Approaches to the Conformational Analysis of Biopharmaceuticals

Roger L. Lundblad

PROTEIN SCIENCE SERIES

SERIES EDITOR

Roger L. Lundblad
Lundblad Biotechnology
Chapel Hill, North Carolina, U.S.A.

PUBLISHED TITLES

Application of Solution Protein Chemistry to Biotechnology
Roger L. Lundblad

Approaches to the Conformational Analysis of Biopharmaceuticals
Roger L. Lundblad

Approaches to the Conformational Analysis of Biopharmaceuticals

Roger L. Lundblad

CRC Press
Taylor & Francis Group
Boca Raton London New York

CRC Press is an imprint of the
Taylor & Francis Group, an **informa** business

Chapman & Hall/CRC
Taylor & Francis Group
6000 Broken Sound Parkway NW, Suite 300
Boca Raton, FL 33487-2742

First issued in paperback 2017

© 2010 by Taylor and Francis Group, LLC
Chapman & Hall/CRC is an imprint of Taylor & Francis Group, an Informa business

No claim to original U.S. Government works

ISBN-13: 978-1-4398-0780-4 (hbk)
ISBN-13: 978-1-138-11494-4 (pbk)

Library of Congress Cataloging-in-Publication Data

Lundblad, Roger L.
    Approaches to the conformational analysis of biopharmaceuticals / Roger L. Lundblad.
        p. ; cm. -- (Protein science)
    Includes bibliographical references and index.
    ISBN 978-1-4398-0780-4 (hardcover : alk. paper)
    1. Protein drugs--Conformation. 2. Pharmaceutical biotechnology. I. Title. II. Series:
Protein science series.
    [DNLM: 1. Biopharmaceutics--methods. 2. Protein Conformation. 3.
Proteins--analysis. QV 38 L962a 2010]

RS431.P75L86 2010
615'.7--dc22   2009024446

Visit the Taylor & Francis Web site at
http://www.taylorandfrancis.com

and the CRC Press Web site at
http://www.crcpress.com

Contents

Series Preface

The universe of biopharmaceutical development revolves around the use of protein science to develop and characterize diagnostic and therapeutic products. Protein Science is a series of books addressed to the application of protein science to biotechnology. Applications of protein science include site-specific chemical modification; spectroscopy and spectrophotometry; electrophoresis; hydrodynamic analytical techniques such as viscosity, light scattering, and analytical ultracentrifugation; chromatographic fractionation including size-exclusion chromatography; and expression and purification systems. This series is directed at the practitioner of commercial biotechnology so there will be volumes on drug product classes including plasma protein products, monoclonal antibodies, cytokines and chemokines, and receptor proteins. To the extent that the series is focused on commercial biotechnology, there will be less discussion of "cutting-edge" and more discussion of direct application of proven new technologies to the study of protein therapeutics. The various volumes will be published in a timely manner to assure immediate value to the biotechnology community.

Preface

This book has been designed to gather as much of the technology for the conformational analysis of biopharmaceutical polymers as possible into a single volume. It is not intended to cover the use of a specific technology in detail. Rather, it is intended to provide sufficient information and references to establish the basis for the selection of a specific experimental approach that would be the most cost-effective in the study of biopolymer conformation. Thus, coverage of some technologies is brief since they have little application to biotechnology products; this is not meant to suggest, however, that such technologies are not of great value in basic science studies. There is also an emphasis to describe studies where multiple technologies were used to address conformational change.

The next several years will see increased interest in the conformational analysis of biopharmaceutical polymers resulting from the development of biosimilar or "follow-on" biological products. The activity of many biopharmaceutical polymers is dependent on conformation. Thus, a comparison of a "generic" (off-patent) biological medicinal product with the originator product includes conformational analysis. A clear understanding of what any differences between products mean or do not mean is critical; thus, the need for independent methods of analysis. It is hoped that this book will emphasize the need for science over hype, the need for rational analysis rather than "smoke and mirrors." Also, renaming a technique does not increase its value in any way.

Finally, I would be remiss if I did not thank Professor Bryce Plapp of the University of Iowa for his continued support of the thermodynamically challenged. Professor Charles Craik of the University of California at San Francisco has also provided some useful advice as has Professor Don Gabriel of the University of North Carolina at Chapel Hill. I am also most indebted to Jill Jurgensen and Barbara Norwitz of Taylor & Francis for their support.

Author

Roger L. Lundblad received his undergraduate education at Pacific Lutheran University, Tacoma, Washington, and his PhD in biochemistry from the University of Washington, Seattle. After postdoctoral work in the laboratories of Stanford Moore and William Stein at the Rockefeller University, New York, he joined the faculty of the University of North Carolina at Chapel Hill in 1968. Dr. Lundblad joined the Hyland division of Baxter Healthcare in 1990. Currently, he is an independent consultant and biotechnology writer based in Chapel Hill, North Carolina. He is an adjunct professor of pathology at the University of North Carolina at Chapel Hill and an editor in chief of the *Internet Journal of Genomics and Proteomics*.

1 Introduction to Biopharmaceutical Conformational Analysis: Issues and Methods

Most biopharmaceuticals are proteins or protein conjugates and are considered to be biopolymers. Proteins have a unique conformation in solution, which is a product of diverse covalent and noncovalent interactions. It is generally accepted that the primary structure of proteins dictates the secondary and tertiary structure of that protein and the final conformation is stabilized by the aforementioned covalent and noncovalent interactions. These interactions can be intramolecular or intermolecular; intramolecular interactions dominate at low protein concentration while intermolecular interactions are more significant at higher protein concentration where such forces are involved in processes such as aggregation. That is not to say that intermolecular interactions are not important at low protein concentrations; however, such interactions are usually driven by specific multivalent interactions.[1]

The study of protein conformation has been of great interest for the study of the relationship between structure and function[3,4] for some time and for the study of protein folding.[5,6] The emergence of biosimilars in commercial biotechnology[7–14] has increased interest in the use of protein conformation study in comparability studies.[15–18] Comparability is also of importance when there are process changes, formulation changes, and change in source material.[19–25]

Protein conformation is the combination of secondary structure (helix, pleated sheet)[26–30] and tertiary structure.[31–37] It is generally accepted that primary structure drives secondary structure, which in turn drives the formation of tertiary structure.[38–40]

The characterization of a protein therapeutic is a critical part of the development and drug approval process. Classic methods such as sequence analysis, compositional analysis, solution behavior with particular emphasis on the formation of aggregates and, more recently, analysis by mass spectrometry are used in the evaluation of protein structure as a biopharmaceutical. The question then is what quality attributes are critical for product performance and what physical/chemical techniques would effectively measure these attributes. It is generally accepted that immunogenicity is a most significant problem. Issues with glycosylation, which influence product half-life and may influence immunogenicity, are also of importance. The problem of immunogenicity is discussed below and elsewhere in detail[41–46] as are techniques for the

1

evaluation of glycosylation.[47–53] Glycosylation presents a little amount of challenge: while glycosylation is important for circulatory half–half (specifically the covering of galactose/galactosamine by sialic acid), there is precious little evidence to suggest a true functional role for glycosylation.

Most solution protein chemistry characterization assays for biologics have focused on chemical structure and biological activity. There is somewhat less interest in the use of conformational analysis. There are several reasons for this. First, to a certain extent, conformational analyses for purposes of identity or comparability only are useful if there is no change: if there is change, it is usually, but not always, difficult to quantitate as compared, for example, to a chemical modification in the peptide chain. However, there are a variety of techniques that can be used to study protein conformation.[54]

Analytical techniques such as amino acid analysis and mass spectrometry provide information regarding the chemical structure of the product. Techniques such as electrophoresis, chromatography, and size exclusion chromatography provide information about purity and can, in selected situations, provide insight into conformation and chemical structure. Hydrophobic interaction chromatography[55–59] can also be useful in the study of conformational changes in proteins.[60–67]

The past 40 years have provided an increase in the sophistication of the technologies available to measure conformational change in proteins; there has not been an increase in the parameters measured. Kauzmann[68] proposed a classification system for the levels of conformation similar to the general classification of primary, secondary, tertiary, and quaternary structure, which separated conformation issues into shape properties and short-range properties. Shape properties (long-range) were parameters dependent on the overall shape (globular, rod, etc.), which might be relatively insensitive to changes in the immediate vicinity of amino acids and peptide bonds. Short-range properties include parameters defined by the immediate environment around individual amino acid residues. Granted that this is an imperfect separation, it does prove useful. Schellman and Schellman[3] reviewed the problem of conformation change in proteins in 1964 and as observed by Cantor and Timascheff,[69] there had been no change in the some 20 years between the two reviews. There has been a marked increase in the sophistication of the instrumentation, and Schellman and Schellman extended Kauzmann's earlier suggestions. Shape properties included hydrodynamic parameters such as frictional coefficient and viscosity changes and solution properties such as fluorescence depolarization and flow birefringence. Also included in shape is electron microscopy, dipole moments, and diffusion through controlled pore membranes (thin film dialysis).[70–72] Short-range properties are, to some extent, "micro" properties as compared to the "macro" properties of shape. Schellman and Schellman include optical properties such as absorbance (IR, UV) and circular dichroism and chemical properties such as side chain reactivity (trace labeling, chemical footprinting), individual pK_a's, hydrogen isotope exchange, biological activity, and immunogenicity as short-range properties. Also included in short-range properties are nuclear magnetic resonance (NMR) and binding of small molecules such as dyes. This division is admittedly imperfect; for example, immunogenicity could be more accurately defined as a shape property but reactivity is dependent on epitopic change.

Most of the techniques used for the conformational analysis of protein were developed either for the study of protein denaturation or, more recently, for the study of protein folding. The focus of this chapter is directed toward the application of solution protein chemistry to the study of conformational change associated with the processing of biotechnology products. These changes can be considered more closely related to denaturation than to protein folding. Denaturation[73] can be considered to be phenomena associated with the change in the spatial arrangement of the polypeptide chains in a protein (tertiary structure) from the native, ordered structure to a more disordered structure in an irreversible process. Denaturation is usually, but not always, associated with loss in solubility. Denaturation is not usually associated with the cleavage of the peptide chain. There are, however, situations that seem to be a slight exception to this; the conversion of fibrinogen to fibrin and the cleavage of peptide chains. Protein denaturation is frequently, but not always, associated with the loss of biological activity as it has long been accepted that configuration is important for biological activity.[74,75] Protein denaturation is not necessarily irreversible[76–81] but there can be a divergence in the quality of structure recovery dependent on measurement.[82–84] The key to renaturation is, in part, dependent on the quality of protein; for example, while some zymogens (e.g., pepsinogen) can be reversibly inactivated under conditions where active enzymes (e.g., pepsin) are irreversibly inactivated.[85–87] On the other hand, trypsin can be reversibly denatured[88,89] and may be more stable than trypsinogen to denaturation.[90–92] Techniques such as light scattering[93–98] and analytical ultracentrifugation[99–104] provide information about the shape and solution behavior of the material (tertiary structure and quaternary structure). These two techniques together with size exclusion chromatography are critical for the evaluation of aggregation in pharmaceutical preparations. There is also reason to consider measurement of the second virial coefficient. The second virial coefficient is a factor used to correct for the nonideal behavior of a particle. Virial coefficients were originally developed as a series of coefficients of inverse powers of V in a polynomial series to approximate the quantity of pV/RT in an equation of state of an ideal gas or similar collection of particles.[105,106] From a practical perspective, the second virial coefficient is related to the excluded volume of a particle[107,108] and is important in accounting for protein–protein interactions and molecular crowding.[109–112] The excluded volume of any particle depends on shape and can be defined as the volume surrounding and including a given object, which is excluded to another object.[107] The second virial coefficient is mentioned most often in the study of the osmotic pressure of proteins but has general use for the study of protein–protein interaction.[113–127] The reader is recommended to articles on protein shape,[128–132] as this attribute is frequently overlooked in favor of the more sophisticated approaches discussed below.

Techniques such as circular dichroism,[133–140] optical rotatory dispersion,[69,141–147] Fourier transform infrared spectroscopy (near infrared [NIR]),[148–157] NMR,[158–170] intrinsic fluorescence,[171–182] binding of fluorescent probes,[183–192] hydrogen–deuterium exchange,[193–207] differential scanning calorimetry,[208–224] Raman spectroscopy,[225–241] protein footprinting,[242–247] limited proteolysis,[248–265] and trace labeling[266–269] can provide information about secondary and tertiary structure. NIR spectroscopy is also useful for noninvasive determination of moisture.[269–276]

One of the major problems with the use of most of these techniques is the requirement for substantial amounts of protein. This can be an issue with therapeutic proteins, which are biologically active at the microgram level and the use of a destructive analytical technique is difficult to justify. However, the use of mass spectrometry for analysis enhances the sensitivity and therefore the value of hydrogen isotope exchange and trace labeling. An analysis of the literature indicates that optical rotatory dispersion is of limited value today as compared to other analytical technologies.

The key issue is—what is the question that you wish to answer? Each of the various techniques has the potential to show changes in conformation secondary to changes in the solvent environment. However, what is the relationship of these changes to biological activity, in vivo clearance, or immunogenicity? In the case of a biopharmaceutical, if you lose activity, you are losing product. Creation of new epitopes (increase or change in immunogenicity; neoantigenicity) either results in an unfortunate immunological response or increased product clearance.[42,44,277] Changes in the immunological properties of biotherapeutic proteins can be identified by established immunoassays.[45,278–281] Changes in glycosylation such as the loss of sialic acid (exposure of galactose/galactosamine) can also increase the rate of product clearance. The demonstration of conformation change in a protein does not necessarily predict a loss of activity or neoantigenicity but can provide insight into the chemistry responsible for such changes. The material below described the relationship between the chemical modification of a protein and changes in secondary or tertiary structure.

First, while the following discussion emphasizes changes in protein conformation (secondary and tertiary structure), primary structure and quaternary structure should be briefly considered. Changes in primary structure can be divided into two categories: first, the modification of individual amino acid residues, which is covered in great detail in Chapter 18 and, second, the cleavage of peptide bonds mostly by proteolytic enzymes or chemical means (i.e., cleavage at asparagine[282–286]). Changes in primary structure will be discussed as such changes in primary structure influence secondary and tertiary structure; changes in quaternary structure are in turn driven by changes in secondary and tertiary structure. As practical note, this author is not aware of any biotherapeutic proteins that were displaying a quaternary structure as an issue; however, general issues of protein–protein interaction, which are important for quaternary structure, are important in the action of most protein biotherapeutics.

Protein conformation can be influenced by both physical and chemical agents. One of the more lively discussions 40 years ago concerned the effect of site-specific modification on protein conformation with respect to the elucidation of the relationship between chemical structure and biological function. Without belaboring the detail, it was generally accepted that it was possible to accomplish the site-specific chemical modification of a protein without gross conformational change, but it was always useful to evaluate such potential changes.[287–289] This has been an active area of interest with the PEGylation of proteins.[290–293]

REFERENCES

1. Stevens, F.J., Analysis of protein–protein interaction by simulation of small-zone size-exclusion chromatography: Application to an antibody–antigen association, *Biochemistry* 25, 981–993, 1986.
2. Sung, M., Poon, G.M.K., and Gariépy, J., The importance of valency in enhancing the import and cell routing potential of protein transduction domain-containing molecules, *Biochim. Biophys. Acta* 1758, 355–363, 2006.
3. Schellman, J.A. and Schellman, C., The conformation of polypeptide chains in proteins, in *The Proteins*, 2nd edn., Vol. 2, ed. H. Neurath, Academic Press, New York, pp. 1–137, Chapter 7, 1964.
4. Fasman, G.D., *Prediction of Protein Structure and the Principles of Protein Conformation*, Plenum Press, New York, 1989.
5. Nall, B.T. and Dill, K.A., *Conformation and Forces in Protein Folding*, AAAS, Washington, DC, 1991.
6. Merz, K.M. and Le Grand, S.M., *The Protein Folding Problem and Tertiary Structure Prediction*, Birkhäuser, Boston, MA, 1994.
7. Schellekens, H., Biosimilar therapeutic agents: Issues with bioequivalence and immunogenicity, *Eur. J. Clin. Invest.* 341, 797–799, 2004.
8. Kessler, M., Goldsmith, D., and Schellekens, H., Immunogenicity of biopharmaceuticals, *Nephrol. Dial. Transplant.* 21(Suppl. 5), v9–v12, 2006.
9. Gerrazani, A.A., Biggio, G., Caputi, A.P. et al., Biosimilar drugs: Concerns and opportunities, *BioDrugs* 21, 351–356, 2007.
10. Roger, S.D. and Mikhail, A., Biosimilars: Opportunity or cause for concern? *J. Pharm. Pharm. Sci.* 10, 405–410, 2007.
11. Pavlovic, M., Girardiin, E., Kapetaneovic, L. et al., Similar biological medicinal products containing recombinant human growth hormone: European regulation, *Horm. Res.* 69, 14–21, 2008.
12. Moran, N., Fractured European market undermines biosimilar launches, *Nat. Biotechnol.* 26, 5–6, 2008.
13. Kawanishi, T., Regulatory perspective from Japan—Comparability of biopharmaceuticals, *Biologicals* 34, 65–68, 2006.
14. Lubiniecki, A.S. and Federici, M.M., Comparability is not just analytical equivalence, *Biologicals* 34, 45–47, 2006.
15. Kuhlmann, M. and Covic, A., The protein science of biosimilars, *Nephrol. Dial. Transplant.* 21(Suppl 5), v4–v8, 2006.
16. Deechongkit, S., Aoki, K.H., Park, S.S., and Kerwin, B.A., Biophysical comparability of the same protein from different manufacturers: A case study using Epoetin alfa from Epogen and Eprex, *J. Pharm. Sci.* 95, 1931–1943, 2006.
17. Heavner, G.A., Arakawa, T., Philo, J.S. et al., Protein isolated from biopharmaceutical formulations cannot be used for comparative studies: Follow-up to a "case study" using Epoetin Alpha Form Epogen and EPREX, *J. Pharm. Sci.* 96, 3214–3225, 2007.
18. ICH 5QE: Comparability of biotechnological/biological products subject to changes in their manufacturing process, International Committee on Harmonisation, http://www.ich.org; http://www.ich.org/cache/compo/276-254-1.html, 2005.
19. DeFelippis, M.R. and Larimore, F.S., The role of formulation in insulin comparability studies, *Biologicals* 34, 49–54, 2006.
20. Petriccciani, J., A global view of comparability concepts, *Dev. Biol.* (Basel) 106, 9–13, 2002.

21. Moos, M. Jr., Regulatory philosophy for comparability protocols, *Dev. Biol.* (Basel) 109, 53–56, 2002.

22. Chirino, A.J. and Mire-Sluis, A., Characterizing biological products and assessing comparability following manufacturing changes, *Nat. Biotechnol.* 22, 1383–1391, 2004.

23. Simek, S.L., Characterization of gene therapy products and the impact of manufacturing changes on product comparability, *Dev. Biol.* (Basel) 122, 139–144, 2005.

24. Robertson, J.S., Changes in biological source material, *Biologicals* 34, 61–63, 2006.

25. Sewerin, K., Shacter, E., Robertson, J., and Wallerius, C., Changes in biological source materials, *Biologicals* 34, 71–72, 2006.

26. Johnson, W.C. Jr., Protein secondary structure and circular dichroism: A practical guide, *Proteins* 7, 205–214, 1990.

27. Yada, R.Y., Jackman, R.L., and Nakai, S., Secondary structure prediction and determination of proteins—A review, *Int. J. Pept. Protein Res.* 31, 98–108, 1988.

28. Andersen, C.A. and Rost, B., Secondary structure assignment, *Methods Biochem. Anal.* 44, 341–363, 2003.

29. Pelton, J.T. and McLean, L.R., Spectroscopic methods for analysis of protein secondary structure, *Anal. Biochem.* 277, 167–176, 2000.

30. Xie, M. and Schowen, R.L., Secondary structure and protein deamidation, *J. Pharm. Sci.* 88, 8–13, 1999.

31. Blow, D.M., Chymotrypsin: Tertiary structure and enzymic activity, *Biochem. J.* 110, 2P, 1968.

32. Crippen, G.M., Correlation of sequence and tertiary structure in globular proteins, *Biopolymers* 16, 2189–2201, 1977.

33. Richardson, J.S., Describing patterns of protein tertiary structure, *Methods Enzymol.* 115, 341–358, 1985.

34. Barton, G.J. and Sternberg, M.J., A strategy for the rapid multiple alignment of protein sequence. Confidence levels from tertiary structure comparisons, *J. Mol. Biol.* 198, 327–337, 1987.

35. Wang, C.X., Shi, Y.Y., and Huang, F.H., Fractal study of tertiary structure of proteins, *Phys. Rev. A* 41, 7043–7048, 1998.

36. Meiler, J. and Baker, D., Coupled prediction of protein secondary and tertiary structure, *Proc. Natl. Acad. Sci. USA* 100, 12105–12110, 2003.

37. Shen, B.W., Spiegel, P.C., Chang, C.H. et al., The tertiary structure and domain organization of coagulation factor VIII, *Blood* 111, 1240–1247, 2008.

38. Sela, M., Anfinsen, C.B., and Harrington, W.F., The correlation of ribonuclease activity with specific aspects of tertiary structure, *Biochim. Biophys. Acta* 26, 502–512, 1957.

39. Anfinsen, C.B., The tertiary structure of ribonuclease, *Brookhaven Symp. Biol.* 15, 184–198, 1962.

40. Anfinesen, C.B., The formation of the tertiary structure of proteins, *Harvey Lect.* 61, 95–116, 1967.

41. Mire-Sluis, A.R., Challenges with current technology for the detection, measurement and characterization of antibodies against biological therapeutics, *Dev. Biol.* (Basel) 109, 59–69, 2002.

42. Hermeling, S., Crommelin, D.J.A., Schellekens, H., and Jiskout, W., Structure–immuonogenicity relationships of therapeutic proteins, *Pharm. Res.* 21, 897–803, 2004.

43. Sampaio, C., Costa, J., and Ferreira, J.J., Clinical comparability of marketed formulations of botulinum toxin, *Mov. Disord.* 19(Suppl 8), S129–S136, 2004.

44. Frost, H., Antibody-mediated side effects of recombinant proteins, *Toxicology* 209, 155–160, 2005.

45. Thorpe, R. and Swanson, S.J., Current methods for detecting antibodies against erythropoietin and other recombinant proteins, *Clin. Diag. Lab. Immunol.* 12, 28–39, 2005.

46. Romer, T., Peter, F., Saenger, P. et al., Efficacy and safety of a new ready-to-use recombinant human growth hormone solution, *J. Endocrinol. Invest.* 30, 578–589, 2007.

47. Henderson, C.J., Holme, M.J., and Aitken, R.J., Analysis of the biological properties of antibodies raised against native and deglycosylated porcine zonae pellucidae, *Gamete Res.* 16, 323–341, 1987.

48. Aouffen, M., Paquin, J., De Grandpre, E. et al., Deglycosylated ceruloplasmin maintains its enzymatic, antioxidant, cardioprotective, and neuronoprotective properties, *Biochem. Cell Biol.*, 79, 489–497, 2001.

49. Raju, T.S., Briggs, J.B., Chamow, S.M. et al., Glycoengineering of therapeutic glycoproteins: In vitro galactoxylation and sialylation of glycoproteins with terminal *N*-acetylglucosamine and galactose residues, *Biochemistry* 31, 8868–8876, 2001.

50. Jefferis, R., Glycosylation of recombinant antibody therapeutics, *Biotechnol. Prog.* 21, 11–16, 2005.

51. Walsh, G. and Jefferis, R., Post-translational modifications in the context of therapeutic proteins, *Nat. Biotechnol.* 24, 1241–1252, 2006.

52. Jefferis, R., Antibody therapeutics: Isotype and glycoform selection, *Expert Opin. Biol. Ther.* 7, 1401–1413, 2007.

53. Temporini, C., Calleri, E., Massolini, G., and Caaccialanza, G., Integrated analytical strategies for the study of phosphorylation and glycosylation in proteins, *Mass Spectrom. Rev.* 27, 207–236, 2008.

54. Crommelin, D.J.A., Storm, G., Verrijk, R. et al., Shifting paradigms: Pharmaceutical versus low molecular weight drugs, *Int. J. Pharmaceut.* 266, 3–16, 2003.

55. Kato, Y., High-performance hydrophobic interaction chromatography of proteins, *Adv. Chromatogr.* 26, 97–115, 1987.

56. Arakawa, T. and Narhi, L.O., Solvent modulation in hydrophobic interaction chromatography, *Biotechnol. Appl. Biochem.* 13, 151–172, 1991.

57. Wu, S.-L. and Karger, B.L., Hydrophobic interaction chromatography of proteins, *Methods Enzymol.* 270, 27–47, 1996.

58. Hemström, P. and Irgum. K., Hydrophilic interaction chromatography, *J. Sep. Sci.* 29, 1784–1821, 2006.

59. Lienqueo, M.E., Mahn, A., Salgado, J.C., and Asenjo, J.A., Current insights on protein behavior in hydrophobic interaction chromatography, *J. Chromatogr. B. Analyt. Technol. Biomed. Life Sci.* 849, 53–68, 2007.

60. Tarvers, R.C., Calcium-dependent changes in properties of human prothrombin: A study using high-performance size-exclusion chromatography and gel-permeation chromatography, *Arch. Biochem. Biophys.* 241, 639–648, 1985.

61. Wu, S.L., Figueroa, A., and Karger, B.L., Protein conformational effects in hydrophobic interaction chromatography. Retention characterization and the role of mobile phase additives and stationary phage hydrophobicity, *J. Chromatogr.* 371, 3–27, 1986.

62. Withka, J., Moncuse, P., Baziotis, A., and Maskiewicz, R., Use of high-performance size-exclusion, ion-exchange, and hydrophobic interaction chromatography for the measurement of protein conformational change and stability, *J. Chromatogr.* 398, 175–202, 1987.

63. Krull, I.S., Stuting, H.H., and Krzysko, S.C., Conformational studies of bovine alkaline phosphatase in hydrophobic interaction and size-exclusion chromatography with linear diode array and low-angle laser light scattering detection, *J. Chromatogr.* 442, 29–52, 1988.

64. Lundblad, R.L., A hydrophobic site in human prothrombin present in a calcium-stabilized conformer, *Biochem. Biophys. Res. Commun.* 157, 295–300, 1988.

65. Haezebrouck, P., Noppe, W., van Dacl, H., and Hanssens, I., Hydrophobic interaction of lysozyme and α-lactalbumin from equine milk whey, *Biochim. Biophys. Acta* 1122, 305–310, 1992.

66. Bjerrun, O.J., Bjerrum, M.J., and Heegaard, N.H., Electrophoretic and chromatographic differentiation of two forms of albumin in equilibrium at neutral pH: New screening techniques for determination of liquid binding to albumin, *Electrophoresis* 16, 1401–1407, 1995.

67. To, B.C.S. and Lenhoff, A.M., Hydrophobic interaction chromatography of proteins. II. Solution thermodynamic properties as a determinant of retention, *J. Chromatogr. A* 1141, 235–243, 2007.

68. Kauzmann, W., Some factors in the interpretation of protein denaturation, *Adv. Prot. Chem.* 14, 1–63, 1959.

69. Cantor, C.R. and Timasheff, S.N., Optical spectroscopy of proteins, in *The Proteins*, 3rd edn., Vol. 5, eds. H. Neurath and R.L. Hill, Academic Press, New York, pp. 145–306, 1982.

70. Craig, L.C. and Chen, H.C., On a theory for the passive transport of solute through semipermeable membranes, *Proc. Natl. Acad. Sci. USA* 69, 702–705, 1972.

71. Chen, H.C., Craig, L.C., and Stoner, E., On the removal of residual carboxylic acid groups from cellulose membranes and Sephadex, *Biochemistry* 11, 3559–3564, 1972.

72. Harris, M.J. and Craig, L.C., A study of the parameters which determine the conformation of linear polypeptides in solution by synthesis of models and determination of thin film dialysis rates, *Biochemistry* 13, 1510–1515, 1974.

73. Putman, F.W., Protein denaturation, in *The Proteins*, eds. H. Neurath and K. Bailey, Academic Press, New York, pp. 808–892, Chapter 9, 1953.

74. Porter, R.R., The relation of chemical structure to the biological activity of proteins, in *The Proteins*, eds. H. Neurath and K. Bailey, Academic Press, New York, pp. 973–1015, Chapter 11, 1973.

75. Putnam, F.W., The chemical modification of proteins, in *The Proteins*, eds. H. Neurath and K. Bailey, Academic Press, New York, pp. 893–972, Chapter 10, 1953.

76. Kodicek, M., Infanzón, A., and Karpenko, V., Heat denaturation of human orosomucoid in water/methanol mixtures, *Biochim. Biophys. Acta* 1246, 10–16, 1995.

77. Pico, G., Thermodynamic aspects of the thermal stability of human serum albumin, *Biochem. Mol. Biol. Int.* 36, 1017–1023, 1995.

78. Herberhold, H. and Winter, R., Temperature- and pressure-induced unfolding and refolding of ubiquitin: A static and kinetic Fourier transform infrared spectroscopy study, *Biochemistry* 41, 2396–2401, 2002.

79. Roychaudhuri, R., Sarath, G., Zeece, M., and Markwell, J., Reversible denaturation of the soybean Kunitz trypsin inhibitor, *Arch. Biochem. Biophys.* 412, 20–26, 2003.

80. Mehta, R., Kundu, A., and Kishore, N., 4-Chlorobutanol induces unusual reversible and irreversible thermal unfolding of ribonuclease A: Thermodynamic, kinetic, and conformational characterization, *Int. J. Biol. Macromol.* 34, 13–20, 2004.

81. Kragh-Hansen, U., Saito, S., Nishi, K. et al., Effect of genetic variation on the thermal stability of human serum albumin, *Biochim. Biophys. Acta* 1747, 81–88, 2005.

82. Ryhähen, L., Zaragoza, E.J., and Uitto, J., Conformational stability of type I collagen triple helix: Evidence for temporary and local relaxation of the protein conformation using a proteolytic probe, *Arch. Biochem. Biophys.* 223, 562–571, 1983.

83. Isaacs, B.S., Brew, S.A., and Ingham, K.C., Reversible unfolding of the gelatin-binding domain of fibronectin: Structural stability in relation to function, *Biochemistry* 28, 842–850, 1989.

84. Mizuno, K. and Hayashi, T., Peculiar effect of urea on the interaction of type I collagen with heparin on chromatography, *J. Biochem.* 116, 1257–1263, 1994.

85. Ahmad, F. and McPhie, P., Thermodynamics of the denaturation of pepsinogen by urea, *Biochemistry* 17, 241–246, 1979.

86. Ahmad, F. and McPhie, P., Characterization of a stable intermediate in the unfolding of diazoacetylglycine ethyl ester—Pepsin by urea, *Biochemistry* 17, 241–246, 1979.

87. Konno, T., Kamatari, Y.O., Tanaka, N. et al., A partially unfolded structure of the alkaline-denatured state of pepsin and its implication for stability of the zymogen-derived protein, *Biochemistry* 39, 4182–4290, 2000.
88. Ruan, K., Lange, R., Meersman, F. et al., Fluorescence and FTIR study of the pressure-induced denaturation of bovine pancreas trypsin, *Eur. J. Biochem.* 265, 79–85, 1999.
89. Brumano, M.H., Rogana, E., and Swaisgood, H.E., Thermodynamics of unfolding of beta-trypsin at pH 2.8, *Arch. Biochem. Biophys.* 382, 57–62, 2000.
90. Hopkins, T.R. and Spikes, J.D., Denaturation of proteins in 8 M urea as monitored by tryptophan fluorescence: Trypsin, trypsinogen and some derivatives, *Biochem. Biophys. Res. Commun.* 30, 540–545, 1968.
91. Delaage, M. and Lazdunski, M., Trypsinogen, trypsin, trypsin-substrate and trypsin-inhibitor complexes in urea solutions, *Eur. J. Biochem.* 4, 378–384, 1968.
92. Otlewski, J., Sywula, A., Kaolasinski, M., and Krowarsch, D., Unfolding kinetics of bovine trypsinogen, *Eur. J. Biochem.* 242, 601–607, 1996.
93. Carlson, F.D., The application of intensity fluctuation spectroscopy to molecular biology, *Annu. Rev. Biophys. Bioeng.* 4, 243–264, 1975.
94. Tinoco, I. Jr., Michols, W., Maestrae, M.F., and Bustamante, C., Absorption, scattering, and imaging of biomolecular structures with polarized light, *Annu. Rev. Biophys. Biophys. Chem.* 16, 319–349, 1987.
95. Eisenberg, H., Thermodynamics and the structure of biological macromolecules. Rozhinkes mit mandolin, *Eur. J. Biochem.* 187, 7–22, 1990.
96. Herding, S.E., Sattelle, D.B., and Bloomfield, V.A. eds., *Laser Light Scattering in Biochemistry*, Royal Society of Chemistry, Cambridge, U.K., 1992.
97. Li-Chain, E.C., Methods to monitor process-induced changes in food proteins. An overview, *Adv. Exp. Med. Biol.* 434, 5–23, 1998.
98. Georgilis, Y. and Saenger, W., Light scattering studies on supersaturated protein solutions, *Sci. Prog.* 82, 271–294, 1998.
99. Aune, K.C., Molecular weight measurements by sedimentation equilibrium: Some common pitfalls and how to avoid them, *Methods Enzymol.* 48, 163–185, 1978.
100. Schuster, T.M. and Toedt, J.M., New revolutions in the evolution of analytical ultracentrifugation, *Curr. Opin. Struct. Biol.* 6, 650–658, 1996.
101. Hensley, P., Defining the structure and stability of macromolecular assemblies in solution: The re-emergence of analytical ultracentrifugation as a practical tool, *Structure* 4, 367–373, 1996.
102. Eisenberg, H., Analytical ultracentrifugation in a Gibbsian perspective, *Biophys. Chem.* 88, 1–9, 2000.
103. Lebowiz, J., Lewis, M.S., and Schuck, P., Modern analytical ultracentrifugation in protein science: A tutorial review, *Protein Sci.* 11, 2067–2079, 2002.
104. Howlett, G.J., Minton, A.P., and Rivas, G., Analytical ultracentrifugation for the study of protein association and assembly, *Curr. Opin. Chem. Biol.*, 10, 430–436, 2006.
105. *Oxford English Dictionary*, Oxford University Press, Oxford, U.K., 2008.
106. Onnes, H.K., 1901, in *Arch. neérlandaises des Sci. exactes & nat.* VI. 874, as cited in *Oxford English Dictionary*, 2008.
107. Flory, P.J., *Principles of Polymer Chemistry*, Cornell University Press, Ithaca, NY, 1953.
108. McCabe, W.C. and Fisher, H.F., Measurement of the excluded volume of protein molecules by differential spectroscopy in the near infra red, *Nature* 207, 1274–1276, 1965.
109. Minton, A.P., Influence of excluded volume upon macromolecular structure and associations in 'crowded' media, *Curr. Opin. Biotechnol.* 8, 65–69, 1997.
110. Minton, A.P., Implications of macromolecular crowding for protein assembly, *Curr. Opin. Struct. Biol.* 10, 34–39, 2000.
111. Despa, F., Orgill, D.P., and Lee, R.C., Molecular crowding effects on protein stability, *Ann. N.Y. Acad. Sci.* 1066, 54–66, 2005.

112. Konopka, M.C., Weisshaar, J.C., and Record, M.T. Jr., Methods of changing biopolymer volume fraction and cytoplasmic solute concentrations for in vivo biophysical studies, *Methods Enzymol.* 428, 487–504, 2007.

113. Nichol, J.W., Janado, M., and Winzor, D.J., The origin and consequences of concentration dependence in gel chromatography, *Biochem. J.* 133, 15–22, 1973.

114. Wan, P.J. and Adams, E.T. Jr., Molecular weights and molecular weight distribution from ultracentrifugation of nonideal solutions, *Biophys. Chem.* 5, 207–241, 1976.

115. Tang, L.H., Powell, D.R., Escott, B.M., and Adams, E.T. Jr., Analysis of various indefinite self-associations, *Biophys. Chem.* 7, 121–139, 1977.

116. Neal, B.L., Asthagiri, D., and Lenhoff, A.M., Molecular origins of osmotic second virial coefficients of proteins, *Biophys. J.* 75, 2469–2477, 1998.

117. Weatherly, G.T. and Pielak, G.J., Second virial coefficients as a measure of protein–osmolyte interactions, *Protein Sci.* 10, 12–16, 2001.

118. Ruppert, S., Sandler, S.L., and Lenhoff, A.M., Correlation between the second virial coefficient and the solubility of proteins, *Biotechnol. Prog.* 17, 182–187, 2001.

119. Tessier, P.M. and Lenhoff, A.M., Measurements of protein self-association as a guide to crystallization, *Curr. Opin. Biotechnol.* 14, 512–516, 2003.

120. Sear, R.P., Solution stability and variability in a simple model of globular proteins, *J. Chem. Phys.* 120, 998–1005, 2004.

121. Valente, J.J., Payne, R.W., Manning, M.C. et al., Colloidal behavior of proteins: Effects of the second virial coefficient on solubility, crystallization and aggregation of proteins in aqueous solution, *Curr. Pharm. Biotechnol.* 6, 427–436, 2005.

122. Paliwal, A., Asthagiri, D., Abras, D. et al., Light-scattering studies of protein solutions: Role of hydration in weak protein–protein interactions, *Biophys. J.* 89, 1564–1573, 2005.

123. Ruckenstein, E. and Shulgin, I.L., Effect of salts and organic additives on the solubility of proteins in aqueous solutions, *Adv. Colloid Interface Sci.* 123–126, 97–103, 2006.

124. Payne, R.W., Nayar, R., Tarantino, R. et al., Second virial coefficient determination of a therapeutic peptide by self-interaction chromatography, *Biopolymers* 84, 527–533, 2006.

125. Bajaj, H., Sharma, V.K., Badkar, A. et al., Protein structural conformation and not second virial coefficient relate to long-term irreversible aggregation of a monoclonal antibody and ovalbumin in solution, *Pharm. Res.* 23, 1382–1394, 2006.

126. Winzor, D.J., Dezczynski, M., Harding, S.E., and Wills, P.R., Nonequivalence of second virial coefficients from sedimentation equilibrium and static light scattering studies of protein solutions, *Biophys. Chem.* 128, 46–55, 2007.

127. Dumetz, A.C., Chockla, A.M., Kaler, E.W., and Lenhoff, A.M., Effects of pH on protein–protein interactions and implications for protein phase behavior, *Biochim. Biophys. Acta* 1784, 600–610, 2008.

128. Blattler, D.P. and Reisthel, F.J., Molecular weight determinations and the influence of gel density and protein shape in polyacrylamide gel electrophoresis, *J. Chromatogr.* 46, 286–292, 1970.

129. Chae, K.S. and Lenhoff, A.M., Computation of the electrophoretic mobility of proteins, *Biophys. J.* 68, 1120–1127, 1995.

130. Røgen, P. and Bohr, H., A new family of global protein shape descriptors, *Math. Biosci.* 182, 167–181, 2003.

131. He, L. and Niemeyer, B., A novel correlation for protein diffusion coefficients based on molecular weight and radius of gyration, *Biotechnol. Prog.* 19, 544–548, 2003.

132. Chang, B.H. and Bae, Y.C., Salting-out in the aqueous single-protein solution: The effect of shape factor, *Biophys. Chem.* 104, 523–533, 2003.

133. Yang, J.T., Wu, C.S.C., and Martinez, H.M., Calculation of protein conformation from circular dichroism, *Methods Enzymol.* 130, 208–270, 1986.

134. Woody, R.W., Circular dichroism, *Methods Enzymol.* 246, 34–71, 1995.

135. Johnson, W.C. Jr., Protein secondary structure and circular dichroism: A practical guide, *Proteins* 7, 205–214, 1990.
136. Bayer, T.S., Booth, L.N., Knudsen, S.M., and Ellington, A.D., Arginine-rich motifs present multiple interfaces for specific binding by RNA, *RNA* 11, 1848–1857, 2005.
137. Harrington, A., Darboe, N., Kenjale, R. et al., Characterization of the interaction of single tryptophan containing mutants of IpaC from *Shigella flexneri* with phospholipid membranes, *Biochemistry* 45, 626–636, 2006.
138. Paramonov, S.E., Jun, H.W., and Hartgerink, J.D., Modulation of peptide–amphiphile nanofibers via phospholipid inclusions, *Biomacromolecules* 7, 24–26, 2006.
139. Miles, A.J. and Wallace, R.A., Synchrotron radiation circular dichroism spectroscopy of proteins and applications in structural and functional genomics, *Chem. Soc. Rev.* 35, 39–51, 2006.
140. Steinberg, I.Z., Circularly polarized luminescence, *Methods Enzymol.* 49, 179–199, 1971.
141. Raghavendra, K. and Ananthanarayanan, V.S., Beta-structure of polypeptides in non-aqueous solutions. I. Spectral characteristics of the polypeptide backbone, *Int. J. Pept. Protein Res.* 17, 412–419, 1981.
142. Hvidt, S., Rodgers, M.E., and Harrington, W.F., Temperature-dependent optical rotatory dispersion properties of helical muscle proteins and homopolymers, *Biopolymers* 24, 1647–1662, 1985.
143. Walji, F., Rosen, A., and Hider, R.C., The existence of conformationally labile (pre-formed) drug binding sites in human serum albumin as evidenced by optical rotatory measurements, *J. Pharm. Pharmacol.* 45, 551–558, 1993.
144. Parkhusrt, L.J., A nanosecond ORD study of hemoglobin, *Biophys. J.* 68, 399–400, 1995.
145. Galvani, M., Hamdan, M., and Righetti, P.G., Probing protein unfolding through monitoring cysteine alkylation by matrix-assisted laser desorption/ionization mass spectrometry, *Rapid Commun. Mass Spectrom.* 14, 1925–1931, 2000.
146. Majewski, A.J., Sanzari, M., Cui, H.L., and Torzilli, P., Effects of ultraviolet radiation on the type-I collagen protein triple helical structure: A method for measuring structural changes through optical activity, *Phys. Rev. E. Stat. Nonlin. Soft Matter Phys.* 65, 031920, 2002.
147. Sakurai, K. and Goto, Y., Dynamics and mechanism of the Tanford transition of bovine beta-lactaglobulin studied using heteronuclear NMR spectroscopy, *J. Mol. Biol.* 356, 483–496, 2006.
148. Susi, H., Infrared spectroscopy—Conformation, *Methods Enzymol.* 26, 455–472, 1972.
149. Susi, H. and Byler, D.M., Resolution-enhanced Fourier-transform infrared spectroscopy of enzymes, *Methods Enzymol.* 130, 290–311, 1986.
150. Siebert, F., Infrared spectroscopy applied to biochemical and biological problems, *Methods Enzymol.* 246, 501–526, 1995.
151. Shaw, R.A. and Mantsch, H.H., Near-IR spectrophotometers, in *Encyclopedia of Spectroscopy and Spectrometry*, eds. J.C. Lindon, G.E. Tauter, and J.L. Holmes, Academic Press, New York, 2000.
152. Yuan, B., Murayama, K., Tsenkova, R. et al., Temperature-dependent near-infrared spectra of bovine serum albumin in aqueous solutions: Spectral analysis by principal component analysis and evolving factor analysis, *Appl. Spectros.* 57, 1223–1229, 2003.
153. Bai, S., Nayar, R., Carpenter, J.F., and Manning, M.C., Noninvasive determination of protein conformation in the solid state using near infrared (NIR) spectroscopy, *J. Pharm. Sci.* 94, 2030–2038, 2005.
154. Izutsu, K., Fujimaki, Y., Kuwabara, A. et al., Near-infrared analysis of protein structure in aqueous solutions and freeze-dried solids, *J. Pharm. Sci.* 95, 781–789, 2006.

155. Bruun, S.W., Holm, J., Hansen, S.I., and Jacobsen, S., Application of near-infrared and Fourier transform infrared spectroscopy in the characterization of ligand-induced conformation changes in folate binding protein purified from bovine milk: Influence of buffer type and pH, *Appl. Spectrosc.* 60, 737–746, 2006.

156. Bruun, S.W., Søndergaard, I., and Jacobsen, S., Analysis of protein structures and interactions in complex food by near-infrared spectroscopy. 2. Hydrated gluten, *J. Agric. Food Chem.* 55, 7244–7251, 2007.

157. Schiro, G. and Cupane, A., Quaternary relaxations in sol-gel encapsulated hemoglobin studied via NIR and UV spectroscopy, *Biochemistry* 46, 11568–11576, 2007.

158. Redfield, A.G., Proton nuclear magnetic resonance in aqueous solutions, *Methods Enzymol.* 49, 253–270, 1978.

159. Crespi, H.L. and Katz, J.J, Preparation of deuterated proteins and enzymes, *Methods Enzymol.* 26, 627–637, 1972.

160. Wagner, G. and Wuthrich, K., Observation of internal mobility of proteins by nuclear magnetic resonance in solution, *Methods Enzymol.* 131, 307–328, 1986.

161. Song, J., Laskowski, M. Jr., Qasim, M.A., and Markley, J.L., NMR determination of pK_a values for asp, glu, his, and lys mutants at each variable contiguous enzyme-inhibitor contact position of the turkey ovomucoid third domain, *Biochemistry* 42, 2847–2856, 2003.

162. Geyer, M., Wilde, C., Selzer, J. et al., Glucosylation of Ras by *Clostridium sordellii* lethal toxin: Consequences for effector loop conformations observed by NMR spectroscopy, *Biochemistry* 42, 11951–11959, 2003.

163. Samson, A.O., Chill, J.H., and Anglister, J., Two-dimensional measurement of protein $T_{1\rho}$ relaxation in unlabeled proteins: mobility changes in α-bungarotoxin upon binding of an acetylcholine receptor peptide, *Biochemistry* 44, 10926–10934, 2005.

164. Mittelmaier, A. and Kay, L.E., New tools provide new insights in NMR studies of protein dynamics, *Science* 312, 224–228, 2006.

165. Jarymowycz, V.A. and Stone, M.J., Fast time scale dynamics of protein backbones: NMR relaxation studies, *Chem. Rev.* 106, 1624–1671, 2006.

166. Zartler, E.R. and Shapiro, M.J., Protein NMR-based screening in drug discovery, *Curr. Pharm. Des.* 12, 3963–3972, 2006.

167. Foster, M.P., McElroy, C.A., and Amero, C.D., Solution NMR of large molecules and assemblies, *Biochemistry* 46, 331–340, 2007.

168. Frieden, C., Protein aggregation processes: In search of the mechanism, *Protein Sci.* 16, 2334–2344, 2007.

169. Pazgier, M., Li, X., Lu, W., and Lubkowski, J., Human defensins: Synthesis and structural properties, *Curr. Pharm. Des.* 13, 3096–3118, 2007.

170. Spiess, H.W., NMR spectroscopy: Pushing the limits of sensitivity, *Angew. Chem. Int. Ed. Engl.* 47, 639–642, 2008.

171. Righetti, P.G. and Verzola, B., Folding/unfolding/refolding of proteins: Present methodologies in comparison with capillary zone electrophoresis, *Electrophoresis* 22, 2359–2374, 2001.

172. Cioni, P. and Strambini, G.B., Tryptophan phosphorescence and pressure effects on protein structure, *Biochim. Biophys. Acta* 1595, 116–130, 2002.

173. Cioni P., Role of protein cavities on unfolding volume change and on internal dynamics under pressure, *Biophys. J.* 91, 3390–3396, 2006.

174. Kumar, A., Tyagi, N.K., and Kinne, R.K., Ligand-mediated conformation changes and positioning of tryptophans in reconstituted human sodium/D-glucose cotransporter1 (hsGLT1) probed by tryptophan fluorescence, *Biophys. Chem.* 127, 69–77, 2007.

175. Harvey, B.J., Bell, E., and Brancaleon, L., A tryptophan rotamer located in a polar environment probes pH-dependent conformational changes in bovine β-lactoglobulin A, *J. Phys. Chem. B.* 111, 2610–2620, 2007.

176. Daly, S.M., Pryzybycien, T.M., and Tilton, R.D., Aggregation of lysozyme and of poly(ethylene glycol)-modified lysozyme after adsorption to silica, *Colloids Surf. B. Biointerfaces* 57, 81–88, 2007.

177. Ghasemi, A., Khajeh, K., and Ranjbar, B., Stabilization of *Bacillus licheniformis* α-amylase by specific antibody which recognizes the N-terminal fragment of the enzyme, *Int. J. Biol. Macromol.* 41, 162–167, 2007.

178. Fan, H., Vitharana, S.N., Chen, T. et al., Effects of pH and polyanions on the thermal stability of fibroblast growth factor 20, *Mol. Pharm.* 4, 232–240, 2007.

179. Benesch, J., Hungerford, G., Shuling, K. et al., Fluorescence probe techniques to monitor protein adsorption-induced conformation changes on biodegradable polymers, *J. Colloid Interface Sci.* 312, 193–200, 2007.

180. Boudier, C., Bonsquet, J.A., Schauinger, S. et al., Reversible inactivation of serpins at acidic pH, *Arch. Biochem. Biophys.* 466, 155–163, 2007.

181. Lee, J. and Tripathi, A., Measurements of label free protein concentration and conformational change using a microfluidic UV-LED method, *Biotechnol. Prog.* 23, 1506–1512, 2007.

182. Ramachander, R., Jiang, Y., Li, C. et al., Solid state fluorescence of lyophilized proteins, *Anal. Biochem.* 376, 173–182, 2008.

183. Kanaoka, Y., Organic fluorescent reagents in the study of enzymes and proteins, *Angew. Chem. Int. Ed. Engl.* 16, 137–147, 1977.

184. Selvin, P.R., The renaissance of fluorescence resonance energy transfer, *Nat. Struct. Biol.* 7, 730–734, 2000.

185. Heyduk, T., Measuring protein conformational changes by FRET/LRET, *Curr. Opin. Biotechnol.* 13, 292–296, 2002.

186. Giepmans, B.N., Adams, S.R., Ellisman, M.H., and Tsien, R.Y., The fluorescent toolbox for assessing protein location and function, *Science* 312, 217–224, 2006.

187. Royer, C.A., Probing protein folding and conformational transitions with fluorescence, *Chem. Rev.* 106, 1769–1784, 2006.

188. Gull, N., Sen, P., and Kabir-Ud-Din, K.R.H., Effect of physiological concentrations of urea on the conformation of human serum albumin, *J. Biochem.* 141, 261–268, 2007.

189. Mazzini, A., Polverini, E., Parisi, M. et al., Dissociation and unfolding of bovine oderant binding protein at acidic pH, *J. Struct. Biol.* 159, 82–91, 2007.

190. Muriznieks, L.D. and Weiss, A.S., Flexibility in solution structure of human tropoelastin, *Biochemistry* 46, 8196–8205, 2007.

191. Vetri, V., Librizzi, F., Leone, M., and Militello, V., Thermal aggregation of bovine serum albumin at different pH: Comparison with human serum albumin, *Eur. Biophys. J.* 36, 717–725, 2007.

192. Rezaei-Ghaleh, N., Ramshini, H., Ebrahim-Habibi, A. et al., Thermal aggregation of α-chymotrypsin: Role of hydrophobic and electrostatic interactions, *Biophys. Chem.* 132, 23–32, 2008.

193. Hvidt, A. and Linderstrom-Lang, K., Exchange of hydrogen atoms in insulin with deuterium atoms in aqueous solutions, *Biochim. Biophys. Acta* 14, 574–575, 1954.

194. Englander, S.W. and Englander, J.J., Hydrogen-tritium exchange, *Methods Enzymol.* 49, 24–39, 1978.

195. Englander, J.J., Rogero, J.R., and Englander, S.W., Identification of an allosterically sensitive unfolding unit in hemoglobin, *J. Mol. Biol.* 169, 325–344, 1983.

196. Englander, J.J., Rogero, J.R., and Englander, S.W., Protein hydrogen exchange studied by the fragment separation method, *Anal. Biochem.* 147, 234–244, 1985.

197. Gregory, R.B. and Rosenberg, A., Protein conformational dynamics measured by hydrogen isotope exchange techniques, *Methods Enzymol.* 131, 448–508, 1986.

198. Rogero, J.R., Englander, J.J., and Englander, S.W., Individual breathing reactions measured by functional labeling and hydrogen exchange methods, *Methods Enzymol.* 131, 508–517, 1986.

199. Bierzyński, A., Methods of peptide conformation studies, *Acta Biochim. Pol.* 48, 1091–1099, 2001.
200. Hoofnagle, A.N., Resing, K.A., and Ahn, N.G., Protein analysis by hydrogen exchange mass spectrometry, *Annu. Rev. Biophys. Biomol. Struct.* 32, 1–23, 2003.
201. Busenlehner, L.S. and Armstrong, R.N., Insights into enzyme structure and dynamics elucidated by amide H/D exchange mass spectrometry, *Arch. Biochem. Biophys.* 433, 34–46, 2005.
202. Wales, T.E. and Enge, J.R., Hydrogen exchange mass spectrometry for the analysis of protein dynamics, *Mass Spectrom. Rev.* 25, 158–170, 2006.
203. Maier, C.S. and Deinzer, M.L., Protein conformations, interactions, and H/D exchange, *Methods Enzymol.* 402, 312–360, 2005.
204. Kaveti, S. and Engen, J.R., Protein interactions probed with mass spectrometry, *Methods Mol. Biol.* 316, 179–197, 2006.
205. Englander, S.W., Hydrogen exchange and mass spectrometry: A historical perspective, *J. Am. Soc. Mass Spectrom.* 17, 1481–1489, 2006.
206. Tsutsui, Y. and Wintrode, P.L., Hydrogen/deuterium exchange-mass spectrometry: A powerful tool for probing protein structure, dynamics and interactions, *Curr. Med. Chem.* 14, 2344–2358, 2007.
207. Li, Y., Williams, T.D., and Topp, E.M., Effects of excipients on protein conformation in lyophilized solids by hydrogen/deuterium exchange mass spectrometry, *Pharm. Res.* 25, 259–267, 2008.
208. Krishnan, K.S. and Brandts, J.F., Scanning calorimetry, *Methods Enzymol.* 49, 3–14, 1978.
209. Biltonen, R.L. and Freire, E., Thermodynamic characterization of conformational states of biological macromolecules using differential scanning calorimetry, *CRC Crit. Rev. Biochem.* 5, 85–124, 1978.
210. Brandts, J.F. and Lin, L.N., Study of strong to ultratight protein interactions using differential scanning calorimetry, *Biochemistry* 29, 6927–6940, 1990.
211. Plum, G.E. and Breslauer, K.J., Calorimetry of proteins and nucleic acids, *Curr. Opin. Struct. Biol.* 5, 682–690, 1995.
212. Weber, P.C. and Salemme, F.R., Applications of calorimetric methods to drug discovery and the study of protein interactions, *Curr. Opin. Struct. Biol.* 13, 115–121, 2003.
213. Matheus, S., Friess, W., and Mahler, H.C., FTIR and nDSC as analytical tools for high-concentration protein formulations, *Pharm. Res.* 23, 1350–1363, 2006.
214. Tavirani, M.R., Moghaddamnia, S.H., Ranjbar, B. et al., Conformational study of human serum albumin in pre-denaturation temperatures by differential scanning calorimetry, circular dichroism and UV spectroscopy, *J. Biochem. Mol. Biol.* 39, 530–536, 2006.
215. Arakawa, T., Kita, Y., Ejima, D. et al., Aggregation suppression of proteins by arginine during thermal unfolding, *Protein Pept. Lett.* 13, 921–927, 2006.
216. de Groot, J., Kosters, H.A., and de Jongh, H.H., Deglycosylation of ovalbumin prohibits formation of a heat-stable conformer, *Biotechnol. Bioeng.* 97, 735–741, 2007.
217. Ejima, D., Tsumoto, K., Fukada, H. et al., Effects of acid exposure on the conformation, stability, and aggregation of monoclonal antibodies, *Proteins* 66, 954–962, 2007.
218. Wakankar, A.A., Lin, J., Vandervelde, D. et al., The effect of cosolutes on the isomerization of aspartic acid residue and conformational stability in a monoclonal antibody, *J. Pharm. Sci.* 96, 1708–1718, 2007.
219. Garber, E. and Demarest, S.J., A broad range of Fab stabilities within a host of therapeutic IgGs, *Biochem. Biophys. Res. Commun.* 355, 751–757, 2007.
220. Kar, K. and Kishore, N., Enhancement of thermal stability and inhibition of protein aggregation by osmolytic effect of hydroxyproline, *Biopolymers* 87, 339–351, 2007.

221. Han, Y., Jin, B.S., and Lee, S.B., Effects of sugar additives on protein stability of recombinant human serum albumin during lyophilization and storage, *Arch. Pharm. Res.* 30, 1124–1131, 2007.

222. Tsybovsky, Y., Shubenok, D.V., Kravchuk, Z.I., and Martsev, S.P., Folding of an antibody variable domain in two functional conformations *in vitro* calorimetric and spectroscopic study of the anti-ferritin antibody VL domain, *Protein Eng. Des. Sel.* 20, 481–490, 2007.

223. Sedlák, E., Zoldák, G., and Wittung-Stafshede, P., Role of copper in thermal stability of human ceruloplasmin, *Biophys. J.* 94, 1384–1391, 2008.

224. Bellezza, F., Cipiciani, A., Quotadamo, M.A. et al., Structure, stability, and activity of myoglobin adsorbed onto phosphate-grafted zirconia nanoparticles, *Langmuir* 23, 13007–13012, 2007.

225. Tobin, M.C., Raman spectroscopy, *Methods Enzymol.* 26, 473–497, 1972.

226. Van Wart, H.E. and Scheraga, H.A., Raman and resonance Raman spectroscopy, *Methods Enzymol.* 49, 67–149, 1978.

227. Williams, R.W., Protein secondary structure analyzed with Raman amide I and amide II spectra, *Methods Enzymol.* 130, 311–331, 1986.

228. Hudson, B. and Mayne, L., Ultraviolet resonance Raman spectroscopy of biopolymers, *Methods Enzymol.* 130, 331–350, 1986.

229. Vass, E., Hollósi, M., Besson, F., and Buchet, R., Vibrational spectroscopic detection of β- and γ-turns in synthetic and natural peptides and proteins, *Chem. Rev.* 103, 1917–1954, 2003.

230. Pimenov, K.V., Bykov, S.V., Mikhonin, A.V., and Asher, S.A., UV Raman examination of α-helical peptide water hydrogen bonding, *J. Am. Chem. Soc.* 127, 2840–2841, 2005.

231. Zhu, F., Issacs, N.W., Hecht, L., and Barron, L.D., Raman optical activity: A tool for protein structure analysis, *Structure* 13, 1409–1419, 2005.

232. Zhu, F., Issacs, N.W., Hecht, L. et al., Raman optical activity of proteins, carbohydrates and glycoproteins, *Chirality* 18, 103–115, 2006.

233. Hédoux, A., Ionov, R., Willart, J.F. et al., Evidence of a two-stage thermal denaturation process in lysozyme: A Raman scattering and differential scanning calorimetry investigation, *J. Chem. Phys.* 124, 14703, 2006.

234. Jaisson, S., Lorimier, S., Ricard-Blum, S. et al., Impact of carbamylation on type I collagen conformational structure and its ability to activate polymorphonuclear neutrophils, *Chem. Biol.* 13, 149–159, 2006.

235. Thawornchinsombut, S., Park, J.W., Meng, G., and Li-Chan, E.C., Raman spectroscopy determines structural changes associated with gelation properties of fish proteins recovered at alkaline pH, *J. Agric. Food Chem.* 54, 2178–2187, 2006.

236. Podstawka, E., Mak, P.J., Kincaid, J.R., and Proniewicz, L.M., Low frequency resonance Raman spectra of isolated alpha and beta subunits of hemoglobin and their deuterated analogues, *Biopolymers* 83, 455–466, 2006.

237. Dàvila, E., Parés, D., and Howell, N.K., Fourier transform Raman spectroscopy study of heat-induced gelation of plasma proteins as influenced by pH, *J. Agric. Food Chem.* 54, 7890–7897, 2006.

238. Xu, M., Shashilov, V.A., Ermolenkov, V.V. et al., The first step of hen egg white lysozyme fibrillation, irreversible partial unfolding, is a two-state transition, *Protein Sci.* 16, 815–832, 2007.

239. Ma, L., Ahmed, Z., Mikonin, A.V., and Asher, S.A., UV resonance Raman measurements of poly-L-lysine's conformational energy landscapes: Dependence on perchlorate concentration and temperature, *J. Phys. Chem. B* 111, 7675–7680, 2007.

240. Strachan, C.J., Rades, T., Gordon, K.C., and Rantanen, J., Raman spectroscopy for quantitative analysis of pharmaceutical solids, *J. Pharm. Pharmacol.* 59, 179–192, 2007.

241. Wen, Z.Q., Raman spectroscopy of protein pharmaceuticals, *J. Pharm. Sci.* 96, 2861–2878, 2007.

242. Heyduk, T., Baichoo, N., and Heyduk, E., Hydroxyl radical footprinting using metal ion complexes, *Met. Ions Biol. Syst.* 38, 255–287, 2001.

243. Kiselar, J.G., Janmey, P.A., Almo, S.C., and Chance, M.R., Structural analysis of gelsolin using synchrotron protein footprinting, *Mol. Cell. Proteomics* 2, 1120–1132, 2003.

244. Guan, J.Q. and Chance, M.R., Structural proteomics of macromolecular assemblies using oxidative footprinting and mass spectrometry, *Trends Biochem. Sci.* 30, 583–592, 2005.

245. Takamoto, K. and Chance, M.R., Radiolytic protein footprinting with mass spectrometry to probe the structure of macromolecular complexes, *Annu. Rev. Biophys. Biomol. Struct.* 35, 251–276, 2006.

246. Gómez, G.E., Cauerhff, A., Craig, P.O. et al., Exploring protein interfaces with a general photochemical reagent, *Protein Sci.* 15, 744–752, 2006.

247. Shchebakova, I., Mitra, S., Beer, R.H., and Brenowitz, M., Fast Fenton footprinting: A laboratory-based method for the time-resolved analysis of DNA, RNA, and proteins, *Nucleic Acids Res.* 34, e48, 2006.

248. Linderstrom-Lange, K., Globular proteins and proteolytic enzymes, *Proc. R. Soc. B* 127, 17, 1939; (a) Wilson, W.D. and Foster, J.F., Conformation dependent limited proteolysis of bovine plasma albumin preparations, *Biochemistry* 10, 1772–1780, 1971.

249. Peters, T. Jr. and Feldhoff, R.C., Fragments of bovine serum albumin produced by limited proteolysis. Isolation and characterization of tryptic fragments, *Biochemistry* 14, 3384–3391, 1975.

250. Reed, R.B., Feldhoff, R.C., Clute, O.L., and Peters, T. Jr., Fragments of bovine serum albumin produced by limited proteolysis. Conformation and ligand binding, *Biochemistry* 14, 4578–4583, 1975.

251. Dombrádi, V., Gergely, P., and Bot, G., Limited proteolysis by subtilisin reveals structural differences between phosphorylase a and b, *Int. J. Biochem.* 15, 1088–1092, 1983.

252. Sheshberadaran, H. and Payne, L.G., Protein antigen-monoclonal antibody contact sites investigated by limited proteolysis of monoclonal antibody-bound antigen: Protein "footprinting," *Proc. Natl. Acad. Sci. USA* 85, 1–5, 1988.

253. Miedel, M.C., Hulmes, J.D., and Pan, Y.C., Limited proteolysis of recombinant human soluble interleukin-2 receptor. Identification of an interleukin-2 binding core, *J. Biol. Chem.* 264, 21097–21105, 1989.

254. Olson, M.O., Kirstein, M.N., and Wallace, M.O., Limited proteolysis as a probe of the conformation and nucleic binding regions of nucleolin, *Biochemistry* 29, 5682–5686, 1990.

255. Wilson, J.E., The use of monoclonal antibodies and limited proteolysis in elucidation of structure–function relationships in proteins, *Methods Biochem. Anal.* 35, 207–250, 1991.

256. Gentile, F. and Salvatore, G., Preferential sites of proteolytic cleavage of bovine, human, and rat thyroglobulin. The use of limited proteolysis to detect solvent-exposed regions of the primary structure, *Eur. J. Biochem.* 218, 603–621, 1993.

257. Fontana, A., Zambonin, M., Polverino de Laureto, P. et al., Probing the conformational state of apomyoglobin by limited proteolysis, *J. Mol. Biol.* 266, 223–230, 1997.

258. Hubbard, S.J., The structural aspects of limited proteolysis of native proteins, *Biochim. Biophys. Acta* 1382, 191–206, 1998.

259. Polverino de Laureto, P., Scaramella, E., Frigo, M. et al., Limited proteolysis of bovine α-lactalbumin: Isolation and characterization of protein domains, *Protein Sci.* 8, 2290–2303, 1999.

260. Yang, S.A. and Klee, C., Study of calcineurin structure by limited proteolysis, *Methods Mol. Biol.* 172, 317–334, 2002.

261. Polverino de Laureto, P., Frare, E., Gottardo, R. et al., Partly folded states of members of the lysozyme/lactalbumin superfamily: A comparative study by circular dichroism spectroscopy and limited proteolysis, *Protein Sci.* 11, 2932–2946, 2002.

262. Fontano, A., de Laureto, P.P., Spolaore, B. et al., Probing protein structure by limited proteolysis, *Acta Biochim. Pol.* 51, 299–321, 2004.

263. Stroh, J.G., Loulakis, P., Lanzetti, A.J., and Xie, J., LC-mass spectrometry analysis of N- and C-terminal boundary sequences of polypeptide fragments by limited proteolysis, *J. Am. Soc. Mass Spectrom.* 16, 38–45, 2005.

264. Williams, J.G., Tomer, K.B., Hice, C.E. et al., The antigenic determinants on HIV p24 for CD4+ T cell inhibiting antibodies as determined by limited proteolysis, chemical modification and mass spectrometry, *J. Am. Soc. Mass Spectrom.* 17, 1560–1590, 2006.

265. Sonoda, S. and Schlamowitz, M., Studies of ^{125}I trace labeling of immunoglobulin G by chloramine-T, *Immunochemistry* 7, 885–898, 1970.

266. Giedroc, D.P., Sinha, S.K., Brew, K., and Puett, D., Differential trace labeling of calmodulin: Investigation of binding sites and conformational states by individual lysine reactivities. Effects of beta-endorphin, trifluoperazine, and ethylene glycol bis(beta-aminoethyl ether)-*N,N,N′,N′*-tetraacetic acid, *J. Biol. Chem.* 260, 13406–13413, 1985.

267. Jackson, A.E., Carraway, K.L. III, Puett, D., and Brew, K., Effects of the binding of myosin light chain kinase on the reactivities of calmodulin lysines, *J. Biol. Chem.* 261, 12226–12232, 1986.

268. Giedroc, D.P., Puett, D., Sinha, S.K., and Brew, K., Calcium effects on calmodulin lysine reactivities, *Arch. Biochem. Biophys.* 252, 136–144, 1987.

269. Lin, T.P. and Hsu, C.C., Determination of residual moisture in lyophilized protein pharmaceuticals using a rapid and non-invasive method: Near infrared spectroscopy, *PDA J. Pharm. Sci. Technol.* 56, 196–205, 2002.

270. Bruun, S.W., Søndergaard, I., and Jacobsen, S., Analysis of protein structures and interaction in complex foods by near-infrared spectroscopy. I. Gluten powder, *J. Agric. Food Chem.* 55, 7234–7243, 2007.

271. Hermida, M., Rodriguez, N., and Rodriguez-Otero, J.L., Determination of moisture, starch, protein, and fat in common beans (*Phaseolus vulgaris* L.) by near infrared spectroscopy, *J. AOAC Int.* 89, 1039–1041, 2006.

272. Berntsson, O., Zackrisson, G., and Ostling, G., Determination of moisture in hard gelatin capsules using near-infrared spectroscopy applications to at-line process control of pharmaceutics, *J. Pharm. Biomed. Anal.* 15, 895–900, 1997.

273. Savage, M., Torres, J., Franks, L. et al., Determination of adequate moisture content for efficient dry-heat viral inactivation in lyophilized factor VIII by loss on drying and by near infrared spectroscopy, *Biologicals* 26, 119–124, 1998.

274. Zheng, Y., Lai, X., Bruun, S.W. et al., Determination of moisture content of lyophilized allergen vaccines by NIR spectroscopy, *J. Pharm. Biomed. Anal.* 46, 592–596, 2008.

275. Brulls, M., Folestad, S., Sparén, A. et al., Applying spectral peak area analysis in near-infrared spectroscopy moisture assays, *J. Pharm. Biomed. Anal.* 44, 127–136, 2007.

276. Nagarajan, R., Singh, P., and Mehrotra, R., Direct determination of moisture in powder milk using near infrared spectroscopy, *J. Autom. Methods Manag. Chem.* 2, 51342, 2006.

277. Konrad, M., Immunogenicity of proteins administered to humans for therapeutic purposes, *Trends Biotechnol.* 7, 175–179, 1989.

278. Wadhwa, M., Bird, C., Dilger, P. et al., Strategies for detection, measurement and characterization of unwanted antibodies induced by therapeutic biologicals, *J. Immunol. Methods* 278, 1–17, 2003.

279. Geng, D., Shankar, G., Schantz, A. et al., Validation of immunoassays used to assess immunogenicity to therapeutic monoclonal antibodies, *J. Pharm. Biomed. Anal.* 39, 364–375, 2005.

280. Doyle, J.W., Johnson, G.L., Eshhar, N., and Hammond, D., The use of rabbit polyclonal antibodies to assess neoantigenicity following viral reduction of an alpha-1-proteinase inhibitor preparation, *Biologicals* 34, 199–297, 2006.

281. Gupta, S., Indelicato, S.R., Jethwa, V. et al., Recommendations for the design, optimization, and qualification of cell-based assays used for the detection of neutralizing antibody responses elicited to biological therapeutics, *J. Immunol. Methods* 321, 1–18, 2007.

282. Brennen, T.V. and Clarke, S., Mechanism of cleavage at Asn 148 during the maturation of jack bean concanavlin A, *Biochem. Biophys. Res. Commun.* 193, 1031–1037, 1993.

283. Capasso, S., Mazzarella, L., Sorrentino, G. et al., Kinetics and mechanism of the cleavage of the peptide bond next to asparagine, *Peptides* 17, 1075–1077, 1996.

284. Mathys, S., Evans, T.C., Chute, I.C. et al., Characterization of a self-splicing mini-intein and its conversion into autocatalytic N- and C-terminal cleavage elements: Facile production of protein building blocks for protein ligation, *Gene* 231, 1–13, 1988.

285. Paulus, H., Protein splicing and related forms of protein autoprocessing, *Annu. Rev. Biochem.* 69, 447–496, 2000.

286. Li, B., Gorman, E.M., Moore, K.D. et al., Effects of acidic N + 1 residues on asparagine deamidation rates in solution and in the solid state, *J. Pharm. Sci.* 94, 666–675, 2005.

287. Kawauchi, H., Bewley, T.A., and Li, C.H., Studies on prolactin. 40. Chemical modification of tyrosine residues in the ovine hormone, *Biochim. Biophys. Acta* 493, 380–392, 1977.

288. Yokosawa, H. and Ishii, S., Anydrotrypsin and trypsin: Subtle differences in the active-site conformations detected by chemical modification and CD spectroscopy, *J. Biochem.* 81, 657–663, 1977.

289. Atassi, M.Z. and Zablocki, W., Conformation, enzymic activity, and immunochemistry of a lysozyme derivative modified at tryptophan 123 by reaction with 2,3-dioxo-5-indolinesulfonic acid, *J. Biol. Chem.* 251, 1653–1658, 1976.

290. Lu, Y., Harding, S.E., Turner, A. et al., Effect of PEGylation on the solution conformation of antibody fragments, *J. Pharm. Sci.* 97, 2062–2097, 2008.

291. Meng, F., Manjula, B.N., Smith, P.K. et al., PEGylation of human serum albumin: Reaction of PEG-phenyl-isothiocyantate with protein, *Bioconjug. Chem.* 19, 1352–1360, 2008.

292. Rodríguez-Martínez, J.A., Solá, R.J., Castillo, B. et al., Stabilization of α-chymotrypsin upon PEG-ylation correlates with reduced structural dynamics, *Biotechnol. Bioeng.* 101, 1142–1149. 2008.

293. Gonnelli, M. and Strambini, G.B., No effect of covalently linked poly(ethylene glycol) chains on protein internal dynamics, *Biochim. Biophys. Acta* 1974, 569–576, 2008.

2 Comparability of Biotechnological/ Biological Products and Biological Generics

Comparability of the final drug product and/or active pharmaceutical ingredient is an issue for at least two situations in the manufacture of therapeutic biopolymers. The first is when there is a change in the manufacturing process[1–5] and the second is with the development of biological generics (follow-on biologicals; biosimilar products).[6–12] It is this area that has attracted the most interest in recent years. In either situation, the concern is that several products are "identical" with respect to bioequivalence and safety. Bioequivalence is driven by pharmacokinetics and therapeutic effect while safety is largely, but not entirely, a product of immunogenicity. These questions can only be satisfactorily addressed by clinical experience. The laboratory challenge is to assure that the several versions of the same biological product, e.g., erythropoietin, will be successful in clinical trials. In other words, the laboratory challenge is to prove that the products are *identical*, The *Oxford Dictionary of the English Language* defines identical as "Agreeing entirely in material, constitution, properties, qualities, or meaning: said of two or more things which are equal parts of one uniform whole, individual examples of one species, or copies of one type, so that any one of them may, for all purposes, or for the purposes contemplated, be substituted for any other." Now, it is relatively easy to show that two forms of aspirin are identical; acetylsalicyclic acid is acetylsalicyclic acid although it might be called acetyl-(2-acetoxy)benzoic acid. The situation gets a little more complex as one considers biological products. Synthetic oligonucleotides can be produced to a high degree of identity within and between production lots[13,14] as can peptide therapeutics such as insulin,[15,16] which do not require glycosylation. However, even insulin preparations are subject to heterogeneity, reflecting aggregation and deamidation.[17–20] Thus, even with a simple peptide, there is heterogeneity, which does not markedly affect product quality. Moving up to a more complex peptide/protein, erythropoietin, the situation becomes more complex because of glycosylation.[21–23] It is clear that there are differences in glycosylation among the several recombinant erythropoietin products,[24] which do result in differences in pharmacokinetics. There are also differences in glycosylation between the native protein and the recombinant protein.[23,25] It is noted that glycosylation of erythropoietin is not required for receptor binding and activity.[26] There are other examples of differences in posttranslational processing with other therapeutic proteins, which do not appear to make a difference in their clinical performance. Factor VIII is a protein whose activity is absent or severely diminished in

19

hemophilia A. A subject with hemophilia A may either be CRM[+] (positive) where an abnormal factor VIII protein is produced, CRM[-] (negative) where factor VIII protein is missing, or CRM[r] (reduced).[27,28] While the majority of these mutations are thought be strictly arising out of changes in the amino acid sequence, there is some evidence[29-31] to suggest that glycosylation may be important in the etiology of hemophilia A. As with most plasma glycoproteins, glycosylation does have an effect on pharmacokinetics[32] but there is no evidence to suggest its importance in function.[33] However, in the case of factor VIII, pharmacokinetics is complicated by in vivo binding to von Willebrand factor.[34] There are three forms of human factor VIII protein, which appear to be considered as identical from a regulatory perspective in that all three derived therapeutic products are licensed globally for the treatment of hemophilia A.[35] There is a fourth form, which is an engineered factor VIII molecule missing in the B-domain[36] and considered bioequivalent to other factor VIII products. The three forms (plasma-derived, derived from baby hamster kidney cells, and derived from Chinese hamster ovary cells) of factor VIII are considered to be identical with respect to amino acid sequence but different with respect to glycosylation.[37-39] This difference with factor VIII is complicated further by the intrinsic heterogeneity of factor VIII with respect to primary structure; while factor VIII is synthesized as a single polypeptide chain, the secreted product is heterogeneous with several peptide bond cleavages.[40-42]

The recombinant factor IX protein product also differs from the plasma-derived material with respect to glycosylation and other posttranslational modifications (β-hydroxylation; sulfation).[43] Heterogeneity is an accepted factor in biological products[44]; it is the management of such heterogeneity that is the challenge, leading to the concept that process drives product.[45,46]

Comparability is the issue with either changes in manufacturing or with biosimilar products. Comparability is defined as the quality of being comparable.[47] Comparable is defined as "A conclusion that products are highly similar before and after manufacturing process changes and that no adverse impact on the quality, safety, or efficacy of the drug product occurred. This conclusion can be based on an analysis of product quality attributes. In some cases, nonclinical or clinical data might be indicated."[1] The International Conference on Harmonisation stresses the importance of the quality attributes of product in establishing comparability.[1] A quality attribute is defined as "A molecular or product characteristic that is selected for its ability to help indicate the quality of the product. Collectively, the quality attributes define the adventitious agent safety, purity, potency, identity, and stability of the product. Specifications measure a selected subset of the quality attributes."[1] The goal of the laboratory/preclinical evaluation of the products is to guide the design of clinical studies that will establish the quality, safety, and efficacy of the drug product. It is relatively easy to establish comparability of amino acid sequence of two protein preparations. It is somewhat more difficult to evaluate the effect of glycosylation as briefly discussed above. It is even more difficult to establish comparability of conformation. Conformation, or protein shape, is likely the most critical factor in the function of a protein; denaturation, for example, destroys protein conformation and function with the retention of primary structure. This book will discuss to a large extent the various technical approaches available for the study of protein conformation as it is a quality attribute of a protein drug products; conformation is less of an issue with oligonucleotide or oligosaccharide products.

There are two technical/regulatory issues that are derived from the manufacturing process. These are both issues of comparability, which are closely related to each other and subject to the same analytical approaches; for all practical purposes, there would be a single problem if only one manufacturer was involved. The first question is whether a product manufactured by A is the same as the product manufactured by A with a significant change in the manufacturing process. For example, a change in media formulation/source materials,[48,49] culture/fermentation temperature, reactor configuration, and agitation conditions has been demonstrated to result in changes in glycosylation pattern.[50–60] The second issue is whether a product made by A is *identical* with the product made by B. such that A can make use of the clinical experience obtained by B to decrease time and expense to market as has been the situation with generic drugs.

The establishment of identity will, of course, depend on what you measure to establish comparability. Establishment of comparability will depend on the definition of the quality attributes of the product concomitant with the methods to measure such attributes. Quality attributes can include characteristics such as color, amino acid sequence, solubility, activity, and immunogenicity, the characteristics that can be clearly related to the safety, efficacy, and quality. We posit that the measurement of conformation is extremely useful if not essential in the establishment of identity between the two products.

The major postulate for the study of product conformation is that conformation is a clear measure of product quality. It is then essential that we have an understanding of which physical and chemical factors can cause conformational change.

Consideration of the limited number of examples in Table 2.1 suggests that there may or may not be conformation secondary to chemical modification of protein primary structure. In general, reactions, which result in charge reversal such as the modification of lysine residues with organic acid anhydrides such as acetic anhydride or maleic anhydride, result in conformational changes while changes in single amino acid residues, which do not involve charge reversal, do not necessarily result in the observed conformational change. An example is provided by the modification of tryptophan in wheat-germ agglutinin (Table 2.1) in which tryptophan residues are modified with no change in circular dichroism (CD) but with total loss of biological activity. Selected studies on the effect of pressure on protein conformation are listed in Table 2.2. Table 2.3 lists some selected studies on the effect of temperature on protein conformation.

Complexity of product[117] has been advanced as a reason for a more rigorous regulatory process[118] for biological generics. Biological generics are designated as biosimlilars or follow-on biologics* because, unlike generic drugs, there is no structural identity[119,120]; it is considered to be comparable to the reference drug but

---

* The nomenclature in this area is evolving. The term biosimilar will be used as used in the European Union, a geography with the most experience in this area with several approved biosimilar drugs. See Rader, R.A., What is a generic biopharmaceutical? biogeneric? follow-on protein? B\biosimilar? follow-on biologic? Part 1: Introduction and basic paradigms, *Bioprocess International*, March 2007; Rader, R.A., What is a generic biopharmaceutical? biogeneric? follow-on protein? B\biosimilar? follow-on biologic? Part 2: Information, nomenclature, perceptions, and the market, *Bioprocess International*, May 2007; Gottlieb, S., Biosimilars: Policy, clinical, and regulatory considerations, *Am. J. Heath Syst. Pharm.* 65 (14 Suppl 6), S2–S8, 2008; Wenzel, R.G., Biosimilars: Illustration of scientific issues in two examples, *Am. J. Health. Syst. Pharm.* 65 (14 Suppl 6), S9–S15, 2008.

TABLE 2.1
Some Studies on the Effect of Chemical Modification on Protein Conformation

Protein	Reagent	Results	Ref.
IgG	Citraconic anhydride	CD[a] showed loss of β-structure (positive CD band at 202 nm became negative; sedimentation coefficient decreased; immunoresponse to goat antiserum lost). All changes reversed by removal of citraconyl groups (dialysis at pH 4.0)	[61][b]
IgG	KCNO/pH 7.7 carbamylation	No change in CD spectra; no change in sedimentation coefficient; no loss of immunoreactivity	[72][b]
Wheat-germ agglutinin	NBS[c]/pH 3.9–8.0 M urea or NBS/pH 6	Two tryptophans modified at pH 6.0 with loss of 85% intrinsic fluorescence (λ_{excit} 280 nm); third tryptophan modified only with pH 3.9/8.0 M urea. No major change in CD spectra (210–260 nm; 260–340 nm); oxidation of one tryptophan resulted in total loss of hemagglutinating activity	[62]
HEWL[d]	I_2	^{13}C NMR[e] used to demonstrate modification of tyrosine residues; modification is pH dependent	[63]
Prolactin	2-Nps-Cl[f] in 70% formic acid	Loss of receptor binding activity; CD shows some loss of α-helical content and some increase in β-sheet content	[64]
Basic pancreatic trypsin inhibitor	Reduced/ carboxymethylated	Sequence-specific assignments for ^{1}H-NMR shifts show changes near modification site and internal hydrogen bonds preserved	[65]
HEWL	Crosslinkage; β-aspartyl formation	Difference in H/D exchange rates of N–1 hydrogens	[66]
Trypsin	Asparagine deamidation	Tertiary structure as a major determinant of asparagine deamidation from neutron crystallographic analysis	[67]
HEWL	Ozone	Oxidation of Trp 62 to kynurenine; little change in CD; decrease in thermal stability	[68]
RNase T$_1$	Ozone	Oxidation of Trp 59 to kynurenine; little change in CD; decrease in thermal stability: this study also used the constant fragment from a λ-immunoglobulin light chain	[68]
Plasma fibronectin	Chloramine-T	Decreased binding of collagen to oxidized (methionine) fibronectin; no change in intrinsic fluorescence or CD spectra	[69]
Amyloid protein	Formic acid	NMR (1D and 2D) demonstrates formation of the formate ester of serine	[70]
Alanine peptides	N/A	Use of CD and amide hydrogen exchange to measure helix-coil transition	[71]
Human IgG$_1$ Fc (recombinant in *Escherichia coli*)	H_2O_2 in phosphate (pH 7.0) or acetate (pH 5.0)	H_2O_2 oxidized methionine residues in recombinant Fc domain resulting in changes in secondary and tertiary structure measured by CD and NMR. There were also changes in DSC analysis	[72]

TABLE 2.1 (continued)
Some Studies on the Effect of Chemical Modification on Protein Conformation

Protein	Reagent	Results	Ref.
BSA, lysozyme, IgG	Succinic anhydride	Results differed with protein studied. In general, there was an increase in Stoke's radius (determined from analytical ultracentrifugation) and a variable difference in the reactivity of disulfide bonds. The sedimentation coefficient for BSA was dependent on sample concentration for the succinylated protein but not for the native protein	[73]
BPTI	Haloacetamide modification of cysteine residue in disulfide bond	NMR measurements detected changes only in the vicinity of the modified residue	[74]
Myoglobin, cytochrome c	Hydroxyl radical	Measurement of rate of γ-ray-mediated oxidation by "on-line" H-D exchange. o-Phosphoric acid increased rate of oxidation while other denaturants such as urea decreases rate by scavenging radical	[75]
Lysozyme[g]	Oxidation	Measurement of the rate of oxidation of native protein and oxidized protein using CD spectroscopy (198 nm). Provides sensitive index of conformational change	[76]
Myoglobin	Hydroxyl radical via γ-radiation; inclusion of O_2	Level of oxidative labeling (oxygen incorporation via mass spectrometry) depends on protein concentration; suggested that oxidative modification does not cause major changes in protein conformation	[77]
SpoOF	Hydroxyl radical via γ-radiation; bicarbonate buffer	Measurement of rate of oxidation and observation of deviation from expected first-order rate constant permits evaluation of conformational change; CD supports loss of some helical structure but bulk of structure is unchanged	[78]
3-Phospho-glycerate kinase	Tetranitromethane	Nitration of single tyrosine inactivates enzyme; CD shows no change 200–260 nm; some small change in environment around aromatic residues on 260–400 nm	[79]
Antithrombin	H_2O_2	Oxidation of two methionine residues (314 and 317) did not results in change in biological activity; oxidation of two more methionine residues (17 and 20) under more rigorous conditions results in some loss of heparin binding affinity; oxidation did result in change in behavior of RP-HPLC (decreased affinity); some change in 180–260 nm (far UV), more pronounced changes in near UV (260–400 nm)	[80]

(continued)

TABLE 2.1 (continued)
Some Studies on the Effect of Chemical Modification on Protein Conformation

Protein	Reagent	Results	Ref.
α-Synculein	4-HNE[h]	Modification of protein with 4-HNE caused major conformation changes (increase β-sheet) as judged by CD and FTIR spectroscopy; decreased aggregation of modified protein	[81]
Catalase	H_2O_2	Oxidation yielded form of enzyme with different catalytic properties and electrophoretic mobility but no gross conformational change (CD, intrinsic fluorescence)	[82]
Calmodulin	H_2O_2	Oxidation of methionine residues is function of solvent exposure; lack of gross conformational changes by CD (small change at 208 nm); oxidized material has somewhat decreased thermal stability	[83]
PAI-1[i]	NClSuc[j]	Oxidation of methionine resulted in CD change which correlated with the loss of activity	[84]
BSA	Iodoacetate, iodoacetamide, DTNB[k]	Modification of the single sulfydryl group in BSA with either iodoacetate or DTNB yielded a derivative which demonstrated β-structure in guandine (>3.0M) while the iodoacetamide derivative demonstrated a random coil under such conditions (measurement by CD). The modification with any of the three reagents did demonstrate a conformational change; a change in conformation was observed with complete reduction of the disulfide bonds of the protein	[85]
Rhodanese	H_2O_2	Oxidized enzyme shows increased susceptibility to proteolysis by either trypsin or chymotrypsin	[86]
N/A	N/A	Review of the use of IR spectroscopy to study conformational change in proteins	[87]

[a]   CD, circular dichroism.

[b]   There is variability in the response of various proteins to different acylating and alkylating agents. See Qasim, M.A. and Salahuddin, A., Changes in conformation and immunological activity of ovalbumin during its modification with different acid anhydrides, *Biochim. Biophys. Acta* 536, 50–63, 1978; Lakkis, J. and Villota, R., Effect of acylation on substructural properties of proteins: A study using fluorescence and circular dichroism, *J. Agric. Food Chem.* 40, 553–560, 1992; Mir, M.M., Fazili, K.M., and Abul Qasim, M., Chemical modification of buried lysine residues of bovine serum albumin and its influence on protein conformation and bilirubin binding, *Biochim. Biophys. Acta* 1119, 261–267, 1992.

[c]   NBS, *N*-bromosuccinimide.

[d]   Hen-egg white lysozyme.

[e]   NMR, nuclear magnetic resonance.

[f]   Nps-Cl; 2-nitrophenylsulfenyl chloride.

[g]   Turkey lysozyme, also β-lactoglobulin, ubiquitin, catalase.

[h]   4-Hydroxy-2-nonenal.

[i]   Plasminogen activator inhibitor-1.

[j]   NClsuc; *N*-chlorosuccinimide.

[k]   DTNB, 5,5-Dithio(bis)nitrobenzoic acid (Ellman's reagent).

TABLE 2.2

Some Studies on the Effect of Pressure on Protein Conformation

Protein	Reaction Conditions	Results	Refs.
Lysozyme	30–2000 bar[a] change[b] 50–100 mM deuterated glycine buffer, pD 2.0	2D NMR; α-helical domain is compressed as the interdomain region; little effect on β-sheet	[88,89]
RNAse and BPTI[c]	1 GPa in 10 mM deuterated Tris-HCl, pD 7.6; 2-mercaptoethanol as reducing agent	Reduction of all disulfide bonds could be accomplished with high pressure. FTIR[d] spectroscopy did not show differences between reduced and unreduced forms under pressure. Amorphous aggregates formed on decompression. It suggested that high pressure results in the formation of aggregation-prone conformers	[90]
N/A	Variable pressure NMR	Review	[91]
N/A	Pressure perturbation calorimetry	Review	[92]
Fibrinogen	Solid-state measurement on KBr pellets; pressure increase to 400 kg/cm²	FTIR spectroscopy demonstrates changes in secondary structure (transition from α-helix to β-sheet) and unfolding/denaturation of fibrinogen[e]	[93]
Canine milk lysozyme	100 MPa; effect on thermal denaturation	UV spectroscopy at pH 4.5 and 2.0	[94]
N/A	Molecular simulation studies	Structural, thermodynamic, and hydration changes as a function of temperature and pressure	[95]
N/A	Review	Pressure changes similar to changes induced by temperature	[96]
Lysozyme	Molecular dynamics simulation at 1 and 3 kbar	Simulation suggests an inversion of hydrophobic and hydrophilic solvent-accessible surface areas; suggests hydrophobic interactions are weaker at higher pressures	[97]
Bacterial inclusion bodies	200 mPa in the presence of reducing agents, pH 8.0[f]	Refolding of inclusion bodies assisted by high hydrostatic pressure. Additives such as arginine prevented aggregation	[98]
BPTI	6000 bar (6 kbar), copper–beryllium high-pressure cell in 50 mM deuterated acetate buffer. Changes in protein structure evaluation by neutron scattering	Increasing pressure reduces radius of gyration; this reduction is not reversible. Slowing down of protein dynamics with increased pressure	[99]
N/A	Simulation studies	Random coil becomes more destabilized with increasing pressure	[100]

(continued)

TABLE 2.2 (continued)
Some Studies on the Effect of Pressure on Protein Conformation

Protein	Reaction Conditions	Results	Refs.
Azurin	0–6 kbar in 2 mM Tris-HCl, pH 7.5	Use of intrinsic fluorescence and phosphorescence to study protein denaturation; decrease in intrinsic fluorescence intensity on denaturation	[101]
Equine serum albumin	0–110 mPa in 0.1 M NaCl at pD 4.4; 60°C	Study of the differential effect of pressure and heat on the denaturation of equine serum albumin using FTIR spectroscopy. Pressure prevents heat aggregation. Heat-induced aggregates with intermolecular β-sheet and pressure-induced aggregates without intermolecular β-sheet	[102]
Thioredoxin	0.1–400 mPa in 50 mM Tris, pH 7.5	Measurement of intrinsic fluorescence as function of pressure. There was an initial decrease in fluorescence reflecting a more compact protein with quenching by a nearby disulfide; at higher pressure, the protein unfolds with an increase in fluorescence	[103]
Ubiquitin	30–3000 atm	NMR studies of water penetration into protein with pressure	[104]

[a] 1 atm = 1.01 bar = 101.3 kPa (kilopascal).

[b] Yamada, H., Pressure-resisting glass cell for high pressure, high resolution NMR measurement, *Rev. Sci. Instrum.* 45, 640–642, 1974.

[c] RNAse, bovine pancreatic ribonuclease; BPTI, bovine pancreatic trypsin inhibitor.

[d] FTIR, Fourier transform infrared.

[e] Fibrinogen is also subject to surface denaturation (Prokopowicz, M., Lukasiak, J., Banecki, B., and Przyjazny, A., In vitro measurement of conformational stability of fibrinogen adsorbed on siloxane, *Biomacromolecules* 6, 39–45, 2005; Santore, M.M. and Wertz, C.F., Protein spreading kinetics at liquid-solid interfaces via an adsorption probe, *Langmuir* 21, 10107–10178, 2005; Xu, L.C. and Siedlecki, C.A., Effects of surface wettability and contact time on protein adhesion to biomaterial surfaces, *Biomaterials* 28, 3273–3283, 2007).

[f] Alkaline pH required for thiolate anion which is required for disulfide reshuffling.

[g] Neutron scattering is a technique for the study of protein conformation (Harroun, T.A., Wignall, G.D., and Kataras, J., Neutron scattering in biology, in *Neutron Scattering in Biology*, eds. J. Fitter, T. Gutberlet, and J. Katsaras, Springer, Berlin, Germany, 2006). This technique requires access to a nuclear reactor or linear accelerator. Elastic scattering provides compositional information while inelastic scattering measures molecular motion (Bucknall, D.G., Neutron scattering in analysis of polymers and rubbers, in *Encyclopedia of Analytical Chemistry*, eds. R.A. Meyers, John Wiley & Sons, Chichester, U.K., 2000). There is a large difference in scattering by hydrogen and deuterium which is quiet useful for the analysis of protein conformation; the ability to use neutron scattering for dry and hydrated samples is of great interest (Marconi, M., Conicchi, E., Onori, G., and Paciaroni, A., Comparative study of protein dynamics in hydrated powders and in solutions: A neutron scattering investigation, *Chem. Phys.* 345, 224–229, 2008; Wood, K., Caronna, C., Fouquet, P. et al., A benchmark for protein dynamics: Ribonuclease A measured by neutron scattering in a large wavevector-energy transfer range, *Chem. Phys.* 345, 305–314, 2008). There have been major technical advances in this field which should facilitate greater use of this technology (Teixeira, S.C.M., Zaccai, G., Ankner, J. et al., New sources and instrumentation for neutrons in biology, *Chem. Phys.* 345, 133–151, 2008).

TABLE 2.3
Some Studies on the Effect of Temperature on Protein Conformation

Protein	Reaction Conditions	Results	Ref.
β-Lactoglobulin	Thermal denaturation	Use of capillary zone electrophoresis (CZE) to measure protein denaturation; CZE performed between 4°C and 95°C	[105]
β-Lactoglobulin	pH 7.0/heating	Synchrotron small-angle x-ray diffraction, Fourier transform IR spectroscopy; above 50°C, IR showed a loss of intramolecular β-sheet and α-sheet	[106]
Horseradish peroxidase	Effect of temperature and pH	CD and intrinsic fluorescence; removal of calcium ions decreases stability	[107]
Lysozyme	^{15}N-labeled human lysozyme (2 mM) in 50 mM KCl/D$_2$O, pH 3.8	Effect of temperature on heteronuclear multidimensional NMR spectroscopy; decrease in volume and surface area with decreasing temperature	[108]
Cutinase (*Fusarium*)	Heating (protein melting) in 0.01 M acetate (pH 4.0), 0.01 M phosphate (pH 6.0, pH 8.0)	A single tryptophan residue; fluorescence quenched in native protein by cystine (disulfide bond); quenching decreased (fluorescence increased) on heating	[109]
Lysozyme and ribonuclease A	Neutral pH (deuterated Tris-HCl, pD 7.6); approximately 3.5 mM protein	Use of FTIR to monitor thermal denaturation. Aggregate formation; dissociation of amorphous aggregates at higher temperatures preceded by conformation change	[110]
N/A	Computer simulation using lattice model	Comparison between chemical (denaturant)- and temperature-induced protein denaturation. Results shows a wider distribution of conformational states than temperature-induced denaturation	[111]
Canine plasminogen	10 mM potassium phosphate–100 mM NaCl, pH 6.5	Large structural changes over the range of 4°C–20°C, Stokes' radius decreases for both the open and closed form as measured by dynamic light scattering or analytical ultracentrifugation	[112]
Bovine serum albumin	Glass transitions of aqueous solutions. Measurement of heat capacity and enthalpy relaxation with adiabatic calorimetry	Relaxation with glass transition is from nonequilibrium to equilibrium state. Enthalpy relaxation rate depends on thermal history. There are three distinguishable glass transitions in subzero range	[113]

(continued)

TABLE 2.3 (continued)

Some Studies on the Effect of Temperature on Protein Conformation

Protein	Reaction Conditions	Results	Ref.
Bovine serum albumin	Dry and hydrated samples	Nuclear magnetic transverse decay; proton second moment. There is a change in water at 170 K; no change with D_2O. Suggested that chains extend from protein in hydrated state but not in dry state	[114]
β-Lactoglobulin	Phenyl-Sepharose® vary salt concentration and temperature	Two-conformation adsorption model; thermodynamic models to predict distribution between various conformers-predict trends in retention strength and stability	[115]
Poly-L-lysine	Perchlorate concentration and temperature	Use of Raman spectroscopy to determine conformational energy landscape; at 1°C, 0.83 M perchlorate, 86% α-helix melting into extended conformation at 60°C	[116]

not necessarily possessing therapeutic equivalence.[121] The term *biosimilar* is really more of a regulatory pathway descriptor than a product designation.[11,122] The laboratory challenge is the clarification of meaningful product quality attributes, which can be subjected to rigorous analysis using established laboratory instrumentation.

Lubiniecki and Federici[44] raise an important point with respect to the consideration of quality attributes. It is not unusual to have a "disconnect" between the structural understanding of a biological drug and the understanding of the mechanism of action. For example, the structure of human growth hormone is well understood but we are still probing the molecular mechanism.[123] Likewise, it can be argued that the structure of blood coagulation factor VIII is understood with greater clarity than is the mechanism action.

Additional issues requiring consideration in comparability exercises include an understanding of the existing product.[44] Some biological products have been in the market for a number of years and the initial review process did not include the technical rigor expected with current products. Lack of a thorough understanding of either the drug product or drug substance confounds any study on comparability. In addition, the choice of reference analyte is an equal challenge: drug substance (active pharmaceutical ingredient) or drug product? It is presumed that there is no difficulty when the comparability exercise is concerned with the effect of a significant change in manufacturing on product attributes. The caveat is the assumption that the product is a "well-characterized biological product" and/or the availability of retention samples of active pharmaceutical ingredient together with an understanding of product stability. The comparability exercise with biosimilars is more of a challenge since it is likely that only drug product (formulated) will be available. Formulation of a biological can result in structural changes without negatively affecting efficacy, quality, or safety.[4] There is also concern that properties can change on storage.[124,125] More recently, it has been argued that a final drug product cannot be used for comparability studies. This poses an issue with biosimilar products when only the final drug product may be available for comparison with another biosimilar product. Generally, the active pharmaceutical ingredient form is used for the evaluation of manufacturing changes; the addition of excipients can unnecessarily complicate the analysis. While this is not a problem within an organization, it is most unlikely the company A (see above) would provide active pharmaceutical ingredient to company B.

It has been suggested that drug cannot be used for comparability exercises.[126] The reader is directed to a useful consideration of the merits of comparability studies.[127]

The remainder of this book contains a discussion of the application of conformational assay methods in the characterization of biological polymers. While structural analysis is useful, conformational analysis is the most useful it determining the properties of a biological macromolecule and guiding the preclinical development of biopharmaceuticals.

REFERENCES

1. ICH 5QE, Comparability of biotechnological/biological products subject to changes in their manufacturing process, International Conference on Harmonisation, http://www.ich.org
2. Schenerman, M.A., Hope, J.N., Kletke, C. et al., Comparability of testing of a humanized monoclonal antibody (Synagis) to support cell line stability, process validation, and scale-up for manufacturing, *Biologicals* 27, 203–215, 1999.
3. Petricciani, J., A global view of comparability concepts, *Dev. Biol.* (Basel), 109, 9–13, 2002.
4. Defelippis, M.R. and Larimore, F.S., The role of formulation in insulin comparability assessments, *Biologicals* 34, 49–54, 2006.
5. Simek, S.L., Characterization of gene therapy products and the impact of manufacturing changes on product comparability, *Dev. Biol.* (Basel), 122, 139–144, 2005.
6. Schellekens H., Biosimilar therapeutic agents: Issues with bioequivalence and immunogenicity, *Eur. J. Clin. Invest.* 34, 797–799, 2004.
7. Kuhlman, M. and Covic, A., The protein science of biosimilars, *Nephrol. Dial. Transplant.* 21(Suppl 5), v4–v8, 2006.
8. Roger, S.S. and Mikhail, A., Biosimilars: Opportunity of cause of concern? *J. Pharm. Pharm. Sci.* 10, 405–410, 2007.
9. Genazzani, A.A., Biggio, G., Caputi, A.P. et al., Biosimilar drugs: Concerns and opportunities, *BioDrugs* 21, 351–356, 2007.
10. Goldsmith, D., Kuhlmann, K., and Covic, A., Through a looking glass: The protein science of biosimilars, *Clin. Exp. Nephrol.* 1, 191–195, 2007.
11. Pavlovic, M., Girardin, E., Kapetanovic, L. et al., Similar biological medicinal products containing human growth hormone: European regulation, *Horm. Res.* 69, 14–21, 2008.
12. Schellekens, H., The first biosimilar epoetin: But how similar is it? *Clin. J. Am. Soc. Nephrol.* 3, 174–178, 2008.
13. Sanghvi, Y.S. and Schulte, M., Therapeutic oligonucleotides: The state-of-the-art in purification technologies, *Curr. Opin. Drug Discov. Dev.* 7, 765–776, 2004.

14. Anon., Oblimersen, BCL-2 antisense oligonucleotide—Genta, G 3139m GC 3139, *Drugs R&D* 8, 321–334, 2007.
15. Bristow, A.F., Recombiant-DNA-derived insulin analogues as potentially useful therapeutic agents, *Trends Biotechnol.* 11, 301–305, 1993.
16. Walsh, G., Therapeutic insulins and their large-scale manufacture, *Appl. Microbiol. Biotechnol.* 67, 151–159, 2005.
17. Ladisch, M.R. and Kohlmann, K.L., Recombinant human insulin, *Biotechnol. Prog.* 8, 469–478, 1992.
18. Branje, J. and Langkjoer, L., Insulin structure and stability, *Pharm. Biotechnol.* 5, 315–350, 1992.
19. Jars, M.U., Hvass, A., and Waaben, D., Insulin aspart (AspB28 human insulin) derivatives formed in pharmaceutical solutions, *Pharm. Res.* 19, 621–628, 2002.
20. Huus, J., Havelund, S., Olsen, H.B. et al., Chemical and thermal stability of insulin: Effects of zinc and ligand binding to the insulin zinc-hexamer, *Pharm. Res.* 23, 2611–2620, 2006.
21. Skibeli, V., Nissen-Lie, G., and Torjesen, P., Sugar profiling proves that human serum erythropoietin differs from recombinant human erythropoietin, *Blood* 98, 3626–2634, 2001.
22. Ohta, M., Kawaski, N., Itoh, S., and Hayakawa, T., Usefulness of glycopeptides mapping by liquid chromatography/mass spectrometry in comparability assessment of glycoprotein products, *Biologicals* 30, 235–244, 2002.
23. Llop, E., Gallego, R.G., Belalcazar, V. et al., Evaluation of protein N-glycosylation in 2-DE: Erythropoietin as a study case, *Proteomics* 7, 4278–4291, 2007.
24. Jurado Garcia, J.M., Torres Sánchez, E., Olmos Hildalgo, D., and Alba Conejo, E., Erythropoietin pharmacology, *Clin. Transl. Oncol.* 9, 715–722, 2007.
25. Amadeo, I., Oggero, M., Zenclussen, M.L. et al., A single monoclonal antibody as probe to detect the entire set of native and partially unfolded rhEPO glycoforms, *J. Immunol. Methods* 293, 191–205, 2004.
26. Sytkowski, A.J., Feldman, L., and Zurbuch, D.J., Biological activity and structural stability of N-glycosylated recombinant human erythropoietin, *Biochem. Biophys. Res. Commun.* 176, 698–704, 1991.
27. Girma, J.P., Fressinaud, E., Houllier, A. et al., Assay of factor VIII antigen (VIII:CAg) in 294 haemophilia A patients by a new commercial ELISA using monoclonal antibodies, *Haemophilia* 4, 98–103, 1998.
28. David, D., Saenko, E.L., Santos, I.M. et al., Stable recombinant expression and characterization of the two haemophilic factor VIII variants C329S (CRM⁻) and G1948D (CRMʳ), *Br. J. Haematol.* 113, 604–615, 2001.
29. Aly, A.M., Higuchi, M., Kasper, C.K. et al., Hemophilia A due to mutations that create new N-glycosylation sites, *Proc. Natl. Acad. Sci. USA* 89, 4933–4937, 1992.
30. Hoyer, L.W., Characterization of dysfunctional factor VIII molecules, *Methods Enzymol.* 222, 169–176, 1993.
31. Amano, K., Sarkar, R., Pemberton, S. et al., The molecular basis for cross-reacting material-positive hemophilia A due to missense mutations within the A2-domain of factor VIII, *Blood* 91, 538–548, 1998.
32. Bovenschen, N., Rijken, D.C., Havekes, L.M. et al., The B domain of coagulation factor VIII interacts with the asialoglycoprotein receptor, *J. Thromb. Haemost.* 3, 1257–1265, 2005.
33. Fay, P.J., Chavin, S.I., Malone, J.E. et al., The effect of carbohydrate depletion on procoagulant activity and *in vivo* survival of highly purified human factor VIII, *Biochim. Biophys. Acta* 800, 152–158, 1984.
34. Denis, C.V., Christophe, O.D., Oortwijn, B.D., and Lenting, P.J., Clearance of von Willebrand factor, *Thromb. Haemost.* 99, 271–278, 2008.
35. Berntorp, E., vWF/FVIII complex and the management of patient with inhibitors: From laboratory to clinical practice, *Haemophilia* 13 (Suppl 5), 69–72, 2007.

36. Kessler, C.M., Gill, J.C., White II, G.C. et al., B-domain deleted recombinant factor VIII preparations are bioequivalent to a monoclonal-antibody purified plasma-derived factor VIII concentrate: A randomized, three-way crossover study, *Haemophilia* 11, 84–91, 2005.
37. Hironaka, T., Furukawa, K., Esmon, P.C. et al., Comparative study of the sugar chains of factor VIII purified from human plasma and from the culture media of recombinant baby hamster kidney cells, *J. Biol. Chem.* 267, 8012–8020, 1992.
38. Kumar, H.P., Hague, C., Haley, T. et al., Elucidation of the N-linked oligosaccharide structures of recombinant human factor VIII using fluorophore-assisted carbohydrate electrophoresis, *Biotechnol. Appl. Biochem.* 24, 207–216, 1996.
39. Schilow, W.F., Schoerner-Burkhardt, E., and Seitz, R., Charge analysis of *N*-glycans from human recombinant coagulation factor VIII and human factor VIII standards, *Thromb. Haemost.* 92, 427–428, 2004.
40. Fulcher, C.A. and Ziumerman, T.S., Characterization of the human factor VIII proco-agulant protein with a heterologous precipitating antibody, *Proc. Natl. Acad. Sci. USA* 79, 1648–1652, 1982.
41. Ganz, P.R., Tackaberry, E.S., Palmer, D.S., and Rock, G.S., Human factor VIII from heparinized plasma, purification and characterization of a single-chain form, *Eur. J. Biochem.* 170, 521–528, 1988.
42. Bihoreau, N., Sauger, A., Yon, J.M., and Van de Pol, H., Isolation and characterization of different activated forms of factor VIII, the human antihemophilic A factor, *Eur. J. Biochem.* 185, 111–118, 1989.
43. Pipe, S.W., Recombinant clotting factors, *Thromb. Haemost.* 99, 840–850, 2008.
44. Lubiniecki, A.S. and Federici, M.M., Comparability is not just analytical equivalence, *Biologicals* 34, 45–47, 2006.
45. Kozlowski, S. and Swann, P., Current and future issues in the manufacturing and development of monoclonal antibodies, *Adv. Drug Deliv. Rev.* 58, 707–722, 2006.
46. Sadick, M., Understanding the puzzle of well-characterized biotechnology products, *Curr. Opin. Biotechnol.* 13, 275–278, 2002.
47. *Oxford English Dictionary,* on-line, Oxford University Press, Oxford, United Kingdom, 2009.
48. Robertson, J.S., Changes in biological source material, *Biologicals* 34, 61–63, 2006.
49. Sewerin, K., Schachter, E., Robertson, J., and Wallerius, C., Changes to biological source materials, *Biologicals* 34, 71–72, 2006.
50. Borys, M.C., Linzer, D.I., and Papoutsakis, E.T., Culture pH affects expression rates and glycosylation of recombinant mouse placental lactogen proteins by Chinese hamster ovary (CHO) cells, *Biotechnology* 11, 720–724, 1993.
51. Watson, E., Shah, B., Leiderman, L. et al., Comparison of N-linked oligosaccharides of recombinant human tissue kallikrein produced by Chinese hamster ovary cells on microcarrier beads and in serum-free suspension culture, *Biotechnol. Prog.* 10, 39–44, 1993.
52. Jenkins, N., Castro, P., Menon, S. et al., Effect of lipid supplements on the production and glycosylation of recombinant interferon-gamma expressed in CHO cells, *Cytotechnology* 15, 209–215, 1994.
53. Gu, X., Xie, L., Harmon, B.J., and Wang, D.I., Influence of Primatone RL supplementation on sialylation of recombinant human interferon-gamma produced by Chinese hamster ovary cell culture using serum-free media, *Biotechnol. Prog.* 56, 353–360, 1997.
54. Nyberg, G.E., Balcarcel, R.R., Follstad, B.D. et al., Metabolic effects on recombinant interferon-gamma glycosylation in continuous culture of Chinese hamster ovary cells, *Biotechnol. Bioeng.* 62, 336–347, 1999.
55. Donaldson, M., Wood, H.A., Kulakosky, P.C., and Shuler, M.L., Glycosylation of a recombinant protein in the Tn5B1-4 insect cell line: Influence of ammonia, time of harvest, temperature, and dissolved oxygen, *Biotechnol. Bioeng.* 63, 255–262, 1999.

56. Bakerm K.N., Rendall, M.N., Hills, A.E. et al., Metabolic control of recombinant protein N-glycan processing in N50 and CHO cells, *Biotechnol. Bioeng.* 73, 188–202, 2001.

57. Joosten, C.E. and Shuler, M.L., Effect of culture conditions on the degree of sialylation of a recombinant glycoprotein expressed in insect cells, *Biotechnol. Prog.* 19, 739–749, 2003.

58. Senger, R.S. and Karim, M.N., Effect of shear stress on intrinsic CHO culture state and glycosylation of recombinant tissue-type plasminogen activator protein, *Biotechnol. Prog.* 19, 1199–1209, 2003.

59. Lipscomb, M.L., Palomares, L.A., Hernández, V. et al., Effect of production method and gene amplification on the glycosylation pattern of a secreted reporter protein in CHO cells, *Biotechnol. Prog.* 21, 40–49, 2005.

60. Serrato, J.A., Hernándex, V., Estrada-Mondaca, S. et al., Differences in the glycosylation profile of a monoclonal antibody produced by hybridoma cultured in serum-supplemented, serum-free or chemical defined media, *Biotechnol. Appl. Biochem.* 47, 113–124, 2007.

61. Nakagawa, Y., Capetillo, S., and Jirgensons, B., Effect of chemical modification of lysine residues on the conformation of human immunoglobulin G, *J. Biol. Chem.* 242, 5703–5708, 1972.

62. Privat, J.P., Lotan, R., Bouchard, P. et al., Chemical modification of the tryptophan residues of wheat-germ agglutinin. Effect on fluorescence and saccharide-binding properties, *Eur. J. Biochem.* 68, 563–572, 1976.

63. Norton, R.S. and Allerhand, A., Studies of chemical modifications of proteins by carbon 13 nuclear magnetic resonance spectroscopy. Reaction of hen egg white lysozyme with iodine, *J. Biol. Chem.* 251, 6522–6528, 1976.

64. Kochman, H., Garnier, J., and Kochman, K., Receptor binding and conformational properties of bovine and ovine prolactins after chemical modification of the two tryptophan residues, *Biochim. Biophys. Acta* 578, 125–134, 1979.

65. Stassinopoulou, C.I., Wagner, G., and Wüthrich, K., Two-dimensional ^{1}H NMR of two chemically modified analogs of the basic pancreatic trypsin inhibitor. Sequence-specific resonance assignments and sequence location of conformation changes relative to the native protein, *Eur. J. Biochem.* 145, 423–430, 1984.

66. Endo, T., Ueda, T., Yamada, H., and Imoto, T., pH dependence of individual tryptophan N–1 hydrogen exchange rates in lysozyme and its chemically modified derivatives, *Biochemistry* 26, 1838–1845, 1987.

67. Kossiakoff, A.A., Tertiary structure is a principal determinant to protein deamidation, *Science* 240, 191–194, 1988.

68. Okajima, T., Kawata, Y., and Hamaguchi, K., Chemical modification of tryptophan residues and stability changes in proteins, *Biochemistry* 29, 9168–9175, 1990.

69. Miles, A.M. and Smith, R.L., Functional methionines in the collagen/gelatin binding domain of plasma fibronectin: Effects of chemical modification by chloramine T, *Biochemistry* 32, 8168–8178, 1993.

70. Klunk, W.E., Xu, C.J., and Pettigrew, J.W., NMR identification of the formic acid-modified residue in Alzheimer's amyloid protein, *J. Neurochem.* 62, 349–354, 1994.

71. Rohl, C.A. and Baldwin, R.L., Comparison of NH exchange and circular dichroism as techniques for measuring the parameters of the helix-coil transition in peptides, *Biochemistry* 36, 8435–8442, 1997.

72. Liu, D., Ren, D., Huang, H. et al., Structure and stability changes on Human IgG1 Fc as a consequence of methionine oxidation, *Biochemistry* 47, 5088–5100, 2008.

73. Habeeb, A.F.S.A., Quantitation of conformation changes on chemical modification of proteins: Use of succinylated proteins as a model, *Arch. Biochem. Biophys.* 121, 652–664, 1967.

74. Yoshioka, S., Abe, H., Noguti, T. et al., Conformational change of a globular protein elucidated at atomic resolution. Theoretical and magnetic resonance study, *J. Mol. Biol.* 170, 1031–1036, 1983.

75. Tong, T., Wren, J.C., and Konermann, L., γ-Ray-mediated oxidative labeling for detecting protein conformational changes by electrospray mass spectrometry, *Anal. Chem.* 80, 2222–2231, 2008.

76. Venkatesh, S., Tomer, K.S., and Sharp, J.S., Rapid identification of oxidation-induced conformational changes by kinetic analysis, *Rapid Commun. Mass Spectrom.* 21, 3927–3936, 2007.

77. Tong, X., Wren, J.C., and Konermann, L., Effects of protein concentration on the extent of γ-ray-mediated oxidative labeling studies by electrospray mass spectrometry, *Anal. Chem.* 79, 6376–6382, 2007.

78. Sharpe, J.S., Sullivan, D.M., and Tomer, K.B., Measurement of multisite oxidation kinetics reveals as active site conformational change in SpoOF as a result of protein oxidation, *Biochemistry* 45, 6260–6266, 2006.

79. Markland, F.S., Bacharach, D.E., Weber, B.H. et al., Chemical modification of yeast 3-phosphoglycerate kinase, *J. Biol. Chem.* 259, 1301–1310, 1975.

80. Van Patten, S.M., Hanson, E., Bernasconi, R. et al., Oxidation of methionine residues in antithrombin. Effects on biological activity and heparin binding, *J. Biol. Chem.* 274, 10268–10276, 1999.

81. Qin, Z., Hu, D., Han, S. et al., Effect of 4-hydroxy-2-nonenal modification on α-synuclein aggregation, *J. Biol. Chem.* 282, 5862–5870, 2007.

82. Diaz, A., Muñoz-Clares, R.A., Rangel, P. et al., Functional and structural analysis of catalase oxidized by singlet oxygen, *Biochimie* 87, 205–214, 2005.

83. Gao, J., Yin, D.H., Yao, Y. et al., Loss of conformational stability in calmodulin upon methionine modification, *Biophys. J.* 74, 1115–1134, 1998.

84. Strandberg, L., Lawrence, D.A., Johansson, L.B., and Ny, T., The oxidative inactivation of plasminogen activator inhibitor type-1 results from a conformational change in the molecule and does not require the involvement of the P1' methionine, *J. Biol. Chem.* 266, 13852–13858, 1991.

85. Batra, P.P., Sasa, K., Ueki, T., and Takeda, K., Circular dichroic study of the conformational stability of sulfydryl-blocked bovine serum albumin *Int. J. Biochem.* 21, 857–862, 1989.

86. Horowitz, P.M. and Bowman, S., Oxidation increases the proteolytic susceptibility of a localized region in rhodanese, *J. Biol. Chem.* 262, 14544–14548, 1987.

87. Barth, A., Infrared spectroscopy of proteins, *Biochim. Biophys. Acta* 1767, 1073–1101, 2007.

88. Akasaka, K., Tezuka, T., and Yamada, H., Pressure-induced changes in folded structure of lysozyme, *J. Mol. Biol.* 271, 671–678, 1997.

89. Reface, M., Tezuka, T., Akasaka, K., and Williamson, M.P., Pressure-dependent changes in the solution structure of hen egg-while lysozyme, *J. Mol. Biol.* 327, 857–865, 2003.

90. Meersman, F., and Heremans, K., High pressure induced the formation of aggregation-prone states of proteins under reducing conditions, *Biophys. Chem.* 104, 297–304, 2003.

91. Akasaka, K., Highly fluctuating protein structures revealed by variable-pressure nuclear magnetic resonance, *Biochemistry* 42, 10875–10885, 2003.

92. Ravindra, R. and Winter, R., Pressure perturbation calorimetry: A new technique provides surprising results on the effects of co-solvents on protein solvation and unfolding behavior, *Chemphyschem* 5, 566–571, 2004.

93. Lin, S.Y., Wei, Y.S., Hsieh, T.F., and Li, M.J., Pressure dependence of human fibrinogen correlated to the conformational alpha-helix to beta-sheet transition: A Fourier transform infrared study microspectroscopic study, *Biopolymers* 75, 393–402, 2004.

94. Watanabe, M., Aizawa, T., Demura, M., and Nitta, K., Effect of hydrostatic pressure on conformational changes of canine milk lysozyme between the native, molten globate and unfolded states, *Biochim. Biophys. Acta* 1702, 129–136, 2004.

95. Paschek, D., Gnanakaran, S., and Garcia, A.E., Simulations of the pressure and temperature unfolding of an alpha-helical peptide, *Proc. Natl. Acad. Sci. USA* 102, 6765–6770, 2005.

96. Marchal, S., Torrent, J., Masson, P. et al., The powerful high pressure tool for protein conformational studies, *Braz. J. Med. Biol. Res.* 38, 1175–1183, 2005.

97. McCarthy, A.N. and Grigera, J.R., Effect of pressure on the conformation of proteins. A molecular dynamics simulation of lysozyme, *J. Mol. Graph. Model.* 24, 254–263, 2006.

98. Chang, B.S., Randolph, T.W., and Kim, Y.S., Effects of solutes on solubilization and refolding of proteins from inclusion bodies with high hydrostatic pressure, *Protein Sci.* 15, 304–313, 2006.

99. Appavou, M.S., Gibrat, G., and Billissent-Funel, M.C., Influence of pressure on structure and dynamics of bovine pancreatic trypsin inhibitor (BPTI) small angle and quasi-elastic neutron scattering studies, *Biochim. Biophys. Acta* 1764, 414–423, 2006.

100. Harano, Y. and Kinoshita, M., Crucial importance of translational entropy of water in pressure denaturation of proteins, *J. Chem. Phys.* 125, 24910, 2006.

101. Cioni, P., Role of protein cavities on unfolding volume change and on internal dynamics under pressure, *Biophys. J.* 91, 3390–3396, 2006.

102. Okuno, A., Kato. M., and Taniguchi, Y., Pressure effects on the heat-induced aggregation of equine serum albumin by FT-IR spectroscopic study: Secondary structure, kinetic and thermodynamic properties, *Biochim. Biophys. Acta* 1774, 652–660, 2007.

103. Ado, K. and Taniguchi, Y., Pressure effects on the structure and function of human thioredoxin, *Biochim. Biophys. Acta* 1774, 813–821, 2007.

104. Day, R. and Garcia, A.E., Water penetration in the low and high pressure native states of ubiquitin, *Proteins* 70, 1175–1184, 2008.

105. Rochu, D., Ducret, G., and Masson, P., Measuring conformational stability of proteins using an optimized temperature-controlled capillary electrophoresis approach, *J. Chromatogr. A* 838, 157–165, 1999.

106. Panick, G., Malessa, R., and Winter, R., Differences between the pressure- and temperature-induced denaturation and aggregation of beta-lactoglobulin A, B, and AB monitored by FT-IR spectroscopy and small-angle X-ray scattering, *Biochemistry* 38, 6512–6519, 1999.

107. Chattopadhyay, K. and Mazumdar, S., Structural and conformational stability of horseradish peroxidase: Effect of temperature and pH, *Biochemistry* 39, 263–270, 2000.

108. Kumeta, H., Miura, A., Kobashigawa, Y. et al., Low-temperature-induced structural changes in human lysozyme elucidated by three-dimensional NMR spectroscopy, *Biochemistry* 42, 1209–1216, 2003.

109. Martinho, J.M., Santos, A.M., Fedorov, A. et al., Fluorescence of the single tryptophan of cutinase: Temperature and pH effect on protein conformation and dynamics, *Photochem. Photobiol.* 78, 15–22, 2003.

110. Meersman, F. and Heremans, K., Temperature-induced dissociation of protein aggregates: Accessing the denatured state, *Biochemistry* 42, 14234–14241, 2003.

111. Choi, H.S., Huh, J., and Jo, W.H., Comparison between denaturant- and temperature-induced unfolding pathways of protein: A lattice Monte Carlo simulation, *Biomacromolecules* 5, 2289–2296, 2004.

112. Kornblatt, J.A. and Schuck, P., Influence of temperature on the conformation of canine plasminogen: An analytical ultracentrifugation and dynamic light scattering study, *Biochemistry* 44, 13122–13131, 2005.

113. Kawai, K., Suzuki, T., and Oguni, M., Low-temperature glass transitions of quenched and annealed bovine serum albumin aqueous solutions, *Biophys. J.* 90, 3732–3738, 2006.

114. Goddard, Y.A., Korb, J.P., and Bryant, R.G., Structural and dynamical examination of the low-temperature glass transition in serum albumin, *Biophys. J.* 91, 3841–4847, 2006.

115. Xiao, Y., Rathore, A., O'Connell, J.P., and Fernandez, E.J., Generalizing a two-conformation model for describing salt and temperature effects on protein retention and stability in hydrophobic interactions chromatography, *J. Chromatogr. A* 1157, 197–206, 2007.

116. Ma, L., Ahmed, Z., Mikhonin, A.V., and Asher, S.A., UV resonance Raman measurements of poly-L-lysine's conformational energy landscapes: Dependence on perchlorate concentration and temperature, *J. Phys. Chem. B* 111, 7675–7680, 2007.

117. Crommelin, D.J.A., Storm, G., Verrijk, R. et al., Shifting paradigms: Biopharmaceuticals versus low molecular weight drugs, *Int. J. Pharm.* 266, 3–16, 2003.

118. Harris, J., Marketing and globalizing biosimilars in a competitive landscape, *J. Generic Med.* 3, 115–120, 2006.

119. Ranke, M.B., New preparations comprising recombinant human growth hormone. Deliberations on the issue of biosimilars, *Horm. Res.* 69, 22–28, 2008.

120. Aldridge, S., Why biosimilars are not true generics? *Pharm. Technol. Europe* June 14, 2007.

121. Mellstedt, H., Niedewieser, D., and Ludwig, H., The challenge of biosimilars, *Ann. Oncol.* 19, 411–419, 2008.

122. Hoppe, W. and Berghout, A., Biosimilar somatotropin: Myths and facts, *Horm. Res.* 69, 29–30, 2008.

123. Lichanska, A.M. and Waters, M.J., New insights into human growth hormone receptor function and clinical implications, *Horm. Res.* 69, 138–149, 2008.

124. Sharma, B., Immunogenicity of therapeutic proteins. Part 1. Impact of product handling, *Biotechnol. Adv.* 25, 310–317, 2007.

125. Sharma, B., Immunogenicity of therapeutic proteins. Part 2. Impact of container closures, *Biotechnol. Adv.* 25, 318–325, 2007.

126. Heavner, G.A., Arakawa, T., Philo, J.S. et al., Protein isolated from biopharmaceutical formulations cannot be used for comparative studies: Follow-up to "A case study using Epoetin Alfa from Epogen and EPREX", *J. Pharm. Sci.* 56, 3214–3225, 2007.

127. Duconge, J., The case of biotech-derived product equivalence: Much ado about nothing, *Curr. Clin. Pharmacol.* 1, 147–156, 2006.

3 Application of Native Electrophoresis for the Study of Protein Conformation

Most biological polymers are present as ions in solution. These macromolecules can be transported relative to solvent in an electric field. The velocity of transport depends on the net electric charge and the size and shape of the macromolecular ion.[1] Size and shape are described by the frictional coefficient.[2-4] Electrophoresis is the method of separation based on the differential migration of molecules in an electric field. Electrophoresis can occur in a fluid phase (moving boundary, zonal using fluid support) or within a matrix (zonal electrophoresis with a polyacrylamide gel or starch gel).[5] Electrophoretic migration within a moving boundary never achieves a complete separation of the various components in a mixture, but a zonal or zone electrophoresis separates various components into discrete regions or zones within the electrophoretogram. Capillary electrophoresis/capillary zone electrophoresis is a microelectrophoretic method providing enhanced speed and sensitivity in analysis.[6-10] Capillary zone electrophoresis was introduced by Jorgenson and Lukcas in 1981[11] and has seen considerable use. CZE (capillary zone electrophoresis) is not to be confused with the earlier technique of continuous zone electrophoresis (CZE).[12] Continuous zone electrophoresis is to be contrasted with multiphasic zone electrophoresis where the support media may require discrete matrices such as in the use of stacking gels for polyacrylamide gel electrophoresis.[13] Capillary zone electrophoresis is also not to be confused with zonal electrophoresis, which is an older technique utilizing a solid support or a gradient system composed of, for example, sucrose.[14] Capillary zone electrophoresis does result in zonal migration of analytes but may not require a specific matrix. Capillary zone electrophoresis is a mature technology for protein analysis,[15,16] which has been demonstrated to be useful for the study of protein conformation.[17,18]

Electrophoresis of proteins can be performed under native (nondenaturing) or denaturing conditions. Polyacrylamide gel electrophoresis in the presence of sodium dodecyl sulfate (SDS-PAGE) is the best known example of electrophoresis under denaturing conditions. Native electrophoresis is actually the earliest form of electrophoresis as represented by moving-boundary electrophoresis.[19-24] Moving-boundary electrophoresis is also known as free-boundary electrophoresis.[25-28] Zonal electrophoresis differs from moving-boundary in that the various species migrate as "zones," which require the use of a supporting medium to restrict diffusion.[29-33] The interaction of analyte with the matrix can provide a complication, resulting in electrochromatography. Moving-boundary electrophoresis, free-boundary electrophoresis, and, capillary zone electrophoresis and zonal electrophoresis are all examples of native electrophoresis.

Blue native electrophoresis[34,35] is a type of native electrophoresis used to separate membrane proteins where Coomassie blue G dye was included in the buffer to induce a charge shift. The electrophoresis usually contains a nondenaturing detergent such as dodecylmaltoside,[36] which is used in a two-dimensional (2D) method with SDS-PAGE as the second dimension. This technique has seen continued development as method for the study of protein aggregates and membrane proteins.[37-42]

All separation technologies that utilize a supporting matrix have the potential of "nonspecific" binding to the matrix. Affinity electrophoresis is a technique where a specific ligand (A) included in the electrophoresis medium alters the mobility of a specific component (B), which binds the ligand (A) in question, thereby forming a complex AB with altered mobility.[43-46] Native electrophoresis has proved useful for the study of interaction of highly charged materials such as heparin with specific proteins.[47-57] The mobility shift on the binding of heparin has been used to detect changes in the conformation of β_2-microglobulin.[56] Mobility shift assays, as the name suggests, are more frequently used for the study of nucleic acid–protein binding.[58-62]

Affinity electrophoresis has been used to estimate the binding constants for antigen–antibody complexes.[63-71] There has been particular interest in the use of affinity electrophoresis for the evaluation of antibodies against polysaccharides[72-74] and for the evaluation of other protein–oligosaccharide interactions.[75-77] It has been possible to capture antibodies by immobilizing antigen in selective areas within a polyacrylamide gel.[78,79] There is no significant migration of the antigen under these conditions and it is possible to selectively capture antibodies to the various antigens. Cibacron blue has been used as ligand for the affinity electrophoresis of albumin.[80] Affinity electrophoresis has been used to isolate monoclonal preparations of anti-hapten antibodies from polyclonal populations. The initial study[81] used dinitrobenzene coupled to an acrylamide–allylamine copolymer (reaction of dinitrofluorobenzene with an acrylamide–allylamine copolymer), which was added to the polyacrylamide slab gel serving as the second dimension in a 2D gel system; the first dimension is an isoelectric-focusing cylindrical gel. It is suggested that this technique was able to separate a polyclonal antibody preparation into monoclonal populations. More recent work from the same research group[82] used fluorescein coupled to a dextran (prepared by reaction of fluorescein isothiocyanate with dextran). The fluorescein-dextran was incorporated into the second dimension of polyacrylamide.

Native electrophoresis has been a useful method for the study of conformation for a number of years. Urea gradient electrophoresis is a powerful tool for the study of protein conformation and denaturation.[83-89] This technique uses a slab gel with a transverse urea gradient (the urea gradient is orthogonal to the direction of the electromigration). The developed electrophoretogram is analogous to a melting curve. Additional variables, such as pressure, can be included.[88]

Traditional moving-boundary electrophoresis and analytical ultracentrifugation are the two the "cleanest" techniques for the investigation of the shape properties of a protein or other macromolecules since a matrix is not involved and wall interactions can generally be discounted because of chamber size and sample concentration. On the other hand, this latter consideration, a comparatively large sample concentration (amount), creates a problem with availability of smaller amounts (microgram) of sample. Nevertheless, the results from early studies on the use of moving-boundary can be instructive in understanding more recent results. Moving-boundary electrophoresis[90–93] was used to study the isomerization (N→F) of albumin[94–96] as can be measured by other physical chemical techniques.[97–101] Clark and colleagues[90] used classic moving-boundary electrophoresis to demonstrate a change in the electrophoretic pattern with albumin when the pH was decreased from 3–4 to 2; the presence of salt had a pronounced effect on the transition from a single peak to two peaks. Subsequent work by Sogami and Foster[91] extended these observations to polyacrylamide gel in demonstrating the N→F transition. More sophisticated electrophoretic techniques have been subsequently used to study the conformational transitions of albumin.[92,93] The work on albumin allows the comparison of several different techniques in the analysis of protein conformation.

Moving boundary electrophoresis has also been used to study the acid transition of α-lactalbumin[102] and the intramolecular oxidation of sulfhydryl groups.[103] The influence of buffer ions on electrophoretic migration was noted by one group[102] and has been discussed by other investigators[22,104,105] for studies on proteins and by other investigators for the electrophoresis of RNA.[106]

Capillary zone electrophoresis[107] is the most suitable of the various modalities of capillary electrophoresis for the study of biological polymer conformation.[108] Capillary zone electrophoresis permits the use of zonal electrophoretic separation without a supporting matrix as convection is not a problem.[109] There are issues with respect to interaction of analytes with the fused capillary wall.[108,110] As the pH of the analytical system increases, the silanol functional groups on the capillary wall dissociate, resulting in the generation of a diffuse negative charge. The interaction of analyte with capillary wall can result in irreversible adsorption or in a reversible interaction, resulting in a peak asymmetry.[111] These interactions can be decreased by modification of the capillary wall[112–115] or by the inclusion of modifying reagents to the electrophoresis media.[116–119] Since capillary zone electrophoresis is a flowing system, the detection systems (flow cells measuring absorbance or fluorescense developed for high-performance liquid chromatography [HPLC]) can be used for analysis in this system.[120] The volume in capillary zone electrophoresis is small, requiring high initial concentration of the analyte or a preanalytical modification of the analyte with a fluorophore or chromophore to provide for detection analyte distribution. Concentration of the sample can be accomplished by stacking or isotachaphoresis. Mass spectrometry can be coupled to capillary zone electrophoresis to enhance analytical sensitivity.[121,122]

In addition to the various articles cited above, there are a number of other excellent articles and books on the development current applications of capillary electrophoresis.[123–131] Examples of the application of capillary zone electrophoresis to the study of biological polymer conformation are shown in Table 3.1. It is also possible to use capillary zone electrophoresis to assess macromolecular interactions such as the interaction of heparin and growth factors (Table 3.2).

TABLE 3.1
Examples of the Application of Capillary Zone Electrophoresis

Polymer	Description of Study	Ref.
Metallothionein	Separation of isoforms[a]	[132]
Ala-Phe-ψ[CS-N]-Pro-Phe-4-nitroanlide	Separation of *cis/trans* isomers of the peptide	[133]
Staphylococcus aureus nuclease A (native and variants)	Use of coated capillaries; separation of variants with the same net charge (K116A and K116G)	[134]
Proline peptides	Separation of *cis/trans* conformers	[135]
Human or salmon calcitonin	Separation of *cis/trans* conformers	[136]
Immunoglobulins	Thermal stability measurements	[137]
Bovine β-lactoglobulin	Measurement of thermal denaturation	[138]
Review	Correlation of capillary zone electrophoresis with other technologies for the study of protein conformation	[17]
Review	Use of capillary zone electrophoresis in the study of protein conformational stability	[18]
Cholinesterase	Detection of ligand stability by protein conformation	[139]
Creatinine kinase	Protein unfolding and refolding	[140]
Albumin, myoglobin, hemoglobin	Determination of nonnative conformations of proteins formed secondary to contact with SDS	[141]

[a] The study notes that capillary zone electrophoresis can be considered a separation technique orthogonal to RP-HPLC.

TABLE 3.2
Examples of the Use of Capillary Zone Electrophoresis to Study Macromolecular Interactions

Polymers	Description of Study	Ref.
α_2-Macroglobulin and protease inhibitors	Used capillary zone electrophoresis to measure interaction with proteases	[142]
Human growth hormone (HGH) and a monoclonal antibody (MAB)	Measurement of interaction of HGH with MAB using CZE; comparison with protein G affinity perfusion chromatography	[143]
Review	Measurement of protein complexes	[144]
Concanavalin A	Interaction of protein and carbohydrate	[145]
Heparin and G-CSF	Interaction of heparin and granulocyte-colony stimulating factors (G-CSF); measurement of binding constants from changes in mobility	[146]
Heparin and IL-1	Interaction of heparin with interleukin-2 (IL-2); measurement of binding constants from changes in mobility	[147]

TABLE 3.2 (continued)

Examples of the Use of Capillary Zone Electrophoresis to Study Macromolecular Interactions

Polymers	Description of Study	Ref.
Heparin and G-CSF	Interaction of heparin oligosaccharide (derived by the action of heparinase) and G-CSF	[148]
Dextran sulfate and G-CSF	Interaction of dextran sulfate (chains of various lengths) and G-CSF	[149]
Heparin and G-MCSF	Comparison of heparin and low-molecular-weight heparin with granulocyte-macrophage colony stimulating factor	[150]
Heparin and PDCD5	Interaction of heparin and PDCD5 (programmed cell death 5) protein and derived peptides; measurement of binding constants from changes in mobility	[151]
Heparin and PDCD5	Interaction of heparin and PDCD5 (programmed cell death 5) protein; measurement of binding constants from changes in mobility	[152]
Heparin and CKLF1	Interaction between heparin and secreted chemokine-like factor 1 (CKFL1); measurement of binding constants from changes in mobility	[153]
Thrombin and some novel natural products	Measurement of interaction between thrombin and natural products derived from *Mellettia*	[154]

REFERENCES

1. Alberty, R.A., Electrochemical properties of proteins and amino acids, in *The Proteins*, Vol. 1, Part A, eds. K. Bailey and H. Neurath, pp. 461–548, Academic Press, New York, 1953.
2. Rothe, G.M., Determination of molecular mass, Stokes' radius, frictional coefficient and isomer-type of non-denatured proteins by time-dependent pore gradient gel electrophoresis, *Electrophoresis* 9, 307–316, 1988.
3. Grossman, P.D. and Soane, D.S., Orientation effects on the electrophoretic mobility of rod-shaped molecules in free solution, *Anal. Chem.* 62, 1592–1596, 1990.
4. Plasson, R. and Cottet, H., Determination of homopolypeptide conformational changes by the modeling of electrophoretic mobilities, *Anal. Chem.* 77, 6047–6054, 2005.
5. Cancalon, P.F., Electrophoresis and isoelectric focusing in food analysis, in *Encyclopedia of Analytical Chemistry*, Vol. 5, ed. R.A. Meyers, John Wiley & Sons, Chichester, U.K., pp. 3929–3955, 2000.
6. Kremser, L., Bilek, G., Blaas, D., and Kenndler, E., Capillary electrophoresis of viruses, subviral particles and virus complexes, *J. Sep. Sci.* 30, 1704–1713, 2007.
7. Ravelet, C., Grosset, C., and Peyrin, E., Liquid chromatography, electrochromatography and capillary electrophoresis applications of DNA and RNA aptamers, *J. Chromatogr. A* 1117, 1–10, 2006.
8. Huang, Y.F., Huand, C.C., Hu, C.C., and Chang, H.T., Capillary electrophoresis-based separation techniques for the analysis of proteins, *Electrophoresis* 27, 3503–3522, 2006.
9. Dolnik, V., Capillary electrophoresis of proteins 2005–2007, *Electrophoresis* 29, 143–156, 2008.
10. Kostal, V., Katzenmeyer, J., and Arriaga, E.A., Capillary electrophoresis in bioanalysis, *Anal. Chem.* 80, 4533–4550, 2008.

11. Jorgenson, J.W. and Lukacs, K.D., Zone electrophoresis in open-tubular glass capillaries, *Anal. Chem.* 53, 1298–1302, 1981.

12. Brattsten, I. and Hedlund, P., Isolation of acute phase protein by means of continuous zone electrophoresis, *Scand. J. Clin. Lab. Invest.* 8, 213–222, 1956.

13. Chen, B., Griffith, A., Catsimpoolas, N. et al., Bandwidth: Comparison between continuous and multiphasic zone electrophoresis, *Anal. Biochem.* 89, 609–615, 1978.

14. Magdoff, B.S., Electrophoresis of proteins in liquid media, in *Analytical Methods of Protein Chemistry*, Vol. 2, *The Composition, Structure, and Reactivity of Proteins*, eds. P. Alexander and R.J. Block, Pergamon Press, New York, 1961.

15. Gebauer, P., Beckers, J.L., and Bocek, P., Theory of system zones in capillary zone electrophoresis, *Electrophoresis* 23, 1779–1785, 2002.

16. Visser, N.F., Lingeman, H., and Irth, H., Sample preparation for peptides and proteins in biological matrices prior to liquid chromatography and capillary zone electrophoresis, *Anal. Biochem. Chem.* 382, 535–558, 2005.

17. Righetti, P.G. and Verzola, B., Folding/unfolding/refolding of proteins: Present methodologies in comparison with capillary one electrophoresis, *Electrophoresis* 22, 2359–2374, 2001.

18. Rochu, D. and Masson, P., Multiple advantages of capillary zone electrophoresis for exploring protein conformational stability, *Electrophoresis* 23, 189–202, 2002.

19. Tseliius, A., A new apparatus for electrophoretic analysis of colloidal mixtures, *Trans. Faraday Soc.* 524–531, 1937.

20. Longworth, L.G., Recent advances in the study of proteins by electrophoresis, *Chem. Rev.* 30, 323–340, 1942.

21. Moore, D.H., Species differences in serum protein patterns, *J. Biol. Chem.* 161, 21–32, 1945.

22. Sendroy, J. Jr., Rodkey, F.L., and MacKenzie, M., Use of tris (hydroxymethyl) aminomethane in moving boundary electrophoresis of serum, *Clin. Chem.* 8, 585–592, 1962.

23. Rilbe, H., Some reminiscences of the history of electrophoresis, *Electrophoresis* 16, 1354–1359, 1995.

24. Righetti, P.G., Bioanalysis: Its past, present, and some future, *Electrophoresis* 25, 2111–2127, 2004.

25. de Nooij, E. and Niemeijer, J.A., Free-boundary electrophoresis of histones, *Biochim. Biophys. Acta* 65, 148–149, 1962.

26. Bookman, R. and Shen, J., Further studies on reaginic antibody by free boundary electrophoresis (electrophoretic behavior of regain), *Int. Arch. Allergy Appl. Immunol.* 24, 158–167, 1964.

27. Takagi, T., Tsujii, K., and Shirahama, K., Binding isotherms of sodium dodecyl sulfate to protein polypeptides with special reference to SDS-polyacrylamide gel electrophoresis, *J. Biochem.* 77, 939–947, 1975.

28. Kubo, K., Isemura, T., and Takagi, T., Interaction between the alpha 1 chain of rat tail collagen and sodium dodecyl sulfate with reference to its behaviour in SDS-polyacrylamide gel electrophoresis, *Biochim. Biophys. Acta* 703, 180–186, 1982.

29. Tselius. A., Electrophoresis, *Methods Enzymol.* 4, 3–20, 1957.

30. Hjérten, S., Agarose as an anticonvection agent in zone electrophoresis, *Biochim. Biophys. Acta* 53, 514–517, 1961.

31. Bier, M., Preparative electrophoresis, *Methods Enzymol.* 5, 33–50, 1962.

32. Altschul, A.M. and Evans, W.J., Zone electrophoresis with polyacrylamide gel, *Methods Enzymol.* 11, 179–186, 1967.

33. Wu, D. and Regnier, F.E., Native protein separations and enzyme microassays by capillary zone and gel electrophoresis, *Anal. Chem.* 65, 2029–2035, 1993.

34. Schägger, H. and von Jagow, G., Blue native electrophoresis for isolation of membrane protein complexes in enzymatically active forms, *Anal. Biochem.* 199, 223–231, 1991.

35. Schägger, H., Cramer, W.A., and von Jagow, G., Analysis of molecular masses and oligomeric states of protein complexes by blue native electrophoresis and isolation of membrane proteins by two-dimensional native electrophoresis, *Anal. Biochem.* 217, 220–230, 1994.

36. Reisinger, V. and Eichacher, L.A., Analysis of membrane protein complexes by blue native PAGE, *Proteomics* 6 (Suppl 2), 6–15, 2006.

37. Poetsch, A., Neff, D., Seelert, H. et al., Dye removal, catalytic activity and 2D crystallization of chloroplast H(+)-ATP synthase purified by blue native electrophoresis, *Biochim. Biophys. Acta* 1466, 339–349, 2000.

38. Fandiño, A.S., Rais, I., Vollmer, M. et al., LC-nanospray-MS/MS analysis of hydrophobic proteins from membrane protein complexes isolated by blue-native electrophoresis, *J. Mass Spectrom.* 40, 1223–1231, 2005.

39. Stegemann, J., Venzki, R., Schrödel, A. et al., Comparative analysis of protein aggregates by blue native electrophoresis and subsequence sodium dodecyl sulfate-polyacrylamide gel electrophoresis in a three-dimensional geometry gel, *Proteomics* 5, 2002–2009, 2005.

40. Nijtmans, L.G., Henderson, N.S., and Holt, I.J., Blue native electrophoresis to study mitochondrial and other protein complexes, *Methods* 26, 327–334, 2002.

41. Reisinger, V. and Eichacker, L.A., Isolation of membrane proteins complexes by blue native electrophoresis, *Methods Mol. Biol.* 424, 423–431, 2008.

42. Swamy, M., Minguet, S., Siegers, G.M. et al., A native antibody-based mobility-shift technique (NAMOS-assay) to determine the stoichiometry of multiprotein complexes, *J. Immunol. Methods* 324, 74–83, 2007.

43. Horejsí, V., Affinity electrophoresis, *Anal. Biochem.* 112, 1–8, 1991.

44. Bøg-Hansen, T.S. and Hau, J., Affinity electrophoresis of glycoproteins, *Acta Histochem.* 71, 47–56, 1982.

45. Hage, D.S., Chiral separations in capillary electrophoresis using proteins as stereoselective binding agents, *Electrophoresis* 18, 2311–2321, 1997.

46. Nakamura, K. and Takeo, K., Affinity electrophoresis and its application to the studies of immune response, *J. Chromatogr. B. Biomed. Sci. Appl.* 715, 125–136, 1998.

47. Varenne, A., Gareil, P., Colliec-Jouault, S., and Daniel, R., Capillary electrophoresis determination of the binding affinity of bioactive sulfated polysaccharides to proteins: Study of the binding properties of fucoidan to antithrombin, *Ann. Biochem.* 315, 152–159, 2003.

48. Grover, M.E. Jr., Heparin and the electrophoretic mobility of β-lipoprotein in coronary disease, *Am. J. Cardiol.* 15, 13–16, 1965.

49. Burstein, M., Interaction of heparin with low density lipoproteins. Formation of soluble high molecular weight complexes, *Biomed. Pharmacother.* 36, 45–47, 1982.

50. Heegaard, N.H., Capillary electrophoresis for the study of affinity interactions, *J. Mol. Recognit.* 11, 141–148, 1998.

51. Wu, Z.L., Zhang, L., Beeler, D.L. et al., A new strategy for defining functional groups on heparin sulfate, *FASEB J.* 16, 539–545, 2002.

52. Vanpouille, C., Denys, A., Carpenter, M. et al., Octasaccharide is the minimal length unit required for efficient binding of cyclophilin B to heparin and cell surface heparan sulphate, *Biochem. J.* 382, 733–740, 2004.

53. McKeon, J. and Holland, L.A., Determination of dissociation constants for a heparin-binding domain of amyloid precursor protein and heparins or heparin sulfate by affinity capillary electrophoresis, *Electrophoresis* 25, 1243–1248, 2004.

54. Moxley, R.A. and Jarrett, H.W., Oligonucleotide trapping method for transcription factor purification systematic optimization using electrophoretic mobility shift assay, *J. Chromatogr. A* 1070, 23–34, 2005.

55. Heegaard, N.N. and De Lorenzi, E., Interactions of charged ligands with β$_2$-microglobulin conformers in affinity capillary electrophoresis, *Biochim. Biophys. Acta* 1753, 131–140, 2005.

56. Heegard, N.H., A novel specific heparin-binding activity of bovine folate-bindning protein characterized by capillary electrophoresis, *Electrophoresis* 27, 1122–1127, 2006.
57. Liu, X., Liu, X., Liang, A. et al., Studying protein–drug interactions by microfluidic chip affinity capillary electrophoresis with indirect laser-induced fluorescence detection, *Electrophoresis* 27, 3125–3128, 2006.
58. Filion, G.J., Fouvry, L., and Defossez, P.A., Using reverse electrophoretic mobility shift assay to measure and compare protein–DNA binding affinities, *Anal. Biochem.* 357, 156–158, 2006.
59. Eguchi, Y., RNA electrophoretic mobility shift assay using a fluorescent DNA sequencer, *Methods Mol. Biol.* 353, 115–123, 2007.
60. Perez-Romero, P. and Imperiale, M.J., Assaying protein–DNA interactions in vivo and in vitro using chromatin immunoprecipitation in the study of transcription in adipose cells, *Methods Mol. Biol.* 131, 123–139, 2007.
61. Hellman, L.M. and Fried, M.G., Electrophoretic mobility shift assay (EMSA) for detection protein–nucleic acid interactions, *Nat. Protoc.* 2, 1849–1861, 2007.
62. Musri, M.M., Gomis, R., and Parrizas, M., Application of electrophoretic mobility shift assay and chromatin immunoprecipitation in the study of transcription in adipose cells, *Methods Mol. Biol.* 456, 231–247, 2008.
63. Caron, M., Faure, A., and Cornillot, P., Affinity electrophoresis. I. Studies to optimize conditions, *J. Chromatogr.* 103, 160–165, 1975.
64. Caron, M., Faure, A., Kedros, G.B., and Cornillot, P., Application of affinity electrophoresis to the study of antigen-association equilibrium, *Biochim. Biophys. Acta* 491, 558–565, 1977.
65. Bøg-Hansen, T.C., Prahl, P., and Løwenstein, H., A set of analytical electrophoresis experiments to predict the results of affinity chromatographic separations: Fractionation of allergens from cow's hair and dander, *J. Immunol. Methods* 22, 293–307, 1978.
66. Breborowicz, J., Gan, J., and Klosin, J., Application of monoclonal antibodies for affinity electrophoresis, *J. Immunol. Methods* 102, 101–107, 1987.
67. Heegaard, N.H. and Bøg-Hansen, T.C., Affinity electrophoresis in agarose gels. Theory and some applications, *Appl. Theor. Electrophor.* 1, 249–259, 1990.
68. Lin, S., Tang, P., and Hsu, S.M., Using affinity capillary electrophoresis to evaluated average binging constant of 18-mer diphosphotyrosine peptide to antiphosphotyrosine Fab, *Electrophoresis* 20, 3388–3395, 1999.
69. Kazama, H., Yamada, K., Aoki, T., and Watabe, H., Application of green fluorescent protein to affinity electrophoresis: Affinity of IgG-binding domain C from streptococcal protein G to mouse IgG1, *Biol. Pharm. Bull.* 25, 168–171, 2002.
70. Sanchez-Muñoz, O.L., Pena, O.C., Ledger, E.D. et al., Zeta potential of staphylococcal protein A (Z-domain) used for determination of complex constant with immunoglobulin Fc fragment by affinity electrophoresis in capillary format, *J. Capill. Electrophor. Microchip Technol.* 7, 31–36, 2002.
71. Lassen, K.S., Bradbury, A.R., Rehfeld, J.F., and Heegaard, N.H., Microscale characterization of the binding specificity and affinity of a monoclonal antisulfotyrosyl IgG antibody, *Electrophoresis* 29, 2557–2564, 2008.
72. Sharon, J., Kabat, E.A., and Morrison, S.L., Association constants of hybridoma antibodies specific for alpha (1 leads to 6) linked dextran determined by affinity electrophoresis, *Mol. Immunol.* 19, 389–397, 1982.
73. Tanaka, T., Nakamura, K., and Takeo, K., A simple method for the determination of anti-dextran IgG in antiserum by means of affinity electrophoresis, *Mol. Immunol.* 19, 389–397, 1989.
74. Mimura, Y., Nakamura, K., and Takeo, K., Analysis of the interaction between an alpha (1–6)dextran-specific mouse hybridoma antibody and dextran B512 by affinity electrophoresis, *J. Chromatogr.* 597, 345–350, 1992.

75. Abou Hachem, M., Nordberg Karlsson, E., Bartonek-Roxâ, E. et al., Carbohydrate-binding modules from a thermostable *Rhodothermus marius* xylanase: Cloning, expression and binding studies, *Biochem. J.* 345, 53–60, 2000.

76. Jamal-Talabani, S., Boraston, A.B., Turkenbrug, J.P. et al., *Ab initio* structure determination and functional characterization of CBM36: A new family of calcium-dependent carbohydrate binding modules, *Structure* 12, 1177–1187, 2004.

77. Ludwiczek, M.L., Heller, M., Kantner, T., and McIntosh, L.P., A secondary xylan-binding site enhances the catalytic activity of a single-domain family 11 glycoside hydrolase, *J. Mol. Biol.* 373, 337–354, 2007.

78. Lee, B.S., Gupta, S., Krisnanchettier, S., and Lateef, S.S., Catching protein antigens by antibody affinity electrophoresis, *Electrophoresis* 25, 3331–3335, 2004.

79. Lee, B.S., Krishnanchettiar, S., Lateef, S.S., and Gupta, S., Capturing sodium dodecyl sulfate-treated protein antigens by antibody affinity electrophoresis, *Electrophoresis* 26, 511–513, 2005.

80. Miller, I. and Gemeiner, M., An electrophoretic study on interactions of albumins of different species with immobilized Cibacron Blue F3G A, *Electrophoresis* 19, 2506–2514, 1998.

81. Takeo, K., Suzuno, R., Tanaka, T., and Nakamura, K., Complete separation of anti-hapten antibodies by two-dimensional affinity electrophoresis, *Electrophoresis* 10, 813–818, 1989.

82. Wang, P. and Nakamura, K., Effects of carrier and hapten array on the production of anti-hapten antibodies analyzed by two-dimensional affinity electrophoresis, *Electrophoresis* 19, 1506–1510, 1998.

83. Creighton, T.E., Electrophoretic analysis of the unfolding of proteins by urea, *J. Mol. Biol.* 129, 235–264, 1979.

84. Creighton, T.E. and Pain, R.H., Unfolding and refolding of *Staphylococcus aureus* penicillinase by urea-gradient electrophoresis, *J. Mol. Biol.* 137, 431–436, 1980.

85. Goldenberg, D.P. and Creighton, T.E., Gel electrophoresis in studies of protein conformation and folding, *Anal. Biochem.* 138, 1–18, 1984.

86. Ewbank, J.J. and Creighton, T.E., Structural characterization of the disulfide folding intermediates of bovine α-lactalbumin, *Biochemistry* 32, 3694–3707, 1993.

87. Creighton, T.E. and Shortle, D., Electrophoretic characterization of the denatured states of staphylococcal nuclease, *J. Mol. Biol.* 232, 670–682, 1994.

88. Paladini, A.A., Weber, G., and Erijman, L., Analysis of dissociation and unfolding of oligomeric proteins using a flat bed gel electrophoresis at high pressure, *Anal. Biochem.* 218, 364–369, 1994.

89. Gainazza, E., Galliano, M., and Miller, I., Structural transitions of human serum albumin: An investigation using electrophoretic techniques, *Electrophoresis* 18, 695–700, 1997.

90. Clark, P., Rachinsky, M.R., and Foster, J.F., Moving boundary electrophoresis behavior and acid isomerization of human mercaptalbumin, *J. Biol. Chem.* 237, 2509–2513, 1962.

91. Sogami, M. and Foster, J.F., Resolution of oligomeric and isomeric forms of plasma albumin by zone electrophoresis on polyacrylamide gel, *J. Biol. Chem.* 237, 2514–2520, 1962.

92. Xu, H., Yu, X.D., and Chen, H.Y., Analysis of conformational change of human serum albumin using chiral capillary electrophoresis, *J. Chromatogr. A* 1055, 209–214, 2004.

93. Jachimska, B., Wasilewska, M., and Adamczyk, Z., Characterization of globular protein solutions by dynamic light scattering, electrophoretic mobility, and viscosity measurements, *Langmuir* 24, 6866–6872, 2008.

94. Winzor, D.J., Ford, C.L., and Nichol, L.W., Thermodynamic nonideality as a probe of macromolecular isomerizations: Application to the acid expansion of bovine serum albumin, *Arch. Biochem. Biophys.* 234, 15–23, 1984.

95. Kuwata, K., Era, S., and Sogami, M., The kinetics studies on the intramolecular SH, S-S exchange reaction of bovine mercaptoalbumin, *Biochim. Biophys. Acta* 1205, 317–324, 1994.

96. Dockal, M., Carter, D.C., and Rüker, F., Conformational transitions of the three recombinant domains of human serum albumin depending on pH, *J. Biol. Chem.* 275, 3042–3050, 2000.

97. Chen, R.F., Fluorescent spectra of human serum albumin in pH region of N–F transition, *Biochim. Biophys. Acta* 120, 169–171, 1966.

98. Etoh, T., Miyazaki, M., Harada, K. et al., Rapid analysis of human serum albumin by high-performance liquid chromatography, *J. Chromatogr. Biomed. Appl.* 578, 292–296, 1992.

99. Sadler, P.J. and Tucker, A., pH-induced transitions of bovine serum albumin–histidine pK_a values and unfolding of the N-terminus during the N–F transition, *Eur. J. Biochem.* 212, 811–817, 1994.

100. Era, S. and Sogami, M., ^{1}H-NMR and CD studies on the structural transition of serum albumin in the acidic region. The N→F transition, *J. Pept. Res.* 52, 431–442, 1998.

101. Ahmad, B., Ankita, and Khah, R.H., Urea induced unfolding of F isomer of human serum albumin: A case study using multiple probes, *Arch. Biochem. Biophys.* 437, 159–167, 2005.

102. Kuwajima, K., Nitta, K., and Sugai, S., Electrophoretic investigations of the acid conformational change of α-lactalbumin, *J. Biochem.* 78, 205–211, 1975.

103. Cann, J.R., Doherty, M.D., and Winzor, D.J., Electrophoresis of reacting systems involving sulfhydryl oxidation of proteins, *Arch. Biochem. Biophys.* 230, 146–153, 1984.

104. Cobb III B.R. and Koenig, V.L., Moving boundary electrophoresis in tris buffers, *Int. J. Protein Res.* 3, 301–311, 1971.

105. Cabenes-Macheteau, M., Chrambach, A., Taverna, M. et al., Resolution of 8-aminonaphthalene-1,3,6-trisulfonic acid-labeled glucose oligomers in polyacrylamide gel electrophoresis at low gel concentration, *Electrophoresis* 25, 8–13, 2004.

106. Hess, E.L., Lagg, S.E., and Utsunomiya, T., Moving boundary electrophoretic behavior of ribonucleic acid, *Biochim. Biophys. Acta* 91, 29–35, 1964.

107. Guzman, N.A. ed. *Capillary Electrophoresis Technology*, Marcel Dekker, New York, 1993.

108. Radka, S.P., Capillary electrophoresis, in *Encyclopedia of Separation Science*, Vol. 9, ed. I.D. Wilson, Academic Press, San Diego, CA, pp. 4009–4044, 2000.

109. Meloan, C.E., *Chemical Separations: Principles, Techniques, and Experiments*, John Wiley & Sons, Inc, New York, Chapter 31, 2000.

110. Trout, A., Muza, M.M., and Landers, J.P., Capillary electrophoresis of proteins and glycoproteins, in *Encylopedia of Analytical Chemistry*, Vol. 7, ed. R.A. Meyers, John Wiley & Sons, Chichester, U.K., pp. 5649–5699, 2000.

111. Stastna, M., Radko, S.P., and Chrambach, A., Discrimination between peak spreading in capillary zone electrophoresis of proteins due to interaction with capillary wall and due to protein microheterogeniety, *Electrophoresis* 22, 66–70, 2001.

112. Chiari, M., Nesi, M., and Righetti, P.G., Surface modification of silica walls: A review of different methodologies, in *Capillary Electrophoresis in Analytical Biotechnology*, ed. P.G. Righetti, CRC Press, Boca Raton, FL, pp. 1–36, Chapter 1, 1996.

113. Cretich, M., Stastna, M., Chrambach, A., and Chiari, M., Decreased protein peak asymmetry and width due to static capillary coating with hydrophilic derivatives of polydimethylacrylamide, *Electrophoresis* 23, 2274–2278, 2002.

114. Gelfi, C., Vigano, A., Ripamonti, M. et al., Protein analysis by capillary zone electrophoresis utilizing a trifunctional diamine for silica coating, *Anal. Chem.* 73, 3862–3868, 2001.

115. Pontoglio, A., Vigano, A., Sebastiano, R. et al., Peptide and protein separations by capillary electrophoresis in the presence of mono- and diquaternarized diamines, *Electrophoresis* 25, 1065–1070, 2004.

116. Lee, K.J. and Heo, G.S., Free solution capillary electrophoresis of proteins using untreated fused-silica capillaries, *J. Chromatogr.* 559, 317–324, 1991.

117. Moring, S.E., Buffers, electrolytes, and additives for capillary electrophoresis, in *Capillary Electrophoresis in Analytical Biotechnology*, ed. P.G. Righetti, CRC Press, Boca Raton, FL, pp. 37–60, Chapter 2, 1996.

118. Corradini, D., Buffer additives other than the surfactant sodium dodecyl sulfate for protein separations by capillary electrophoresis, *J. Chromatogr. B Biomed. Sci. Appl.* 699, 221–256, 1997.

119. Dolnik, V., Capillary zone electrophoresis of proteins, *Electrophoresis* 18, 2353–2361, 1997.

120. Weber, P.L., Capillary electrophoresis in peptide and protein analysis, Detection modes for, in *Encyclopedia of Analytical Chemistry*, Vol. 7, ed. R.A. Meyers, John Wiley & Sons, Chichester, U.K., pp. 5614–5628, 2000.

121. Tomlinson, A.J. and Naylor, S., Capillary electrophoresis/mass spectrometry in peptide and protein analysis, in *Encyclopedia of Analytical Chemistry*, Vol. 7, ed. R.A. Meyers, John Wiley & Sons, Chichester, U.K., pp. 5649–5699, 2000.

122. Strege, M.A. and Lager, A.L., eds. *Capillary Electrophoresis of Proteins and Peptides*, Humana Press, Totowa, NJ, 2005.

123. Currell, G., *Analytical Instrumentation Performance, Characteristics and Quality*, John Wiley & Sons, Ltd., Chichester, U.K., 2002.

124. Aboul-Enein, H.Y. and Ali, I., Capillary electrophoresis, in *Ewing's Analytical Instrumentation Handbook*, 3rd edn., ed. J. Cazes, Marcel Dekker, New York, pp. 803–806, Chapter 25, 2005.

125. Barttle, K.D. and Myers, P., *Capillary Electrochromatography*, Royal Society of Chemistry, Cambridge, U.K., 2001.

126. Deyl, Z. and Šveč, F. eds. *Capillary Electrochromatography*, Elsevier, Amsterdam, the Netherlands, 2001.

127. Camilleri, P. ed., *Capillary Electrophoresis Theory and Practice*, CRC Press, Boca Raton, FL, 1993.

128. Landers, J.P. ed. *Capillary Electrophoresis*, CRC Press, Boca Raton, FL, 1994.

129. Khaledi, M.G. ed. *High-Performance Capillary Electrophoresis Theory, Techniques, and Applications*, John Wiley & Sons, New York, 1998.

130. Begley, D.J., Free zone capillary electrophoresis, in *The Proteins Protocols Handbook*, ed. J.M. Walker, Humana Press, Totowa, NJ, 1996.

131. Righetti, P.G. ed., *Capillary Electrophoresis in Analytical Biotechnology*, CRC Press, Boca Raton, FL, 1996.

132. Richards, M.P. and Beattie, J.H., Characterization of metallothionein isoforms. Comparison of capillary electrophoresis with reversed-phase high-performance liquid chromatography, *J. Chromatogr.* 648, 459–468, 1993.

133. Meyer, S., Joles, A., Schutkowski, M., and Fischer, G., Separation of *cis/trans* of a prolyl peptide bond by capillary zone electrophoresis, *Electrophoresis* 15, 1151–1157, 1994.

134. Kálmán, F., Ma, S., Hodel, A. et al., Charge and size effects in the capillary zone electrophoresis of nuclease A and its variants, *Electrophoresis* 16, 595–603, 1995.

135. Ma, S., Kálmán, F., Kálmán, A. et al., Capillary zone electrophoresis at subzero temperature. I. Separation of the *cis* and *trans* conformers of small peptides, *J. Chromatogr. A* 716, 167–182, 1995.

136. Thunecke, F. and Fischer, G., Separation of *cis/trans* conformers of human and salmon calcitonin by low temperature capillary electrophoresis, *Electrophoresis* 19, 288–294, 1998.

137. Dai, H.J. and Krull, I.S., Thermal stability studies of immunoglobulins using capillary isoelectric focusing and capillary zone electrophoretic methods, *J. Chromatogr. A* 807, 121–128, 1998.

138. Rochu, D., Ducret, G., and Masson, P., Measuring conformational stability of proteins using an optimized temperature-controlled capillary electrophoresis approach, *J. Chromatogr. A* 838, 157–165, 1999.

139. Rochu, D., Rnault, F., and Masson, P., Detection of unwanted protein-bound ligands that stabilize cholinesterase conformation, *Electrophoresis* 23, 930–937, 2002.

140. Deng, B., Ru, Q.H., and Luo, G.A., Monitoring the unfolding and refolding of creatinine kinase by capillary zone electrophoresis, *Capill. Electrophor. Microchip Technol.* 8, 7–10, 2003.

141. Stutz, H., Wallner, M., Mallisa, H. Jr. et al., Detection of coexisting protein conformers in capillary zone electrophoresis subsequent to transient contact with sodium dodecyl sulfate, *Electrophoresis* 26, 1089–1105, 2005.

142. Reif, O.W. and Freitag, R., Studies of complexes between proteases, substrates, and the protease inhibitor α_2-macroglobulin using capillary electrophoresis with laser-induced fluorescence detection, *J. Chromatogr. A* 716, 363–369, 1995.

143. Zou, H., Zhang, Y., Lu, F., and Krull, I.S., Characterization of immunochemical reaction for human growth hormone with its monoclonal antibody by perfusion protein G affinity chromatography and capillary zone electrophoresis, *Biomed. Chromatogr.* 10, 78–82, 1996.

144. Tulp, A., Verwoerd, D., and Neefjes, J., Electromagnetism for separation of protein complexes, *J. Chromatogr. B Biomed. Sci. Appl.* 722, 141–151, 1999.

145. Al-Arhabi, M., Mrestani, Y., Richter, H., and Neubert, R.N., Characterization of interaction between protein and carbohydrate using CZE, *J. Pharm. Biomed. Anal.* 29, 555–560, 2002.

146. Liang, A., He, X., Du, Y. et al., Capillary zone electrophoresis investigation of the interaction between heparin and granulocyte-colony stimulating factor, *Electrophoresis* 25, 870–875, 2004.

147. Liang, A., He, X., Du, Y. et al., Capillary zone electrophoresis characterization of low molecular weight heparin binding to interleukin 2, *J. Pharm. Biomed. Anal.* 38, 408–413, 2004.

148. Liang, A., Chao, Y., Du, Y. et al., Separation, identification, and interaction of heparin oligosaccharides with granulocyte-colony stimulating factor using capillary electrophoresis and mass spectrometry, *Electrophoresis* 26, 3460–3467, 2005.

149. Liang, A., Zhou, X., Wang, Q. et al., Interactions of dextran sulfate with granulocyte-colony stimulating factor and their effects on leukemia cells, *Electrophoresis* 27, 3195–3201, 2006.

150. Liang, A., Du, Y., Wang, K., and Lin, B., Quantitative investigation of the interaction between granulocyte-macrophage colony-simulating factor and heparin by capillary zone electrophoresis, *J. Sep. Sci.* 29, 1637–1641, 2006.

151. Ling, X., Liu, Y., Fan, H. et al., Studies on interactions of programmed cell death 5 (PDCD5) and its related peptides with heparin by capillary zone electrophoresis, *Anal. Bioanal. Chem.* 387, 909–916, 2007.

152. Yi, L., Ziamol, L, Hui, F. et al., Determination of binding constant and binding region of programmed cell death 5 and heparin by capillary zone electrophoresis, *J. Chromatogr. A* 1143, 284–287, 2007.

153. Liu, Y., Zhang, S., Liang, X. et al., Analysis of the interactions between the peptides from secreted human CKLF1 and heparin using capillary zone electrophoresis, *J. Pept. Sci.* 14, 984–988, 2008.

154. Zhang, S., Yin, T., Ling, X. et al., Interactions between thrombin and natural products of *Millettia* species Champ. using capillary zone electrophoresis, *Electrophoresis* 29, 3391–3397, 2008.

4 Affinity Chromatography Including Hydrophobic Interaction Chromatography in the Study of Biopolymer Conformation

Affinity chromatography[1-5] can, for the specific purpose of this chapter, be divided into three areas: "active-site" affinity chromatography, "exosite" affinity chromatography, and immunoaffinity chromatography. Immunoaffinity chromatography will be discussed in Chapter 19. "Active-site" affinity chromatography uses substrate-related ligands such as inhibitors bound to matrices for the purification of specific enzymes.[6] "Exosite" affinity chromatography includes hydrophobic affinity chromatography and chromatography on heparin matrices. The term "exosite" refers to a site outside of the active site, which has specific binding characteristics such as the exosite in thrombin.[7] Exosite interactions also tend to be "nonspecific" compared to "active-site" affinity. For example, hydrophobic interaction sites tend to be nonspecific with respect to the affinity ligand; the same can be true for heparin-binding sites.

The effect of "nonspecific" factors in chromatography fractionation has been known for many years and form, for example, the basis for separation of amino acids on ion-exchange columns.[8-10] The tight binding of proteins and peptides to early ion-exchange resins required that chromatography be performed in the presence of urea.[11-13] The inclusion of urea was required because of forces other than ion exchange are involved in the binding of protein to the resin. It is of interest that the chromatography of trypsin on IRC-50 in 8.0 M urea required performance in the cold to avoid autolysis.[12]

The use of change in functional properties at the enzyme active site as measure of protein conformational change is more difficult and, as such, used far less than "exosite" chromatography. Quantitative affinity chromatography[14-18] has the potential to examine changes in active site binding to ligands, resulting from conformational change but has not yet seen direct application. The key question is how much does affinity for substrate change relative to other changes in protein conformation? Consideration of the discussions in Chapter 19 would suggest that, at least with low-molecular-weight substrates, there can be considerable conformational change before there are significant changes in catalytic parameter such as K_m. It is possible that this approach could be more useful for enzymes that work on complex carbohydrates such as those missing in lysosomal storage diseases; affinity chromatography has been used to purify enzymes in this class.[19–23] Despite the above caveats, it is not prudent to discard the concept of quantitative affinity chromatography in the evaluation of protein conformation.

Quantitative affinity chromatography has the potential to be useful in the evaluation of the integrity of an enzyme based on the binding of a specific cofactor or substrate where the term "biospecific" is sometimes used to describe the process. Quantitative affinity chromatography can be used to measure binding constants in crude extracts as shown by Brodelius and Mosbach for various dehydrogenases.[24,25] The K_{diss} values of porcine (H_4) and rabbit (M_4) lactate dehydrogenases were closed, if not identical, for the crystalline and crude enzyme extract.[25] It is necessary to have values for the purified enzyme to estimate the values in the crude tissue extracts. Nevertheless, the technique does have the potential for application for in-process analysis use. Denatured enzymes do not bind to affinity columns.[24,26,27] Some applications of quantitative affinity chromatography are presented in Table 4.1.

The process of "nonspecific" or "exosite" affinity chromatography has the potential of being more useful that biospecific chromatography in the study of protein conformation. The term patch has also been used to describe used to describe regions of unique chemical characteristic on a protein such as exosites; a PUBMED search revealed equal use of the terms as descriptor of a protein domain with little overlap between the terms. The term exosite seems to be used to describe secondary binding sites on various proteolytic enzymes. The term patch has a broader use with acidic patches[45–51] and basic patches,[52–59] and hydrophobic patches[60–69] described in a variety of proteins.

The hypothesis underlying the use of "nonspecific" affinity chromatography is that a conformational change in protein will expose a buried or cryptic patch or exosite, which will then bind to an attractive matrix. It is also possible to use the concepts of quantitative affinity chromatography with exosite affinity chromatography (Table 4.2) as, for example, with the use of heparin as a ligand.[70,71,73] Heparin is a highly charged polyanionic molecule, which can act as an ion-exchanger;[70,71] cation-exchange matrices such as sulfoalkyl-polysaccharide matrices (e.g., Sulfopropyl-Sephadex®). More frequently, however, heparin is used as a ligand for a specific site on a protein (Table 4.3) and has been used to study conformational change in proteins. Heparin–agarose has been used to separate conformational forms of antithrombin,[91,99,100] mutant forms for antithrombin,[101] and chemically modified antithrombin,[102] inactive and active forms of bovine trypsin,[103] mutant forms of receptor-associated protein,[104] and conformational forms of fibronectin.[105] In these cases, a cryptic patch containing basic amino acids is exposed, permitting binding to an acid matrix such as heparin–agarose. In some of these situations, the exosite is exposed on the native protein. This is case with thrombin where the isoelectric point is approximately 5.6, yet it behaves as a cation on chromatography at pH 6.5.[106]

There is extensive use of hydrophobic affinity chromatography (hydrophobic interaction chromatography [HIC]) both for protein purification and to study

TABLE 4.1
Application of Quantitative Affinity Chromatography to the Study of Enzyme–Ligand Binding Constants

Protein	Ligand/Study	Ref.
Staphylococcal nuclease	Use of thymidine-5′-phosphate-3′-aminophenylphosphate-Sepharose to determine binding constants using competitive inhibitors	[28]
Ribonuclease	Use of uridine-5′-(Sepharose-4-aminophenylphosphoryl)-2(3′)-phosphate for analysis of nucleotide–RNAse interactions	[29]
Staphylococcal nuclease	Comparison of quantitative affinity chromatography and equilibrium dialysis	[30]
Myeloma protein TEPC 15	Phosphorylcholine-Sepharose; evaluation of multivalency of antibody	[31]
Carboxypeptidase B	Evaluation of multiple binding sites using D-phenylalanine bound to agarose matrix via a hexyl spacer. The bound ligand bound to the enzyme more tightly than the free amino acid suggesting the importance of the spacer group in the interaction with the enzyme	[32]
A-chymotrypsin	Matrix of 4-phenylbutylamine linked to succinylated polyacrylic hydrazide agarose. Binding is enhanced by the presence of a competitive inhibitor, carbobenzoxy-phenylalaine and decreased by the presence of benzyloxycarbonyl-alanine-alanine. It is suggested that selected solutes can specifically modulate the binding of enzyme to a matrix.	[33]
Adenosine deaminase	Matrix inosine bound to agarose via a 6-aminohexyl spacer	[16]
Phosphorylase b	AMP-agarose matrix used to determine binding constants for IMP, AMP and AMP analogues	[34]
Immunoglobulins	Dinitrophenyl-aminohexyl-Agarose. Measurement of binding affinities for detergent-solubilized surface membrane IgA, comparison with reduced carboxymethylated secreted immunoglobulins	[35]
Myosin subfragment-1	Periodate-oxidized ATP coupled to hydrazide agarose	[36]
Lactate dehydrogenase	Blue-Sepharose[a]; evaluation of the effect of gel filtration on quantitative affinity chromatography[b] on agarose matrices	[37]
Plasminogen (Glu-plasminogen; lys-plasminogen; miniplasminogen)	Used aminohexyl agarose to measure binding affinity or various ligands such as lysine and lysine derivatives	[38]
Lectins	Use of unmodified agarose and other polymerized polysaccharide matrices as affinity matrices for carbohydrate-binding proteins such as lectins	[39]
DNA-binding proteins	Used DNA-cellulose to measure nonspecific binding of proteins to DNA	[40]
Cytochrome c	Photosynthetic reaction centers immobilized in liposomes (containing biotinylated phosphatidylethanolamine), which are encapsulated in streptavidin-coupled gel (Sephadex S-100 or Sephacryl)	[41]

(*continued*)

TABLE 4.1 (continued)

Application of Quantitative Affinity Chromatography to the Study of Enzyme–Ligand Binding Constants

Protein	Ligand/Study	Ref.
Prothrombin and thrombin	Hirudin-derived peptide coupled to agarose used to identify a "proexosite" of prothrombin which binds to hirudin; establish that binding affinity increased 100-fold on activation to thrombin	[42]
Glutamate dehydrogenase	Perphrenazine-Sepharose used to measure interaction of drugs with glutamate dehydrogenase	[43]
Glutamate dehydrogenase isoenzymes	Perphrenazine-Sepharose	[44]
Review	Review	[18]

[a] Blue Sepharose refers to an affinity matrix prepared by the coupling of Cibacron blue F-GA to agarose yielding a versatile matrix. See Travis, J., Bower, J., Tewksbury, D. et al., Isolation of albumin from whole human plasma and fractionation of albumin-depleted plasma, *Biochem. J.* 157, 301–306, 1976; Lamkin, G.E. and King, E.E., Blue Sepharose: A reusable affinity chromatography medium for purification of alcohol dehydrogenase, *Biochem. Biophys. Res. Commun.* 72, 560–565, 1976.

[b] Some authors use the term bioaffinity chromatography (cf. Turková, J., *Bioaffinity Chromatography*, 2nd edn., Elsevier, Amsterdam, 1993) or biospecific chromatography (cf. Irvine, G.B. and Murphy, R.F., Biospecific chromatography with macro-ligands, *Biochem. Soc. Trans.* 9, 284–285, 1981; Harakas, N.K., Protein purification process engineering. Biospecific affinity chromatography, *Bioprocess Technol.* 18, 259–316, 1994).

TABLE 4.2

Exosite Affinity Chromatography of Biological Macromolecules

Protein	Ligand/Study	Ref.
Thrombin	Use of heparin-agarose to study solvent effects on binding to thrombin; comparison with effect on the extrinsic fluorescence probe, *p*-aminobenzamidine	[70]
Thrombin and antithrombin	Affinity partitioning on heparin-agarose	[71]
Human C4b-binding protein	Serum amyloid protein coupled (CNBr) to agarose	[72]

conformational change in protein. There is a strong attraction when two hydrophobic surfaces approach each other, which is likely due changes in solvation, reflecting the removal of water.[107–114] It is recognized that the driving force is the attraction of water to itself and the self-attraction of nonpolar groups is less significant.[115–117] Hydrophobicity and the qualitative measurement of hydrophobicity will be a recurring theme is this volume as the exposure of cryptic hydrophobic sites is a frequent consequence of conformational change in proteins.[118–118j]

TABLE 4.3
The Use of Heparin-Agarose for Analysis of Protein Conformation

Protein	Ligand/Study	Ref.
Type I collagen	Correlation of binding of type I collagen and fragments to heparin-agarose with changes in electrophoretic mobility	[74]
Lipoprotein lipase	Used heparin-agarose to study conformational change (dimer to monomer transition)	[75]
Lipoprotein lipase	Used heparin-agarose to study conformational changes; correlation with reaction with monoclonal antibodies	[76]
Follistatin	Use of heparin-agarose and sulfated cellulose[a] to identify heparin binding site of follistatin using *Staphylococcus aureus* V8 protease digestion	[77]
Lipoprotein lipase	Evaluation of mutant forms (replacement of basic residues by alanine) of lipoprotein lipase by chromatography of heparin-agarose	[78]
Type I collagen	Use of heparin-agarose to study conformation of type I collagen. Native type I collagen binds to heparin-agarose while heat-denatured type I collagen does not bind to heparin-agarose. Urea (3 M) reduced the amount of type I collagen binding to heparin-agarose while 4 M urea has no effect on CD spectra	[79]
Lipoprotein lipase	Native dimeric lipoprotein lipase binds to heparin-agarosel lipoprotein lipase produced in the presence of tunicamycin (which blocks *N*-glycosylation) is aggregated and binds poorly to heparin-agarose	[80]
Asymmetric acetylcholinesterase	Use of heparin-agarose to characterize the collagen-tailed form of acetylcholinesterase	[81]
Vitronectin	Characterization of plasma vitronectin, which does not bind to heparin-agarose, and platelet vitronectin, which does bind heparin-agarose	[82]
Human extracellular superoxide dismutase	Isolation of the heparin-binding domain of extracellular superoxide dismutase	[83]
Human complement factor H	Use of heparin-agarose to evaluate truncated and mutant forms of human complement factor H including the role of short consensus repeats.	[84]
Heliobacter pylori urease	*H. pylori* urease purification using a heparinoid (Cellulfine sulfate[a])	[85]
Papain	Conformation change (CD, enzyme activity) in the presence of heparin; interaction demonstrated by binding to heparin-agarose column	[86]
Human selenium-dependent thioredoxin reductase	Separation of different forms (varied oxidation states) by heparin-agarose	[87]
Fibronectin	Modification with glucose in the presence of oxygen decreases affinity to heparin-agarose. Conformational change (CD, electrophoresis) was also observed under these conditions of modification (glycoxidative)	[88]

(*continued*)

TABLE 4.3 (continued)
The Use of Heparin-Agarose for Analysis of Protein Conformation

Protein	Ligand/Study	Ref.
Kallistatin	Use of heparin-agarose to determined effect of amino acid changes (alanine substitution for lysine) on binding	[89]
Human heparin/heparan sulfate interacting protein/ L29 (human HIP/L29)	Use of heparin-agarose to evaluated mutant and truncated forms of HIP/L29. Comparison with CD studies on heparin binding	[90]
Antithrombin	Separation of latent and prelatent antithrombin from native commercial and heat-treated antithrombin preparations evaluation of anti-angiogenic active of prelatent antithrombin	[91]
Tenascin-X	Use heparin-agarose to demonstrated conformational dependence of tenascin binding to heparin	[92]
Receptor-associated protein (RAP)	Use of heparin-agarose to evaluate effect of mutation on binding affinity of RAP. Correlation with effect on binding to low-density lipoprotein receptor family	[93]
Minicollagen XII	Use of heparin-agarose to demonstrate conformation of minicollagen expressed in insect cells.	[94]
Human extracellular superoxide dismutase	Use of heparin-agarose to evaluated effect of substitution of basic amino acids (alanine scanning) for binding to heparin and heparan sulfate proteoglycans	[95]
Co1Q (collagen tail subunit of synaptic acetylcholinesterase)	Use of heparin-agarose to identify heparin-binding domains in the triple helix domain of Co1Q	[96]
Human extracellular superoxide dismutase	Use of heparin-agarose to show importance of an inter-subunit disulfide bond in binding to heparin	[97]
Antithrombin	Separation of perlatent antithrombin and native antithrombin	[98]

[a] Matrex® Cellufine™ sulfate is a beaded cellulose with substitution of the 6-hydroxyl group with sulfate.

HIC, hydrophobic chromatography, and hydrophobic affinity chromatography represent an application of reverse-phase (RP) chromatography (reverse-phase high-performance liquid chromatography [RP-HPLC]). RP chromatography is distinguished from conventional or normal-phase chromatography in that the stationary phase is nonpolar; in normal-phase chromatography, the stationary phase is polar as, for example, with ion-exchange chromatography. Polar analytes may be studied in RP chromatography by a technique called ion-pair (Figure 4.1) chromatography.[119–121] The use of hydrophobic affinity chromatography for conformational change is based on the availability of hydrophobic patches on the proteins surface and the interaction of such domains with the hydrophobic stationary phase. The matrix for HIC is usually modest in being composed of a phenyl group or a C_4-hydrocarbon chain while RP-HPLC uses a more hydrophobic C_8 or C_{18} groups; matrices include silica (Figure 4.2) or various organic materials (Figure 4.3). One advantage of hydrophobic affinity chromatography is the coupling with other technologies such as the use of

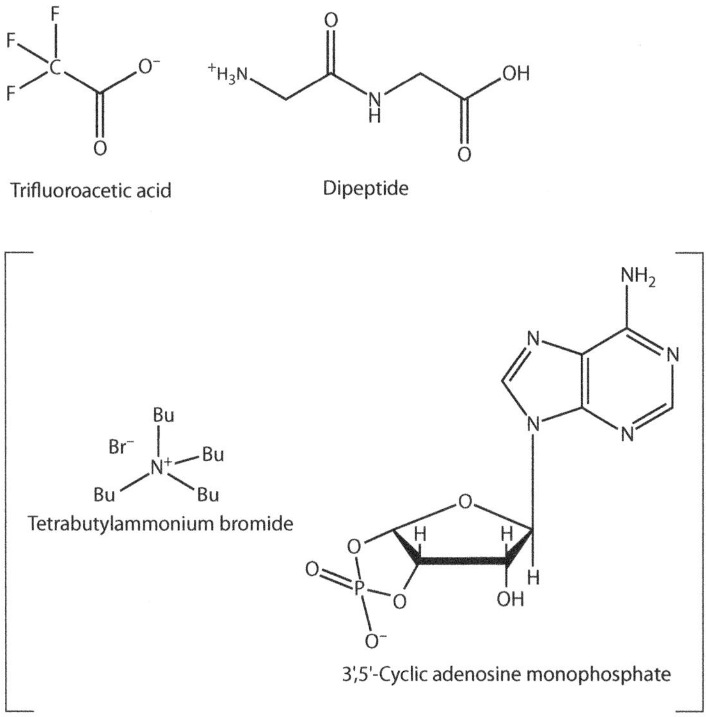

FIGURE 4.1 Examples of ion-pair reagents used in RP-HPLC (see Pipkorn, R., Boenke, C., Gehrke, M., and Hoffman, R., High-throughput peptide synthesis and peptide purification strategy at the low micromole-scale using the 96-well format, *J. Pept. Res.* 59, 105–114, 2002; Witters, E., Van Dongen, W., Esmans, E.L., and Van Onckelen, H.A., Ion-pair liquid chromatography-electrospray mass spectrometry for the analysis of cyclic nucleotides, *J. Chromatogr. B.* 694, 55–63, 1997; Hearn, M.T.W., ed., *Ion-Pair Chromatography: Theory and Biological and Pharmaceutical Applications*, Marcel Dekker, New York, Chapter 1, 1995).

fluorescent probes.[118g,122–125] The RP-HPLC matrix is usually more densely packed that the HIC matrix.[126] As would be expected, there is difference in the performance of proteins in the two systems.[127] The nature of HIC matrix has a major effect on the binding and separations of proteins.[128–130]

Thiophilic chromatography is a special case of hydrophobic interaction chromatography where a matrix is produced by the sequential modification of agarose by divinylsulfone and 2-mercaptoethanol (Figure 4.4).[131] Thiophilic chromatography is highly specific for immunoglobulins,[132] but has been useful for other proteins including human transferrin.[133,134] A complement-activating component from Cobra venom,[135] amyloid-beta protein precursor,[136] prostate-specific antigen,[137] and serum albumin.[138] It is noted that immunoglobulins bind more tightly than other plasma proteins[138] but thiophilic matrix is promiscuous.[139] The mechanism of binding to thiophilic matrices is not clear.[140,141] While the matrix has polar characteristics, solvent

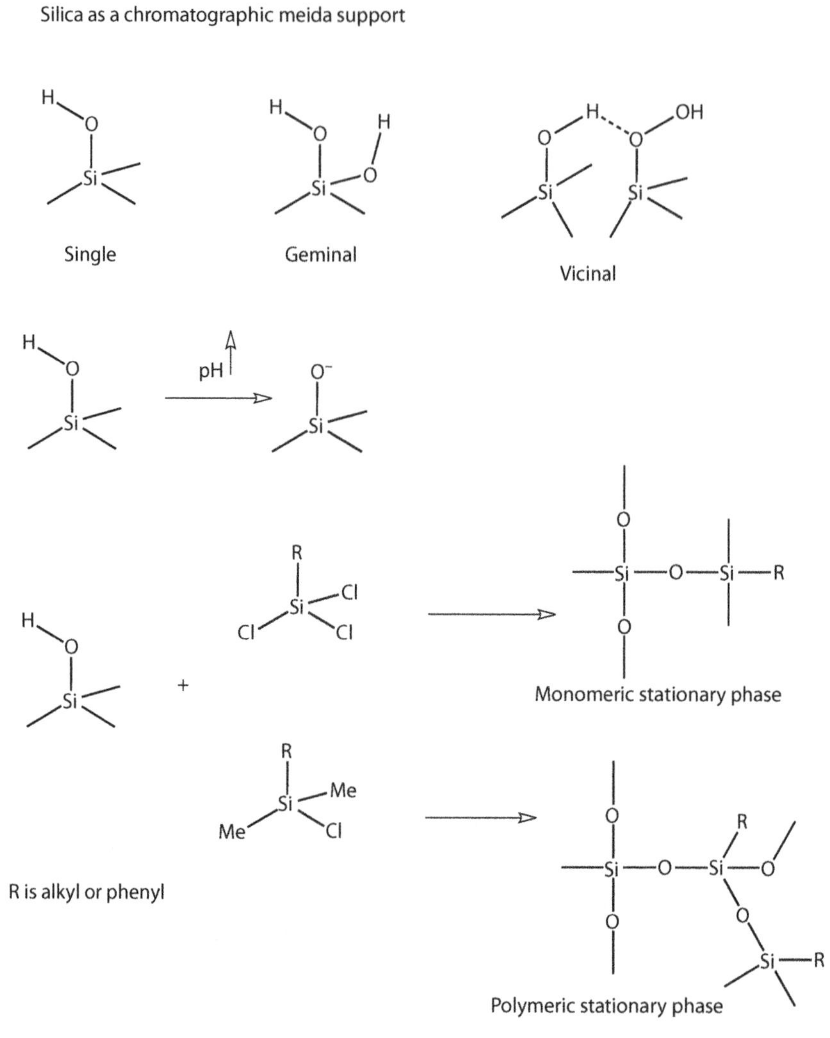

Silica as a chromatographic meida support

FIGURE 4.2   Properties of silica and modification for RP-HPLC and HIC (see Unger, K.K., Silica as a support, in *HPLC of Biomacromolecules. Methods and Applications*, eds. K.M. Gooding and F.E. Regnier, Marcel Dekker, New York, 1990). Silica is a solid hydrous oxide which can be represented as SiO_2. There are Brønsted acids sites at the silica surface; the pK_a is approximately 7; the isoelectric point is 2.2. Organosilanes are used to modify surface hydroxyl groups with alkyl or phenyl functions to provide a hydrophobic matrix (see Doyle, C.A. and Dorsey, J.G., Reversed-phase HPLC: Preparation and characterization of the reversed-phase stationary phases, in *Handbook of HPLC*, eds. E. Katz, R. Eksteen, D. Schoenmakers, and N. Miller, Marcel Dekker, New York, 1998; Fargáca, E. and Cserbáti, T., *Molecular Basis of Chromatographic Separation*, CRC Press, Boca Raton, FL, 1997; Cazes, J. ed., *Encyclopedia of Chromatography*, Marcel Dekker, New York, 2001).

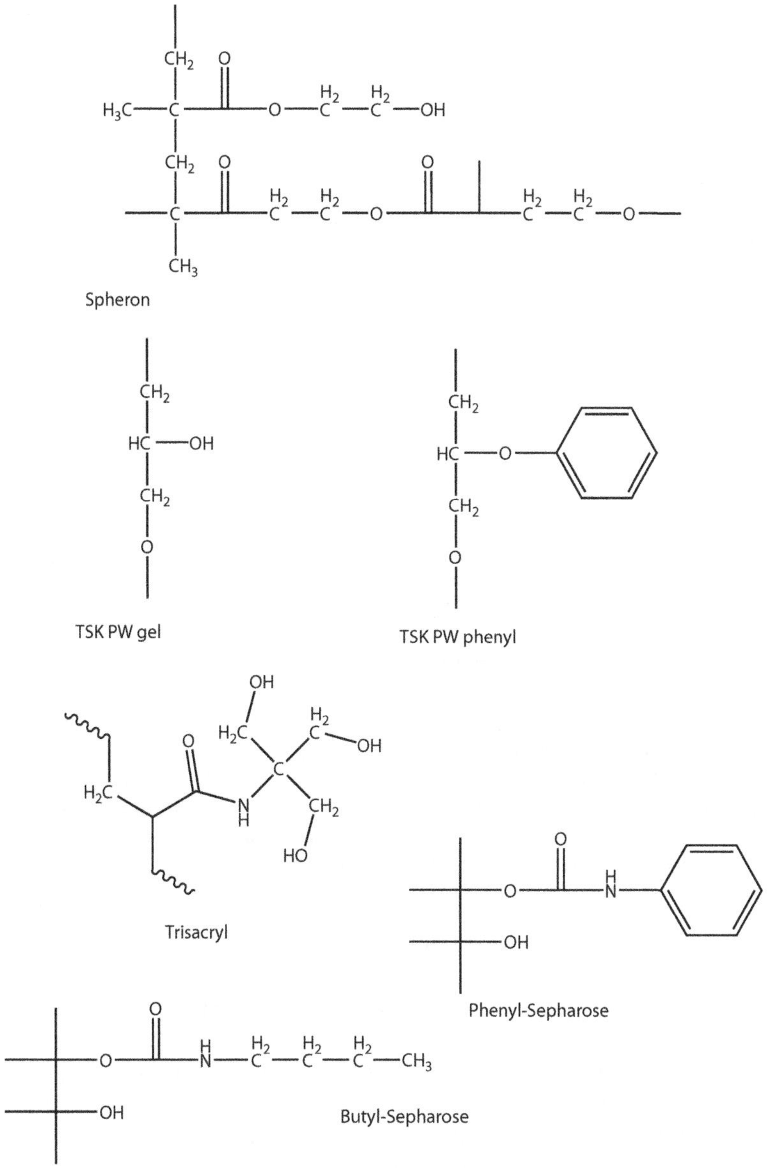

FIGURE 4.3 Some hydrophobic matrices used RP-HPLC and HIC (see Mikes, O. and Coupek, J., Organic supports, in *HPLC of Biomacromolecules. Methods and Applications*, eds. K.M. Gooding and F.E. Regnier, Marcel Dekker, New York, Chapter 2, 1990; Shansky, R.E., Wu, S.-L., Figueroa, A., and Karger, B.L., Hydrophobic interaction chromatography of biopolymers, in *HPLC of Biomacromolecules. Methods and Applications*, eds. K.M. Gooding and F.E. Regnier, Marcel Dekker, New York, Chapter 5, 1990; Jennissen, H.P., Hydrophobic interaction chromatography, in *Encyclopedia of Separation Science*, Vol. 1, ed. I.D. Wilson, Academic Press, San Diego, CA, pp. 265–277, 2000).

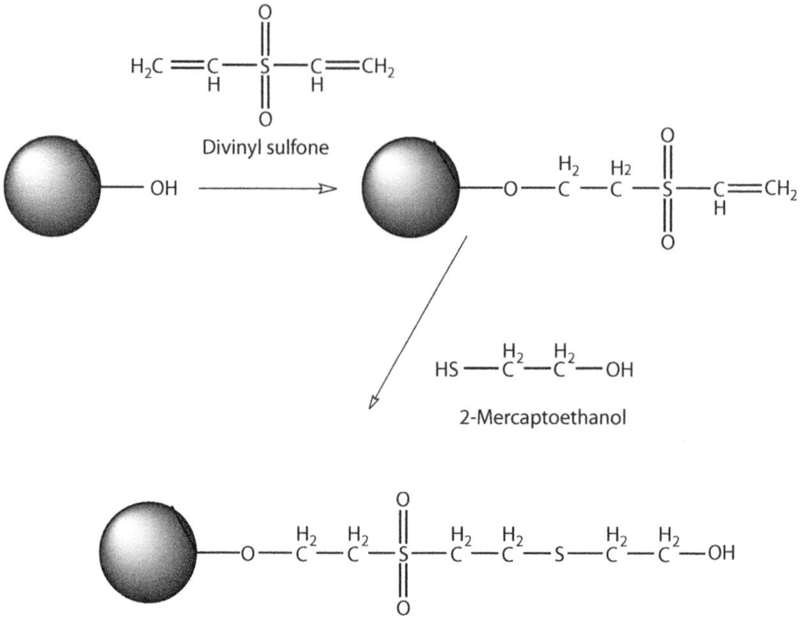

FIGURE 4.4 Scheme for the preparation of a thiophilic gel (see Porath, J., Maisano, F., and Belew, M., Thiophilic adsorption—A new method for protein fractionation, *FEBS Lett.* 185, 306–310, 1985; Hutchens, T.W. and Porath, J., Thiophilic adsorption of immunoglobulins— Analysis of conditions optimal for selective immobilization and purification, *Anal. Biochem.* 159, 217–226, 1986; Boschetti, E., The use of thiophilic chromatography for antibody purification: A review, *J. Biochem. Biophys. Methods* 49, 361–389, 2001).

condition for adsorption and elution are similar to those using for HIC in that binding is promoted by hydrophilic salts.

Hydrophobic affinity chromatography has been used to study the conformation of the vitamin K-dependent coagulation factors. The vitamin K-dependent coagulation factors such as factor IX require calcium ions for function; some divalent cations can replace calcium in function.[142] Proteins such as prothrombin and factor IX have been demonstrated to have a calcium-stabilized conformation where cryptic epitopes are expressed.[143,144] In addition to the exposure of cryptic epitopes, the conformation observed in the presence of calcium ions include changes in the environment around aromatic residues[145,146] and such changes appear to be associated with phospholipid binding.[145–148] Conformation-specific antibodies have been useful for the purification of the vitamin K-dependent coagulation factors.[148–152]

Prothrombin is retarded on size exclusion chromatography in the presence of calcium ions,[153,154] suggesting a decrease in molecular size (Stokes' radius). Subsequent work[155] showed that the retention of prothrombin or prothrombin fragment 1 on chromatography on phenyl-Sepharose is retarded in the presence of calcium ions, but not magnesium ions, suggesting exposure of a cryptic hydrophobic site; the oxidation

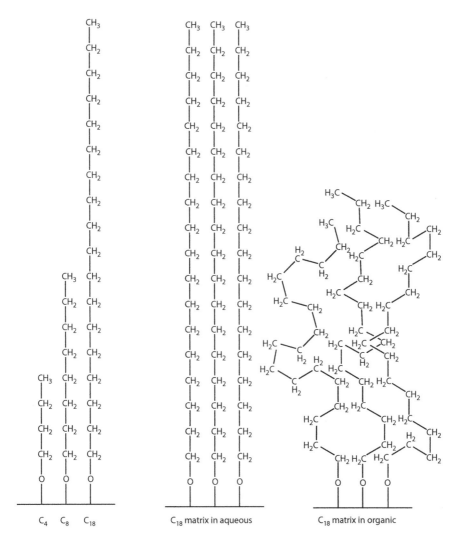

FIGURE 4.5 The structure of some alkyl hydrophobic matrices. There is likely a difference in structure in the alkyl chains dependent on solvent as shown by the C_{18} matrices in water and organic solvents. (Adapted from Pesek, J.J. and Matyska, M.T., Reversed-phase stationary phases, in *Encyclopedia of Chromatography*, ed. J. Cazes, Marcel Dekker, New York, 2001.) See also Shatiel, S., Hydrophobic chromatography, *Methods Enzymol.* 34, 126–140, 1974; Shatiel, S., Hydrophobic chromatography, *Methods Enzymol.* 104, 69–96, 1984; Fargács, E. and Cserháti, T., *Molecular Basis of Chromatographic Separation*, CRC Press, Boca Raton, FL, 1997.

of a tryptophan residue in prothrombin fragment 1 by *N*-bromosuccinimide reduced the extent of retardation. This is consistent with the observations of Liebmann[146] on the oxidation of factor IX by *N*-bromosuccinimide, which suggested a role for tryptophan in expression of a phospholipid-binding site. In earlier studies, the migration of bovine Factor IXa on G-200 Sephadex was retarded by the presence of calcium ions.[156] The conformational change of vitamin K–dependent proteins in the presence of calcium ions is critical for biological function and it is suggested that the measure of the conformational integrity as measured by binding of protein to HIC or gel filtration media is required for biological function. This interaction is unique as high salt is not required to promote adsorption to the matrix as discussed below. In this case, HIC or size-exclusion chromatography in the presence and absence of calcium ions can be used as a surrogate marked for the biological activity of Factor IX. The binding of proteins to a hydrophobic matrix is promoted by the presence of high salt concentration (high salts decrease the solubility of hydrocarbons in water). Conventional HIC requires the presence of high salt and elution from the matrix is promoted by decreasing salt or increased pH.[156a,156b] It is noted that there is interest in performed HIC at lower ionic strength (0.5 M $(NH_4)SO_4$).[156c]

HIC can be used to separate vitamin K–dependent proteins in the absence of calcium ions.[157] The protein sample is applied to the column in 3.0 M NaCl and elution is accomplished with a decreasing salt gradient. Factor X was not released by a phenyl-Sepharose column while prothrombin was eluted before Factor IX. Other HIC media were evaluated: prothrombin was retained by octyl(C_8) but not by a butyl(C_4) agarose matrix. Thrombin is separated from prothrombin on a phenyl-Sepharose column[158]; this matrix can be used for the large-scale purification of thrombin.[159]

The migration of hydrophobic proteins on polysaccharide-based chromatographic matrices (agarose) is retarded.[160] Similar phenomenon is observed with basic proteins as low ionic strength.[160] The interaction of hydrophobic proteins with the polysaccharide matrix is enhanced by high salt concentrations.[160] The "nonspecific" binding of protein to gel filtration media via hydrophobic interaction has been discussed above and is seen with a variety of matrices.[161–164] Analyte charge also has a role in gel filtration.[165–168]

α-Lactalbumin is a small, acidic protein (14,200, 123 amino acids) found at high concentrations in milk where it is involved in the action of lactose synthase.[169,170] The binding of calcium ions to α-lactabumin is associated with a conformational change in the protein. A decrease in fluorescence quantum yield and shift to shorter wavelength ("blue shift"); both of these measures are associated with a change in the environment of tryptophanyl residues.[171] The binding of calcium ions to α-lactabumin also results in an increase thermal stability. α-Lactalbumin has been used as model protein for a large number of studies on protein conformation.[172–179] Some of these studies have used chromatography on hydrophobic matrices to study conformation change in α-lactalbumin. Wu and coworkers[180] evaluated the thermal behavior of α-lactalbumin on HIC (methyl and ethyl matrices) and observed that retention increased with increasing temperature. Subsequent work by this group[181] showed that the addition of calcium ions to the mobile phase resulted in an increase in the temperature required to result in an increase in column retention. To and Lenhoff[182] studied the chromatography of α-lactalbumin on phenyl-Sepharose in an isocratic elution system with ammonium sulfate. Increasing ammonium sulfate concentration results in an increasing retention; in the absence of calcium ions, there is considerable tailing; in the presence of 10 mM Ca+, the peaks are asymmetric but with considerably less tailing. Fogle and Fernandez[183] used hydrogen–deuterium exchange to study the solvent exposure of proteins adsorbed to phenyl-Sepharose. The extent of exchange was determined by mass spectrometry. There was no difference in the extent of hydrogen–deuterium exchange with lysozyme in solution and bound to the matrix. There was an increase in the solvent exposure of α-lactalbumin bound to the phenyl-Sepharose matrix as compared to α-lactalbumin in solution. This is consistent with a conformational change in α-lactalbumin bound to the hydrophobic matrix. Subsequent work from this group[184] showed that increased protein loading decreased the amount protein unfolding on the phenyl-Sepharose matrix. Adsorption to butyl-Sepharose was demonstrated to be more destabilizing than phenyl-Sepharose. Ueberbacher and colleagues[185] studied the chromatography of α-lactalbumin on butyl-Sepharose under isocratic solutions using ammonium sulfate as the mobile phase. As with the studies using phenyl-Sepharose described above, increasing ammonium sulfate concentration increases the retention of α-lactalbumin; there was no retention when 0.5 M ammonium sulfate was used and a compete retention when 1.0 M ammonium sulfate was used. Lysozyme was not retained on the column while the behavior of bovine serum albumin was similar to that of α-lactabumin. Studies with Fourier transform IR spectroscopy confirmed the conformational changes occurring the presence of ammonium sulfate, which are likely responsible for the binding to the hydrophobic matrices.

While the primary thrust of this chapter is the description of the use of HIC to study protein conformation, it is necessary to briefly discuss the use of RP-HPLC.[186–190] RP-HPLC is used for both protein purification and characterization. Some selected examples are presented in Table 4.4 with an emphasis on issues related to commercial biotechnology. Most of these studies use an alkyl chain bonded to a matrix such as silica or an organic polymer (Figure 4.3). The most common matrices are butyl (C_4), octyl (C_8), octadecanyl (C_{18}), or phenyl (Figure 4.5). The RP-HPLC of proteins is usually performed under conditions where the protein sample undergoes denaturation.[207,208] Failure to obtain complete denaturation prevents a correlation of retention time with protein hydrophobicity and possible lack of reproducibility. Conformational change can occur during RP-HPLC as demonstrated by Cohen and workers for RNAse[209] where there is an equilibrium between native and denatured protein at 23°C; if the chromatography is performed at 37°C, only a single peak corresponding to the denatured protein is observed.

In addition to effective sample composition (protein quality) and sample pretreatment, a variety of other factors contribute to the RP-HPLC of proteins including matrix, solvent, and temperature.[210]

Some selected studies on the use of HIC to study protein conformation are presented in Table 4.5. It must be recognized that these are not static measurement of the quality of the sample. It is possible that there is significant change in protein conformation in HIC as a result of the high salt conditions or binding to the matrix.[220–222] It is also possible that there is hysteresis in binding to hydrophobic surfaces as well as irreversible binding.[223,224]

TABLE 4.4

Selected Examples of the Use of RP-HPLC and Hydrophobic Interaction Chromatography for the Purification and Characterization of Proteins

Protein	Matrix and Conditions	Refs.
Human monocyte chemoattractant protein (MCP-1)	RP-HPLC (C_{18}); TFA/ACN[a]; Separation of three forms of MCP-1, which differ the extent of glycosylation	[191]
Recombinant leech-derived tryptase inhibitor	RP-HPLC (C_{18})	[192]
Rat ribosomal proteins	RP-HPLC (C_{18}). TFA/CAN; Separation of various posttranslationally modified rat ribosomal proteins (e.g., truncation, methylation); analysis of separated proteins by mass spectrometry	[193]
^{125}I-EGF	RP-HPLC (C_{18}) TEA-HOAc[b], pH 7.1/ACN Separation of various species of iodinated EGF (chloromine T iodination);	[194]
Surface glycoprotein from *Trypanosoma brucei*	RP-HPLC, TFA/2-propanol; resolution improved by changing to 1-propanol; degradation of sample in TFA eliminated by change to ammonium formate	[195]
High mobility groups 1 and 1 proteins (HMG 1 and HMG 2)	Separation of oxidized and reduced forms of HMG 1 and HMG 2 proteins	[196]
Monoclonal antibody (MAB)	RP-HPLC (various alklyl matrices). Monitor MAB stability, evaluation of mobile phases, column temperature, and column matrices. The mobile phase solvent was ACN and columns developed with an organic solvent; optimal resolution was obtained with 1-propanol or 2-propanol	[197]
Human albumin	RP-HPLC; Formic acid/2-propanol	[198]
Variant IL-2 proteins (muteins[c]) including oxidized forms	RP-HPLC (C_4), TFA/ACN; separation of various variant forms of IL-2.	[199]
Human insulin-like growth factor I (hIGF-I)	RP-HPLC, (C_8) pH 2.0 sodium phosphate with propanesulfonic acid, sodium salt/CAN. Separation of native and oxidized forms of the protein	[200,201]
Recombinant methionyl human granulocyte colony stimulating factor (Neupogen®)	RP-HPLC (C_4), sodium perchlorate, pH 2.0/ACN; separation of oxidized products formed by reaction with hydrogen peroxide	[202]
Recombinant human interleukin 11	RP-HPLC (C_4) TFA/ACN; separation of oxidized (hydrogen peroxide) protein from native protein. Samples stored in polypropylene tubes were oxidized more rapidly than those stored in polystyrene tubes. Polypropylene tubes exposed to light promoted oxidation. The small difference in hydrophobicity provided by methionine oxidation required a shallow ACN gradient for resolution of native and oxidized protein	[203]
Monoclonal antibody	RP-HPLC (polystyrene-divinyl benzene); TFA/CAN; separation of MAB with oxidized tryptophan. RP-HPLC was found to be more reproducible that size-exclusion chromatography[d]	[204]

TABLE 4.4 (continued)
Selected Examples of the Use of RP-HPLC and Hydrophobic Interaction Chromatography for the Purification and Characterization of Proteins

Protein	Matrix and Conditions	Refs.
Human insulin	RP-HPLC (C_4); TFA/ACN; evaluation of human insulin degradation by gamma irradiation or accelerated electrons	[205]
Recombinant human stem cell factor	RP-HPLC (C_4), TFA/ACN; analysis of oxidation (hydrogen peroxide) products.	[206]

a TFA, trifluoroacetic acid; ANC, acetonitrile; TFA is usually the mobile phase solvent with a gradient of ACN used for column development.

b TEA, triethanolamine; HOAc, acetic acid.

c The term "mutein" refers to a protein obtained from the expression of a gene sequence altered by site-specific mutagenesis (Kunitani, M., Hirtzer, P., Johnson, D. et al., Reversed-phase chromatography of interleukin-2 muteins, *J. Chromatogr.* 359, 391–402, 1986).

d The initial observation was that the oxidized material eluted from a size exclusion column (TSK) prior to the native material reflecting non-specific adsorption of protein to the SEC matrix (see Edwards, S.L. and Dubin, J., pH effects on nonideal protein size-exclusion chromatography on Superose 6, *J. Chromatogr.* 648, 3–7, 1993; Pujar, N.S. and Zydney, A.L., Electrostatic effects on protein partitioning in size-exclusion chromatography and membrane ultrafiltration, *J. Chromatogr. A.* 796, 229–238, 1998).

TABLE 4.5
HIC and Protein Conformation

Protein	Matrix/Conditions/Study	Ref.
Phosphorylase b	Cooperative binding of phosphorylase b to butyl-Sepharose matrix	[211]
Soluble interleukin I receptor type II	Isopropyl-Sepharose; Reverse ammonium sulfate gradient; use for identifying degraded forms of formulated protein; use in stability indicating assay	[212]
N/A	Correlation of second virial coefficient with retention on HIC columns	[213]
Albumin	Phenyl-Sepharose; separation of fatty acid-free albumin from liganded albumin	[214]
Urea-denatured proteins	Correlations of HIC with SEC, fluorescence, ion-exchange	[215]
A-lactalbumin	Phenyl-Sepharose; apo-α-lactalbumin binds to phenyl-Sepharose and is eluted with Ca^{2+} or Na^+	[216]
Calbindin D (28K)	Phenyl-Sepharose; Reverse NaCl gradient, change in hydrophobic site in he presence of calcium ions; not as pronounced as with calmodulin. Correlation with fluorescent probe (ANS) binding; intrinsic fluorescence, CD, UV spectroscopy	[118]
Troponin C	Phenyl-Sepharose; binds in the presence of calcium ions (25 mM MOPS, pH 7.1), eluted with EDTA	[217]
Equine lysozyme	Phenyl-Sepharose; binds in the absence of calcium ions, eluted with 1 mM Ca^{2+}; correlation with ANS fluorescence	[218]
Myoglobin, cytochrome *c*	Phenyl-TSK; Reverse ammonium sulfate gradient. Evaluation of protein denaturation with urea and guanidine thiocyanate.	[219]

REFERENCES

1. Lowe, C.R. and Dean, P.D.G., *Affinity Chromatography*, John Wiley & Sons, London, U.K., 1974.
2. Gribnau, T.C.S., Visser, J., and Nivard, R.J.F. eds., *Affinity Chromatography and Related Techniques*, Elsevier, Amsterdam, the Netherlands, 1982.
3. Dean, P.D.G., Johnson, W.S., and Middle, F.A. eds., *Affinity Chromatography. A Practical Approach*, IRL Press, Oxford, U.K., 1985.
4. Chaiken, I.M., *Analytical Affinity Chromatography*, CRC Press, Boca Raton, FL, 1987.
5. Hage, D.S. ed., *Handbook of Affinity Chromatography*, 2nd edn., Taylor & Francis, Boca Raton, FL, 2005.
6. Cuatrecasas, P., Wilchek, M., and Anfinsen, C.B., Selective enzyme purification by affinity chromatography, *Proc. Natl. Acad. Sci. USA* 61, 636–643, 1968.
7. Fenton II J.W., Witting, J.I., Pouliott, C., and Fareed, J., Thrombin anion-binding exosite interactions with heparin and various polyanions, *Ann. N.Y. Acad. Sci.* 556, 158–165, 1989.
8. Moore, S. and Stein, W.H., Chromatography of amino acids on starch columns: Solvent mixtures for the fractionation of protein hydrolysates, *J. Biol. Chem.* 178, 53–57, 1949.
9. Moore, S. and Stein, W.H., Chromatography of amino acids on sulfonated polystyrene resins, *J. Biol. Chem.* 192, 663–681, 1951.
10. Moore, S. and Stein, W.H., Procedures for the chromatographic determination of amino acids on four per cent cross-linked sulfonated polystyrene resins, *J. Biol. Chem.* 211, 893–906, 1954.
11. Cole, R.D., The chromatography of insulin in urea-containing buffer, *J. Biol. Chem.* 235, 2294–2299, 1960.
12. Cole, R.D. and Kinkade, J.M. Jr., A chromatographic study of trypsin, *J. Biol. Chem.* 236, 2443–2445, 1961.
13. Cole, R.D., Ion exchange chromatography of prolactin in urea-containing buffers, *J. Biol. Chem.* 236, 1369–1371, 1961.
14. Dunn, B.M. and Gilbert, W.A., Quantitative affinity chromatography of alpha-chymotrypsin, *Arch. Biochem. Biophys.* 198, 533–540, 1979.
15. Danner, J., Somerville, J.E, Truner, J., and Dunn, B.M., Multiple binding sites of carboxypeptidase B: The evaluation of dissociation constants by quantitative affinity chromatography, *Biochemistry* 18, 3039–3045, 1979.
16. Rosemeyer, H. and Seela, F., Quantitative affinity chromatography of adenosine deaminase on polymer-bound inosine: The assessment of binding constants by biospecific elution, *Anal. Biochem.* 115, 339–346, 1981.
17. Bergman, D.A. and Winzor, D.J., Quantitative affinity chromatography: Increased versatility of the technique for studies of ligand binding, *Anal. Biochem.* 153, 380–386, 1986.
18. Winzor, D.J., Determination of binding constants by affinity chromatography, *J. Chromatogr. A* 1037, 351–367, 2004.
19. Nakasone, S., Ohshita, T., and Iwamasa, T., Heterogeneity of pig lysosomal acid alpha-glucosidase. Affinity to Sephacryl S-200 gel and tissue distribution, *Biochem. J.* 279, 719–726, 1991.
20. Weber, B., Hopwood, J.J., and Yogalingam, G., Expression and characterization of human recombinant and α-*N*-acetylglucosaminidase, *Protein Expr. Purif.* 21, 251–259, 2001.
21. Yasuda, K., Chang, H.H., Wu, H.L. et al., Efficient and rapid purification of recombinant human α-galactosidase A by affinity column chromatography, *Protein Expr. Purif.* 37, 499–506, 2004.

22. Matsushita-Oikawa, H., Komatsu, M., Lida-Tanaka, N. et al., Novel carbohydrate-binding activity of bovine liver β-glucuronidase toward lactose/N-acetyllactosamine sequences, *Glycobiology* 16, 891–901, 2006.

23. Samoylova, T.I., Martin, D.R., Morrison, N.E. et al., Generation and characterization of recombinant feline β-galactosidase for preclinical enzyme replacement therapy study in GM1 gangliosidosis, *Metab. Brain Dis.* 23, 161–173, 2008.

24. Brodelius, P. and Mosbach, K., Separation of the isoenzymes of lactate dehydrogenase by affinity chromatography using immobilized AMP-analogues, *FEBS Lett.* 35, 223–226,1972.

25. Brodelius, P. and Mosbach, K., Determination of dissociation constants for binary dehydrogenase-coenzyme complexes by (bio)affinity chromatography on an immobilized AMP-analogue, *Anal. Biochem.* 72, 629–636, 1976.

26. Parikh, I. and Cuatrecasas, P., Affinity chromatography, *Vox Sang.* 23, 141–145, 1972.

27. Hoffman, D.L., Purification and large-scale preparation of antithrombin III, *Am. J. Med.* 87, 23S–26S, 1989.

28. Dunn, B.M. and Chaiken, I.M., Quantitative affinity chromatography. Determination of binding constants by elution with competitive inhibitors, *Proc. Natl. Acad. Sci. USA* 71, 2382–2385, 1974.

29. Chaiken, I.M. and Taylor, H.C., Analysis of ribonuclease-nucleotide by quantitative affinity chromatography, *J. Biol. Chem.* 251, 2044–2048, 1976.

30. Dunn, B.M. and Chaiken, I.M., Evaluation of quantitative affinity chromatography by comparison with kinetic and equilibrium dialysis methods for the analysis of nucleotide binding to staphylococcal nuclease, *Biochemistry* 14, 2343–2349, 1975.

31. Eliat, D., Chaiken, I.M., and McCormick, W.M., Expression of multivalency in the affinity chromatography of antibodies. Appendix. Derivation and evaluation of equations for independent bivalent interacting systems in quantitative affinity chromatography, *Biochemstry* 18, 790–795, 1979.

32. Somerville, D.J., Truner, J.E., and Dunn, B.M., Multiple bonding sites of carboxypeptidase B: The evaluation of dissociation constants by quantitative affinity chromatography, *Biochemistry* 18, 3039–3045, 1979.

33. Dunn, B.M. and Gilbert, W.A., Quantitative affinity chromatography of α-chymotrypsin, *Arch. Biochem. Biophys.* 198, 533–540, 1979.

34. Koayashi, M., Soman, G., and Graves, D.J., A comparison of the activator sites of liver and muscle glycogen phosphorylase *b*, *J. Biol. Chem.* 257, 14041–14047, 1982.

35. Jarvis, M.R. and Voss, E.W. Jr., Determination of dissociation constants and ligand specificity of detergent solubilized surface membrane immunoglobulin A from MOPC-315, *Mol. Immunol.* 20, 125–136, 1983.

36. Schaub, M.C., Watterson, J.G., Loth, K., and Foletta, D., The role of magnesium in binding of the nucleotide polyphosphate chain to the active site of myosin subfragment-1, *Eur. J. Biochem.* 134, 197–204, 1983.

37. Hogg, P.J. and Winzor, D.J., Quantitative affinity chromatography: Further developments in the analysis of experimental results from column chromatography and partition equilibrium studies, *Arch. Biochem. Biophys.* 234, 55–60, 1984.

38. Christensen, U., The AH-site of plasminogen and two C-terminal fragments. A weak lysine-preferring ligands not carrying a free carboxylate function, *Biochem. J.* 223, 413–421, 1984.

39. Hogg, P.J. and Winzor, D.J., Studies of lectin-carbohydrate interactions by quantitative affinity chromatography: Systems with galactose and ovalbumin as saccaridic ligand, *Anal. Biochem.* 163, 331–338, 1987.

40. Jenuwine, E.S. and Shaner, S.L., Equilibrium constants for nonspecific binding of proteins to DNA may be obtained by quantitative zonal DNA affinity chromatography, *Anal. Biochem.* 242, 228–233, 1996.

41. Yang, Q., Liu, X.Y., Hara, M. et al., Quantitative affinity chromatographic studies of mitochondrial cytochrome *c* binding to bacterial photosynthesis reaction center, reconstituted in liposome membranes and immobilized by detergent dialysis and avidin–biotin binding, *Anal. Biochem.* 280, 945–102, 2000.

42. Anderson, P.J., Nessett, A., Dhamawardana, K.R., and Bock, P.E., Characterization of proexosite I on prothrombin, *J. Biol. Chem.* 275, 16428–16434, 2000.

43. Veronese, F.M., Bevilacqua, R., and Chaiken, I.M., Drug–protein interactions: Evaluation of the binding of antipsychotic drugs to glutamate dehydrogenase by quantitative affinity chromatography, *Mol. Pharmacol.* 15, 313–321, 1979.

44. Yoon, H.Y., Hwang, S.H., Lee, E.Y. et al., Effects of ADP on different inhibitory properties of brain glutamate dehydrogenase isoproteins by perphrenazine, *Biochemie* 63, 907–913, 2001.

45. Simpson, R.J., Moritz, R.L., Nice, E.C. et al., Complete amino acid sequence of plastocyanin from a green alga, *Enteromorpha prolifera*, *Eur. J. Biochem.* 157, 497–506, 1986.

46. Bushman, F.D. and Ptashne, M., Turning lambda Cro into a transcriptional activator, *Cell* 54, 191–197, 1988.

47. Modi, S., Nordling, M., Lundberg, L.G. et al., Reactivity of cytochrome *c* and *f* with mutant forms with spinach plastocyanin, *Biochim. Biophys. Acta* 1102, 85–90, 1992.

48. Young, S., Sigfridsson, K., Olesen, K., and Hansson, O., The involvement of the two acidic patches of spinach plastocyanin in the reaction with photosystem I, *Biochim. Biophys. Acta* 1322, 106–114, 1997.

49. Sato, K., Kohzuma, T., and Dennison, C., Pseudospecificity of the acidic patch of plastocyanin for the interaction with cytochrome *f*, *J. Am. Chem. Soc.* 126, 3028–3029, 2004.

50. Koch, M., Velarde, M., Harrison, M.D. et al., Crystal structures of oxidized and reduced stallacyanin from horseradish roots, *J. Am. Chem. Soc.* 127, 158–166, 2005.

51. Smits, C., Czabotar, P.E., Hinds, M.G., and Day, C.L., Structural plasticity underpins promiscuous binding of the prosurvival protein A1, *Structure* 16, 818–829, 2008.

52. Fairbrother, W.J., Hall, L., Littlechild, J.A. et al., Site-directed mutagenesis of histidine 52 in the "basic patch" region of yeast phosphoglycerate kinase, *FEBS Lett.* 258, 247–250, 1989.

53. Kulish, D., Lee, J., Lomakin, I. et al., The functional role of basic patch, a structural element of *Escherichia coli* transcript cleavage factors GreA and GreB, *J. Biol. Chem.* 275, 12789–12798, 2000.

54. Gong, X.S., Wne, J.Q., Fisher, N.E. et al., The role of individual lysine residues in the basic patch on turnip cytochrome *f* for electrostatic interactions with plastocyanin in vitro, *Eur. J. Biochem.* 267, 3461–3468, 2000.

55. Edlich, C., Stier, G., Simon, B. et al., Structure and phosphatidylinositol-(3,4)-bisphosphate binding of the C-terminal PH domain of human pleckstrin, *Structure* 13, 277–286, 2005.

56. Gennadios, H.S. and Christianson, D.W., Binding of uridine 5′-diphosphate in the "basic patch" of the zinc deacetylase LpxC and implications for substrate binding, *Biochemistry* 45, 15216–15223, 2006.

57. Zuo, Y., Zhend, H., Wang, Y. et al., Crystal structure of RNAse T, an exonuclease involved in tRNA maturation and end turnover, *Structure* 15, 417–428, 2007.

58. Tarendeau, F., Crepin, T., Guillgay, D. et al., Host determinant residue lysine 627 lies on the surface of a discrete, folded domain of influenza virus polymerase PB2 subunit, *PLoS Pathog.* 4(8), e1000136, 2008, doi:10.1371/journal.ppat.1000136.

59. Luo, D., Xu, T., Hunke, C. et al., Crystal structure of the NS3 protease from dengue virus, *J. Virol.* 82, 173–183, 2008.

60. Huang, M., Furie, B.C., and Furie, B., Crystal structure of the calcium-stabilized human factor IX Gla domain bound to a conformation-specific anti-factor IX antibody, *J. Biol. Chem.* 279, 14338–14346, 2004.

61. Pantoja-Uceda, D., Shewry, P.R., Bruiz, M. et al., Solution structure of a methionine-rich 2S albumin from sunflower seeds: Relationship to its allergenic and emulsifying properties, *Biochemistry* 43, 6976–6986, 2004.
62. Janz, J.M. and Farrens, D.L., Rhodopsin activation exposes a key hydrophobic binding site for the transducin α-subunit C terminus, *J. Biol. Chem.* 279, 29767–29773, 2004.
63. Geszvain, K., Tanja, M., Mooney, R.A. et al., A hydrophobic patch on the flap-tip helix of *E. coli* RNA polymerase mediates σ70 region 4 function, *J. Mol. Biol.* 343, 569–587, 2004.
64. Borucki, B., Kyndt, J.A., Joshi, C.P. et al., Effect of salt and pH on the activation of photoactive yellow protein and gateway mutants Y98Q and Y98F, *Biochemistry* 44, 13650–13663, 2005.
65. Johnsborg, O., Kristiansen, P.E., Blomqvist, T. et al., A hydrophobic patch in the competence-stimulating peptide, a pneumococcal competence phenomena, is essential for specificity and biological activity, *J. Bacteriol.* 188, 1744–1749, 2006.
66. Macek-Keber, M. and Roman, J., Structural similarity between the hydrophobic fluorescent probe and lipid A as a ligand of MD-2, *FASEB J.* 20, 1836–1842, 2006.
67. Dikeakos, J.D., Lacombe, M.-J., Chantal, M. et al., A hydrophobic patch in a charged α-helix is sufficient to target proteins to dense core secretory granules, *J. Biol. Chem.* 282, 1136–1143, 2007.
68. Lijnzaad, P., Feenstra, K.A., Heninga, J. et al., On defining the dynamics of hydrophobic patches on protein surfaces, *Proteins* 72, 105–114, 2008.
69. Gratacos-Cubarsi, M. and Lametsch, R., Determination of charges in protein conformation caused by pH and temperature, *Meat Sci.* 80, 545–549, 2008.
70. Olson, S.T., Halvorson, H.R., and Björk, I., Quantitative characterization of the thrombin-heparin interaction. Discrimination between specific and nonspecific binding models, *J. Biol. Chem.* 266, 6342–6352, 1991.
71. Olson, S.T., Bock, P.E., and Sheffer, R., Quantitative evaluation of solution equilibrium binding interactions by affinity partitioning: Application to specific and nonspecific protein-heparin interactions, *Arch. Biochem. Biophys.* 286, 533–545, 1991.
72. García de Frutos, P., Hardig, Y., and Dahlback, B., Serum amyloid P component binding to C4b-binding protein, *J. Biol. Chem.* 270, 26950–26955, 1995.
73. Hogg, P.J., Jackson, C.M., and Winzor, D.J., Use of quantitative affinity chromatography for characterizing high-affinity interactions: Binding of heparin to antithrombin III, *Anal. Biochem.* 192, 303–311, 1991.
74. Keller, M.M., Keller, J.M., and Kühn, K., The C-terminus of type I collagen is a major binding site for heparin, *Biochim. Biophys. Acta* 882, 1–5, 1986.
75. Chajek-Shaul, T., Friedman, G., Ziv, E. et al., Fate of lipoprotein lipase taken up by the rat liver. Evidence for a conformational change with loss of catalytic activity, *Biochim. Biophys. Acta* 963, 183–191, 1988.
76. Peterson, J., Fujimoto, W.Y., and Brunzell, J.D., Human lipoprotein lipase: Relationship of activity, heparin affinity, and conformation as studied with monoclonal antibodies, *J. Lipid Res.* 33, 1165–1170, 1992.
77. Inouye, S., Ling, N., and Shimasaki, S., Localization of the heparin binding site of follistatin, *Mol. Cell. Endocrinol.* 90, 1–6, 1992.
78. Hata, A., Ridinger, D.N., Sutherland, S. et al., Binding of lipoprotein lipase to heparin. Identification of five critical residues in two distinct segments of the amino-terminal domain, *J. Biol. Chem.* 268, 8447–8457, 1993.
79. Mizuno, K. and Hayashi, T., Peculiar effect of urea on the interaction of type I collagen with heparin on chromatography, *J. Biochem.* 116, 1257–1263, 1994.
80. Park, J.W., Oh, M.S., Yang, J.Y. et al., Glycosylation, dimerization, and heparin affinity of lipoprotein lipase in 3T3-L1 adipocytes, *Biochim. Biophys. Acta* 1254, 45–50, 1995.

81. Deprez, P.N. and Inestrosa, N.C., Two heparin-binding domains are present on the collagenic tail of asymmetric acetylcholinesterase, *J. Biol. Chem.* 270, 11043–11046, 1995.

82. Seiffert, D. and Schleef, R.R., Two functionally distinct pools of vitronectin (Vn) in the blood circulation: Identification of a heparin-binding competent population of Vn within platelet α-granules, *Blood* 88, 552–560, 1996.

83. Tibell, L.A., Sethson, I., and Buevich, A.V., Characterization of the heparin-binding domain of human extracellular superoxide dismutase, *Biochim. Biophys. Acta* 1340, 21–32, 1997.

84. Blackmore, T.K., Hellwage, J., Sadon, T.A. et al., Identification of the second heparin-binding domain in human complement factor H, *J. Immunol.* 160, 3342–3348, 1998.

85. Icatlo, F.C. Jr., Kuroki, M., Kobayashi, C. et al., Affinity purification of *Helicobacter pylori* urease. Relevance to gastic mucin adherence by urease protein, *J. Biol. Chem.* 273, 18130–18138, 1998.

86. Almeida, P.C., Nantes, I.L., Rizzi, C.C. et al., Cysteine proteinase activity regulation. A possible role of heparin and heparin-like glycosaminoglycans, *J. Biol. Chem.* 274, 30433–30438, 1999.

87. Gorlatov, S.N. and Stadtman, T.C., Human selenium-dependent thioredoxin reductase from HeLa cells: Properties of forms with differing heparin affinities, *Arch. Biochem. Biophys.* 369, 133–142, 1999.

88. Sakata, N., Sasatomi, Y., Ando, S. et al., Causal relationship between conformational change and inhibition of domain functions of glycoxidative fibronectin, *Connect. Tissue Res.* 41, 117–129, 2000.

89. Chen, V.C., Chao, L., Pimenta, D.C. et al., Identification of a major heparin-binding site in kallistatin, *J. Biol. Chem.* 276, 1276–1284, 2001.

90. Hoke, D.E., LaBrenz, S.R., Höök, M., and Carson, D.D., Multiple domain contributes to heparin/heparan sulfate binding by human HIP/L29, *Biochemistry* 39, 15686–15694, 2000.

91. Larsson, H., Akerud, P., Nordling, K. et al., A novel anti-angiogenic form of antithrombin with retained proteinase binding ability and heparin affinity, *J. Biol. Chem.* 276, 11996–12002, 2001.

92. Lethias, C., Elefteriou, F., Parsiegla, G. et al., Identification and characterization of a conformational heparin-binding site involving two fibronectin type III modules of bovine tenascin-X, *J. Biol. Chem.* 276, 16432–16438, 2001.

93. Melman, L., Cao, Z.F., Rennke, S. et al., High affinity binding of receptor-associated protein to heparin and low density lipoprotein receptor-related protein requires similar basic amino acid sequence motifs, *J. Biol. Chem.* 276, 29338–29346, 2001.

94. Aubert-Foucher, E., Goldschmidt, D., Jaquinod, M., and Mazzorana, M., Processing in the C-terminal domain of minicollagen XII removes a heparin-binding site, *Biochem. Biophys. Res. Commun.* 286, 1131–1139, 2001.

95. Stenlund, P., Lindberg, M.J., and Tibell, L.A., Structural requirements for high affinity heparin binding: Alanine scanning analysis of charged residues in the *C*-terminal domain of human extra-cellular superoxide dismutase, *Biochemistry* 41, 3168–3175, 2002.

96. Deprez, P., Inestrosa, N.C., and Krejci, E., Two different heparin-binding domains in the triple-helical domain of ColQ, the collagen tail subunit of synaptic acetylcholinesterase, *J. Biol. Chem.* 278, 23233–23242, 2003.

97. Ookawara, T., Eguchi, H., Kizaki, T. et al., An inter-subunit bond affects affinity of human lung extracellular superoxide dismutase to heparin, *Free Radic. Res.* 37, 823–827, 2003.

98. Richard, B., Swanson, R., Schedin-Weiss, S. et al., Characterization of the conformational alterations, reduced anticoagulant activity, and enhanced anticoagulant activity of prelatent antithrombin, *J. Biol. Chem.* 283, 14417–14429, 2008.
99. Chang, W.S. and Harper, P.L., Commercial antithrombin concentrate contains inactive L-forms of antithrombin, *Thromb. Haemost.* 77, 323–328, 1997.
100. Heger, A., Grunert, T., Schulz, P. et al., Separation of active and inactive forms of human antithrombin by heparin affinity chromatography, *Thromb. Res.* 106, 157–164, 2002.
101. Owen, M.C., Shaw, G.J., Grau, E. et al., Molecular characterization of antithrombin Barcelona-2: 47 arginine to cysteine, *Thromb. Res.* 55, 451–457, 1989.
102. Scully, M.F., Kakkar, V.V., and Goodwin, C.A., Non-specific influence of chemical modification upon the properties of antithrombin III: Modification of carboxyl groups, *Thromb. Res.* 67, 447–456, 1992.
103. Finotti, P., Separation by heparin-affinity chromatography of catalytically active and inactive forms of trypsin which retain the (Na,K) ATPase stimulating property, *Clin. Chim. Acta* 256, 37–51, 1996.
104. Migliorini, M.M., Behre, E.H., Brew, S. et al., Allosteric modulation of ligand binding to low density lipoprotein receptor-related protein by the receptor-associated protein requires critical lysine residues within its carboxyl-terminal domain, *J. Biol. Chem.* 278, 17986–17992, 2003.
105. Patel, S., Chaffotte, A.F., Amana, B. et al., In vitro denaturation–renaturation of fibronectin. Formation of multimers disulfide-linked and shuffling of intramolecular disulfide bonds, *Int. J. Biochem. Cell Biol.* 38, 1547–1560, 2006.
106. Lundblad, R.L., A rapid method for the purification of bovine thrombin and the inhibition of the purified enzyme with phenylmethylsulfonyl fluoride, *Biochemistry* 10, 2501–2506, 1971.
107. Tanford, C., *The Hydrophobic Effect Formation of Micelles and Biological Membranes*, Wiley-Interscience, New York, 1973.
108. Reynolds, J.A., Gilbert, D.B., and Tanford, C., Empirical correlation between hydrophobic free energy and aqueous cavity surface area, *Proc. Natl. Acad. Sci. USA* 71, 2925–2927, 1974.
109. Tanford, C., Interfacial free energy and the hydrophobic effect, *Proc. Natl. Acad. Sci. USA* 76, 4175–4176, 1979.
110. Abraham, M.H., Free energies of solution of rare gases and alkanes in water and non-aqueous solvents: A quantitative assessment of the hydrophobic effect, *J. Am. Chem. Soc.* 101, 5477–5484, 1979; (a) Garbassi, F., Morra, M., and Occhiello, E., *Polymer Surfaces from Physics to Technology*, John Wiley & Sons, Chichester, U.K., 1994.
111. Kyte, J., *Structure in Protein Chemistry*, Garland Publishing, New York, 1995.
112. Tsai, C.J., Maizel, J.V. Jr., and Nussinov, R., The hydrophobic effect, a new insight from cold denaturation and two-stage water structure, *Crit. Rev. Biochem. Mol. Biol.* 37, 55–69, 2002.
113. Lee, C.Y., McCammon, J.A., and Rossky, P.J., The structure of liquid water at an extended hydrophobic surface, *J. Chem. Phys.* 80, 4448–4452, 2008.
114. Ewell, J., Gibb, B.C., and Rick, S.W., Water inside a hydrophobic cavity and molecule, *J. Phys. Chem. B* 112, 10272–10279, 2008.
115. Kyte, J., The basis of the hydrophobic effect, *Biophys. Chem.* 100, 193–203, 2003.
116. Dill, K.A., Truskett, T.M., Vlachy, V., and Huber-Lee, B., Modeling water, the hydrophobic effect, and its salvation, *Annu. Rev. Biophys. Biomol. Struct.* 34, 173–199, 2005.
117. Yu, H. and Dill, K.A., Water's hydrogen bonds in the hydrophobic effect: A simple model, *J. Phys. Chem. B.* 109, 23611–23617, 2005.

118. Tall, A.R., Shipley, G.G., and Small, D.M., Conformational and thermodynamic properties of apo A-1 of human plasma high density lipoproteins, *J. Biol. Chem.* 251, 3749–3755, 1976; (a) Chollet, R. and Anderson, L.L., Conformational changes associated with the reversible cold inactivation of ribulose-1,5-bisphosphate carboxylase-oxygenase, *Biochim. Biophys. Acta* 482, 228–240, 1977; (b) Erlanson-Albertsson, C. and Aakerlund, H.E., Conformational change in pancreatic lipase induced by colipase, *FEBS Lett.* 144, 38–42, 1982; (c) Follenius, A. and Gerard, D., Fluorescence investigations of calmodulin hydrophobic sites, *Biochem. Biophys. Res. Commun.* 199, 1154–1560, 1984; (d) Buerkler, J. and Krebs, J., Structural studies of calmodulin and related calcium-binding proteins by hydrophobic labeling, *FEBS Lett.* 182, 167–170, 1985; (e) Horowitz, P.M. and Criscimagna, N.L., Sulfhydryl-directed triggering of conformational changes in the enzyme rhodanase, *J. Biol. Chem.* 263, 10278–10283, 1988; (f) Dukhanina, E.A., Dukhanin, A.S., Lomonsov, M.Y. et al., Calcium-binding characteristics and conformational changes in metastatin, a member of the S-100 protein family, *Biochemistry (Moscow)* 62, 530–536, 1997; (g) Berggård, T., Silow, M., Thulin, E. et al., Ca^{2+}- and H^+-dependent conformational changes of calbindin D_{28k}, *Biochemistry* 39, 6864–6873, 2000; (h) Weers, P.M.M., Kay, C.M., and Ryan, R.O., Conformational changes of an exchangeable apolipoprotein, apolipophorinIII from *Locusta migratoria* at low pH, *Biochemistry* 40, 7754–7760, 2001; (i) Braga, C.A.C.A., Pinto, J.R., Valenta, A.P. et al., Ca^{2+} and Mg^{2+} binding to weak sites of TnC C-domain induces exposure of a large hydrophobic surface that leads to loss of TnC from the thin filament, *Int. J. Biochem. Cell Biol.* 38, 110–122, 2006; (j) Sharp, J.S. and Tomer, K.B., Analysis of the oxidative damage-induced conformational changes of apo- and holocalmodulin by dose-dependent protein oxidative surface mapping, *Biophys. J.* 92, 1682–1692, 2007.

119. *Ion-Pair Chromatography. Theory and Biological and Pharmaceutical Applications*, ed. M.T.W. Hearn, Marcel Dekker, New York, 1985.

120. Wang, X. and Carr, P.W., An unexpected observation concerning the effect of anionic additives on the retention behavior of basic drugs and peptides in reversed-phase liquid chromatography, *J. Chromatogr. A* 1154, 165–173, 2007.

121. McCalley, D.V., Comparison of an organic polymeric column and a silica-based reversed-phase column for the analysis of basic peptides by high-performance liquid chromatography, *J. Chromatogr. A* 1073, 137–145, 2005.

122. Jayat, D., Gaudin, J.C., Chobert, J.M. et al., A recombinant C121S mutant of bovine β-lactoglobulin is more susceptible to peptic digestion and to denaturation by reducing agents and heating, *Biochemistry* 43, 6312–6321, 2004.

123. Flores-Flores, C., Martinez-Martinez, A., Muñoz-Delgado, E., and Vidal, C.J., Conversion of acetylcholinesterase hydrophilic tetramers into amphiphilic dimers and monomers, *Biochem. Biophys. Res. Commun.* 219, 53–58, 1996.

124. Fitzgerald, R.J. and Swaisgood, H.E., Binding of ions and hydrophobic probes to α-lactalbumin and κ-casein as determined by analytical affinity chromatography, *Arch. Biochem. Biophys.* 268, 239–248, 1989.

125. Shanker, V., Naziruddin, B., Reyes de la Rocha, S., and Sachdev, G.P., Evidence of hydrophobic domains in human respiratory mucins. Effect of sodium chloride on hydrophobic binding properties, *Biochemistry* 29, 5856–5864, 1990.

126. Doyle, C.A. and Dorsey, J.G., Reversed-phase HPLC: Preparation and characterization of reversed-phase stationary phase, in *Handbook of HPLC*, eds. E. Katz, R. Eksteen, D. Schoenmakers, and N. Miller, Marcel Dekker, New York, 1998.

127. Fausnaugh, J.L., Kennedy, L.A., and Regnier, F.E., Comparison of hydrophobic-interaction and reversed phase chromatography of proteins, *J. Chromatogr.* 317, 141–155, 1984.

128. Shatiel, S., Hydrophobic chromatography, *Methods Enzymol.* 34, 126–140, 1974; (a) Jennissen, H.P. and Heilmeyer, L.M.G. Jr., General aspects of hydrophobic chromatography. Adsorption and elution characteristics of some skeletal muscle enzymes, *Biochemistry* 14, 754–760, 1975; (b) Homcy, C.J., Wrenn, S.M., and Haber, E., Demonstration of the hydrophobic character of adenylate cyclase following hydrophobic resolution on immobilized alkyl residues, *J. Biol. Chem.* 252, 8957–8964, 1977.

129. Shatiel, S., Hydrophobic chromatography, *Methods Enzymol.* 104, 69–96, 1984.

130. Dias-Cabral, A.C., Ferreira, A.S., Phillips, J. et al., The effects of ligand chain length, salt concentration and temperature on the adsorption of bovine serum albumin onto polypropyleneglycol-Sepharose, *Biomed. Chromatogr.* 19, 606–616, 2005.

131. Porath, J., Maissano, F., and Belew, M., Thiophilic adsorption—A new method for protein fractionation, *FEBS Lett.* 185, 306–310, 1985.

132. Hutchens, T.W. and Porath, J., Thiophilic adsorption of immunoglobulins—Analysis of conditions optimal for selective immobilization and purification, *Anal. Biochem.* 159, 217–226, 1986.

133. Srikrishnan, T., MacKenzie, J.T., and Sulkowski, E., Thiophilic interaction chromatography of human transferrin, *J. Chromatogr. Sci.* 44, 634–638, 2006.

134. MacKenzie, J.T., Srikrishnan, T., and Sulkowski, E., Thiophilic interaction chromatography of mammalian and avian transferrins, *J. Chromatogr. Sci.* 45, 710–713, 2007.

135. Kölln, J., Braren, I., Bredehorst, R., and Spillner, E., Purification of native and recombinant corbra venom factor using thiophilic adsorption chromatography, *Protein Pept. Lett.* 14, 475–480, 2007.

136. Bourhim, M., Johnson, T., Oyston, D., and Srikrishnan, T., Thiophilic interaction chromatography (TIC) of amyloid-beta precursor, *J. Alzheimers Dis.* 12, 143–149, 2007.

137. Sathaeesh, B.H.K., Vijayalakshmi, M.A., Smith, G.J., and Chadha, K.C., Thiophilic-interaction chromatography of enzymatically active tissue prostate-specific antigen (T-PSA) and its modulation by zinc ions, *J. Chromatogr. B.* 861, 227–235, 2007.

138. Bourhim, M., Rajendran, A., Ramos, V. et al., Thiophilic interaction chromatography of serum albumin, *J. Chromatogr. Sci.* 46, 574–576, 2008.

139. Hardouin, J., Duchateau, M., Canelle, L. et al., Thiophilic adsorption revisited, *J. Chromatogr. B.* 845, 226–231, 2006.

140. Lakhiari, H. and Muller, D., Purification of IgG and insulin on supports grafted by sialic acid developing "thiophilic-like" interactions, *J. Chromatogr. B.* 818, 53–59, 2005.

141. Boschetti, F., The use of thiophilic chromatography for antibody purification: A review, *J. Biochem. Biophys. Methods* 49, 361–389, 2001.

142. Amphlett, G.W., Kisiel, W., and Castellino, F.J., The interaction of Ca^{2+} with human factor IX, *Arch. Biochem. Biophys.* 208, 576–585, 1981.

143. Lewis, R.M., Furie, B.C., and Furie, B., Conformation-specific monoclonal antibodies directed against the calcium-stabilized structure of human prothrombin, *Biochemistry* 22, 948–954, 1983.

144. Poort, S.R., van den Linden, I.K., Krommenkak-van Es, C. et al., Rabbit polyclonal antibodies against the calcium-dependent conformation of factor IX and their applications in solid phase immunoradiometric assays, *Thromb. Haemost.* 55, 122–128, 1986.

145. Nelsestuen, G.L., Role of γ-carboxylglutamic acid. An unusual protein transition required for the calcium-dependent binding of prothrombin to phospholipid, *J. Biol. Chem.* 251, 5648–5656, 1976.

146. Liebman, H.A., The metal-dependent conformational changes in Factor IX associated with phospholipid binding. Studies using antibodies against a synthetic peptide and chemical modification of Factor IX, *Eur. J. Biochem.* 212, 339–345, 1993.

147. Freedman, S.J., Blostein, M.D., Baleja, J.D. et al., Identification of the phospholipid binding site in the vitamin K-dependent blood coagulation protein Factor IX, *J. Biol. Chem.* 271, 16227–16236, 1996.

148. Lecompte, M.F., Interaction of an amphitropic protein (factor Xa) with membrane models in a complex system, *Biochim. Biophys. Acta* 1724, 307–314, 2005.

149. Bessos, H. and Prowse, C.V., Immunopurification of human coagulation Factor IX using monoclonal antibodies, *Thromb. Haemost.* 56, 86–89, 1986.

150. Liebman, H.A., Limentani, S.A., Furie, B.C., and Furie, B., Immunoaffinity purification of factor IX (Christmas Factor) by using conformation-specific antibodies directed against the factor IX-metal complex, *Proc. Natl. Acad. Sci. USA* 82, 3879–3883, 1985.

151. Yan, B.S., Review of conformation-specific affinity purification methods for plasma vitamin K-dependent proteins, *J. Mol. Recogn.* 9, 211–218, 1996.

152. Wajih, N., Hutson, S.M., Owen, J., and Wallin, R., Increased production of functional recombinant clotting factor IX by baby hamster kidney cells engineered to overexpress VKORC1, the vitamin K 2,3-epoxide-reducing enzyme of the vitamin K cycle, *J. Biol. Chem.* 280, 31603–31607, 2005.

153. Tarvers, R.C., Calcium-dependent changes in properties of human prothrombin: A study using high-performance size-exclusion chromatography and gel permeation chromatography, *Arch. Biochem. Biophys.* 241, 639–648, 1985.

154. Church, F.C., Lundblad, R.L., Noyes, C.M., and Tarvers, R.C., Effect of divalent cations on the limited proteolysis of prothrombin by thrombin, *Arch. Biochem. Biophys.* 240, 607–612, 1985.

155. Lundblad, R.L., A hydrophobic site in human prothrombin present in a calcium-stabilized conformers, *Biochem. Biophys. Res. Commun.* 157, 295–300, 1988.

156. Chuang, T.F., Sargeant, R.B., and Hougie, C., The effect of calcium ions on the properties of Factor IX and its activated form, *Br. J. Haematol.* 27, 281–287, 1974; (a) Hjerten, S., Some general aspects of hydrophobic interaction chromatography, *J. Chromatogr.* 87, 325–331, 1973; (b) Shansky, R.E., Wu, S.-L., Figuero, A., and Karger, B.L., Hydrophobic interaction chromatography of biopolymers, in *HPLC of Biological Macromolecules. Methods and Applications*, eds. K.M. Gooding and F.E. Regnier, Marcel Dekker, New York, Chapter 5, 1990; (c) Chen, J., Luo, Q., Breneman, C.M., and Cramer, S.M., Classification of protein adsorption and recovery at low salt concentration in hydrophobic interaction chromatographic systems, *J. Chromatogr. A.* 1139, 236–246, 2007.

157. Husi, H. and Walkinshaw, M.D., Separation of human vitamin K-dependent coagulation proteins using hydrophobic chromatography, *J. Chromatogr. B.* 736, 77–88, 1999.

158. Karlsson, G., Analysis of human α-thrombin by hydrophobic interaction high-performance liquid chromatography, *Protein Expr. Purif.* 27, 171–174, 2003.

159. Aizawa, P., Winge, S., and Karlsson, G., Large-scale preparation of thrombin from human plasma, *Thromb. Res.* 122, 560–567, 2008.

160. Lee, S.-C. and Whitaker, J.R., Are molecular weights of proteins determined by Superose 12 column chromatography, *J. Agric. Food Chem.* 52, 4948–4952, 2004.

161. Adachi, K., Kim, J.Y., and Asakura, T., Effects of difference in shape and hydrophobicity of surface amino acids on high performance gel-permeation chromatography, *J. Chromatogr.* 428, 247–254, 1988.

162. Golovchenko, N.P., Kataeva, I.A., and Akimento, V.K., Analysis of the pH-dependent interactions with gel filtration media, *J. Chromatogr.* 591, 121–128, 1992.

163. Joyce, J.G., Cook, J.C., Przysiecki, C.T., and Lehman, E.D., Chromatographic separation of low-mass recombinant proteins and peptides on Superdex 30 prep grade, *J. Chromatogr. B.* 662, 325–334, 1994.

164. Arakawa, T., Tsumoto, K., Kita, Y. et al., Biotechnology applications of amino acids in protein purification and formulations, *Amino Acids* 33, 587–605, 2007.

165. Romano, V.A., Ebeyer, T., and Dubin, P.L., Influence of net protein charge and stationary phase charge on protein retention in size exclusion chromatography, in *Strategies in Size Exclusion Chromatography*, eds. H. Potschcke and P.L. Dubin, American Chemical Society, Washington, DC, 1996.

166. Pujar, N.S. and Zydney, A.L., Electrostatic effects of protein partitioning in size-exclusion chromatography and membrane ultrafiltration, *J. Chromatogr. A.* 796, 299–238, 1998.

167. Zhu, Y.X., Poschka, M., Dubin, P.L., and Cai, C., A method for the quantitation of charge by size exclusion chromatography demonstrated with components of ficoll 400, *Macromol. Chem. Phys.* 202, 61–72, 2001.

168. Irvine, G.B., High-performance size-exclusion chromatography of peptides, *J. Biochem. Biophys. Methods* 56, 233–242, 2003.

169. McKenzie, H.A. and White, F.H. Jr., Lysozyme and α-lactalbumin: Structure, function, and interrelationships, *Adv. Protein Chem.* 41, 173–315, 1991.

170. Permyalkov, E.A. and Berliner, L.J., α-Lactalbumin: Structure and function, *FEBS Lett.* 423, 269–274, 2000.

171. Jung, H.J., Lee, J.Y., Kim, S.H. et al., Solution structure and lipid membrane partitioning of VSTxl, an inhibitor of the KvAP potassium channel, *Biochemistry* 44, 6015–6023, 2005.

172. Shortle, D.R., Structural analysis of non-native states of proteins by NMR methods, *Curr. Opin. Struct. Biol.* 6, 24–30, 1996.

173. Fontana, A., de Laureto, P.P., Spolaore, B. et al., Probing protein structure by limited proteolysis, *Acta Biochim. Pol.* 51, 299–321, 2004.

174. Redfield, C., Using nuclear magnetic resonance spectroscopy to study molten globule states of proteins, *Methods* 34, 121–132, 2004.

175. Huppertz, T., Fox, P.F., de Kruif, K.G., and Kelly, A.L., High pressure-induced changes in bovine milk proteins: A review, *Biochim. Biophys. Acta* 1764, 593–598, 2006.

176. Munishkina, L.A., Ahmad, A., Fink, A.L., and Uversky, V.N., Guiding protein aggregation with macromolecular crowding, *Biochemistry* 47, 8993–9006, 2008.

177. Athamneh, A.I., Griffin, M., Whaley, M., and Barone, J.R., Conformational changes and molecular mobility in plasticized proteins, *Biomacromolecules*, 9, 3181–3187, 2008.

178. Guan Sekhar, P.M. and Prakash, V., Interaction of selected cosolvents with bovine α-lactalbumin, *Int. J. Biol. Macromol.* 42, 348–355, 2008.

179. Halskau, O. Jr., Perez-Jimenez, R., Ibarra-Molero, B. et al., Large-scale modulation of thermodynamic protein folding barriers linked to electrostatic, *Proc. Nat. Acad. Sci. USA* 105, 8625–8630, 2008.

180. Wu, S.-L., Benedek, K., and Karger, B.L., Thermal behavior of proteins in high-performance hydrophobic-interaction chromatography. On-line spectroscopic and chromatographic characterization, *J. Chromatogr.* 359, 3–17, 1986.

181. Wu, S.-L., Figueroa, A., and Karger, B.L., Protein conformational effects in hydrophobic interaction chromatography. Retention characterization and the role of mobile phase additives and stationary phase hydrophobicity, *J. Chromatogr.* 371, 3–27, 1986.

182. To, B.C.S. and Lenhoff, A.M., Hydrophobic interaction chromatography of proteins. I. The effects of protein and adsorbent properties on retention and recovery, *J. Chromatogr.* 1141, 191–205, 2007.

183. Fogle, J.L. and Fernandez, E.J., Amide hydrogen-deuterium exchange: A fast tool for screening protein stabilities in chromatography, *LCGC North America*, June, 2006.

184. Fogle, J.L., O'Connell, J.P., and Fernandez,. E.J., Loading, stationary phase, and salt effects during hydrophobic interaction chromatography: α-Lactabumin is stabilized at high loading, *J. Chromatogr.* 1121, 209–218, 2006.

185. Ueberbacher, R., Haimer, E., Hahn, R., and Jungbauer, A., Hydrophobic interaction chromatography of proteins. V. Quantitative assessment of conformational changes, *J. Chromatogr. A* 1198–1199, 154–163, 2008.

186. Gooding, K.M. and Regnier, F.E. eds., *HPLC of Biological Macromolecules*, Marcel Dekker, New York, 1990.
187. Neue, U.D., *HPLC Columns. Theory, Technology, and Practice*, Wiley-VCH, New York, 1997.
188. Cunico, R.L., Gooding, K.M., and Wehr, T., *Basic HPLC and GC of Biomolecules*, Bay Analytical Laboratory, Richmond, CA, 1998.
189. Katz, E., Eksteen, R., Schoenmakers, D., and Miller, N. eds., *Handbook of HPLC*, Marcel Dekker, New York, 1998.
190. Meyer, V.R., *Practial High-Performance Liquid Chromatography*, John Wiley & Sons, Chichester, U.K., 2004.
191. Ishii, K., Yamagami, S., Tanaka, T. et al., Full active baculovirus-expressed human monocyte chemoattractant protein 1 with the intact *N*-terminus, *Biochem. Biophys. Res. Commun.* 206, 959–961, 1995
192. Pohling, G., Fendrich, G., Knecht, R. et al., Purification, characterization and biological evaluation of recombinant leech-derived tryptase inhibitor (rLDTI) expressed at high levels in the yeast *Saccharomyces cerevesiae*, *Eur. J. Biochem.* 241, 619–626, 1996.
193. Williamson, N.A., Raliegh, J., Morrice, N.A, and Wettenhall, R.E.H., Post-translational processing of rat ribosomal proteins. Ubiquitous methylation of Lys 22 within the zinc finger motif of Rl-40 (carboxy-terminal extension protein 52) and tissue-specific methylation of Lys 4 in RL29, *Eur. J. Biochem.* 246, 786–793. 1997.
194. Matrisian, L.M., Planck, S.R., Finch, J.S., and Magun, B.E., Heterogeniety of ^{125}I-labeled epidermal growth factor, *Biochim. Biophys. Acta* 839, 139–146, 1985.
195. Tetaert, D., Gomes, V., Huet-Duvielier, G. et al., Heterogeneity in high-performance liquid chromatography of a variant surface glycoprotein of *Trypanosoma brucei*, *Biochem. Biophys. Res. Commun.* 144, 1222–1228, 1987.
196. Elton, J.S. and Reeves, R., The effects of oxidation of the reverse-phase high-performance chromatography characteristics of the high mobility group 1 and 2 proteins, *Anal. Biochem.* 149, 316–321, 1985.
197. Dillon, T., Bandarenko, P.V., Rehder, D.S. et al., Optimization of a reversed-phase high-performance liquid chromatography/mass spectrometry method for characterizing recombinant antibody heterogeneity and stability, *J. Chromatogr. A.* 1120, 112–120, 2006.
198. Vidal, P., Nielsen, E., and Welinder, B.S., Effect of glycation on the heterogeneities of human serum albumin analyzed by reversed-phase high-performance liquid chromatography in a solvent containing formic acid, *J. Chromatogr.* 573, 201–206, 1992.
199. Kunitani, M., Reversed-phase HPLC separation of homologous proteins: Interleukin-2 muteins, in *High-Performance Liquid Chromatography of Peptides and Proteins*, eds. C.F. Mant and R.S. Hodges, CRC Press, Boca Raton, FL, 1991.
200. Fransson, J., Florin-Robertsson, E., Axelsson, K., and Nyhlén, C., Oxidation of human insulin-like growth factor I in formulation studies: Kinetics of methionine oxidation in aqueous solution and in solid state, *Pharm. Res.* 13, 1252–1257, 1996.
201. Fransson, J. and Hagman, A., Oxidation of human insulin-like growth factor I in formulation studies. II. Effects of oxygen, visible light, and phosphate on methionine oxidation in aqueous solution and evaluation of possible mechanisms, *Pharm. Res.* 13, 1476–1481, 1996.
202. Léon, J., Reubsaet, E., Beijnen, J.H. et al., Oxidation of recombinant methionyl human granulocyte colony stimulating factor, *J. Pharmaceut. Biomed. Anal.* 17, 283–289, 1998.
203. Yokota, H., Saito, H., Masuoka, K. et al., Reversed phase HPLC of Met^{58} oxidized rhIL-11: Oxidation enhanced by plastic tubes, *J. Pharmaceut. Biomed. Anal.* 24, 317–324, 2000.

204. Yang, J., Wang, S., Liu, J., and Ragani, A., Determination of tryptophan oxidation of monoclonal antibody by reversed phase high performance liquid chromatography, *J. Chromatogr. A*. 1156, 174–182, 2007.
205. Terryn, H., Maquille, A., Houée-Levin, C., and Tilquin, B., Irradiation of human insulin in aqueous solution, first step towards radiosterilization, *Int. J. Pharm*. 343, 4–11, 2007.
206. Hsu, Y.-R., Narhi, L.O., Spahr, C. et al., In vitro methionine oxidation of *Escherichia coli*-derived human stem cell factor: Effects on the molecular structure, biological activity, and dimerization, *Protein Sci*. 5, 1165–1173, 1996.
207. Nugent, K.D., Burton, W.G., Slattery, T.K., and Johnson, B.F., Separation of proteins by reversed-phase high performance liquid chromatography. II. Optimizing sample pretreatment and mobile phase composition, *J. Chromatogr*. 443, 381–397, 1988.
208. Mant, C.T., Zhou, N.E., and Hodges, R.S., Correlation of protein retention times in reversed-phase chromatography with polypeptide chain length and hydrophobicity, *J. Chromatogr*. 476, 363–375, 1989.
209. Cohen, S.A., Benedek, K., Tapuhi, Y. et al., Conformational effects in the reversed-phase liquid chromatography of ribonuclease A, *Anal. Biochem*. 144, 275–284, 1985.
210. Burton, W.G., Nugent, K.D, Slattery, T.K., and Summers, B.R., Separation of proteins by reversed-phase high-performance liquid chromatography. I. Optimizing the column, *J. Chromatogr*. 443, 363–379, 1988.
211. Jennissen, H.P., Evidence for negative cooperativity in the adsorption of phosphorylase b on hydrophobic agaroses, *Biochemistry* 15, 5683–5692, 1976.
212. Zhang, Y., Martinez, T., Woodruff, B. et al., Hydrophobic interaction chromatography of soluble interleukin I receptor type II to reveal chemical degradation resulting in loss of potency, *Anal. Chem*. 80, 7022–7028, 2008.
213. To, B.C. and Lenhoff, A.M., Hydrophobic interaction chromatography of proteins. II. Solution thermodynamic properties as a determinant of retention, *J. Chromatogr. A*. 1141, 235–243, 2007,
214. Bjerrum, O.J., Bjerrum, M.J., and Heegaard, N.H., Electrophoretic and chromatographic differentiation of two forms of albumin in equilibrium at neutral pH: New screening technique for determination of lipid binding to albumin, *Electrophoresis* 16, 1401–1407, 1995.
215. Withka, J., Moncuse, P., Baziatis, A., and Maskiewicz, R., Use of high-performance size-exclusion, ion-exchange, and hydrophobic interaction chromatography for the measurement of protein conformational change and stability, *J. Chromatogr*. 398, 175–202, 1987.
216. Desmet, J., Hanssens, I., and van Caulwelaert, F., Comparison of the binding of Na^+ and Ca^{2+} to bovine α-lactalbumin, *Biochim. Biophys. Acta* 912, 211–219, 1987.
217. Iio, T., Melittin-binding of troponin C, *J. Biochem*. 114, 773–728, 1993.
218. Haezebrouck, P., Nappe, W., Van Dael, H., and Hanssons, I., Hydrophobic interactions of lysozyme and α-lactalbumin from equine milk whey, *Biochim. Biophys. Acta* 1122, 305–310, 1992.
219. Bramanti, E., Ferri, F., Sortino, C. et al., Characterization of denatured proteins by hydrophobic interaction chromatography: A preliminary study, *Biopolymers* 69, 293–300, 2003.
220. Geng, X., Guo, L., and Change, J., Study of the retention mechanism of proteins in hydrophobic interaction chromatography, *J. Chromatogr*. 507, 1–23, 1990.
221. Jungbauer, A., Machold, C., and Hahn, R., Hydrophobic interaction chromatography of proteins. III. Unfolding of proteins on adsorption, *J. Chromatogr. A*. 1079, 221–228, 2005.

222. To, B.C.S. and Lenoff, A.M., Hydrophobic interaction chromatography of proteins, *J. Chromatogr. A* 1205, 46–59, 2008.
223. Hearn, M.T.W. and Zhao, G., Investigations into the thermodynamics of polypeptide interaction with nonpolar ligands, *Anal. Chem.* 71, 4874–4886, 1999.
224. Lyklema, J., Interfacial thermodynamics with special reference to biological systems, in *Physical Chemistry of Biological Interfaces*, ed. A. Baszkin, Marcel Dekker, New York, 2008.

5 Size-Exclusion Chromatography and Biomolecular Conformation

Size-exclusion chromatography refers to a technology that separates molecules on the basis of hydrodynamic volume (Stokes' radius) using a cylindrical column with laminar fluid flow. Gel permeation chromatography is taken to mean the separation of organic compounds in apolar solvents[1] while gel filtration means the separation of materials in aqueous solvents systems.[2–6] The material in this chapter will discuss two applications of gel filtration in the study of biomolecular conformation. The first is the use of gel filtration to measure changes in hydrodynamic radius/Stokes' radius as a measure of conformational change; the second is the "nonspecific" adsorption of biomolecules to the matrix as a result of change in structure.

Stokes' radius of a biomolecule is the radius of a sphere, which has the same diffusion coefficient as the biomolecule.[7]

$$f_o = 6\pi\eta r \tag{5.1}$$

where

f_o is the frictional coefficient
η is the viscosity
r the radius of a sphere (Stokes' radius)

Extending this to a protein (by definition asymmetric), the Stokes' radius of a protein would be

$$r_{protein} = \frac{k_B T}{6\pi\eta D protein} \tag{5.2}$$

where $r_{protein}$ is the effective radius or Stokes' radius of the protein.

The value for D (translational diffusion coefficient) is obtained from dynamic light scattering or analytical ultracentrifugation. The reader is directed to Kyte's excellent book[7] for a further discussion the shape of protein and other biomolecules.

The hydrodynamic radius is the radius of a sphere with the same volume as the hydrodynamic particle.[7] The hydrodynamic particle contains the solvation shell.

While Stokes' radius and hydrodynamic radius are not synonymous, both refer to the volume of a macromolecule in solution; both are approximation of the space that a molecule occupies in solution. The primary measurement of the hydrodynamic radius is most frequently by light scattering.[8–12] The theory of gel filtration has been reviewed by Ackers and others.[13–22] The primary process in size-exclusion chromatography is the partitioning of solute between the mobile phase (outside the pores) and inside the porous particle (the stationary phase). A partition coefficient can derived from a consideration of the elution volume and the total volume of the system.[16]

$$K_D = (V_e - V_o)/V_t \qquad (5.3)$$

where

V_e is the elution volume for analyte
V_o is the excluded column volume or void volume
V_t is total column volume

A substance, which diffuses into all pores and is not adsorbed to the matrix, has a K_D of 1.0. An analytic, which is excluded from the matrix, had a K_D of less that 1.0 while a material that is adsorbed to the matrix will have a K_D of greater than 1.0. Gelotte[23] studied the interaction of a variety of simple compounds (adsorption) with the Sephadex matrix in water or 0.05 M NaCl. Sodium hydroxide and potassium hydroxide had K_D values greater than 2 in water, reflecting the interaction of hydroxide with the hydroxyl groups of the matrix. Solvent can have a significant effect; the K_D value for 2,4,6-trimethylpyridine (γ-collidine) in water is >5.0 while it is 0.0 in 0.05 M NaCl. It would appear that 2,4,6-trimethylpyrisine interacts with the column matrix by hydrophobic forces. Gelotte reported a K_D for arginine of >13 in water while a K_D of 1.0 was determined for arginine in 0.05 M NaCl; in this case, there was a combination of ionic interaction and hydrogen bonding. There are other examples of mixed-mode retention such as reported for (–)-epigallocatechin gallate on an agarose matrix.[24]

Gel filtration is frequently used to determine the molecular weight of a protein or other macromolecule. The accuracy of the value obtained for the molecular weight of biopolymers in turns depends on accuracy of the measurement of the partition coefficient and the relation of the partition coefficient to the molecular properties of the biopolymer.[16] The latter consideration provides the greatest challenge in that it is related to the symmetry of the particle; thus, the partition coefficient is related to the hydrodynamic volume/Stokes' radius and not necessary to molecular weight. Nevertheless, it possible to calibrate a gel filtration with proteins of known molecular weight and obtained reasonable data[16]; however, it considered more useful to calibrate a gel filtration column with proteins of known Stokes' radius.[25,26] Accurate values for Stokes' radius may be obtained from dynamic light scattering or analytical ultracentrifugation[27–34] and are used for validation of standards for gel filtration or to validate data obtained from gel filtration with standards for which Stokes' radii are not available. Some values for Stokes' radii and selected other hydrodynamic parameters for some native proteins are shown in Table 5.1. Other tabulations of hydrodynamic parameters have been compiled by Edsall,[44] Damm and coworkers,[1] Sober,[45] Kumonski and Pessen,[46] and Uversky.[4]

TABLE 5.1

Some Values for Stokes' Radii, Molecular Weight, Axial Ratio, and Frictional Ratio for Some Proteins

Protein	MW (kDa)	Stokes' Radius (nm)	Axial Ration (a/b)	f/f_0^a	Refs.
RNase	12.4	1.90		1.05	[35,36]
		2.14c			
RCM-RNaseb		2.91			[36]
		3.14c			
Lysozyme	14.1	1.9		1.21	[35]
Trypsinogen	23.7	1.96		1.15	[35,37]
Chymotrypsinogen	25.0	2.17		1.19	[35]
denatured		4.78			[38]
Corn zein	50.0			2.17	[35]
Human albumin	72.3	3.6		1.30	[35]
Succinylated albumin		6.3			
Aldolase denatured	147	4.66		1.31	[35,38]
		6.08			
Bovine fibrinogen	333			2.34	[35]
Human fibrinogen	277	10.7d	7e		[32]
Human factor IX	65	4.08		1.54	[39]
Ovalbumin Denatured	43.5	3.02			[38]
ovalbumin		5.20			[38]
Plectin	1200	27	~50	4	[40]
Apoferritin	482	6.7			[26]
	480	6.1		1.2	[41]
Γ-Globulin	160	5.2		1.45	[41]
Waxy maize amylopectin	560,000	348			[42]
Soluble extracellular	65	3.9f			[43]
domain of amyloid		5.1g			
precursor protein		Anisometric protein			

a Frictional ratio.

b Reduced, carboxymethylated protein.

c In the presence of 6.0 M guanidine hydrochloride.

d Determined by sedimentation equilibrium and sedimentation velocity measurement. A value of 7.1 nm was obtained by native gel filtration. The authors suggest end-on insertion of this asymmetric protein resulting in an artifactually low value for the molecular weight. A value of 10.1 for the Stokes' radius of human fibrinogen was established by light scattering measurements (Hantgan, R.R., Braaten, J.V., and Rocco, M., Dynamic light scattering studies of $\alpha_{IIb}\beta_3$ solution conformation, *Biochemistry* 32, 3935–3941, 1993).

e Determined for porcine fibrinogen (Serallach, E., Hofmann, V., Zulnauf, M. et al., Conformation of the fibrinogen molecule in solution: Dynamic light scattering spectroscopy, *Schweiz. Med. Wochenschr.* 106, 1380, 1976).

f Determined with small-angle x-ray scattering.

g Gel filtration.

Burnett and coworkers[47] studied the movement of proteins from blood to amniotic fluid and suggested that Stokes' radius should be used in preference to molecular weight in studies on protein filtration systems. The reader is directed to other studies on Stokes' radius and membrane permeability.[48–51]

A consideration of the above methods suggests that when there is no significant change in molecular mass of a protein or other biomacromolecule, a change in hydrodynamic radius/Stokes' radius as measured by gel filtration can be a measure of conformational change (increase or decrease in asymmetry as measured by changes in frictional coefficient). As we will see below, there are conformational changes in proteins, which result in the exposure of surfaces which physically interact with the gel filtration matrix. In general, the conformational changes observed in biopharmaceuticals result in an apparent increase in Stokes' radius similar to those observed with the reduction and alkylation of a protein (see RNase in Table 5.1). Some selected examples of changes in conformation as measured by gel filtration are presented in Table 5.2.

Ideal gel filtration (size-exclusion chromatography) as described above separates proteins and other biomolecules on the basis size (Stokes' radius). Nonideal gel filtration combine the size partitioning with interaction with the matrix, yielding K_D values

---

TABLE 5.2

Alterations in Biopolymer Conformation as Determined by Changes in Stokes' Radius Measured by Gel Filtration

Protein	Chromatographic Media and Change in Migration	Refs.
Fc fragment of rabbit IgG	Thin-layer gel filtration on G-200; pH-dependent conformational change as expressed relative to migration of Fab fragment. The conformational change in the Fc domain has been demonstrated by limited proteolysis and scanning microcalorimetry[a]	[52]
Human transferring and apotransferrin	Gel filtration on G-100 Sephadex demonstrated changes in Stokes' radius on binding of iron to apotransferrin	[53]
Succinylated ovalbumin	Used gel filtration on G-100 and G-200 Sephadex to measure conformational changes in ovalbumin on reaction with succinic anhydride. There was a increase in axial ratio, f/f_o, and Stokes' radius as results of modification of ovalbumin with succinic anhydride	[54]
Disulfide-reduced ovalbumin	Use of gel filtration to study conformational changes in ovalbumin resulting from reduction of disulfide bonds. There was no difference in the Stokes' radius between intact and disulfide-bond reduced. The reduced protein was more susceptible to proteolysis by pepsin	[55]
Transferrin and apotransferrin	Use of G-100 Sephadex to measure changes in Stokes' radius occurring on the binding of iron to apotransferrin. The extent of change depends on species of protein and nature of counterion (oxalate or bicarbonate)	[56]
Cyclic nucleotide phosphodiesterase	Use of gel filtration to study changes in conformation using partially purified proteins	[57]

TABLE 5.2 (continued)
Alterations in Biopolymer Conformation as Determined by Changes in Stokes' Radius Measured by Gel Filtration

Protein	Chromatographic Media and Change in Migration	Refs.
Thyroglobulin	Ca^{2+} caused a decrease in migration (increase in elution volume) on Sepharose 6B consistent with a decrease in Stokes' radius. There was also a general decrease (thyroglobulin, egg albumin, N-acetyltryptophanamide) in migration (increase in elution volume) in the presence of 10% PEG 6000. Changes in the apparent molecular size of thyroglobulin in the presence of calcium ions were also observed on sucrose density gradient centrifugation	[58]
Radioiodinated polypeptide hormones[b]	Iodination (^{125}I) of the various proteins (except PRL) was accomplished with chloramine T; the radiolabeling of PRL was accomplished with lactoperoxidase. Changes in Stokes' radii were evaluated by gel filtration. It is noted that gel filtration behavior of HGH is influenced by buffer and ionic strength (Stokes' radius 2.22 in 0.05 M phosphate; 2.69 in 0.025 M Tris)	[59,60]
Monoclonal IgM cryoglobulin	Analytical ultracentrifugation gave a S020, W of 20.1 compared to 17–19 for normal cryoglobulin. Bio-Gel A1.5 M gave MW of 995 kDa for monoclonal IgM cryoglobulin while a value of 870 kDa was obtained for normal cryoglobulin. This difference was a function of the Fabμ region where gel filtration provide a MW value of 65 kDa for the monoclonal IgM cryoglobulin and a value 48 kDa for the normal cryoglobulin; analysis of the fragments under denaturing conditions (6.0 M GuCl) yielded a value of 50 and 52 kDa, respectively	[61]
α2-Macroglobulin	Use of HPLC gel filtration (TSK G4000 SW) to study conformational change of α2-macroglobulin on binding to protease (chymotrypsin). There was a decrease in Stokes' radius from 8.67 to 7.85 nm; this is also the change observed on treatment with methylamine. The Stokes' radius of ovomacroglobulin (chicken and crocodilian) also decreased on binding of chymotrypsin but no change was observed with methylamine. It is noted that other investigators[c] have noted nonideal behavior of α2-macroglobulin on gel filtration	[62]
Lysozyme	Use of size-exclusion chromatography to measure hydrodynamic volume changes during refolding after denaturation in urea/dithiothreitol	[63]
Bovine serum albumin	Use of gel filtration (Sephacryl S-300) to determine changes in Stokes' radius resulting from modification with succinic anhydride. Stokes' radius increases with increased modification with succinic anhydride. Concomitant changes in conformation were observed with intrinsic fluorescence measurement, difference UV-VIS spectroscopy and immunological reactivity	[64]

(continued)

TABLE 5.2 (continued)
Alterations in Biopolymer Conformation as Determined by Changes in Stokes' Radius Measured by Gel Filtration

Protein	Chromatographic Media and Change in Migration	Refs.
PEGylated[d] human serum albumin	Use of gel filtration (Superose 12; FPLC) to measure changes in hydrodynamic radius, resulting from the modification of human serum albumin with PEG-phenyl isothiocyanate. Stokes' radius increased from 3.95 to 6.57 nm with the hexapegylated derivative. There was also an increase in viscosity and colloid osmotic pressure	[65]
Linear PEGylated proteins and branched PEGylated proteins	Use of gel filtration to measure changes in Stokes' radius following modification of proteins with linear PEG chains or branched PEG chains	[66]
Metallothionein	Use of gel filtration (G-75 Sephadex) to assess effect of zinc on the conformation of metallothionein. The Stokes' radius of the apoprotein is 2.08 nm, which decreases to 1.54 nm on saturation with zinc	[67]
Villin	Conformational change observed in the presence of calcium ions studied by gel filtration on Sephacryl S-300. Stokes' radius increases from 4.4 to 4.9 nm in the presence of $20 \mu M$ Ca^{2+}. Conformational change also observed with analytical ultracentrifugation, UV-VIS difference spectroscopy, and CD	[68]
Bovine cardiac troponin C	Use of Sephacryl S-300 to measure Ca^{2+}-dependent conformational changes in bovine cardiac troponin C. The Stokes' radius decrease from 2.63 nm in the absence of calcium ions to 2.43 nm in the presence of 2 mM Ca^{2+}	[69]
Prorenin and renin	G-100 Sephadex used to show an apparent increase in Stokes' radius with acid-activated renin	[70]
Bovine pancreatic DNAse	Use of G-100 used to demonstrate pH-dependent conformational change in DNase and the effect of calcium ions on that change	[71]

[a] Stewart, G.A. and Stenworth, D.R., Effect of acid treatment upon susceptibility of rabbit IgG to proteolytic cleavage with various enzymes, *Immunochemistry* 12, 713–721, 1975; Tischenko, V.M., Metastable state of the Fc fragment, *J. Thermal Anal. Calorimetry* 62, 63–68, 2000.

[b] Human growth hormone (HGH); luteotropin (LH); follicle-stimulating hormone (FSH); thyrotropin (TSH); prolactin (PRL); corticotropin (ACTH).

[c] Gonias, S.L., Roche, P.A., and Pizzo, S.V., Purification and characterization of human alpha-2-macroglobulin conformational variant by non-ideal high performance size-exclusion chromatography, *Biochem. J.* 235, 559–567, 1986.

[d] PEGylated—Modification with poly(ethylene glycol).

generally larger (increased retention) than predicted by Stokes' radius. Examples of such interaction were described by Gelotte[23] and briefly described above. Nonideal gel filtration most commonly results from interaction with the matrix via electrostatic, hydrophobic, or hydrogen bonding interactions.[72–76] Some examples of nonideal gel filtration for the study of protein conformation have been presented in Chapter 4 with particular reference to the vitamin K–dependent coagulation factors. It is necessary to have other measures of conformational change to validate the conclusions from changes in gel filtration properties.[77]

REFERENCES

1. Sharma, J., Principles of gel permeation chromatography, *J. AOAC Int.* 91, 113A–118A, 2008.
2. Wood, G.C. and Cooper, P.F., The application of gel filtration to the study of protein-binding of small molecules, *Chromatogr. Rev.* 1, 88–107, 1970.
3. Andrews, P., Estimation of molecular size and molecular weights of biological compounds by gel filtration, *Methods Biochem. Anal.* 18, 1–53, 1970.
4. Williams, K.W., Solute–gel interactions in gel filtration, *Lab. Pract.* 21, 667–670, 1972.
5. Winzor, D.J., From gel filtration to biosensor technology: The development of chromatography for the characterization of protein interactions, *J. Mol. Recognit.* 13, 279–298, 2000.
6. Porath, J., From gel filtration to adsorptive size exclusion, *J. Protein Chem.* 16, 463–468, 1997.
7. Kyte, J., *Structure in Protein Chemistry*, 2nd edn., Garland Science, New York, 2007.
8. Yarmush, D.M., Murphy, R.M., Colton, C.K. et al., Quasi-elastic light scattering of antigen complexes, *Mol. Immunol.* 25, 17–32, 1988.
9. Wang, Q., Huang, X., Nakamura, A. et al., Molecular characterisation of soybean polysaccharides: An approach by size exclusion chromatography, dynamic and static light scattering methods, *Carbohydr. Res.* 340, 2637–2644, 2005.
10. Stopa, B., Rybarska, J., Drozd, A. et al., Albumin binds self-assembling dyes as specific polymorphonuclear ligands, *Int. J. Biol. Macromol.* 40, 1–8, 2006.
11. Flammer, I. and Ri Ka, J., Dynamic light scattering with single-mode receivers: Partial heterodyning regime, *Appl. Opt.* 36, 7508–7517, 1997.
12. Pons, T., Uyeda, H.T., Medintz, I.L., and Mattoussi, H., Hydrodynamic dimensions, electrophoretic mobility, and stability of hydrophilic quantum dots, *J. Phys. Chem. B.* 110, 20308–20317, 2006.
13. Edmond, E., Farquhar, S., Dunstone, J.R., and Ogston, A.G., The osmotic behavior of Sephadex and its effects on chromatography, *Biochem. J.* 108, 755–763, 1968.
14. Polson, A. and Katz, W., A quantitative theory for gel-exclusion chromatography, *Biochem. J.* 112, 387–388, 1969.
15. Rodbard, D. and Chrambach, A., Unified theory for gel electrophoresis and gel filtration, *Proc. Natl. Acad. Sci. USA* 65, 970–977, 1970.
16. Ackers, G.K., Molecule sieve methods of analysis, in *The Proteins*, 3rd edn., Vol. I, eds. H. Neurath and R.L. Hill, Academic Press, New York, Chapter 1, 1975.
17. Cann, J.R. and Hinman, N.D., Hummel-Dreyer gel chromatographic procedure as applied to ligand-mediated association, *Biochemistry* 15, 4614–4622, 1976.
18. Minton, A.P., Thermodynamic nonideality and the dependence of partition coefficient upon solute concentration in exclusion chromatography. II. An improved theory of equilibrium partitioning of concentrated protein solutions. Application to hemoglobin, *Biophys. Chem.* 18, 139–143, 1983.
19. le Maire, M., Ghazi, A., Martin, M., and Brochard, F., Calibration curves for size-exclusion chromatography: Description of HPLC gels in terms of porous fractals, *J. Biochem.* 106, 814–817, 1989.

20. Bedani, F., Kok, W.T., and Janssen, H.G., A theoretical basis for parameter selection and instrument design in comprehensive size-exclusion chromatography × liquid chromatography, *J. Chromatogr. A*. 1133, 126–134, 2006.
21. Winzor, D.J. and De Jersey, J., Biospecific interactions: Their quantitative characterization and use of solute purification, *J. Chromatogr.* 492, 377–430, 1989.
22. Funasaki, N., Gel filtration chromatographic study on the self-association of surfactants and related compounds, *Adv. Colloid Interface Sci.* 43, 87–136, 1993.
23. Gelotte, B., Studies on gel filtration Sorption properties of the bed material sephadex, *J. Chromatogr.* 3, 330–342, 1960.
24. Xu, J., Tan, T., and Janson, J.-C., Mixed-mode retention mechanisms for (–)-epigallocatechan gallate on a 12% cross-linked agarose gel media, *J. Chromatogr.* 1137, 49–55, 2006.
25. le Maire, M., Aggerbeck, L.P., Monteilhet, C. et al., The use of high-performance liquid chromatography for the determination of size and molecular weights of proteins: A caution and a list of membrane proteins suitable as a standards, *Anal. Biochem.* 154, 525–535, 1986.
26. de Haën, C., Molecular weight standards for calibration of gel filtration and sodium dodecyl sulfate-polyacrylamide gel electrophoresis: Ferritin and apoferritin, *Anal. Biochem.* 166, 235–245, 1987.
27. Schmitz, K.S., *An Introduction to Dynamic Light Scattering by Macromolecules*, Academic Press, Boston, MA, 1990.
28. Damaschun, G., Damaschun, H., Gast, H. et al., Acid denatured apo-cytochrome c is a random coil: Evidence from small angle X-ray scattering and dynamic light scattering, *Biochim. Biophys. Acta* 24, 1078, 289–295, 1991.
29. Gast, K., Zirwer, D., Müller-Frohne, M., and Damschun, G., Compactness of the kinetic molten globule of bovine α-lactalbumin: A dynamic light scattering study, *Protein Sci.* 7, 2004–2011, 1998.
30. Kornblatt, J.A. and Schuck, P., Influence of temperature on the conformation of canine plasminogen: An analytical ultracentrifugation and dynamic light scattering study, *Biochemistry* 44, 13122–13131, 2005.
31. d'Orlyé, F., Varenne, A., and Gareil, P., Determination of nanoparticle diffusion coefficients by Taylor dispersion analysis using a capillary electrophoresis instrument, *J. Chromatogr.* 1204, 226–232, 2008.
32. Nozaki, Y., Schechter, N.M., Reynolds, J.A., and Tanford, C., Use of gel chromatography for the determination of the Stokes radii of proteins in the presence of and absence of detergents. A reexamination, *Biochemistry* 15, 3884–3890, 1976.
33. Gaulfetti, P.J., Iwakura, M., Lee, C. et al., Apparent radii of the native, stable intermediates and unfolded conformers of the α-subunit of tryptophan synthase from *E. coli*, a TIM barrel protein, *Biochemistry* 38, 13367–13378, 1999.
34. Brown, P.H. and Schuck, P., Macromolecular size-and-shape distributions by sedimentation velocity analytical ultracentrifugation, *Biophys. J.* 90, 4651–4661, 2006.
35. Damm, H.C., Besch, P.K., and Goldwyn, A.J. eds., *Handbook of Biochemistry and Biophysics*, World Publishing Company, Cleveland, OH, 1966.
36. Nöopert, A., Gast, K., Müller-Frohne, M. et al., Reduced-denatured ribonuclease A is not in a compact state, *FEBS Lett.* 380, 179–182, 1996.
37. al-Obeidi, A.M. and Light, A., Size-exclusion high performance liquid chromatography of native trypsinogen, the denatured protein, and partially refolded molecules. Further evidence that non-native disulfide bonds are dominant in refolding the completely reduced protein, *J. Biol. Chem.* 263, 8642–8645, 1988.
38. Uversky, V.N., Use of fast protein size-exclusion liquid chromatography to study the unfolding of proteins which denature through the molten globule, *Biochemistry* 32, 13288–13298, 1993.

39. Suomela, H., Human coagulation factor IX. Isolation and characterization, *Eur. J. Biochem.* 71, 145–154, 1976.

40. Foisner, R. and Wiche, G., Structure and hydrodynamic properties of plectin molecule, *J. Mol. Biol.* 198, 515–531, 1987.

41. Demassieux, S. and Lachance, J.P., Tamisage moléculaire sur Sepharose 6. Relations enpiriques entre le coefficient de distribution et le rayon de Sotkes d'une port, et le poids, moléculaire d'autre port de proteins de degrees d'asymétrée varies, *J. Chromatogr.* 89, 251–258, 1974.

42. Millard, M.M., Wolf, W.J., Dintzis, F.R., and Wilett, J.L., The hydrodynamic characteristics of waxy maize amylopectin in 90% dimethylsulfoxide–water by analytical ultracentrifugation, dynamic, and static light scattering, *Carbohydr. Poly.* 39, 315–320, 1999.

43. Gralle, M., Botello, M.M., Oliviera, C.L. et al., Solution studies and structural model of the extracellular domain of the human myeloid precursor protein, *Biophys. J.* 83, 3513–3524, 2002.

44. Edsall, J.T., The size, shape, and hydration of protein molecules, in *The Proteins*, Vol. 1, eds. H. Neurath and K. Bailey, Academic Press, New York, Chapter 1, 1953.

45. Sober, H.H. ed., *Handbook of Biochemistry*, CRC Press, Cleveland, OH, 1968.

46. Kumonski, T.F. and Pessen, H., Estimation of sedimentation coefficients of globular proteins: An application of small-angle x-ray scattering, *Arch. Biochem. Biophys.* 219, 89–100, 1982.

47. Burnett, D., Wood, S.M., and Bradwell, A.R., Estimation of the Stokes radii of serum proteins for a study of protein movement from blood to amniotic fluid, *Biochim. Biophys. Acta* 427, 231–237, 1976.

48. Clauss, M.A. and Jain, R.K., Interstitial transport of rabbit and sheep antibodies in normal and neoplastic tissues, *Cancer Res.* 50, 3487–3492, 1990.

49. Palassini, M. and Remuzzi, A., Numerical analysis of viscous flow through fibrous media: A model for glomerular basement membrane permeability, *Am. J. Physiol.* 274, F223–F231, 1998.

50. Curcio, E., De Bartolo, L., Barbieri, G. et al., Diffusive and convective transport though hollow fiber membranes for liver cell culture, *J. Biotechnol.* 117, 309–321, 2005.

51. Kolin, D.L., Ronis, D., and Wiseman, P.W., k-Space image correlation spectroscopy: A method for accurate transport measurements independent of fluorophore photophysics, *Biophys. J.* 91, 3061–3075, 2006.

52. Charlwood, P.A. and Utsumi, S., Conformational changes and dissociation of Fc fragments of rabbit immunoglobulin G as a function of pH, *Biochem. J.* 112, 357–365, 1969.

53. Charlwood, P.A., Differential sedimentation velocity and gel-filtration measurements on human apotransferrin and iron-transferrin, *Biochem. J.* 125, 1019–1026, 1971.

54. Kidwai, S.A., Ansari, A.A., and Salahuddin, A., Effect of succinylation (3-carboxypropionylation) on the conformation and immunological activity of ovalbumin, *Biochem. J.* 155, 171–180, 1976.

55. Tatsumi, E., Yoshimatsu, D., and Hirose, M., Conformational state of disulfide-reduced ovalbumin at acidic pH, *Biosci. Biotechnol. Biochem.* 63, 1285–1290, 1999.

56. Jarritt, P.H., Effect of iron on sedimentation-velocity and gel filtration behaviour of transferrins from several vertebrates, *Biochim. Biophys. Acta* 453, 332–343, 1976.

57. Francis, S.H., Chu, D.M., Thomas, M.K. et al., Ligand-induced conformation changes in cyclic nucleotide phosphodiesterases and nucleotide-dependent protein kinases, *Methods* 14, 81–92, 1998.

58. Formisano, S., Di Jeso, B., Aquarviva, R. et al., Calcium-induced changes in thyroglobulin conformation, *Arch. Biochem. Biophys.* 227, 351–357, 1983.

59. Ribela, M.T.C.P. and Bartolini, P., Stokes radius determination of radioidinated polypeptide hormones by gel filtration, *Anal. Biochem.* 174, 693–697, 1988.

60. Bartolini, P. and Ribel, M.T.C.P., Influence of chloramine T iodination on the biological and immunological activity of the molecular radius of the human growth hormone molecule, *J. Immunoassay* 7, 129–138, 1986.

61. Middaugh, C.R., Oshman, R.G., and Litman, W., Localization of a conformational anomaly to the Fabμ region of a monoclonal IgM cryoglobulin, *Clin. Exp. Immunol.* 31, 126–130, 1978.

62. Nishigai, M., Osada, T., and Ikai, A., Structural changes in alpha-2- and ovomacroglobulins studied by gel chromatography and electron microscopy, *Biochim. Biophys. Acta* 831, 236–241, 1985.

63. Batas, B., Jones, H.R., and Chaudhuri, J.B., Studies of the hydrodynamic volume changes that occur during refolding of lysozyme using size-exclusion chromatography, *J. Chromatogr. A* 766, 109–119, 1997.

64. Tayyah, S. and Qasim, M.A., A correlation between changes in conformation and molecular properties of bovine serum albumin upon succinylation, *J. Biochem.* 100, 1125–1136, 1986.

65. Meng, F., Manjula, B.N., Smith, P.K., and Acharya, S.A., PEGylation of human serum albumin: Reaction of PEG-phenyl-isothiocyanate with protein, *Bioconjug. Chem.* 19, 1352–1360, 2008.

66. Fee, C.J., Size comparison between protein PEGylated with branched and linear poly (ethylene glycol) molecules, *Biotechnol. Bioeng.* 98, 725–731, 2007.

67. Phillips, J.L., Effect of zinc on the Stokes' radius of metallothionein, *Biochem. Biophys. Res. Commun.* 114, 998–1000, 1983.

68. Hesterberg, L.K. and Weber, K., Ligand-induced conformational changes in villin, a calcium-controlled actin-modulating protein, *J. Biol. Chem.* 258, 359–364, 1983.

69. Byers, D.M. and Kay, C.M., Hydrodynamic properties of bovine cardiac troponin C, *Biochemistry* 21, 229–233, 1982.

70. Hsueh, W.A., Carlson, E.J., and Israel-Hagman, M., Mechanism of acid-activation of rennin: Role of kallikrein in renin activation, *Hypertension* 3(3Pt 2), I22–I29, 1981.

71. Lizárraga, B., Sánchez-Romero, D., Gil, A., and Melgar, E., The role of Ca^{2+} on pH-induced hydrodynamic changes of bovine pancreatic deoxyribonuclease A, *J. Biol. Chem.* 253, 3191–3195, 1978.

72. Determann, H., *Gel Chromatography. A Laboratory Handbook*, Springer-Verlag, New York, 1969.

73. Dubin, P.L. ed., *Aqueous Size-Exclusion Chromatography*, Elsevier, Amsterdam, the Netherlands, 1988.

74. Wu, C.-S. ed., *Handbook of Size-Exclusion Chromatography*, Marcel Dekker, New York, 1995.

75. Petschka, M. and Dubin, P.L. eds., *Strategies in Size-Exclusion Chromatography*, American Chemical Society, Washington, DC, 1996.

76. Mori, S. and Barth, H.G., *Size-Exclusion Chromatography*, Springer-Verlag, Berlin, 1999.

77. Dekker, C., Agianian, B., Weik, M. et al., Biophysical characterization of the influence of salt on tetrameric SecB, *Biophys. J.* 81, 455–462, 2001.

6 Use of Analytical Ultracentrifugation to Study Biomolecular Conformation

The past 40 years have provided an increase in the sophistication of the technologies available to measure conformational change in proteins; there has not been an increase in the parameters measured. It is important to understand that while there have been some advances in experimental approaches, most of the improvement has resulted from more sophistication in instrumentation and data analysis including computer technology. Analytical ultracentrifugation is an example of a technique that has benefited from improvement in instrument technology and data analysis.[1-6]

Kauzmann[7] proposed a classification system for the levels of conformation similar to the general classification of primary, secondary, tertiary, and quaternary structure, which separated conformation issues into shape properties and short-range properties. Shape properties (long-range) were parameters dependent on the overall shape (globular, rod, etc.), which might be relatively insensitive to changes in the immediate vicinity of amino acids and peptide bonds. Short-range properties include parameters defined by the immediate environment around individual amino acid residues. Granted that this is an imperfect separation, it does prove useful. Schellman and Schellman reviewed the problem of conformation change in proteins in 1964[8] and as observed by Cantor and Timasheff,[9] there had been no change in the some 20 years between the two reviews. There has been a marked increase in the sophistication of the instrumentation that Schellman and Schellman used extending Kauzmann's earlier suggestions. Shape properties included hydrodynamic parameters such as frictional coefficient and viscosity changes and solution properties such as fluorescence depolarization and flow birefringence. Also included in shape is electron microscopy, dipole moments, and diffusion through controlled pore membranes (thin film dialysis).[10-12] Short-range properties are, to some extent, "micro" properties as compared to the "macro" properties of shape. Schellman and Schellman include optical properties such as absorbance (IR, UV) and circular dichroism and chemical properties such as side chain reactivity (trace labeling, chemical footprinting), individual pK_a's, hydrogen isotope exchange, biological activity, and immunogenicity as short-range properties. Also included in short-range properties are NMR and binding of small molecules such as dyes. This division is admittedly is imperfect; for example, immunogenicity could be more accurately defined as shape property but reactivity is dependent on epitopic change.

Techniques such as light scattering[13–18] and analytical ultracentrifugation[19–24] provide information about the shape and solution behavior of the material (tertiary and quaternary structures). These two techniques together with size-exclusion chromatography are critical for the evaluation of aggregation in pharmaceutical preparations. There is also reason to consider measurement of the second virial coefficient. The second virial coefficient is a factor used to correct for the nonideal behavior of a particle. Virial coefficients were originally developed as a series of coefficients of inverse powers of V in a polynomial series to approximate the quantity of pV/RT in an equation of state of an ideal gas or similar collection of particles.[25,26] From a practical perspective, the second virial coefficient is related to the excluded volume of a particle[27,28] and is important in accounting for protein–protein interactions and molecular crowding.[29–32] The excluded volume of any particle depends on shape and can be defined as the volume surrounding and including a given object, which is excluded to another object.[27] The second virial coefficient is mentioned most often in the study of the osmotic pressure of proteins but has general use for the study of protein–protein interaction.[33–47] The reader is recommended to articles on protein shape[48–52] for additional information.

Analytical ultracentrifugation is a powerful tool for the analysis of the size and shape of biological macromolecules. Analytical ultracentrifugation may include sedimentation velocity or sedimentation measurements.[53,54] As the name suggests, sedimentation velocity measures the rate of movement of a particle in a gravitational field while sedimentation equilibrium allows a system to go to equilibrium under centrifugal force. Sedimentation velocity uses higher speeds for a shorter period of time while sedimentation equilibrium uses lower speeds for longer periods of time. Sedimentation equilibrium allows the direct determination of molecular weight. Both approaches require the knowledge of the partial specific volume of the analyte. Sedimentation velocity measures the rate of transport of a solute across a given surface (dx/dt) in a centrifugal field ($\omega^2 x$), the rate of transport is defined by

$$s = \frac{dx/dt}{\omega^2 x} \qquad (6.1)$$

where s (cm/s/dyne/g of s^{-1}) is Svedberg unit (10^{-13} s), and

$$M = RTs/D\left(1 - v\rho\right) \qquad (6.2)$$

where
M is the molecular weight
s is Svedberg unit
D is the diffusion constant
v is the partial specific volume
ρ is the solution density

It is possible to obtain an accurate molecular weight from sedimentation equilibrium.

We consider two concepts here: size and shape. Size refers to magnitude, bulk, "bigness," or dimensions of anything.[54] Shape refers to the external form or contour of an object or geometrical figure, which depends on the constant relations of position and proportionate distance among the various points comprising the outline or external surface. The other quality that influences solution behavior is surface properties such as charge and hydrophobicity. Density does not play a major role as there is not much difference between biological macromolecules; partial specific volume is a descriptor for the density of proteins. Shape (conformation) is a macroproperty as described above and is measured by techniques such as ultracentrifugation and light scattering.[55]

The shape of a protein (or other macromolecule) can be described by several measurements. Stokes' radius and the hydrodynamic radius define the volume of a molecule in solution. These have been described in Chapter 5. Briefly, the hydrodynamic particle is defined as the biological macromolecule and any molecules of water, which occupy cervicular areas and/or are bound to the molecule surface. The hydrodynamic volume is defined by frictional ratio, which is defined as the product obtained as a result of dividing the observed frictional coefficient (f) (Equation 6.3) by the frictional coefficient for a hard sphere of the same partial specific volume; the diffusion constant is a measure of the frictional coefficient (Equation 6.4).[56] An approximation to can be obtained

$$f = 6\pi\eta r \tag{6.3}$$

where
  η is the solvent viscosity
  r is the hydrodynamic radius

$$D = k_B T / f \tag{6.4}$$

The reader is directed to Kyte[57] and others for a consideration of hydrodynamic properties of proteins and other biological macromolecules.

The results from analytical ultracentrifugation are not influenced by matrix interactions such as those observed with gel filtration; the data obtained from analytical ultracentrifugation can, however, reflect the pressure induced by the high centrifugal forces.[58–65] Protein conformation can be influenced by hydrostatic pressure.[66–69] Some examples of the application of analytical ultracentrifugation in the study of biopharmaceutical polymers is shown in Table 6.1.

Analytical ultracentrifugation also can provide a direct determination of the molecular weight of a protein or other biomacromolecules.[82–85] Analytical ultracentrifugation is gold standard for identifying aggregates in protein preprations[86–88] and for examining protein–protein interactions.[89–93] Analytical ultracentrifugation is also used as a primary method for the determination of molecular weight.[94–100]

TABLE 6.1

Use of Ultracentrifugation to Measure Conformational Change in Biological Macromolecules

Macromolecule	Description of Study	Ref.
Antibody fragments	Effect of PEGylation on solution conformation of antibody fragments. Modification of Fab' with poly(ethylene glycol) decreases S, increases frictional ratio, and Stokes' radius	[70]
Recombinant G-MCSF[a]	Use of sedimentation velocity and sedimentation equilibrium to compare human and murine proteins and evaluation of structure changes after reduction. Comparison with other techniques such as spectroscopy and PAGE	[71]
Ricin	Modest decrease is $S_{20,w}$ at acid pH; larger decrease at alkaline pH consistent with unfolding from native compact form ($f/f_o = 1.28$); correlation with viscometry, CD, and fluorescence	[72]
Equine β-lactoglobulin	Use of sedimentation velocity and sedimentation equilibrium to study conformational change during cold denaturation. Increase in Stokes' radius with cold denaturation. Comparison of analytical ultracentrifugation data with CD data, dynamic light scattering, and small-angle x-ray	[73]
Canine plasminogen	Use of sedimentation to measure conformational changes in canine plasminogen as a function of temperature, ligand binding. Shape changes confirmed by light scattering	[74]
EPEC Tir[b] protein	Use of sedimentation velocity to assess conformational changes occurring on phosphorylation; also CD measurements	[75]
Recombinant porcine growth hormone	Used of sedimentation equilibrium to measure conformational change during acid denaturation. Self-association was prominent at pH 2.0; self-association reduced by 4.0 M urea	[76]
Calmodulin	Use of sedimentation equilibrium to measure the pH-dependence of monomer–dimer equilibrium; correlation with ESI-FTICR[c] mass spectrometry	[77]
Murine interleukin-6	Use of sedimentation velocity and sedimentation equilibrium to assess conformational change in murine interleukin-6 (IL-6)	[78]
Interleukin-8 (IL-8)	Use of sedimentation equilibrium analysis to measure monomer–dimer equilibrium for IL-8 and IL-8 mutants: Correlation with NMR and in vitro biological activity	[79]
Myrosinase	Use of sedimentation equilibrium to characterize protein. Decrease in $S_{20,w}$ in 6.0 M guanidine	[80]
Acid phosphatase from Pinto beans	Used analytical ultracentrifugation to measure conformational change associated with loss of activity at acid pH ($S_{20,w}$ of 4.7 in Tris, pH 7.5; 2.5 in acetate at pH 5.2)	[81]

[a] G-MCSF; granulocyte-macrophage colony-stimulating factors.

[b] Tir, translocated intimin receptor.

[c] ESI-FTICR, electrospray ionization Fourier-transform ion cyclotron resonance.

REFERENCES

1. Lebowitz, J., Lewis, M.S., and Schuck, P., Modern analytical ultracentrifugation in protein science: A novel review, *Protein Sci.* 11, 2667–2679, 2002.
2. Schuck, P., On the analysis of protein self-association by sedimentation velocity analytical ultracentrifugation, *Anal. Biochem.* 320, 104–124, 2003.
3. Lechner, M.D. and Borger, I. eds., *Analytical Ultracentrifugation VII: Progress in Colloid and Polymer Science*, Vol. 127, Springer-Verlag, Berlin, Germany, 2003.
4. Ebel, C., Analytical ultracentrifugation for the study of biological macromolecules, in *Analytical Ultracentrifugation VII: Progress in Colloid and Polymer Science*, Vol. 127, eds. M.D. Lechner and I. Borger, Springer-Verlag, Berlin, Germany, pp. 72–82, 2003.
5. Brown, P.H. and Schuck, P., Macromolecular size-and-shape by sedimentation velocity analytical ultracentrifugation, *Biophys. J.* 90, 4651–4661, 2006.
6. Cole, J.L., Lary, J.W., Moody, T., and Lane, T.M., Analytical ultracentrifugation: Velocity and sedimentation equilibrium, *Methods Cell Biol.* 84, 143–179, 2008.
7. Kauzmann, W., Some factors in the interpretation of protein denaturation, *Adv. Protein Chem.* 14, 1–63, 1959.
8. Schellman, J.A. and Schellman, C., The conformation of polypeptide chains in proteins, in *The Proteins*, 2nd edn., ed. H. Neurath, Academic Press, New York, pp. 1–137, Chapter 7, 1964.
9. Cantor, C.R. and Timasheff, S.N., Optical spectroscopy of proteins, in *The Proteins*, 3rd edn., Vol. 5, eds. H. Neurath and R.L. Hill, Academic Press, New York, pp. 145–306, 1982.
10. Craig, L.C. and Chen, H.C., On a theory for the passive transport of solute through semipermeable membranes, *Proc. Natl. Acad. Sci. USA* 69, 702–705, 1972.
11. Chen, H.C., Craig, L.C., and Stoner, E., On the removal of residual carboxylic acid groups from cellulose membranes and Sephadex, *Biochemistry* 11, 3559–3564, 1972.
12. Harris, M.J. and Craig, L.C., A study of the parameters which determine the conformation of linear polypeptides in solution by synthesis of models and determination of thin film dialysis rates, *Biochemistry* 13, 1510–1515, 1974.
13. Carlson, F.D., The application of intensity fluctuation spectroscopy to molecular biology, *Annu. Rev. Biophys. Bioeng.* 4, 243–264, 1975.
14. Tinoco, I. Jr., Michols, W., Maestrae, M.F., and Bustamante, C., Absorption, scattering, and imaging of biomolecular structures with polarized light, *Annu. Rev. Biophys. Biophys. Chem.* 16, 319–349, 1987.
15. Eisenberg, H., Thermodynamics and the structure of biological macromolecules. Rozhinkes mit mandolin, *Eur. J. Biochem.* 187, 7–22, 1990.
16. Herding, S.E., Sattelle, D.B., and Bloomfield, V.A., eds., *Laser Light Scattering in Biochemistry*, Royal Society of Chemistry, Cambridge, U.K., 1992.
17. Li-Chain, E.C., Methods to monitor process-induced changes in food proteins. An overview, *Adv. Exp. Med. Biol.* 434, 5–23, 1998.
18. Georgilis, Y. and Saenger, W., Light scattering studies on supersaturated protein solutions, *Sci. Prog.* 82, 271–294, 1998.
19. Aune, K.C., Molecular weight measurements by sedimentation equilibrium: Some common pitfalls and how to avoid them, *Methods Enzymol.* 48, 163–185, 1978.
20. Schuster, T.M. and Toedt, J.M., New revolutions in the evolution of analytical ultracentrifugation, *Curr. Opin. Struct. Biol.* 6, 650–658, 1996.
21. Hensley, P., Defining the structure and stability of macromolecular assemblies in solution: The re-emergence of analytical ultracentrifugation as a practical tool, *Structure* 4, 367–373, 1996.
22. Eisenberg, H., Analytical ultracentrifugation in a Gibbsian perspective, *Biophys. Chem.* 88, 1–9, 2000.

23. Lebowiz, J., Lewis, M.S., and Schuck, P., Modern analytical ultracentrifugation in protein science: A tutorial review, *Protein Sci.* 11, 2067–2079, 2002.

24. Howlett, G.J., Minton, A.P., and Rivas, G., Analytical ultracentrifugation for the study of protein association and assembly, *Curr. Opin. Chem. Biol.* 10, 430–436, 2006.

25. *Oxford English Dictionary*, Oxford University Press, Oxford, U.K., 2008.

26. Onnes, H.K., 1901, in *Arch. neérlandaises des Sci. exactes & nat.* VI. 874, as cited in *Oxford English Dictionary*, 2008.

27. Flory, P.J., *Principles of Polymer Chemistry*, Cornell University Press, Ithaca, NY, 1953.

28. McCabe, W.C. and Fisher, H.F., Measurement of the excluded volume of protein molecules by differential spectroscopy in the near infra red, *Nature* 207, 1274–1276, 1965.

29. Minton, A.P., Influence of excluded volume upon macromolecular structure and associations in "crowded" media, *Curr. Opin. Biotechnol.* 8, 65–69, 1997.

30. Minton, A.P., Implications of macromolecular crowding for protein assembly, *Curr. Opin. Struct. Biol.* 10, 34–39, 2000.

31. Despa, F., Orgill, D.P., and Lee, R.C., Molecular crowding effects on protein stability, *Ann. N.Y. Acad. Sci.* 1066, 54–66, 2005.

32. Konopka, M.C., Weisshaar, J.C., and Record, M.T. Jr., Methods of changing biopolymer volume fraction and cytoplasmic solute concentrations for in vivo biophysical studies, *Methods Enzymol.* 428, 487–504, 2007.

33. Nichol, J.W., Janado, M., and Winzor, D.J., The origin and consequences of concentration dependence in gel chromatography, *Biochem. J.* 133, 15–22, 1973.

34. Wan, P.J. and Adams, E.T. Jr., Molecular weights and molecular weight distribution from ultracentrifugation of nonideal solutions, *Biophys. Chem.* 5, 207–241, 1976.

35. Tang, L.H., Powell, D.R., Escott, B.M., and Adams, E.T. Jr., Analysis of various indefinite self-associations, *Biophys. Chem.* 7, 121–139, 1977.

36. Neal, B.L., Asthagiri, D., and Lenhoff, A.M., Molecular origins of osmotic second virial coefficients of proteins, *Biophys. J.* 75, 2469–2477, 1998.

37. Weatherly, G.T. and Pielak, G.J., Second virial coefficients as a measure of protein–osmolyte interactions, *Protein Sci.* 10, 12–16, 2001.

38. Ruppert, S., Sandler, S.L., and Lenhoff, A.M., Correlation between the second virial coefficient and the solubility of proteins, *Biotechnol. Prog.* 17, 182–187, 2001.

39. Tessier, P.M. and Lenhoff, A.M., Measurements of protein self-association as a guide to crystallization, *Curr. Opin. Biotechnol.* 14, 512–516, 2003.

40. Sear, R.P., Solution stability and variability in a simple model of globular proteins, *J. Chem. Phys.* 120, 998–1005, 2004.

41. Valente, J.J., Payne, R.W., Manning, M.C. et al., Colloidal behavior of proteins: Effects of the second virial coefficient on solubility, crystallization and aggregation of proteins in aqueous solution, *Curr. Pharm. Biotechnol.* 6, 427–436, 2005.

42. Paliwal, A., Asthagiri, D., Abras, D. et al., Light-scattering studies of protein solutions: Role of hydration in weak protein–protein interactions, *Biophys. J.* 89, 1564–1573, 2005.

43. Ruckenstein, E. and Shulgin, I.L., Effect of salts and organic additives on the solubility of proteins in aqueous solutions, *Adv. Colloid Interface Sci.* 123–126, 97–103, 2006.

44. Payne, R.W., Nayar, R., Tarantino, R. et al., Second virial coefficient determination of a therapeutic peptide by self-interaction chromatography, *Biopolymers* 84, 527–533, 2006.

45. Bajaj, H., Sharma, V.K., Badkar, A. et al., Protein structural conformation and not second virial coefficient relate to long-term irreversible aggregation of a monoclonal antibody and ovalbumin in solution, *Pharm. Res.* 23, 1382–1394, 2006.

46. Winzor, D.J., Dezczynski, M., Harding, S.E., and Wills, P.R., Nonequivalence of second virial coefficients from sedimentation equilibrium and static light scattering studies of protein solutions, *Biophys. Chem.* 128, 46–55, 2007.

47. Dumetz, A.C., Chockla, A.M., Kaler, E.W., and Lenhoff, A.M., Effects of pH on protein–protein interactions and implications for protein phase behavior, *Biochim. Biophys. Acta* 1784, 600–610, 2008.
48. Blattler, D.P. and Reisthel, F.J., Molecular weight determinations and the influence of gel density and protein shape in polyacrylamide gel electrophoresis, *J. Chromatogr.* 46, 286–292, 1970.
49. Chae, K.S. and Lenhoff, A.M., Computation of the electrophoretic mobility of proteins, *Biophys. J.* 68, 1120–1127, 1995.
50. Røgen, P. and Bohr, H., A new family of global protein shape descriptors, *Math. Biosci.* 182, 167–181, 2003.
51. He, L. and Niemeyer, B., A novel correlation for protein diffusion coefficients based on molecular weight and radius of gyration, *Biotechnol. Prog.* 19, 544–548, 2003.
52. Chang, B.H. and Bae, Y.C., Salting-out in the aqueous single-protein solution: The effect of shape factor, *Biophys. Chem.* 104, 523–533, 2003.
53. Schachman, H.K., *Ultracentrifugation in Biochemistry*, Academic Press, New York, 1954.
54. Haschemeyer, R.H. and Haschemeyer, A.E.V., *Proteins. A Guide to Study by Physical and Chemical Methods*, John Wiley & Sons, New York, 1973.
55. Errington, N. and Rowe, A.J., Probing conformation and conformational change in proteins is optimally undertaken in relative modes, *Eur. Biochem. J.* 32, 511–517, 2003.
56. Christensen, L.L.H., Theoretical analysis of protein concentration determination using biosensor technology under conditions of partial mass transport limitation, *Anal. Biochem.* 245, 153–164, 1997.
57. Kyte, J., *Structure in Protein Chemistry*, 2nd edn., Garland Science, New York, 2007.
58. Hauge, J.G., Pressure-induced dissociation of ribosomes during ultracentrifugation, *FEBS Lett.* 17, 168–172, 1971.
59. Balerlein, R. and Infante, A.A., Pressure-induced dissociation of ribosomes, *Methods Enzymol.* 30, 328–345, 1975.
60. Dicamelli, R.F., Balbinder, E., and Lebowitz, J., Pressure effects on the association of the α and β_2 subunits of tryptophan synthetase from *Escherichia coli* and *Salmonella typhimurium*, *Arch. Biochem. Biophys.* 155, 315–324, 1973.
61. van Diggelen, O.P., Oostrom, H., and Bosch, L., The association of ribosomal subunits of *Escherichia coli*. 2. Two types of association products differing in sensitivity to hydrostatic pressure generated during centrifugation, *Biophys. Chem.* 3, 21–34, 1975.
62. Powell, J.T. and Brew, K., A comparison of the interactions of galactosyltransferase with a glycoprotein substrate (ovalbumin) and α-lactalbumin, *J. Biol. Chem.* 251, 3653–3663, 1976.
63. Poto, E.M. and Wood, H.G., Association–dissociation of transcarboxylase, *Biochemistry* 16, 1949–1955, 1977.
64. Marcum, J.M. and Borisy, G.G., Sedimentation velocity analyses of the effect of hydrostatic pressure on the 30 S microtubule protein oligomer, *J. Biol. Chem.* 253, 2852–2857, 1978.
65. Potschka, M. and Schuster, T.M., Determination of reaction volumes and polymer distribution characteristics of tobacco mosaic virus coat protein, *Anal. Biochem.* 161, 70–79, 1987.
66. Kim, Y.S., Randolph, T.W., Seefeldt, M.B., and Carpenter, J.F., High-pressure studies on protein aggregates and amyloid fibrils, *Methods Enzymol.* 413, 237–253, 2006.
67. Stanicová, J., Sedlák, E., Musatov, A., and Robinson, N.C., Differential stability of dimeric and monomeric cytochrome c oxidase exposed to elevated hydrostatic pressure, *Biochemistry* 46, 7146–7152, 2007.
68. Rocha, C.B., Suarez, M.C., and Yu, A., Volume and free energy of folding for troponin C C-domain: Linkage to ion binding and N-domain interaction, *Biochemistry* 47, 5047–5058, 2008.

69. Barstow, B., Ando, N., Kim, C.U., and Gruner, S.M., Alteration of citrine structure by hydrostatic pressure explains the accompanying spectral shift, *Proc. Natl. Acad. Sci. USA* 105, 13362–13366, 2008.

70. Lu, Y., Harding, S.E., Turner, A. et al., Effect of pegylation on the solution conformation of antibody fragments, *J. Pharm. Sci.* 97, 2062–2079, 2008.

71. Wingfield, P., Graber, P., Moonen, P. et al., The conformation and stability of recombinant-derived granulocyte-macrophage colony stimulating factors, *Eur. J. Biochem.* 173, 65–72, 1988.

72. Frénoy, J.P., Effect of physical environment on the conformation of ricin. Influence of low pH, *Biochem. J.* 240, 221–226, 1986.

73. Yamada, Y., Yajima, T., Fujiwara, K. et al., Helical and expanded conformation of equine β-lactoglobulin in the cold-denatured state, *J. Mol. Biol.* 350, 338–348, 2005.

74. Kornblatt, J.A. and Schuck, P., Influence of temperature on the conformation of canine plasminogen: An analytical ultracentrifugation and dynamic light scattering study, *Biochemistry* 44, 13122–13131, 2005.

75. Race, P.R., Solovyova, A.S., and Banfield, M.J., Conformation of the EPEC Tir protein in solution: Investigating the impact of serine phosphorylation at positions 434/463. *Biophys. J.* 93, 586–596, 2007.

76. Parkinson, E.J., Morris, M.B., and Bastiras, S., Acid denaturation of recombinant porcine growth hormone: Formation and self-association of folding intermediates, *Biochemistry* 39, 12345–12356, 2000.

77. Lafitte, D., Heck, A.J., Hill, T.J. et al., Evidence of noncovalent dimerization of calmodulin, *Eur. J. Biochem.* 261, 337–344, 1999.

78. Ward, L.D., Zhang, J.G., Checkley, G. et al., Effect of pH and denaturants on the folding and stability of murine interleukin-6, *Protein Sci.* 2, 1291–1300, 1993.

79. Lowman, H.B., Fairbrother, W.J., Slagel, P.B. et al., Monomeric variants of IL-8: Effects of side chain substitutions and solution conditions upon dimer formation, *Protein Sci.* 6, 598–608, 1997.

80. Pessina, A., Thomas, R.M., Palmieri, S., and Luisi, P.L., An improved method for the purification of myrosinase and its physicochemical characterization, *Arch. Biochem. Biophys.* 280, 383–389, 1990.

81. Staples, R.C., McCarthy, W.J., and Stahmann, M.A., Heat stabilities of acid phosphatases from pinto bean leaves, *Science* 149, 1248–1249, 1965.

82. Lustig, A., Engel, A., Tsiotis, G. et al., Molecular weight determination of membrane proteins by sedimentation equilibrium at the sucrose or nycodenz-adjusted density of the hydrated detergent micelle, *Biochim. Biophys. Acta* 1464, 199–206, 2000.

83. Pavlov, G., Finet, S., Tatarenko, K. et al., Conformation of heparin studied with macromolecular hydrodynamic methods and x-ray scattering, *Eur. Biophys. J.* 32, 437–449, 2003.

84. van Holde, K.E., Sedimentation equilibrium and the foundations of protein chemistry, *Biophys. Chem.* 108, 5–8, 2004.

85. Burgess, N.K., Stanley, A.M., and Fleming, K.G., Determination of membrane protein molecular weights and association equilibrium constants using sedimentation equilibrium and sedimentation velocity, *Methods Cell Biol.* 84, 181–211, 2008.

86. Philo, J.S., Is any measurement method optimal for all aggregate sizes and types?, *AAPS J.* 8, E564–E571, 2006.

87. Liu, J., Andya, J.D., and Shire, S.J., A critical review of analytical ultracentrifugation and field flow fractionation methods for measuring protein aggregation, *AAPS J.* 8, E580–E589, 2006.

88. Berkowitz, S.A., Role of analytical ultracentrifugation in assessing the aggregation of protein biopharmaceutical, *AAPS J.* 8, E590–E605, 2006.

89. Behlke, J. and Ristau, O., Analysis of protein self-association under conditions of the thermodynamic non-ideality, *Biophys. Chem.* 87, 1–13, 2000.
90. Behlke, J. and Ristau, O., Sedimentation equilibrium: A valuable tool to study homologous and heterologous interactions of proteins or proteins and nucleic acids, *Eur. J. Biochem.* 32, 427–431, 2003.
91. Stafford, W.F. and Sherwood, P.J., Analysis of heterologous interacting systems by sedimentation velocity: Curve fitting algorithms for estimation of sedimentation coefficients, equilibrium and kinetic constants, *Biophys. Chem.* 108, 231–243, 2004.
92. Balbo, A., Minor, K.H., Velikovsky, C.A. et al., Studying multiprotein complexes by multisignal sedimentation velocity analytical ultracentrifugation, *Proc. Natl. Acad. Sci. USA* 102, 81–86, 2005.
93. Brown, P.H., Balbo, A., and Schuck, P., Characterizing protein–protein interactions by sedimentation velocity analytical ultracentrifugation, *Curr. Protoc. Immunol.* 81, 18.15.1–18.15.39, 2008.
94. Gerwing, J., Dolman, C.E., Reichmann, M.E., and Bains, H.S., Purification and molecular weight determination of *Clostridium botulinum* type E toxin, *J. Bacteriol.* 88, 216–219, 1964.
95. Varley, P.G., Brown, A.J., Dawkes, H.C., and Burns, N.R., A case study and use of sedimentation equilibrium analytical ultracentrifugation as a tool for biopharmaceutical development, *Eur. Biophys. J.* 25, 437–443, 1997.
96. Fowler, S., Byron, O., Jumel, K. et al., Novel configurations of high molecular weight species of the pertussis toxin vaccine component, *Vaccine* 21, 2678–2688, 2003.
97. Sukumar, M., Doyle, B.L., Combs, J.L., and Pekar, A.H., Opalescent appearance of an IgG1 antibody of high concentrations and its relationships to noncovalent associations, *Pharm. Res.* 21, 1087–1093, 2004.
98. Harding, S.E., Challenges for the modern analytical ultracentrifuge analysis of polysaccharides, *Carbohydr. Res.* 340, 811–826, 2005.
99. Koschella, A., Inngjerdingen, K., Paulsen, B.S. et al., Unconventional methyl galactan synthesized via the thexyldimethylsilyl intermediate: Preparation, characterization, and properties, *Macromol. Biosci.* 8, 96–105, 2008.
100. Demeler, B., Brookes, E., and Nagel-Steger, L., Analysis of heterogeneity in molecular weight and shape by analytical ultracentrifugation using parallel distributed computing, *Methods Enzymol.* 454, 87–113, 2009.

7 Use of Differential Scanning Calorimetry to Measure Conformational Change in Proteins and Other Biomacromolecules

Differential scanning calorimetry (DSC) is a technique used to measure heat flow in or out of a sample undergoing a physical transition with heat flow output based on differences in temperatures between sample and reference.[1-7] A midpoint of a thermal transition (T_M) can be obtained; this point is analogous to the melting point of nucleic acids. The transition can be seen as a sharp peak or a broad peak, reflecting the kinetic pathway of denaturation; temperature scan speed can also have an effect. The area under the transition peak yields the transition enthalpy. The peak can reflect an endothermic process or an exothermic process; "melting" (thermal denaturation or unfolding) is endothermic while crystallization is exothermic. The process of thermal denaturation is usually but not always irreversible[8-13] and hysteresis is not uncommon.[14-18] DSC is a method for quantifying the thermal denaturation of proteins and a measure of thermal stability.[19] As such, it has proved useful in the development of formulation strategies for biopharmaceuticals.[20-31] DSC has been used on complex protein mixtures such as human blood plasma,[32,33] complex biological structures,[34,35] and food products.[36,37] It is noted that the scanning rate (rate of temperature increase) can influence the quality of results, reflecting the various kinetic pathways for the unfolding of the biomacromolecules.[38-44]

The application of DSC for evaluation of biopharmaceutical formulation and stability has been mentioned in the discussion previously. The specific application of DSC to the characterization of monoclonal antibodies is presented in Table 7.1. It should be noted that DSC can be applied to solid and liquid samples. Table 7.2 provides some examples of the application of DSC to the study of conformational change in biomacromolecules such as nucleic acids, proteins, and carbohydrates. Table 7.3 provides some examples of the use of DSC as a stability-indicating assay.[112-115]

TABLE 7.1

Use of Differential Scanning Calorimetry for the Characterization of Monoclonal Antibodies

Monoclonal Antibody	Description of Study	Ref.
Deacetylvinblastine coupled to a monoclonal antibody	A cytotoxic vinca alkaloid is coupled to antibody via periodate oxidized carbohydrate; formulation at various moisture levels evaluated via DSC, RP-HPLC, size-exclusion chromatography. DSC demonstrated samples were above glass transition temperature at 40°C	[45]
Monoclonal antibody	Evaluation of sucrose formulation and trehalose formulations using DSC and dielectric relaxation spectroscopy to determine glass transition temperature using solid samples. Aggregation evaluated by size-exclusion chromatography	[46]
Anti-fluorescein monoclonal antibody	Use of DSC to evaluate importance of conformational changes outside of the variable region	[47]
Murine Mab against HIV type 1 capsid protein p24	Use of DSC, CD, and intrinsic fluorescence to evaluate acid-induced conformation changes in intact antibody, Fab, and Fc regions	[48]
Human anti-IL8 monoclonal antibody	Study to evaluate excipients including histidine on stability of monoclonal antibody. DSC used to determine glass transition temperature of solid sample and melting temperature of liquid samples. Histidine was found to have beneficial effects on formulation stability	[49,50]
IgG1 recombinant monoclonal antibody	DSC used to demonstrate stabilizing effect of posttranslational modification with oligosaccharides with monoclonal antibody; use of derivative fragments suggest stabilization effect is through CH(2) domains	[51]
Concentrated monoclonal antibody solutions	DSC and CD suggest stabilization at higher protein concentrations while spectroscopic techniques suggest decrease in stability. Suggest that decrease in stability due to aggregation in concentrated solution not to intrinsic stability	[52]
Humanized monoclonal antibodies	Use of DSC and CD to evaluation effect of low pH on conformation. Limited conformational change is observed at low pH (2.7–3.9). Adjustment to pH 6.0 from 3.5 restored native conformation; return to native conformation is not observed with pH 2.7	[53]
Mab I	DSC used to evaluate conformational stability of Mab I. Glycerol or sucrose increased thermal stability but had a negative effect on the isomerization of Asp 32	[54]
Monoclonal antibody	DSC used for development of lyophilization cycle for monoclonal antibody formulated without a crystalline bulking agent	[55]
Humanized IgG1 monoclonal antibodies	DSC used to evaluate thermal stability of three humanized IgG1 monoclonal antibodies and their Fab and Fc fragments. The measured enthalpy for the three Fab fragments was similar; the melting point showed considerable variability. Stability of Fab fragment may be critical for long-term storage	[56]

TABLE 7.2

Use of Differential Scanning Calorimetry to Measure Conformational Change in Biological Macromolecules

Macromolecule	Description of Study	References
DNAse and lactate dehydrogenase	Heating enzyme samples below T_M showed structural alterations as measured by Fourier transform-Raman spectroscopy and loss of enzyme activity	[12]
Various proteins subjected to γ-irradiation	γ-Irradiation of solid phase and solution proteins at ambient temperature and −80°C decreases onset temperature and T_M	[57]
Albumin, fibrinogen	Use of DSC to evaluate conformational changes of proteins adsorbs on low-temperature isotropic carbon[a]	[58]
Cytochrome C	Maleylation decreased T_M and ΔH (enthalpy change). The effect on T_M was observed at 50% modification while the effect of enthalpy was observed only a higher (77%) modification	[59]
Fibronectin	Little effect of Ca^{2+} or ionic strength on thermal transitions: There was no effect of pH from 7.0 to 10.0; at pH 10.5 and 11, the transition peaks were mostly missing; transitions were missing below pH 5.0 as well	[60]
Various	DSC demonstrated a conformational change in the presence of 1-propanol under conditions used for RP-HPLC. This conformational change was also demonstrated with CD, UV-VIS spectroscopy, and fluorescence	[61]
Protoxin from *Bacillus thuringiensis*	DSC demonstrates conformational change on activation of protoxin by limited proteolysis	[62]
Streptokinase	Demonstration of two-domain structure by DSC; domain structure lost below pH and in the presence of urea	[63]
RNase T1	DSC used to evaluate effect of pH on the conformational stability of ribonuclease T1; the stabilization of the protein by NaCl and $MgCl_2$	[64]
Glucoamylase 1 from *Aspergillus niger*	Use of DSC to evaluate the effect of β-cyclodextrin on the thermal denaturation pathway of glucoamylase 1; evaluate to the effect of catalytic domain on thermal stability of binding domain	[65]
HMW-Kininogen[b]	Evaluation of conformational stability of factor XI-binding domains and prekallikrein using large peptides prepared by condensation of long-chain peptides using orthogonal chemistry protocols	[66]
Adrenodoxin	DSC studies of holo adrenodoxin and apo adrenodoxin demonstrate importance of iron–sulfur cluster for the conformational stability of adrenodoxin	[67]
α-Crystallin	DSC results dependent on scan rate	[68]
Double-stranded DNA	Use of DSC to study thermal denaturation of double-stranded DNA; use of Raman spectroscopy revealed three separate transition: base unstacking, rearrangement of deoxyribose-phosphate moiety, and change in environment of phosphate groups	[69]

(continued)

TABLE 7.2 (continued)

Use of Differential Scanning Calorimetry to Measure Conformational Change in Biological Macromolecules

Macromolecule	Description of Study	References
Human serum albumin	Comparison of DSC, intrinsic fluorescence, and hydrolase activity as measures of thermal unfolding of protein	[70]
Variant lysozyme	Use of DSC to study variant lysozyme proteins; water molecule in a cavity created in the interior of a protein increases stability	[71]
β-Lactoglobulin	Aggregation process of β-lactoglobulin studied by DSC, binding of fluorescent probe (1-anilino-6-naphthalenesulphonic acid, ANS) and reaction with 2,2′-dithio-bis-(2-nitrobenzoic acid (DTNB))	[72]
Human factor VIIa	DSC used to study stability of factor VIIa and derivatives. Ca^{2+} stabilizes factor VIIa. Calcium ion effect also studied with intrinsic fluorescence and guanidine denaturation	[73]
Lysozyme variants	Use of DSC to study the relationship between changes in conformational stability and changes in accessible surface area of hydrophobic residues due to mutations	[74]
Monoclonal antibodies	Use of DSC to study the differences in monoclonal antibody stability; demonstration of subclass-independent structural variations associated with different thermodynamic stability (binding of antigen induced conformational change in Fc domain demonstrated by protein A-binding affinity	[75]
Metmyoglobin and heparin	DSC used to characterize complex formation between metmyoglobin and heparin; complex formation lowers transition temperature; heparin also prevents aggregation at the isoelectric point resulting in an increase in reversibility of denaturation	[76]
Bovine serum albumin	DSC used to measure conformational change on pyridoxal phosphate	[77]
α-Lactalbumin	Use of DSC to measure the effect of calcium ions and sodium ions on the stability of lactalbumin; also use of CD and isothermal titration microcalorimetry	[78]
Monoclonal antibody and Fab and Fc fragments	DSC used to study pH-induced conformational changes in a monoclonal antibody and its Fab and Fc fragments	[48]
RNase	Use of DSC to determine heat capacity changes during the folding of ribonuclease A	[79]
B-Lactoglobulin	Modification with carboxymethylated dextran increases thermal stability as evaluation by DSC. The derivatives also demonstrated reduced immunogenicity	[80]
Bovine serum albumin	DSC used to evaluate conformation changes secondary to albumin binding to silica or polystyrene surfaces	[81]
Cytochrome *c*	Development of temperature-gradient gel electrophoresis as a technique to study the thermal unfolding of proteins and comparison with DSC	[82]

TABLE 7.2 (continued)

Use of Differential Scanning Calorimetry to Measure Conformational Change in Biological Macromolecules

Macromolecule	Description of Study	References
Human lysozyme variants	Use of DSC to evaluate effect of deletion or replacement mutations on the stability of human lysozyme	[83,84]
Yeast enolase 1	Use of DSC to evaluate effect of metal ions, substrates, substrate analogues, and chaotropic agents on thermal stability of yeast enolase 1	[85]
Glucose oxidase from *Aspergillus niger*	DSC used to evaluate the effect of monovalent cations on the stability of glucose oxidase from *Aspergillus niger*. The results area compared with those obtained from limited proteolysis, spectroscopy, and size-exclusion chromatography. Monovalent cations stabilized against thermal denaturation and denaturation but destabilized pH denaturation	[86]
IgG	DSC used to evaluate the effect of glycosylation in the Fc domain on IgG stability and binding to Fc gamma receptor	[87]
Spinach Rubisco[c]	DSC used to show that conformational change (and activity loss) was reversible with heating to 45°C, partially reversible from 45°C to 60°C, and irreversible above 60°C. Results confirmed with CD and fluorescence showed a three-stage thermal unfolding for Rubisco	[88]
Plasmid DNA-cationic polymer complex	DSC used to evaluate stability of complexes between plasmid DNA and poly(dimethylamino)ethyl methacrylate; complex formation raises thermal transition temperature of plasmid DNA	[89]
Glycosidases	Use of DSC to evaluate conformational stability of cold-adapted (psychrotolerant) and heat-adapted (thermotolerant) glycosidases; combined with intrinsic fluorescence, susceptibility to chaotropic agents, and fluorescence quenching. The cold-adapted enzyme has low stability while the thermotolerant enzyme has higher stability	[90]
Staphylococcal nuclease	Evaluation of stability as function of temperature and pressure	[91]
Trypsin	Use of DSC to evaluate thermal stability in the presence of various stabilizers or chaotropic agents	[92]
Phenylalanine hydroxylase	DSC used to evaluate the effect of phosphorylation on the conformation of phenylalanine hydroxylase	[93]
Hens egg white lysozyme (HEWL)	Used DSC to evaluated structural changes in HEWL absorbed and desorbed from magnetic particles	[94]
Hens egg white lysozyme (HEWL)	Used DSC to evaluate thermal denaturation of lysozyme in combination with Raman spectroscopy	[95]
Bovine serum albumin; α-lactalbumin	Use of DSC to evaluation binding of protein to Triton X-100; monomer form of detergent does not affect stability while micellar form destabilizes proteins	[96]

(continued)

TABLE 7.2 (continued)
Use of Differential Scanning Calorimetry to Measure Conformational Change in Biological Macromolecules

Macromolecule	Description of Study	References
Human serum albumin (HSA)	Use of DSC, CD, and UV-VIS spectroscopy to study reversible, "pre-denaturation" thermal conformational changes in HSA. The transition from 22°C to 55°C induces reversible conformation changes	[97]
Monoclonal antibody	Exposure to low pH causes limited conformational change as assessed by CD and sedimentation velocity. DSC decreased melting temperature at low pH (2.7–3.9)	[53]
β-Lactoglobulin	Use of DSC and CD to evaluation conformational changes in β-lactoglobulin resulting from glycationation (maltopentaose) and phosphorylation (pyrophosphate) during dry heating	[98]
Trypsin	Use of DSC and CD to evaluation conformational change caused by mutagenesis in the activation domain of trypsin	[99]
α_2-Macroglobulin	Use of DSC to evaluate conformational change occurring on the "activation" of α_2-macroglobulin, Both cleavage of the bait region and covalent incorporation of protein ligands into thioester linkage have important roles in stability of complexes	[100]
Lysozyme	Use of T (ZERO) modulated temperature differential scanning calorimetry (MDSC); enables separation of overlapping transitions	[101]
Myoglobin	DSC demonstrates modification of myoglobin by 4-hydroxy-2-nonenal (4-HNE) results in decreased melting temperature	[102]
Lactoferrins	DSC studies on lactoferrins derived from various species; iron stabilized apolactoferrins from all species; human lactoferrin demonstrated the greatest heat stability	[103]

[a] There are several studies on the use of DSC to study protein adsorption to surfaces (see Welzel, P.B., Investigation of adsorption-induced structural changes of proteins at solid/liquid interfaces by differential scanning calorimetry, *Thermochim. Acta* 382, 175–188, 2002; Serro, A.P., Bastos, M., Pessoa, J.C., and Saramago, B., Bovine serum albumin conformational changes upon adsorption on titania and on hydroxyapaptite and their relation with biomineralization, *J. Biomed. Mater. Res. A* 70A, 420–427, 2004; Brandes, N., Weizel, P.B., Werner, C., and Kroh, L.W., Adsorption-induced conformational changes of proteins onto ceramic particles: Differential scanning calorimetry and FTIR analysis, *J. Colloid Interface Sci.* 299, 56–69, 2006).

[b] HMW-kininogen, high-molecular weight kininogen.

[c] Rubisco, ribulose bisphosphate carboxylase/oxygenase.

TABLE 7.3

Some Examples of the Use of Differential Scanning Calorimetry as a Stability Indicating Assay

Biopharmaceutical	Description of Study	References
LADH and LDH[a]	Use of DSC to evaluate stability of enzymes in solution and bound to agarose; in the presence and absence of coenzymes and coenzyme fragments	[104]
Human serum transferrin (HST), human lactoferrin (HLF), and rabbit serum transferring (RST)	Removal of iron lead to thermal destabilization of HST, HLF, and RST; secondary structure not altered by iron binding or release; Fourier transform infrared spectroscopy demonstrates lost of secondary structure on thermal denaturation	[105]
Recombinant factor XIII	Use of DSC to elucidate the thermal denaturation of factor XIII and various derived domain structures	[106]
α-Chymotrypsin	Use of DSC to identify metal-stabilizing binding sites and metal-destabilizing binding sites in α-chymotrypsin	[107]
Recombinant DNase	Use of DSC to measure denaturation temperature and enthalpy and the effect of additives such as chaotropic agents, divalent cations, and disaccharides. Ca^{2+} stabilizes DNase while Mg^{2+}, Mn^{2+}, and Zn^{2+} destabilize the protein	[108]
Serum albumins	DSC used to compare the thermal stability of five serum albumins (human, bovine, dog, rabbit, and rat); there are species differences in conformational stability and mechanism of unfolding pathway	[109]
Blood coagulation factor VIIa	DSC measures the effect of Ca^{2+} on thermal stability of factor VIIa	[73]
Bovine serum albumin (BSA)	DSC used to measure the effect of various sugars (sucrose, trehalose, dextrans) on the structural stability (glass transition temperature) of BSA in the solid state. This is an example of the direct application of DSC to the study of the stability of formulated final drug product	[110]
Insulin	DSC used to study the mobility of insulin in lyophilized preparations [trehalose and poly(vinylpyrrolidone)]	[111]
Human serum albumin	DSC used to measure effect of NaCl on albumin formulations contains mannitol using different freeze-drying protocols	[22]
N/A	DSC study of frozen solutions of excipients [maltose, trehalose, sucrose, dextran 40, poly(vinylpyrrolidone)]	[25]

[a] LADH and LDH, liver alcohol dehydrogenase and lactate dehydrogenase.

REFERENCES

1. Bershtein, V.A. and Egorov, V.M., *Differential Scanning Calorimetry of Polymers: Physics, Chemistry, Analysis, Technology*, Ellis Horwood, New York, 1994.
2. Höhne, G., Hemminger, W., and Fllamershim, H.-J., *Differential Scanning Calorimetry*, Springer-Verlag, Berlin, Germany, 1996.
3. Haines, P.J., *Principles of Thermal Analysis and Calorimetry*, Royal Society of Chemistry, Cambridge, U.K., 2002.
4. Rigo, A. and Collins, R., Differential scanning calorimetry and differential thermal analysis, in *Encyclopedia of Analytical Chemistry. Applications, Theory, and Instrumentation*, Vol. 15, ed. R.A. Meyers, John Wiley & Sons, Ltd., Chichester, U.K., pp. 12147–13179, 2000.
5. Brandts, J.F. and Lin, L.N., Study of strong to ultratight protein interactions using differential scanning calorimetry, *Biochemistry* 29, 6927–6940, 1990.
6. Sanchez-Ruiz, J.M., Differential scanning calorimetry of proteins, *Subcell. Biochem.* 24, 133–176, 1995.
7. Spink, C.H., Differential scanning calorimetry, *Methods Cell Biol.* 84, 115–141, 2004.
8. Velicelebi, G. and Sturtevant, J.M., Thermodynamics of the denaturation of lysozyme in alcohol–water mixtures, *Biochemistry* 18, 1180–1186, 1979.
9. Kasimova, M.R., Milstein, S.J., and Freire, E., The conformational equilibrium of human growth hormone, *J. Mol. Biol.* 277, 409–418, 1998.
10. Lin, S.Y., Hsieh, T.F., Wei, Y.S., and Li, M.J., Mechanical compression affecting the thermal-induced conformational stability and denaturation temperature of human fibrinogen, *Int. J. Biol. Macromol.* 37, 127–133, 2005.
11. Forbes, R.T., Barry, B.W., and Elkordy, A.A., Preparation and characterization of spray-dried and crystallized trypsin: FT-Raman study to detect protein denaturation after thermal stress, *Eur. J. Pharm. Sci.* 30, 315–323, 2007.
12. Elkordy, A.A., Forbes, R.T., and Barry, B.W., Study of protein conformational stability and integrity using calorimetry and FT-Raman spectroscopy correlated with enzymatic activity, *Eur. J. Pharm. Sci.* 33, 177–190, 2008.
13. Sawano, M., Yamamoto, H., Ogasahara, K. et al., Thermodynamic basis of the stabilities of three CutAls from *Pyrococcus horikoshii*, *Thermus thermophilius*, and *Oryza sativa*, with unusually high denaturation temperatures, *Biochemistry* 47, 721–730, 2008.
14. Na, G.C., Monomer and oligomer of type I collagen: Molecular properties and fibril assembly, *Biochemistry* 28, 7161–7167, 1989.
15. Fidanza, M., Dentini, M., Crescenzi, V., and Del Vecchio, P., Influence of charged groups on the conformational stability of succinogycan in dilute aqueous solution, *Int. J. Biol. Macromol.* 11, 372–376, 1989.
16. Mazen, F., Milas, M., and Rinaudo, M., Conformational transition of native and modified gellan, *Int. J. Biol. Macromol.* 26, 109–118, 1999.
17. Benitez-Cardoza, C.G., Rojo-Dominguez, A., and Hernández-Arana, A., Temperature-induced denaturation and renaturation of triosephosphate isomerase from *Saccharomyces cerevisiae*: Evidence of dimerization coupled to refolding of the thermally unfolded protein, *Biochemistry* 40, 9059–9058, 2001.
18. Kocherbitov, V. and Arnebrant, T., Hydration of thermally denatured lysozyme studied by sorption calorimetry and differential scanning calorimetry, *J. Phys. Chem. B* 110, 10144–10150, 2006.
19. Yadav, S. and Ahmad, F., A new method for the determination of stability parameters of proteins from their heat-induced denaturation curves, *Anal. Biochem.* 283, 207–213, 2000.

20. Liao, Y.H., Brown, M.B., and Martin, G.P., Investigation of the stabilization of freeze-dried lysozyme and the physical properties of the formulations, *Eur. J. Pharm. Biopharm.* 58, 15–24, 2004.
21. Tang, X.C. and Pikal, M.J., The effect of stabilizers and denaturants on the cold denaturation temperatures of proteins and implications for freeze-drying, *Pharm. Res.* 22, 1167–1175, 2005.
22. Hawe, A. and Friess, W., Physico-chemical lyophilization behavior of mannitol, human serum albumin formulations, *Eur. J. Pharm. Sci.* 28, 224–232, 2006.
23. Mosharraf, M., Malmberg, M., and Fransson, J., Formulation, lyophilization and solid-state properties of a pegylated protein, *Int. J. Pharm.* 336, 215–232, 2007.
24. Kawai, K. and Suzuki, T., Stabilizing effect of four types of disaccharide on the enzymatic activity of freeze-dried lactate dehydrogenase: Step by step evaluation from freezing to storage, *Pharm. Res.* 24, 1883–1890, 2007.
25. Nesarikar, V.V. and Nassar, M.N., Effect of cations and anions on glass transition temperatures in excipient solutions, *Pharm. Dev. Technol.* 12, 259–264, 2007.
26. Pikal, M.J., Rigsbee, D.R., and Roy, M.L., Solid state chemistry of proteins: I. Glass transition behavior in freeze dried disaccharide formulations of human growth hormone (hGH), *J. Pharm. Sci.* 96, 2765–2776, 2007.
27. Liu, W. and Guo, R., Effects of Triton X-100 nanoaggregates on dimerization and antioxidant activity of morin, *Mol. Pharm.* 5, 588–597, 2008.
28. Salnikova, M.S., Middaugh, C.R., and Rytting, J.H., Stability of lyophilized human growth hormone, *Int. J. Pharm.* 358, 108–113, 2008.
29. Schüle, S., Schulz-Fademrecht, T., Garidel, P. et al., Stabilization of IgG1 in spray-dried powders for inhalations, *Eur. J. Pharm. Biopharm.* 69, 793–807, 2008.
30. Chang, B.S. and Randall, C.S., Use of subambient thermal to optimize protein lyophilization, *Cryobiology* 29, 632–656, 1992.
31. Biltonen, R.L. and Freire, E., Thermodynamic characterization of conformational states of biological macromolecules using differential scanning calorimetry, *CRC Crit. Rev. Biochem.* 5, 85–124, 1978.
32. Garbett, N.C., Miller, J.C., Jenson, A.B. et al., Calorimetric analysis of the plasma proteome, *Semin. Nephrol.* 27, 621–626, 2007.
33. Garbett, N.C., Miller, J.C., Jenson, J.J. et al., Calorimetry outside the box: A new window into the plasma proteome, *Biophys. J.* 94, 1377–1383, 2008.
34. Tzou, D.L., Lee, S.M., and Yeung, H.N., Temperature dependence and phase transition of proton relaxation of hydrated collagen in intact beef tendon specimens via cross-relaxation spectroscopy, *Magn. Reson. Med.* 37, 359–365, 1997.
35. Giannini, S., Buda, R., Di Caprio, F. et al., Effects of freezing on the biomechanical and structural properties of human posterior tibial tendons, *Int. Orthop.* 32, 145–151, 2008.
36. Xie, F., Dowell, F.E., and Sun, X.S., Using visible and near-infrared reflectance spectroscopy and differential scanning calorimetry to study starch, protein, and temperature effects on bread staling, *Cereal Chem.* 81, 249–254, 2004.
37. Wang, X.-S., Tang, C.-H., Li, B.-S. et al., Effects of high pressure treatment on some physicochemical and functional properties of soy protein isolates, *Food Hydrocoll.* 22, 560–567, 2008.
38. Grinberg, V.Y., Burova, T.V., Haertlé, T., and Tolstoguzov, V.E., Interpretation of DSC data on protein denaturation complicated by kinetic and irreversible effects, *J. Biotechnol.* 79, 269–280, 2000.
39. Weijers, M., Barneveld, P.A., Stuart, M.A.C., and Visschers, R.W., Heat-induced denaturation and aggregation of ovalbumin at neural pH described by irreversible first-order kinetics, *Protein Sci.* 12, 2693–2703, 2003.

40. Ding, Y.W., Ye, X.D., and Zheng, G.Z., Microcalorimetric investigation on aggregation and dissolution of poly(*N*-isopropylacrylamide) chains in water, *Macromolecules* 38, 904–908, 2005.

41. Remmele, R.L., Erik, J.Z.V., Diarmavaram, V. et al., Scan-rate-dependent melting transitions of interleukin-1 receptor (type II): Elucidation of meaningful thermodynamic and kinetic parameters of aggregation acquired from DSC simulations, *J. Am. Chem. Soc.* 127, 8328–8339, 2005.

42. Bao, L.D., Chatterjee, S., Lohmer, S. et al., An irreversible and kinetically controlled process: Thermal induced denaturation of L-2-hydroxyisocaproate dehydrogenase from *Lactobacillus confuses*, *Protein J.* 26, 143–151, 2007.

43. Bodnar, M.A. and Britt, B.M., Transition state characterization of the low- to physiological-temperature nondenaturational conformational change in bovine adenosine deaminase by slow scan rate differential scanning calorimetry, *J. Biochem. Mol. Biol.* 39, 167–170, 2006.

44. Davodi, J., Wakarchuk, W.W., Surewicz, W.K., and Carey, P.R., Scan-rate dependence in protein calorimetry: The reversible transitions of *Bacillus circulans* zylanase and a disulfide-bridge mutant, *Protein Sci.* 7, 1538–1544, 1998.

45. Roy, M.L., Pikal, M.J., Rickard, E.C., and Maloney, A.M., The effects of formulation and moisture on the stability of a freeze-dried monoclonal antibody-vinca conjugate: A test of the WLF glass transition temperature, *Dev. Biol. Stand.* 74, 323–339, 1992.

46. Duddu, S.P. and Dal Monte, P.R., Effect of glass transition temperature on the stability of lyophilized formulations containing a chimeric therapeutic monoclonal antibody, *Pharm. Res.* 14, 591–595, 1997.

47. Mummert, M.E., and Voss, E.W. Jr., Effects of secondary forces on the ligand binding and conformational state of antifluorescein monoclonal antibody 9–40, *Biochemistry* 36, 11918–11922, 1997.

48. Welfle, K., Misselwitz, R., Hausdorf, G. et al., Conformation, pH-induced conformational changes, and thermal unfolding of anti-p24 (HIV-1) monoclonal antibody CB4-1 and its Fab and Fc fragments, *Biochim. Biophys. Acta* 1431, 120–131, 1999.

49. Chen, B., Bautista, R., Yu, K. et al., Influence of histidine on the stability and physical properties of a fully human antibody in aqueous and solid forms, *Pharm. Res.* 20, 1952–1960, 2003.

50. Tian, F., Sane, S., and Rytting, J.H., Calorimetric investigation of protein/amino acid interactions in the solid state, *Int. J. Pharm.* 310, 175–186, 2006.

51. Liu, H., Bulseco, G.G., and Sun, J., Effect of posttranslational modifications of the thermal stability of a recombinant monoclonal antibody, *Immunol. Lett.* 106, 144–153, 2006.

52. Harn, N., Allan, C., Oliver, C., and Middaugh, C.R., Highly concentrated monoclonal antibody solutions: Direct analysis of physical structure and thermal stability, *J. Pharm. Sci.* 96, 532–546, 2007.

53. Ejima, D., Tsumoto, K., Fukada, H. et al., Effects of acid exposure on the conformation, stability, and aggregation of monoclonal antibodies, *Proteins* 66, 954–962, 2007.

54. Wakankar, A.A., Liu, J., Vandervelde, D. et al., The effect of cosolutes on the isomerization of aspartic acid residues and conformation stability in a monoclonal antibody, *J. Pharm. Sci.* 96, 1708–1718, 2007.

55. Colandene, J.D., Maldonado, L.M., Creagh, A.T. et al., Lyophilization cycle development for a high-concentration monoclonal antibody formulation lacking a crystalline bulking agent, *J. Pharm. Sci.* 96, 1598–1608, 2007.

56. Ionescu, R.M., Vlasak, J., Price, C., and Kirchmeier, M., Contribution of variable domains to the stability of humanized IgG1 monoclonal antibodies, *J. Pharm. Sci.* 97, 1414–1426.

57. Ciesta, K., Ros, Y., and Giuszewski, W., Denaturation processes in gamma irradiated proteins studied by differential scanning calorimetry, *Radiat. Phys. Chem.* 58, 233–243, 2000.
58. Feng, L. and Andrade, J.D., Protein adsorption on low-temperature isotropic carbon: I. Protein conformational change probed by differential scanning calorimetry, *J. Biomed. Mater. Res.* 28, 735–743, 1994.
59. Ismond, M.A., Murray, E.D., and Arntfield, S.D., Thermal properties of chemical modified cytochrome *c*, *Biochem. Int.* 15, 245–253, 1987.
60. Franceschi de Carreira, P. and Castellino, F.J., The thermotropic properties of human plasma fibronectin, *Arch. Biochem. Biophys.* 243, 284–291, 1985.
61. Sadler, A.J., Micanovic, R., Katzenstein, G.E. et al., Protein conformation and reversed-phase high-performance liquid chromatography, *J. Chromatogr.* 317, 93–101, 1984.
62. Choma, C.T., Surewicz, W.K., and Kaplan, H., The toxic moiety of the *Bacillus thurigi-nesis* protoxin undergoes a conformational change upon activation, *Biochem. Biophys. Res. Commun.* 179, 933–938, 1991.
63. Welfle, K., Pfeil, W., Misselwitz, R. et al., Conformational properties of streptokinase— Differential scanning calorimetric investigations, *Int. J. Biol. Macromol.* 14, 19–22, 1992.
64. Hu, C.Q., Sturtevant, J.M., Thomson, J.A. et al., Thermodynamics of ribonuclease T1 denaturation, *Biochemistry* 31, 4876–4872, 1992.
65. Williamson, G., Belshaw, N.J., Noel, T.R. et al., *O*-Glycosylation and stability. Unfolding of glucoamylase induced by heat and guanidine hydrochloride, *Eur. J. Biochem.* 207, 661–670, 1992.
66. You, J.L., Page, J.D., Scarsdale, J.N. et al., Conformational analysis of synthetic peptides encompassing the factor XI and prekallikrein overlapping binding domains of high molecular weight kininogen, *Peptides* 14, 867–876, 1993.
67. Burova, T.V., Bernhardt, R., and Pfeil, W., Conformational stability of bovine holo and apo adrenodoxin—A scanning calorimetric study, *Protein Sci.* 4, 909–916, 1995.
68. Gresierich, U. and Pfeil, W., The conformational stability of α-crystallin is rather low: Calorimetric results, *FEBS Lett.* 393, 151–154, 1996.
69. Duguid, J.G., Bloomfield, V.A., Benevides, J.M., and Thomas, G.J. Jr., DNA melting investigated by differential scanning calorimetry and Raman spectroscopy, *Biophys. J.* 71, 3350–3360, 1996.
70. Picó, G.A., Thermodynamic features of the thermal unfolding of human serum albumin, *Int. J. Biol. Macromol.* 20, 63–73, 1997.
71. Takano, K., Funahashi, J., Yamagata, Y. et al., Contribution of water molecules in the interior of a protein to the conformational stability, *J. Mol. Biol.* 274, 132–142, 1997.
72. Relkin, P., Reversibility of heat-induced conformational changes and surface exposed hydrophobic clusters of β-lactoglobulin: Their role in heat-induced sol–gel state transition, *Int. J. Biol. Macromol.* 22, 59–66, 1998.
73. Freskgård, P.G., Petersen, L.C., Gabriel, D.A. et al., Conformational stability of factor VIIa: Biophysical studies of thermal and guanidine hydrochloride-induced denaturation, *Biochemistry* 37, 7203–7212, 1998.
74. Takano, K., Yamagata, Y., and Yutani, K., A general rule for the relationship between hydrophobic effect and conformational stability of a protein: Stability and structure of a series of hydrophobic mutants of human lysozyme, *J. Mol. Biol.* 280, 749–761, 1998.
75. Kravchuk, Z.I., Chumanevich, A.A., Viasov, A.P., and Martsev, S.P., Two high-affinity monoclonal IgG2a antibodies with differing thermodynamic stability demonstrate distinct antigen-induced changes in protein A-binding affinity, *J. Immunol. Methods* 217, 131–141, 1998.
76. Fedunová, D. and Antalik, M., Studies on interactions between metmyoglobin and heparin, *Gen. Physiol. Biophys.* 17, 117–131, 1998.

77. Zhang, F., Thottananiyil, M., Martin, D.L., and Chen, C.H., Conformational alteration in serum albumin as a carrier for pyridoxal phosphate: A distinction from pyridoxal phosphate-dependent glutamate dehydrogenase, *Arch. Biochem. Biophys.* 364, 195–202, 1999.

78. Griko, Y.V. and Remeta, D.P., Energetics of solvent and ligand-induced conformational changes in α-lactalbumin, *Protein Sci.* 8, 554–561, 1999.

79. Pace, C.W., Grimsley, G.R., Thomas, S.T., and Makhatadze, G.I., Heat capacity change for ribonuclease A folding, *Protein Sci.* 8, 1500–1504, 1999.

80. Hattori, M., Nagasawa, K., Ohgata, K. et al., Reduced immunogenicity of β-lactoglobulin by conjugation with carboxymethyl dextran, *Bioconjug. Chem.* 11, 84–93, 2000.

81. Norde, W. and Giacomelli, C.E., BSA structural changes during homomolecular exchange between the adsorbed and the dissolved states, *J. Biotechnol.* 79, 259–268, 2000.

82. Viglaský, V., Antalik, M., Bagel'ová, J. et al., Heat-induced transition of cytochrome c observed by temperature gradient gel electrophoresis at acidic pH, *Electrophoresis* 21, 850–858, 2000.

83. Takano, K., Yamagata, Y., and Yutani, K., Role of amino acid residues at turns in the conformational stability and folding of human lysozyme, *Biochemistry* 39, 8655–8665, 2000.

84. Fanahashi, J., Takano, K., Yamagata, Y., and Yutani, K., Role of surface hydrophobic residues in the conformational stability of human lysozymes at three different positions, *Biochemistry* 39, 14448–14456, 2000.

85. Brewer, J.M. and Wampler, J.E., A differential scanning calorimetric study of the effect of metal ions, substrate/products, substrate analogues and chaotropic anions on the thermal denaturation of yeast enolase 1, *Int. J. Biol. Macromol.* 28, 213–218, 2001.

86. Ahmad, A., Akhtar, M.S., and Shakuni, V., Monovalent cation-induced conformational change in glucose oxidase leading to stabilization of the enzyme, *Biochemistry* 40, 1945–1955, 2001.

87. Mimura, Y., Sondemann, P., Ghirlando, R. et al., Role of oligosaccharide residues of IgG1-Fc in Fc gamma RIIb binding, *J. Biol. Chem.* 276, 45539–45547, 2001.

88. Li, G., Mao, H., Ruan, X. et al., Association of heat-induced conformational change with activity loss of Rubisco, *Biochem. Biophys. Res. Commun.* 290, 1128–1132, 2002.

89. Kuo, J.H., Lo, Y.L., Shau, M.D., and Cherng, J.Y., A thermodynamic study of cationic polymer-plasmid DNA complexes by highly-sensitive differential scanning calorimetry, *J. Control. Release* 81, 321–325, 2002.

90. Collies, T., Meuwis, M.A., Gerday, C., and Feller, G., Activity, stability, and flexibility in glycosidases adapted to extreme thermal environments, *J. Mol. Biol.* 328, 419–428, 2003.

91. Ravindra, R. and Winter, R., On the temperature–pressure free-energy landscape of proteins, *Chem. Phys. Chem.* 4, 359–365, 2003.

92. Bittar, E.R., Caldeira, F.R., Santos, A.M. et al., Characterization of trypsin at acid pH by differential scanning calorimetry, *Braz. J. Med. Biol. Res.* 36, 1621–1627, 2003.

93. Miranda, F.F., Thórólfsson, M., Teigen, K. et al., Structural and stability effects of phosphorylation: Localized structural changes in phenylalanine hydroxylase, *Protein Sci.* 13, 1219–1226, 2004.

94. Peng, Z.G., Hidajat, K., and Uddin, M.S., Adsorption and desorption of lysozyme on nano-sized magnetic particles and its conformational changes, *Colloids Surf. B. Biointerface* 35, 169–174, 2004.

95. Hédoux, A., Ionov, R. Willart, J.F. et al., Evidence for a two-stage thermal denaturation process in lysozyme: A Raman scattering and differential scanning calorimetry investigation, *J. Chem. Phys.* 124: 14703, 2006.

96. Singh, S.K. and Kisore, N., Thermodynamic insights into the binding of Triton X-100 to globular proteins: A calorimetric and spectroscopic investigation, *J. Phys. Chem. B* 110, 9728–9737, 2006.

97. Tavirani, M.R., Moghaddamnia, S.H., Ranjbar, B. et al., Conformational study of human serum albumin in pre-denaturation temperatures by differential scanning calorimetry, circular dichroism and UV spectroscopy, *J. Biochem. Mol. Biol.* 39, 530–536, 2006.

98. Enomoto, H., Li, C.P., Morizane, K. et al., Glycation and phosphorylation of β-lactoglobulin by dry-heating: Effect on protein structure and some properties, *J. Agric. Food Chem.* 55, 2392–2398, 2007.

99. Gombos, L., Kardos, J., Patthy, A. et al., Probing conformational plasticity of the activation domain of trypsin: The role of glycine hinges, *Biochemistry* 47, 1675–1684, 2008.

100. Kaczowka, S.J., Madding, L.S., Epting, K.L. et al., Probing the stability of native and activated forms of α_2-macroglobulin, *Int. J. Biol. Macromol.* 42, 62–67, 2008.

101. Badkar, A., Yohannes, P., and Banga, A., Application of TZERO calibrated modulated temperature differential scanning calorimetry to characterize model protein formulation, *Int. J. Pharm.* 309. 146–156, 2006.

102. Alderton, A.L., Faustman, C., Liebler, D.C., and Hill, D.W., Induction of redox instability of bovine myoglobin by adduction with 4-hydroxy-2-nonenal, *Biochemistry* 42, 4398–4405, 2003.

103. Coness, C., Sámchez, L., Rota, C. et al., Isolation of lactoferrin from milk of different species: Calorimetric and antimicrobial studies, *Comp. Biochem. Physiol. B Biochem. Mol. Biol.* 150, 131–139, 2008.

104. Koch-Schmidt, A.C. and Mosbach, K., Studies on conformation of soluble and immobilized enzymes using differential scanning calorimetry 1. Thermal stability of nicotinamide adenine dinucleotide dependent dehydrogenases, *Biochemistry* 16, 2101–2105, 1977.

105. Hadden, J.M., Bloemendal, M., Harris, P.I. et al., Fourier transform infrared spectroscopy and differential scanning calorimetry of transferrins: Human serum transferrin, rabbit serum transferring, and human lactoferrin, *Biochim. Biophys. Acta* 1205, 59–67, 1994.

106. Kurochkin, I.V., Procyk, R., Bishop, P.D. et al., Domain structure, stability and domain-domain interactions in recombinant factor XIII, *J. Mol. Biol.* 248, 414–430, 1995.

107. Sagar, S.L. and Domach, M.M., Using differential scanning calorimetry to elucidate metal-protein binding sites in α- and γ-chymotrypsin, *Bioseparations* 5, 289–294, 1995.

108. Chan, H.K., Au-Yeung, K.L., and Gonda, I., Effects of additives on heat denaturation of rhDNase in solutions, *Pharm. Res.* 13, 756–761, 1996.

109. Kosa, T., Maruyama, T., and Otagiri, M., Species differences of serum albumins: II. Chemical and thermal stability, *Pharm. Res.* 15, 449–454, 1998.

110. Imamura, K., Ogawa, T., Sakiyama, T., and Nakanishi, K., Effects of type of sugars on the stabilization of protein in the dried state, *J. Pharm. Sci.* 92, 266–274, 2003.

111. Yoshioka, S. and Aso, Y., A quantitative assessment of the significance of molecule mobility as a determinant for the stability of lyophilized insulin formulations, *Pharm. Res.* 22, 1358–1364, 2005.

112. Taylor, R.B. and Shivji, A.S., A critical appraisal of drug stability testing methods, *Pharm. Res.* 4, 177–180, 1987.

113. Bakshi, M. and Singh, S., Development of validated stability-indicating assay methods—Critical review, *J. Pharm. Biomed. Anal.* 28, 1011–1140, 2002.

114. Bhutani, H., Singh, S., Vir, S. et al., LC and LC-MS study of stress decomposition behaviour of isoniazid and establishment of validated stability-indicating assay method, *J. Pharm. Biomed. Anal.* 43, 1213–1220, 2007.

115. Guidance for Industry. Quality of biotechnological products: Stability testing of biotechnological/biological products, http://www.fda.gov/cder/guidance/ichq5c.pdf

8 Light Scattering and Biomacromolecular Conformation

Diffusion can be separated into lateral or translational diffusion and rotation diffusion. Translational diffusion is movement of a macromolecule through a solution in response to a concentration gradient provided there is no barrier to diffusion[1-3]; there has been considerable recent interest in measuring the translational diffusion of proteins at high concentration and/or in solutions of high viscosity to "mimic" the molecular crowding inside the cell.[4-6] Rotational diffusion is the rotational motions of molecule or particle relative to the mass center of the molecule or particle.[7-9]

The translation diffusion constant for a macromolecule can be obtained by several techniques; light scattering is one of the more popular techniques. Other molecular characteristics, which can be obtained from light scattering, include molecular weight, radius of gyration, and the second virial coefficient. The translational diffusion constant for a macromolecule may also be determined by other techniques including a porous diaphragm cell, free diffusion, ultracentrifugation, and flow field fractionation.[1,10,11] The rotational diffusion constant can be measured by several techniques including fluorescence anisotropy and NMR.[8,12,13]

Incident light is modified by contact with an object in light scattering. In elastic light scattering, which has found considerable use for cells,[14-20] the energy of the incident light is only slightly decreased by contact with the target molecule or the cell while some of the incident energy is absorbed with inelastic light scattering (Raman scattering), resulting in a change in the wavelength of the scattered light.[21-23] The adsorption of light by a molecule has been considered as an extreme form of inelastic light scattering. Elastic and inelastic light scattering as described above is considered to be classical or static light scattering (also known as Rayleigh scattering) and the measurement of scattered light averages over a relative long (2 s) period of time. Dynamic light scattering (also known as photon correlation spectroscopy [PCS] and quasielastic light scattering [QELS]) measures the time-dependent fluctuations in the intensity of scattered light that occurs because the particles undergo Brownian motion. The velocity of this Brownian motion is measured and is called the translational diffusion coefficient D. This diffusion coefficient can be converted into particle size (Stokes' radius, hydrodynamic radius) using the Stokes–Einstein equation

$$D_{trans} = kT/6\pi\eta_0 R_h \qquad (8.1)$$

where

k is the Boltzmann constant

D_{trans} is the translational diffusion constant

η_0 is the viscosity of the solution

R_h is hydrodynamic radius

The diffusion constant is related to the frictional coefficient f by the Einstein–Sutherland equation:

$$D_{trans} = kT/f \qquad (8.2)$$

The reader is directed to several excellent works on theory and instrumentation for light scattering.[1, 24–30]

It is also important to differentiate light scattering from turbidity and nephelometry. Turbidity is the measure of the light scattered by a particle or suspension; it is a measure of light scattered rather than adsorbed by a solution. In this sense, light scattering is used to evaluate the final quality of final drug product preparations such as albumin, intravenous immunoglobulin, and other biopharmaceuticals.[31–40] Turbidity is used as measure of protein denaturation.[41–44] A correction for light scattering should be made for the determination of protein concentration by UV-VIS spectroscopy.[45–47] Nephelometry is a technique that measures the intensity of light scattered by an object in a different direction such as 90° relative to the incident light; nephelometry is widely used for the estimation of antigen–antibody complexes.[48–51] Multiangle light scattering is related to nephelometry in that the intensity of the scattered light is measured; it differs from nephelometry in that the scattering is measured at several angles relative the incident light[51–57] and is used with in-line analysis of chromatographic fractionation[58–66] and asymmetric flow field flow fractionation.[67–73]

Dynamic light scattering can provide information about changes in protein conformation and protein aggregation. Our focus here is not on applications in biotechnology, for which there is a considerable body of literature, which we will not discuss; the reader is directed to other sources for wider coverage of the application of this technology to a variety of biological problems.[47–53] Dynamic light scattering can also provide information on the size and homogeneity of larger particles such as vaccine formulations and gene therapy delivery vector and vehicles. These materials have formulation issues similar to those for protein drug products.

One of quality attributes of any parenteral drug product is the absence of particulated material such as aggregated protein, which can responsible for an adverse reaction.[74] While the detection of such materials is critical, studies on development of such aggregates during the processing of a therapeutic protein are critical for the understanding of the process.[75] A listing of some studies using dynamic light scattering to characterize changes in proteins under conditions used for the processing and formulation of proteins is presented in Table 8.1. Studies with model proteins are also included. Stress conditions are used similar to those used in accelerated stability studies for biological products.[96–98] The body of data suggest that dynamic light scattering (DLS) is most useful for detection of aggregates and elucidation of pathways from denaturation to aggregation. DLS may be less useful as a stand-alone technique for the assessment of denaturation as there are other techniques such as circular dichorism (CD) and hydrogen–deuterium exchange are available. The use of DLS for the characterization of vaccine and vaccine-related products is provided in Table 8.2. Table 8.3 lists some

TABLE 8.1

Application of Dynamic Light Scattering for the Evaluation of Biopharmaceuticals during Processing and Formulation

Protein	Description of Study	Ref.
Hydrophobic cytokine	DLS used to evaluate the ability of excipients such NaCl to prevent aggregation/precipitation of a recombinant hydrophobic cytokine formulated with human serum albumin and mannitol	[76]
Interferon alpha2b	DLS used to evaluate metal-catalyzed aggregate formation	[74]
Human albumin	DLS used to evaluate the effect of lyophilization on aggregate formation; aggregate formation; conformational change detected by DLS and SDS-PAGE but not by CD or infrared spectroscopy	[77]
IgG	DLS used to study aggregation of IgG; capillary zone electrophoresis and two-phase partitioning also used	[78]
Monoclonal antibody	DLS used to evaluate aggregate formation in acid-stressed and unstressed monoclonal antibody preparations; comparison with size-exclusion chromatography (SEC), sedimentation velocity analytical ultracentrifugation (SV-AUC) and asymmetrical flow field flow fractionation (AF_4) for detection of aggregate formation. DLS, SV-AUC, and AF_4 detected aggregates missed by SEC	[10]
Glucagon-like peptide-1 analogue	DLS used to evaluate aggregate formation occurring on lyophilization of a glucagon-like peptide-1 analogue; comparison with FTIR, CD, and intrinsic fluorescence	[79]
Human growth hormone	DLS used to evaluate the effect of organic solvents on human growth hormone: comparison with CD and fluorescence	[80]
IgG	DLS used to study opalescence of concentrated IgG solutions; also used static light scattering, nephelometry, and analytical ultracentrifugation; opalescence was a consequence of Rayleigh light scattering	[81]
IgG	DLS used to demonstrate aggregate formation	[82]
Insulin glargine	DLS used to characterize concentration-dependent aggregation of insulin glargine	[83]
IgG	DLS used to determine aggregate composition and extent of aggregate formation for IgG in various process steps during manufacture. SEC was not useful for large aggregates; utility limited to dimers and oligomers	[84]
Bovine serum albumin	Use of DLS to study thermal aggregation of BSA combined with the use of FITR spectroscopy to measure changes in protein conformation	[85]

(*continued*)

TABLE 8.1 (continued)

Application of Dynamic Light Scattering for the Evaluation of Biopharmaceuticals during Processing and Formulation

Protein	Description of Study	Ref.
IgG	Use of DLS and DSC to evaluate the thermal stability of IgG during formulation development	[86]
Lysozyme as model protein	Use of DLS to study the effect of pressure on proteins	[87]
Bovine serum albumin	DLS used to measure the effect of dextran sulfate on the thermal denaturation and aggregation of bovine serum albumin. At pH 6.2, dextran sulfate did not bind to native BSA but did bind to partially unfolded BSA CD studies suggested that dextran sulfate facilitated denaturation of BSA but suppressed aggregation	[88]
Lens protein	DLS used to measure aggregation of lens protein after glycation with ascorbic acid	[89]
Insulin; alcohol dehydrogenase	DLS used to evaluate the effect of arginine on protein aggregation. There was a concentration-dependent effect of arginine on protein aggregation (heat-induced aggregation with alcohol dehydrogenase; dithiothreitol-induced aggregation with insulin)	[90]
Monoclonal antibody	DLS used to the aggregation of a IgG1 monoclonal antibody at two temperatures (5°C and 25°C) under two different mechanical stress conditions (shaking, stirring). Shaking was more stressful; headspace in vials had a major effect on aggregation with shaking	[91]
Bovine serum albumin; human serum albumin	Combination of dynamic viscosity and DLS for detecting onset of protein aggregation and form/composition of aggregated forms	[92]
Bovine serum albumin; staphylococcal nuclease	DLS used to assess protein aggregation following thermal/chaotrope-induced denaturation of protein	[93]
Lysozyme	DLS used to measure effect of benzyl alcohol on thermal denaturation of lysozyme; other techniques included CD, biological assay, DSC, and native electrophoresis. Benzyl alcohol increased susceptibility to denaturation but decreased aggregation	[94]
Lysozyme	DLS used to evaluated structural changes of lysozyme on lyophilization in the presence of sucrose; studies combined with internal reflectance infrared spectroscopy	[95]

of the studies on the use of DLS for the characterization of gene therapy delivery vehicles and vectors. Tables 8.4 through 8.6 contain a partial listing of studies on the application of DLS for the characterization of proteins, polysaccharides, and nucleic acids, respectively.

TABLE 8.2

The Use of Dynamic Light Scattering for the Characterization of Vaccines and Vaccine-Related Products

Product	Description of Study	Ref.
Clostridium difficile toxins and toxoids	DLS used with CD, intrinsic and extrinsic fluorescence, high-resolution UV spectroscopy and turbidity to characterize toxoids derived from toxins via formaldehyde cross-linkage including pH and thermal stability studies permitting the development of phase diagrams	[99]
Hepatitis-B surface antigen (HbsAg)	DLS used to evaluate the formulation of a HbsAg dry powder vaccine for epidermal powder immunization; use of spray freeze-drying	[100]
Norwalk virus-like particles	Use of DLS to evaluate the stability of recombinant virus-like particles for use as a vaccine antigen; also use of intrinsic and extrinsic fluorescence, derivative UV spectroscopy, CD and DSC	[101]
Engineered ricin for use as vaccine	DLS used to characterize a truncated A-chain form for ricin to be used for vaccine	[102]
Strepococcal pyrogenic exotoxin A (SPEA)	DLS used to evaluate stability of engineered forms of SPEA in combination with antigenicity studies: suggestion that protein conformation has significant role in generating effective neutralizing antibodies	[103]
Hepatitis A virus	DLS used to characterize the size and stability of hepatitis A virus used to prepare VAQTA vaccine	[104]
Virus-like particles (VLP) derived from murine polyomavirus VP-1 in insect cells or *Escherichia coli*	DLS compared with asymmetric flow field flow fractionation for size homogeneity characterization of VLP (vaccine candidate)	[105]
Cationic supported lipid bilayer	DLS used to characterize an antigen-presentation vehicle consisting of sulfated polystyrene particles coated with a cationic lipid (dioctadehyldimethylammonium bromide) bilayer. *Taenia crassiceps* protein was the model antigen	[106]
Human influenza A virus	DLS used to evaluate cell culture (MDCK cells, serum-free) process for the production of human influenza A virus to be used for vaccine production (β-propiolactone)	[107]
Autographa californica M nucleopolyhedrovirus	DLS used to measure effect of thermal stress on *Autographa californica* M nucleopolyhedrovirus; an increase in temperature (45°C) reduced infectivity while DLS showed increase in particle size from 150 to 249 nm (55°C)	[108]
Human respiratory syncytial virus (RSV)	DLS, intrinsic and extrinsic fluorescence, second derivative UV spectroscopy, and CD used to define effect of pH (3–8) and temperature (10°C–85°C) on RSV; DLS and turbidity specifically used for evaluation of integrity of viral particles. Data used to develop an empirical phase diagram	[109]

(continued)

TABLE 8.2 (continued)
The Use of Dynamic Light Scattering for the Characterization of Vaccines and Vaccine-Related Products

Product	Description of Study	Ref.
Polyurethane particles for protein-based vaccines	DLS and transmission electron microscopy (TEM) used to characterize acid-degradable polyurethane particles containing ovalbumin as a model antigen	[110]
Polycaprolactone/maltodextrin capsules for protein and vaccine delivery	DLS used to define size determining formulation process variables (SDFPV) of polycaprolactone/maltodextrin capsules	[111]
Stabilizers for RSV	DLS used to evaluate stabilizers for RSV as method for developing formulated product; the initial screening of stabilizers was based on turbidity at 56°C. Evaluation also included second derivative UV spectroscopy	[112]
Cationic liposomes	DLS used to characterize liposomes consisting of dimethyldioctadecylammonium and trehalose 6,6′-dibehenate (synthetic cord factor from *Mycobacterium tuberculosis*) to be used as adjuvant	[113]
Human papillomavirus-like particles	DLS used to characterize human papillomavirus-like particles produced by recombinant technology	[114]
Diphtheria toxoid-loaded elastic vesicles	Use of DLS and measurement of zeta potential for characterization of surfactant-based vesicle formulations for diphtheria toxoid including stability studies	[115]

TABLE 8.3
The Use of Dynamic Light Scattering for the Characterization of Gene Therapy Viral Vectors and Other Vehicles for the Delivery of Nucleic Acids and Related Derivative Products

Vehicle/Vector	Description of Study	Ref.
Adenovirus type 5	DLS, intrinsic and extrinsic fluorescence. second derivative UV spectroscopy used to evaluate thermal stability of adenovirus type 5 as a function of pH (3–8) at low (0.075 M NaCl) and high (1.0 M NaCl) ionic strength	[116]
Adenovirus type 2	DLS, intrinsic and extrinsic fluorescence, second derivative spectroscopy, and CD used to evaluate thermal stability of adenovirus type 2. An empirical phase diagram was developed and used to define "phase boundaries"	[117]
Adenovirus 5	Use of DLS for the development of a sorption process for virus purification via ion-exchange	[118]

TABLE 8.3 (continued)
The Use of Dynamic Light Scattering for the Characterization of Gene Therapy Viral Vectors and Other Vehicles for the Delivery of Nucleic Acids and Related Derivative Products

Vehicle/Vector	Description of Study	Ref.
Copolymer of amine methacrylate with poly(ethylene glycol)	Use of DLS and transmission electron microscopy (TEM) to characterize poly(ethylene glycol) copolymer with poly (2-dimethylamino)-ethylmethacrylate as vehicle to condense DNA for gene therapy; TEM shows uniformly discrete spheres having a diameter of 80–100 nm as determined by DLS	[119]
Reducible polycation (obtained by condensing $Cys(Lys)_{10}Cys$ complex with DNA	DLS used to characterize a polyplex of reducible polycation–DNA	[120]
Galactosylated chitosan/DNA nanoparticle	Galactose was coupled to chitosan to provide specificity for hepatocytes; the galactosylated chitosan was complexed with plasmid DNA and the resulting complexes characterized with DLS to determine particle size	[121]
DNA-polymethacrylate-Penetratin	DNA is condensed with polymethacrylate and then combined with Penetratin to provide for increased transfection efficiency. DLS is used to establish composition and stability of this ternary complex	[122]
Supercoiled plasmids	Use of DLS, CD, FTIR, and DSC to evaluate supercoiled plasmids by various purification protocols	[123]
Thioplex	DLS was used to characterize the complex formed between DNA and a polyaspartamimde polymer (α,β-poly-N-2-hydroxyethyl-D,L-asparamide) containing amine groups (introduced with 3-carboxypropyl trimethylammonium chloride) and sulfhydryl groups introduced with N-succinimidyl-2-(2-pyridyldithio)propionate	[124]
Hydrophobic derivatives of polyethyleneimine (PEI)	DLS, freeze-fracture microscopy, and fluorescence microscopy used to characterize DNA complexes (polyplexes) with hydrophobic derivatives of polyethyleneimine	[125]
Cationic liposome–DNA complexes (lipoplexes)	DLS used to assess stability of a series of cationic liposome–DNA complexes (lipoplexes)	[126]
Polyion complex micelles (PICMs)	PICMs are obtained by the mixing of copolymers of hydrophilic nonionic and cationic blocks with oligonucleotides (e.g., antisense oligonucleotides; siRNA). The particles were characterized by gel electrophoresis and dynamic light scattering	[127]
Complex of DNA and hydrophobically modified low molecular weight chitosan	Chitosan is modified by N-/2(3)-(dodec-2-enyl)succinoyl groups and complexed with DNA. These complexes are characterized by DLS, electrophoresis, and fluorescence	[128]

(*continued*)

TABLE 8.3 (continued)

The Use of Dynamic Light Scattering for the Characterization of Gene Therapy Viral Vectors and Other Vehicles for the Delivery of Nucleic Acids and Related Derivative Products

Vehicle/Vector	Description of Study	Ref.
Calcium phosphate–DNA nanocomposites	Preparations of calcium phosphate entrapping DNA were prepared for use in gene therapy. These composites are characterized by DLS, transmission electron microscopy, and scanning electron microscopy	[129]
Complexes of small RNA molecules (siRNA) and aminoglycoside derivatives	Small double-stranded RNA (siRNA) was complexed with aminoglycosides, which were characterized by fluorescence and DLS; the complexes are intended for use in delivery of siRNA for gene silencing	[130]
Lyophilized pegylated polyplex micelles	DLS and atomic force microscopy used to characterize a freeze-dried formulation of a polyplex consisting of a poly(ethylene glycol) shell prepared by polyion complex of plasmid DNA and thiolated PEG–poly(lysine) block copolymers	[131]
Polymethacrylate–plasmid nanoparticles	Complex formed between a plasmid and poly(2-dimethylamino)ethyl methacrylate forms nanoparticles which are characterized dynamic light scattering and electrophoresis	[132]

TABLE 8.4

Characterization of Some Proteins and Derivatives by Dynamic Light Scattering

Biomacromolecule	Description of Study	Ref.
α-Glucan and glycogen from *Mycobacterium bovis*	DLS and analytical ultracentrifugation used to the glycogen-like α-glucan in the capsule from *M. bovis*	[133]
Antibody-labeled gold particles	DLS used to characterize antibody-labeled gold particles for use in a sol particle immunoassay to used to quantify murine IgG antibodies	[134]
Human serum albumin (HSA) and bovine serum albumin (BSA)	The effect of pH and protein concentration on the properties of solutions of two model proteins, HSA and BSA. Properties included DLS, dynamic viscosity, and electrophoretic mobility	[92]
Lysozyme	DLS and static light scattering studies of lysozyme in guanidinium chloride; second virial coefficient is positive indicating protein–protein repulsion; unfolding was not associated with increase in hydrodynamic radius	[135]
Lysozyme	DLS and static light scattering used to study nucleating lysozyme in $NaCl/NH_4SO_4$ solutions as function of lysozyme concentration	[136]

TABLE 8.4 (continued)
Characterization of Some Proteins and Derivatives by Dynamic Light Scattering

Biomacromolecule	Description of Study	Ref.
Human serum albumin (HSA)	Use of DLS, FRET, and CD to characterized conformational changes in HAS as a function of pH and ligand binding	[137]
RSV NS1 protein	His-tagged human respiratory syncytial virus nonstructural protein NS1 (RSV NS1 protein) was expressed in *E. coli* and characterized by DLS, analytical ultracentrifugation, CD, and SEC	[138]
Adiponectin	Use of analytical ultracentrifugation, DLS, and gel electrophoresis to characterize the various oligomeric forms of adiponectin isolated from bovine serum or 3T3-L1 adipocytes	[139]
Gelatin	DLS and shear rheology were used to validated pulsed field gradient NMR for determination of the gel temperature of gelatin	[140]
Jararassin-I	DLS and SDS-PAGE were used to characterize jarrassin-I, a thrombin-like enzyme purified from *Bothrops jararaca* venom	[141]
Spinach plastocyanin	Use of DLS, DSC, NMR, UV spectroscopy, and mass spectrometry to study of the role of copper site oxidation state on the thermal denaturation of spinach plastocyanin	[142]
Barstar	DLS, fluorescence, and far-UV CD used to compare two unfolded forms of barstar (pH 12) or guanidine hydrochloride	[143]
Plakophilin 1	DLS, CD, limited proteolysis, analytical ultracentrifugation, and electron microscopy used to evaluate the effect of zinc on the solution structure of plakophilin 1	[144]
Human ferrochelatase	Gel filtration and DLS used to characterize human ferrochelatase in detergent solution as a homodimer	[145]
Human uroporphyrinogen decarboxylase (with His tag)	His-tagged human uroporphyrinogen decarboxylase was expressed in *E. coli*, purified to homogeneity and characterized with DLS, analytical ultracentrifugation, and crystallographic analysis (3.0 Å)	[146]
Human recombinant cyclooxygenase-2	Crystallographic analysis showed dimer structure consistent with DLS and SEC	[147]
Glycosylated insulin	Glycosylated insulin derivatives were prepared from insulin by reaction with *p*-succinimidylphenyl glucopyranoside. The resulting derivative forms were isolated and characterized by DLS, UV spectroscopy, and CD spectroscopy; the thermal stability of the various forms was evaluated	[148]
Acidic fibroblast growth factor	Use of static light scattering, DLS, and analytical ultracentrifugation to study interaction of heparin with acidic fibroblast growth factor	[149]
Ca^{2+}-Mg^{2+}-ATPase from sarcoplasmic reticulum	DLS was used to study the delipidated Ca^{2+} + Mg^{2+}-ATPase after detergent removal to evaluate aggregation and conformational change	[150]

TABLE 8.5

Characterization of Some Polysaccharides and Derivatives by Dynamic Light Scattering

Polysaccharide	Description of Study	Ref.
Hyperbranched β-D-glucan	Microwave treatment is used to solubilize a water-insoluble hyperbranched β-D-glucan(TM3a) from sclerotia of *Pleurotus*. SEC and DLS were used to characterize the solubilized material	[151]
Heparin	DLS used to study the self-association of heparin; the self-association of heparin is thought to have a role in the gelation of heparin-functionalized hydrogels	[152]
Hyaluronic acid	DLS and viscosity measurements used to evaluate the role of chemical cross-linking (water-soluble carbodiimide) in the formation and stability of gels	[153]
Starch	NMR and DLS used to measure the rate of starch dissolution in dimethylsulfoxide	[154]
Hydroxyethyl starch (HES)	HES was esterified with lauric, palmitic, and stearic acid; resulting derivative forms self-assembled into micelles/vesicles which were characterized by DLS and TEM	[155]
Heparin–deoxycholic acid conjugates	Heparin–deoxycholic acid conjugates were synthesized and evaluated with ^{1}H NMR, TEM, and DLS	[156]
Starch and amylose	Use of DLS, static light scattering, SEC with multi-angle light scattering were used to characterize various amyloses including starches in dimethylsulfoxide	[157]
An exocellular polysaccharide derived from *Lactococcus lactis*	DLS was used to characterize a polysaccharide produced by the lactic acid bacterium *L. lactis*	[158]
Various polysaccharides	Comparison of asymmetrical flow field flow fractionation with multiangle light scattering for determination of temperature on mass distribution with prior data from DLS and static light scattering	[159]
Low-molecular-weight heparin (LMW heparin)	DLS, NMR, Raman spectroscopy, and SEC used to characterize commercial preparations of low-molecular-weight heparin	[160]
Heparin	DLS used to characterize a commercial preparation of heparin	[161]
Pectin	DLS, static light scattering and SEC used to characterize several sharp fractions from low methoxyl citrus pectin	[162]
Acetylated chitosans	DLS, analytical ultracentrifugation, and viscometry used to characterized various acetylated chitosan preparations	[163]

TABLE 8.6

The Characterization of Some Nucleic Acids and Derivatives by Dynamic Light Scattering

Nucleic Acid or Derivative Form	Description of Study	Ref.
Cylindrical poly (oligo-DNA)	Oligonucleotides (5′-TCCATGACGTTC-3′) were modified at the 5′ end with an amine function, which was modified with *N*-methacryloyloxysuccinimide to generate a derivative, which could be polymerized to a comb polymer, forming a cylinder; characterized by DLS and atomic force microscopy	[164]
Self-assembled peptide nucleic acid amphiphiles	A 6-mer peptide nucleic acid amphiphiles with a 12-carbon *n*-alkane tail forms ellipsoidal micelles which are characterized by DLS and small-angle x-ray scattering	[165]
Supercoiled plasmids	DLS, CD, FTIR, and DSC are used to characterize supercoiled plasmids produced by different production techniques; stability studies and biological assays were included. The results provide guidance for the manufacture of supercoiled plasmids for gene delivery vehicles	[123]
Short oligonucleotides	Depolarized dynamic light scattering is useful for distinguishing between different conformations (DNA double helix versus hairpin structure) of oligonucleotides	[166]
tRNA(Phe)	Use of depolarized dynamic light scattering to characterize solution structure (internal dynamics) of brewer's yeast tRNA(Phe)	[167]

REFERENCES

1. Haschemeyer, R.H. and Haschemeyer, A.E.V., *Proteins. A Guide to Study by Physical and Chemical Methods*, Wiley-Interscience, New York, 1973.
2. Kuttner, Y.Y., Kozer, N., Segal, E. et al., Separating the contribution of translation and rotational diffusion to protein association, *J. Am. Chem. Soc.* 127, 15138–15144, 2005.
3. Peters, R. and Kubitscheck, U., Scanning microphotolysis: Three-dimensional diffusion measurement and optical single-transporter recording, *Methods* 18, 508–517, 1999.
4. Bernadó, P., García de la Torre, J., and Pons, M., Macromolecular crowding in biological systems: Hydrodynamics and NMR methods, *J. Mol. Recognit.* 17, 397–407, 2004.
5. Kozer, N., Kuttner, Y.Y., Haran, G., and Schreiber, G., Protein–protein association in polymer solutions: From dilute to semidilute to concentrated, *Biophys. J.* 92, 2139–2149, 2007.
6. Zorrilla, S., Hink, M.A., Visser, A.J., and Lillo, M.P., Translational and rotational motions of protein in a protein crowded environment, *Biophys. Chem.* 125, 298–305, 2007.
7. Lakowicz, J.R., Fluorescence spectroscopic investigations of the dynamic properties of proteins, membranes, and nucleic acids, *J. Biochem. Biophys. Methods* 2, 91–119, 1980.
8. Case, D.A., Molecular dynamics and NMR spin relaxation in proteins, *Acc. Chem. Res.* 35, 325–331, 2002.

9. Bax, A. and Grishaev, A., Weak alignment NMR: A hawk-eyed view of biomolecular structure, *Curr. Opin. Struct. Biol.* 15, 563–570, 2005.

10. Gabrielson, J.P., Brader, M.L., Pekar, A.H. et al., Quantitation of aggregate levels in a recombinant humanized monoclonal antibody formulation by size-exclusion chromatography, asymmetrical flow field flow fractionation, and sedimentation velocity, *J. Pharm. Sci.* 96, 268–279, 2007.

11. Hawe, A., Fress, W., Sutter, M., and Jiskoot, W., Online fluorescent dye detection method for the characterization of immunoglobulin G aggregation by size-exclusion chromatography and asymmetrical flow field flow fractionation, *Anal. Biochem.* 378, 115–122, 2008.

12. Moncrieffe, M.C., Eaton, S., Bajzer, Z. et al., Rotational and translational motion of troponin C, *J. Biol. Chem.* 274, 17464–17470, 1999.

13. Blake-Hall, J., Walker, O., and Fushman, D., Characterization of the overall rotational diffusion of a protein from ^{15}N relaxation measurements and hydrodynamic calculations, *Methods Mol. Biol.* 278, 139–160, 2004.

14. Kerker, M., Chew, H., McNulty, P.J. et al., Light scattering and fluorescence by small particles having internal structure, *J. Histochem. Cytochem.* 27, 250–263, 1979.

15. Mourant, J.R., Campolet, M., and Brocker, C., Light scattering from cells: The contribution of the nucleus and the effect of proliferative status, *J. Biomed. Opt.* 5, 131–137, 2000.

16. Ferri, F., Greco, M., Arcovito, G. et al., Structure of fibrin gels studied by elastic light scattering techniques: Dependence of fractal dimension, gel crossover length, fiber diameter, and fiber density on monomer concentration, *Phys. Rev. E. Stat. Nonlin. Soft Matter Phys.* 66, 11913, 2002.

17. Allen, J., Liu, Y., and Kim, Y.L. et al., Spectroscopic translation of cell–material interactions, *Biomaterials* 28, 162–174, 2007.

18. Prosperi, D., Morasso, C., Tortora, P. et al., Avidin decorated core-shell nanoparticles for biorecognition studies by elastic light scattering, *Chembiochem* 8, 1021–1028, 2007.

19. Mustafi, D., Smith, C.D., Makinen, M.W., and Lee, R.C., Multi-block poloxamer surfactants suppress aggregation of denatured proteins, *Biochim. Biophys. Acta* 1780, 7–15, 2008.

20. Orlova, D.Y., Yurkin, M.A., Hoekstra, A.G., and Maltsev, V.P., Light scattering by neutrophils: Model, simulation, and experiment, *J. Biomed. Opt.* 13:054057, 2008.

21. Dubin, S.B., Lunacek, J.H., and Benedek, G.B., Observation of the spectrum of light scattered by solutions of biological macromolecules, *Proc. Natl. Acad. Sci. USA* 57, 1164–1171, 1967.

22. Cummins, H.Z., Carlson, F.D., Herbert, T.J., and Woods, G., Translational and rotational diffusion constants of tobacco mosaic virus from Rayleigh linewidths, *Biophys. J.* 9, 518–546, 1969.

23. Lin, S.H.C., Dewan, R.K., Bloomfield, V.A., and Morr, C.V., Inelastic light scattering study of the size distribution of bovine milk casein micelles, *Biochemistry* 10, 4788–4793, 1971.

24. Fabelinskii, I.L., *Molecular Scattering of Light*, Plenum Press, New York, 1968.

25. Chu, B., *Laser Light Scattering*, Academic Press, New York, 1974.

26. Bohren, C.F. and Huffman, D.R., *Absorption and Scattering of Light by Small Particles,* Wiley, New York, 1983.

27. Schuerman, D.W. ed., *Light Scattering by Irregularly Shaped Particles*, Plenum Press, New York, 1989.

28. Schmitz, K.S., *An Introduction to Dynamic Light Scattering by Macromolecules*, Academic Press, Boston, MA, 1990.

29. Harding, S.E., Satelle, D.B., and Bloomfield, V.A. eds., *Laser Light Scattering in Biochemistry*, Royal Society of Chemistry, Cambridge, U.K., 1992.

30. Mishchenko, M.L. and Hovenier, J.W., *Light Scattering by Nonspherical Particles: Theory, Measurements, and Applications,* Academic Press, San Diego, CA, 2000.

31. Hawe, A. and Freiss, W., Physico-chemical lyophilization behavior of mannitol, human serum albumin formulations, *Eur. J. Pharm. Sci.* 28, 224–232, 2006.

32. Hawe, A. and Friess, W., Development of HSA-free formulations for a hydrophobic cytokine with improved stability, *Eur. J. Pharm. Biopharm.* 68, 169–182, 2008.

33. McCarthy, D., Goddard, D.H., Pell, B.K., and Holborow, E.J., Intrinsically stable IgG aggregates, *J. Immunol. Methods* 41, 63–74, 1981.

34. Lundblad, J.L., Mitra, G., Sternberg, M.M., and Schroeder, D.D., Comparative studies of impurities in intravenous immunoglobulin preparations, *Rev. Infect. Dis.* 8 (Suppl 4), S382–S390, 1986.

35. Lundblad, J.L. and Londeree, N., The effect of processing methods on intravenous immune globulin preparations, *J. Hosp. Infect.* 12 (Suppl D), 3–15, 1988.

36. Garcia, M., Monge, M., Leon, G. et al., Effect of preservatives of IgG aggregation, complement-activating effect and hypotensive activity of horse polyvalent antivenom used in snakebite envenomation, *Biologicals* 30, 143–151, 2002.

37. Mahler, H.C., Müller, R., Friess, W. et al., Induction and analysis of aggregates in a liquid IgG1-antibody formulation, *Eur. J. Pharm. Biopharm.* 59, 407–417, 2005.

38. Wang, W., Wang, Y.J., and Wang, D.Q., Dual effects of Tween 80 on protein stability, *Int. J. Pharm.* 347, 31–38, 2008.

39. Rojas, G., Espinoza, M., Lomonte, B., and Gutiérrez, J.M., Effect of storage temperature on the stability of the liquid polyvalent antivenom, *Toxion* 28, 101–105, 1990.

40. Usami, A., Ohtsu, A., Takahama, S., and Fujii, T., The effect of pH, hydrogen peroxide and temperature on the stability of human monoclonal antibody, *J. Pharm. Biomed. Anal.* 14, 1133–1140, 1996.

41. Suzuki, K., Miyosawa, Y., and Suzuki, C., Protein denaturation by high pressure. Measurements of turbidity of isoelectric ovalbumin and horse serum albumin under high pressure, *Arch. Biochem. Biophys.* 101, 225–228, 1963.

42. Ramzi, A., Sutter, M., Hennink, W.E., and Jiskoot, W., Static light scattering and small-angle neutron scattering study on aggregated recombinant gelatin in aqueous solution, *J. Pharm. Sci.* 95, 1703–1711, 2006.

43. Vetri, V., Librizzi, F., Leone, M., and Militello, V., Thermal aggregation of bovine serum albumin at different pH: Comparison with human serum albumin, *Eur. Biophys. J.* 36, 717–725, 2007.

44. Choi, S.J., Lee, S.E., and Moon, T.W., Influence of sodium chloride and glucose on acid-induced gelation of heat-denature ovalbumin, *J. Food Sci.* 73, C312–C322, 2008.

45. Wetlaufer, D.B., Ultraviolet spectra of proteins and amino acids, *Adv. Prot. Chem.* 17, 303–390, 1962.

46. Gill, S.C. and von Hippel, P.H., Calculation of protein extinction coefficients from amino acid composition, *Anal. Biochem.* 182, 319–326. 1989.

47. Colon, W., Analysis of protein structure by solution optical spectroscopy, *Methods Enzymol.* 309, 605–632, 1999.

48. Kaaden, O.R., Frenzel, B., and Moennig, V., Comparative evaluation of kinetic nephelometry for antibody detection, *Comp. Immunol. Microbiol. Infect. Dis.* 2, 477–484, 1979.

49. Höffken, K., Bestek, U., Sperber, U., and Schmidt, C.G., Quantitation of immune complexes by nephelometry, *J. Immunol. Methods* 29, 237–244, 1979.

50. Feldkamp, C.S., Levinson, S.S., Perry, M., and Amin, V., Anti-IgG combined with rate nephelometry for measure polyethylene glycol-precipitated circulating immune complexes, *Clin. Chem.* 31, 2024–2027, 1985.

51. Lee, C.J., Quality control of polyvalent pneumococcal polysaccharide-protein conjugate vaccine by nephelometry, *Biologicals* 30, 97–103, 2002.

52. Jones, A., Young, D., Taylor, J. et al., Quantification of microbial productivity via multi-angle light scattering and supervised learning, *Biotechnol. Bioeng.* 59, 131–143, 1998.
53. Girod, S., Baldet-Dupy, P., Maillols, H., and Devoisselle, J.M., On-line direct determination of the second virial coefficient of a natural polysaccharide using size-exclusion chromatography and multi-angle laser light scattering, *J. Chromatogr. A* 943, 147–152, 2002.
54. Mendichi, R., Rizzo, V., Gigli, M., and Schieroni, A.G., Fractionation and characterization of a conjugate between a polymeric drug-carrier and the antitumor drug camptothecin, *Bioconjug. Chem.* 13, 1253–1258, 2002.
55. White, D.R. Jr., Hudson, P., and Adamson, J.T., Dextrin characterization by high-performance anion-exchange chromatography—Pulsed amperometric detection and size-exclusion chromatography–multi-angle light scattering–refractive index detection, *J. Chromatogr. A* 997, 79–85, 2003.
56. Barackman, J., Prado, I., Karunatilake, C., and Furuya, K., Evaluation of on-line high-performance size-exclusion chromatography, differential refractometry, and multi-angle laser light scattering analysis of the monitoring of the oligomeric state of human immunodeficient virus vaccine protein antigen, *J. Chromatogr. A* 1043, 57–64, 2004.
57. MacNair, J.E., Desai, T., Teyral, J. et al., Alignment of absolute and relative molecular size specifications for a polyvalent pneumococcal polysaccharide vaccine (PNEUMOVAX 23), *Biologicals* 33, 49–58, 2005.
58. Wahlund, K.-G. and Giddings, J.R., Properties of an asymmetrical flow field-flow fractionation channel having a permeable wall, *Anal. Chem.* 59, 1332–1339, 1987.
59. van Bruijnsvoort, M., Wahlund, K.G., Nilsson, G., and Kok, W.T., Retention behavior of amylopectins in asymmetrical flow field-flow fractionation studied by multi-angle light scattering detection, *J. Chromatogr. A* 925, 171–182, 2001.
60. Frankema, W., van Bruijnsvoort, M., Tijssen, R., and Kot, W.T., Characterisation of core-shell latexes by flow field flow fractionation with multi-angle light scattering detection, *J. Chromatogr. A* 943, 251–261, 2002.
61. Wolf, M., Buckau, G., and Chanel, V., Asymmetrical flow field-flow fractionation of humic substances: Comparison of polyacrylic acids and polystyrene sulfonate as molar mass standards, in *Humic Substances Molecular Details and Applications in Land and Water Conservation*, eds. E.A. Ghabboor and G. Davis, Taylor & Francis, New York, Chapter 2, 2004.
62. Luo, J., Leeman, M., Ballagi, A. et al., Size characterization of green fluorescent protein inclusion bodies in *E. coli* using asymmetrical flow field-flow fractionation-multi-angle light scattering, *J. Chromatogr. A* 1120, 158–164, 2006.
63. Mao, S., Augsten, C., Mäder, K., and Kissel, T., Characterization of chitosan and its derivatives using asymmetrical flow field-flow-fractionation: A comparison with traditional methods, *J. Pharm. Biomed. Anal.* 45, 736–741, 2007.
64. Leeman, M., Islam, M.T., and Haseltine, W.G., Asymmetrical flow field-flow fractionation coupled with multi-angle light scattering and refractive index detections for characterization of ultra-high molar mass poly(acrylamide) flocculants, *J. Chromatogr. A* 1172, 194–203, 2007.
65. Augsten, C. and Mäder, K., Characterization molar mass distributions and molecule structure of different chitosans using asymmetrical flow field-flow fractionation combined with multi-angle light scattering, *Int. J. Pharm.* 351, 23–30, 2008.
66. Pease III L.F., Lipin, D.I., Tsai, D.H. et al., Quantitative characterization of virus-like particles by asymmetrical flow field flow fractionation, electrospray differential mobility analysis, and transmission electron microscopy, *Biotechnol. Bioeng.* 102, 845–855, 2009.
67. Fujime, S., Quasi-elastic scattering of laser light. A new tool for the dynamic study of biological macromolecules, *Adv. Biophys.* 3, 1–43, 1972.

68. Schurr, J.M., Dynamic light scattering of biopolymers and biocolloids, *CRC Crit. Rev. Biochem.* 4, 371–431, 1977.

69. Langowski, J., Kremer, W., and Kapp, U., Dynamic light scattering for study of solution conformation and dynamics of superhelical DNA, *Methods Enzymol.* 211, 430–448, 1992.

70. Fincham, A.G. and Moradian-Oldak, J., Recent advances in amelogenin biochemistry, *Connect. Tissue Res.* 32, 119–124, 1998.

71. Pérez, S., Rodriguez-Carvajal, M.A., and Doco, T., A complex plant cell wall polysaccharide: Rhamnogalacturonan II. A structure in quest of a function, *Biochimie* 85, 109–121, 2003.

72. Sebag, J., To see the invisible: The quest of imaging vitreous, *Dev. Ophthalmol.* 42, 5–28, 2008.

73. Rodríguez, H.B. and San Román, E., Excitation energy transfer and trapping in dye-loaded solid particles, *Ann. N.Y. Acad. Sci.* 1130, 247–252, 2008.

74. Hermeling, S., Schellekens, H., Mass, C. et al., Antibody response to aggregated human interferon α2b in wild-type and transgenic immune tolerant mice depends on type and level of aggregation, *J. Pharm. Sci.* 95, 1084–1096, 2006.

75. Nobbmann, U., Connah, M., Fish, B. et al., Dynamic light scattering as a relative tool for assessing the molecular integrity and stability of monoclonal antibodies, *Biotechnol. Genet. Eng. Rev.* 24, 117–128, 2007.

76. Hawe, A. and Friess, W., Stabilization of a hydrophobic recombinant cytokines by human serum albumin, *J. Pharm. Sci.* 96, 2987–2999, 2007.

77. Lin, J.J., Meyer, J.D., Carpenter, J.F., and Manning, M.C., Stability of human serum albumin during bioprocessing: Denaturation and aggregation during processing of albumin paste, *Pharm. Res.* 17, 391–396, 2000.

78. Bermudez, O. and Faciniti, D., Aggregation and denaturation of antibodies: A capillary electrophoresis, dynamic light scattering, and aqueous two-phase partitioning study, *J. Chromatogr. B* 807, 17–24, 2004.

79. Doyle, B.L., Pollo, M.J., and Pekar, A.H., Biophysical signatures of noncovalent aggregates formed by a glucagonlike peptide-1 analogue: A prototypical example of biopharmaceutical aggregation, *J. Pharm. Sci.* 94, 2749–2763, 2005.

80. Sukumar, M., Storms, J.M., and De Filippis, M.R., Non-native intermediate conformational states of human growth hormone in the absence of organic solvents, *Pharm. Res.* 22, 789–796, 2005.

81. Sukumar, M., Doyle, B.L., Combs, J.L., and Pekar, A.H., Opalescent appearance of an IgG1 antibody at high concentration and its relationship to noncovalent association, *Pharm. Res.* 21, 1087–1093, 2004.

82. Hartmann, W.K., Saptharishi, N., Yang, X.Y. et al., Characterization and analysis of thermal denaturation of antibodies by size-exclusion high-performance liquid chromatography with quadrupole detection, *Anal. Biochem.* 325, 227–239, 2004.

83. Coppolino, R., Coppolino, S., and Villari, V., Study of the aggregation of insulin glargine by light scattering, *J. Pharm. Sci.* 95, 1029–1034, 2006.

84. Ahrer, K., Buchacher, A., Iberer, G. et al., Analysis of aggregates of human immunoglobulin G using size-exclusion chromatography, static and dynamic light scattering, *J. Chromatogr. A* 1009, 89–96, 2003.

85. Militello, V., Casarino, C., Emanuele, A. et al., Aggregation kinetics of bovine serum albumin studied by FTIR spectroscopy and light scattering, *Biophys. Chem.* 107, 175–187, 2004.

86. Ahrer, K., Buchacher, A., Iberer, G., and Jungbauer, A., Thermodynamic stability and formation of aggregates of human immunoglobulin G characterized by differential scanning calorimetry and dynamic light scattering, *J. Biochem. Biophys. Methods* 66, 73–86, 2006.

87. Banachowicz, E., Light scattering studies of proteins under compression, *Biochim. Biophys. Acta* 1764, 405–413, 2006.
88. Chung, K., Kim, J., Cho, B.K. et al., How does dextran sulfate prevent heat induced aggregation of proteins? The mechanism and its limitation as aggregation inhibitor, *Biochim. Biophys. Acta* 1774, 249–257, 2007.
89. Linetsky, M., Shipova, E., Cheng, R., and Ortwerth, B.J., Glycation by ascorbic acid oxidation products leads to the aggregation of lens proteins, *Biochim. Biophys. Acta* 1782, 22–34, 2008.
90. Lyutova, E.M., Kasakov, A.S., and Gurvitz, B.Y., Effects of arginine on kinetics of protein aggregation studied by dynamic laser light scattering and turbidimetry techniques, *Biotechnol. Prog.* 23, 1411–1416, 2007.
91. Kiese, S., Pappenberger, A., Friess, W., and Mahler, H.C., Shaken, not stirred: Mechanical stress testing of an IgG1 antibody, *J. Pharm. Sci.* 97, 4347–4366, 2008.
92. Jachimska, B., Wasilewska, M., and Adamczyk, Z., Characterization of globular protein solutions by dynamic light scattering, electrophoretic mobility and viscosity measurements, *Langmuir* 24, 6866–6872, 2008.
93. Chodanker, S., Aswal, V.K., Kohlbrecher, J. et al., Structural evolution during protein denaturation as induced by different methods, *Phys. Rev. E Stat. Nonlin. Soft Matter Phys.* 77, 031901, 2008.
94. Goyal, M.K., Roy, I., Banerjee, U.C. et al., Role of benzyl alcohol in the prevention of heat-induced aggregation and inactivation of hen egg white lysozyme, *Eur. J. Pharm. Biopharm.* 71, 367–376, 2009.
95. Remmele, R.L. Jr., Stushnoff, C., and Carpenter, J.F., Real-time in situ monitoring of lysozyme during lyophilization using infrared spectroscopy: Dehydration stress in the presence of sucrose, *Pharm. Res.* 14, 1548–1555, 1997.
96. Cowdery, S., Frey, M., Orlowski, S., and Grey, A., Stability characteristics of freeze-dried human live virus vaccines, *Dev. Biol. Stand.* 36, 297–303, 1976.
97. Mariner, J.C., House, J.A., Sollod, A.E. et al., Comparison of the effect of various chemical stabilizers and lyophilization cycles on the thermostability of a Vero cell-adapted rinderpest vaccine, *Vet. Microbiol.* 21, 195–209, 1990.
98. Magari, R.T., Munoz-Antoni, I., Baker, J., and Flagler, D.J., Determining shelf life by comparing degradations at elevated temperatures, *J. Lab. Clin. Anal.* 18, 159–164, 2004.
99. Salnikova, M.S., Joshi, S.B., and Rytting, J.H., Physical characterization of clostridium difficile toxins and toxoids: Effect of the formaldehyde crosslinking on thermal stability, *J. Pharm. Sci.* 97, 3735–3752, 2008.
100. Maa, Y.F., Ameri, M., Shu, C. et al., Hepatitis-B surface antigen (HbsAg) powder formulation: Process and stability assessment, *Curr. Drug Deliv.* 4, 57–67, 2007.
101. Ausar, S.F., Foubert, T.R., Hudson, M.H. et al., Conformational stability and disassembly of Norwalk virus-like particles. Effect of pH and temperature, *J. Biol. Chem.* 281, 19478–19488, 2006.
102. McHugh, C.A., Tammariello, R.F., Millard, C.B., and Carra, J.H., Improved stability of a protein vaccine through elimination of a partially unfolded state, *Protein Sci.* 13, 2736–2743, 2004.
103. Carra, J.H., Welcher, B.C., Shokman, R.D. et al., Mutational effects on protein folding stability and antigenicity: The case of streptococcal pyrogenic exotoxin A, *Clin. Immunol.* 108, 60–68, 2003.
104. Volkin, D.B., Burke, D.J., Marfie, K.E. et al., Size and conformational stability of the hepatitis A virus used to prepare VAQTA, a highly purified inactivated vaccine, *J. Pharm. Sci.* 86, 666–673, 1997.
105. Chuan, Y.P., Fan, Y.Y., Lua, L., and Middleberg, A.P., Quantitative analysis of virus-like particle size and distribution by field-flow fractionation, *Biotechnol. Bioeng.* 99, 1425–1433, 2008.

106. Lincopan, N., Espindola, N.M., Vas, A.J., and Carmona-Ribeiro, A.M., Cationic supported lipid bilayers for antigen presentation, *Int. J. Pharm.* 340, 216–222, 2007.
107. Kalbfuss, B., Genzel, Y., Wolff, M. et al., Harvesting and concentration of human influenza virus produced in serum-free mammalian cell culture for the production of vaccines, *Biotechnol. Bioeng.* 97, 73–85, 2007.
108. Michalsky, R., Pframm, P.H., Czernak, P. et al., Effects of temperature and shear force on infectivity of the baculovirus *Autographa californica* M nucleopolyhedrovirus, *J. Virol. Methods* 153, 90–96, 2008.
109. Ausar, S.F., Rexroad, J., Frolor, V.G. et al., Analysis of the thermal and pH stability of human respiratory syncytial virus, *Mol. Pharm.* 2, 491–499, 2005.
110. Bachelder, E.M., Beaudette, T.T., Broaders, K.E. et al., Acid-degradable polyurethane particles for protein-based vaccines: Biological evaluation and in vitro analysis of particle degradation products, *Mol. Pharm.* 5, 876–884, 2008.
111. Devineni, D., Ezekwudo, D., and Palaniappan, R., Formulation of maltodextrin entrapped in polycaprolactone microparticles for protein and vaccine delivery: Effect of size determining formulation process variables of microparticles on the hydrodynamic diameter of BSA, *J. Microencapsul.* 24, 358–370, 2007.
112. Ausar, S.F., Espina, M., Brock, J. et al., High-throughput screening of stabilizers for respiratory syncytial virus: Identification of stabilizers and their effects on the conformational thermostability of viral particles, *Hum. Vaccin.* 3, 94–103, 2007.
113. Davidsen, J., Rosenkrands, I., Christensen, D. et al., Characterization of cationic liposomes based on dimethyldioctadecylammonium and synthetic cord factor from *M.tuberculosis* (trehalose 6,6′-debehenate)—A novel adjuvant inducing both strong CMI and antibody responses, *Biochim. Biophys. Acta* 1718, 22–31, 2005.
114. Shi, L., Sanyal, G., Ni, A. et al., Stabilization of human papilloma virus-like particles by non-ionic surfactants, *J. Pharm. Sci.* 94, 1538–1551, 2005.
115. Ding, Z., Rivas-Benita, M., Hischberg, H. et al., Preparation and characterization of diphtheria toxoid-loaded elastic vesicles for transcutaneous immunization, *J. Drug Target.* 16, 555–563, 2008.
116. Rexroad, J., Evans, R.K., and Middaugh, C.R., Effect of pH and ionic strength on the physical stability of adenovirus type 5, *J. Pharm. Sci.* 95, 237–247, 2006.
117. Rexroad, J., Martin, T.T., McNeilly, D. et al., Thermal stability of adenovirus type 2 as a function of pH, *J. Pharm. Sci.* 95, 1469–1479, 2006.
118. Trilisky, E.I. and Lenhoff, A.M., Sorption processes in ion-exchange chromatography of viruses, *J. Chromatogr. A* 1142, 2–12, 2007.
119. Rungsardthong, U., Deshpande, M., Bailey, L. et al., Copolymers of amine methacrylate with poly(ethylene glycol) as vectors for gene therapy, *J. Control. Release* 73, 359–380, 2001.
120. Read, M.L., Bremner, K.H., Oupický, D. et al., Vectors based on reducible polycations facilitate intracellular release of nucleic acids, *J. Gene Med.* 5, 232–245, 2003.
121. Kim, T.H., Park, I.K., Nah, J.W. et al., Galactosylated chitosan/DNA nanoparticles prepared using water-soluble chitosan as a gene carrier, *Biomaterials* 25, 3783–3792, 2004.
122. Christiaens, B., Dubruel, P., Grooten, J. et al., Enhancement of polymethacrylate-mediated gene delivery by Penetratin, *Eur. J. Pharm. Sci.* 24, 525–537, 2005.
123. Tumanova, I., Boyer, J., Ausar, S.F. et al., Analytical and biological characterization of supercoiled plasmids purified by various chromatographic techniques, *DNA Cell Biol.* 24, 819–831, 2005.
124. Cavallaro, G., Campiai, M., Licciardi, M. et al., Reversibly stable thioplexes for intracellular delivery of genes, *J. Control. Release* 115, 322–334, 2006.
125. Masotti, A., Moretti, F., Mancini, F. et al., Physicochemical and biochemical study of selected hydrophobic polyethyleneimine-based polycationic liposomes and their complexes with DNA, *Bioorg. Med. Chem.* 15, 1504–1515, 2007.

126. Masotti, A., Mossa, G., Cametti, C. et al., Comparison of different commercially available cationic liposome–DNA lipoplexes: Parameters influencing toxicity and transfection efficiency, *Colloids Surf. B. Biointerfacs* 68, 136–144, 2009.

127. Dufresne, M.-H., Eisabahy, E., and Leroux, J.-C., Characterization of polyion complex micelles designed to address the challenges of oligonucleotide delivery, *Pharm. Res.* 25, 2083–2093, 2008.

128. Zhang, X., Ercelen, S., Duportali, G. et al., Hydrophobically modified low molecular weight chitosans as efficient and nontoxic gene delivery vehicles, *J. Gene Med.* 10, 527–539, 2008.

129. Singh, R., Saxena, A., and Moxumdar, S., Calcium phosphate–DNA nanocomposites morphological studies and their bile duct infusion for liver-directed gene therapy, *Int. J. Appl. Ceramic Technol.* 5, 1–10, 2008.

130. Desigaux, L., Sainlos, M., Lambert, O. et al., Self-assembled lamellar complexes of siRNA with lipidic aminoglycoside derivatives promote efficient siRNA delivery and interference, *Proc. Natl. Acad. Sci. USA* 104, 16534–16539, 2007.

131. Miyata, K., Kakizawa, Y., Nishiyama, N. et al., Freeze-dried formulations for in vivo gene delivery of PEGylated polyplex micelles with disulfide crosslinked cores to the liver, *J. Control. Release* 109, 15–23, 2005.

132. Cherng, J.-Y., van de Wetering, P., Taisma, H. et al., Effect of size and serum proteins on transfection efficiency of poly((2-dimethylamino)ethyl methacrylate)-plasmid nanoparticles, *Pharm. Res.* 13, 1038–1042, 1996.

133. Dinadayala, P., Sambou, T., Daffé, M., and Lemassu, A., Comparative structural analysis of the α-glucan and glycogen from *Mycobacterium bovis*, *Glycobiology* 18, 502–508, 2008.

134. Martin, J.M., Pâques, M., van der Velden-de Groot, T.A., and Beuvery, E.C., Characterization of antibody labelled colloidal gold particles and their applicability in a sol particle immuno assay (SPIA), *J. Immunoassay* 11, 31–47, 1990.

135. Liu, W., Cellmer, T., Keerl, D. et al., Interaction of lysozyme in guanidinium chloride solutions from static and dynamic light scattering measurements, *Biotechnol. Bioeng.* 90, 482–490, 2005.

136. Georgalis, Y., Umbach, P., Raptis, J., and Saenger, W., Lysozyme aggregation studied by light scattering. II. Variations of protein concentration, *Acta Crystallogr. D Biol. Cyrstallogr.* 53, 703–712, 1997.

137. Shaw, A.K. and Pal, S.K., Resonance energy transfer and ligand binding studies on pH-induced folded states of human serum albumin, *J. Photochem. Photobiol. B* 90, 187–197, 2008.

138. Ling, Z., Tran, K.C., Arnold, J.J., and Teng, M.N., Purification and characterization of recombinant human respiratory syncytial virus nonstructural protein NS1, *Protein Expr. Purif.* 57, 261–270, 2008.

139. Suzuki, S., Wilson-Kubalek, E.M., Wart, D. et al., The oligomeric structure of high molecular weight adiponectin, *FEBS Lett.* 581, 809–814, 2007.

140. Brand, T., Richter, S., and Berger, S., Diffusion NMR as a new method for the determination of the gel point of gelatin, *J. Phys. Chem. B* 110, 15853–15857, 2006.

141. Vierira, D.F., Watanabe, I., Sant'ana, C.D. et al., Purification and characterization of jararassin-I, a thrombin-like enzyme from *Bothrops jararaca* snake venom, *Acta Biochim. Biophys. Sin.* 34, 798–802, 2004.

142. Sandberg, A., Harrison, D.J., and Karlsson, B.G., Thermal denaturation of spinach plastocyanin: Effect of copper site oxidation state and molecular oxygen, *Biochemistry* 42, 10301–10310, 2003.

143. Rami, B.R. and Udogaonkar, J.B., pH-jump-induced folding and unfolding studies of barstar: Evidence for multiple folding and unfolding pathways, *Biochemistry* 40, 15267–15279, 2003.

144. Hofmann, I., Mücke, N., Read, J. et al., Physical characterization of plakophilin 1 reconstituted with and without zinc, *Eur. J. Biochem.* 267, 4381–4389, 2000.
145. Burden, A.E., Wu, C., Dailey, T.A. et al., Human ferrochelatase: Crystallization of the [2FE-2S] cluster and determination that the enzyme is homodimer, *Biochim. Biophys. Acta* 1453, 191–197, 1999.
146. Phillips, J.D., Whitby, F.G., Kushner, J.P., and Hill, C.P., Characterization and crystallization of human uroporphyrinogen decarboxylase, *Protein Sci.* 6, 1343–1346, 1997.
147. Di Marco, J.S., Priestle, J.P., Grütter, M.G. et al., Characterization, crystallization and preliminary crystallographic analysis of human recombinant cyclooxygenase-2, *Acta Crystallogr. D Biol. Crystallogr.* 53, 224–226, 1997.
148. Baudys, M., Uchio, T., Mix, D. et al., Physical stabilization of insulin by glycosylation, *J. Pharm. Sci.* 84, 28–33, 1995.
149. Mach, H., Volkin, D.B., Burke, C.J. et al., Nature of the interaction of heparin with acidic fibroblast growth factor, *Biochemistry* 32, 5480–5489, 1993.
150. Yen, Y., Seizer, J.C., and Baskin, R.J., Dynamic light scattering characterization of the detergent-free, delipidated $(Ca^{2=} + Mg^{2+})$-ATPase from sarcoplasmic reticulum, *Biochim. Biophys. Acta* 509, 78–89, 1979.
151. Tao, Y. and Xu, W., Microwave-solubilization and solution properties of hyperbranched polysaccharide, *Carbohydr. Res.* 343, 3071–3078, 2008.
152. Spinelli, F.J., Kiick, K.L., and Furst, E.M., The role of heparin self-association in the gelation of heparin-functionalized polymers, *Biomaterials* 29, 1299–1306, 2008.
153. Maleki, A., Kjøniksen, A.L., and Nyström, B., Characterization of the chemical degradation of hyaluronic acid during chemical gelation in the presence of different cross-linker agents, *Carbohydr. Res.* 342, 2775–2792, 2007.
154. Dona, A., Yuen, C.W., Peate, J. et al., A new NMR method for directly monitoring and quantifying the dissolution kinetics of starch in DMSO, *Carbohydr. Res.* 342, 2604–2610, 2007.
155. Besoeer, A., Hause, G., Kressler, J., and Mäder, K., Hydrophobically modified hydroxyethyl starch: Synthesis, characterization, and aqueous self-assembly into nano-sized polymeric micelles and vesicles, *Biomacromolecules* 8, 359–367, 2007.
156. Park, K., Kim, K., Kwon, I.C. et al., Preparation and characterization of self-assembled nanoparticles of heparin-deoxycholic acid conjugates, *Langmuir* 20, 11726–11731, 2004.
157. Radosta, S., Haberer, M., and Vorwarg, W., Molecular characterization of amylose and starch in dimethyl sulfoxide, *Biomacromolecules* 2, 970–978, 2001.
158. Tuinier, R., Zoon, P., Olieman, C. et al., Isolation and physical characterization of an exocellular polysaccharide, *Biopolymers* 49, 1–9, 1999.
159. Viebka, C. and Williams, P.A., The influence of temperature on the characterization of water-soluble polymers using asymmetrical flow field-flow-fractionation coupled to multiangle laser light scattering, *Anal. Chem.* 72, 3896–3901, 2000.
160. Atha, D.H., Coxon, B., Reipa, V., and Gaigalas, A.K., Physicochemical characterization of low molecular weight heparin, *J. Pharm. Sci.* 84, 360–364, 1995.
161. Gaigalas, A.K., Hubbard, J.B., LeSage, R. and Atha, D.H., Physical characterization of heparin by light scattering, *J. Pharm. Sci.* 84, 355–359, 1995.
162. Ousalem, M., Busnel, J.P., and Nicolai, T., A static and dynamic light scattering study of sharp pectin fractions in aqueous solution, *Int. J. Biol. Macromol.* 15, 209–213, 1993.
163. Errington, N., Harding, S.E., Vårum, K.M., and Illum, L., Hydrodynamic characterization of chitosans varying in degrees of acetylation, *Int. J. Biol. Macromol.* 15, 113–117, 1993.
164. Fluegel, S. and Maskos, M., Cylindrical poly(oligo DNA), *Biomacromolecules* 8, 700–702, 2007.

165. Lau, C., Bitton, R., Bianco-Peled, H. et al., Morphological characterization of self-assembled peptide nucleic acid amphiphiles, *J. Phys. Chem. B* 110, 9027–9033, 2006.
166. Eimer, W., Williamson, J.R., Boxer, S.G., and Percora, R., Characterization of the overall and internal dynamics of short oligonucleotides by depolarized dynamic light scattering and NMR relaxation measurements, *Biochemistry* 29, 799–811, 1990.
167. Patkowski, A., Eimer, W., and Dorfmüller, T., Internal dynamics of tRNA(Phe) by depolarized dynamic light scattering, *Biopolymers* 30, 975–983, 1990.

9 Use of Luminescence to Measure Conformational Change in Biopharmaceuticals with Emphasis on Protein and Protein Drug Products

Fluorescence is the emission of energy from a molecule following adsorption of electromagnetic radiation and is a useful tool in the study of protein structure. Fluorescence derived from a protein (or other biomacromolecules) can be separated into intrinsic fluorescence and extrinsic fluorescence. Intrinsic fluorescence is the emission of light from a protein after excitation by monochromatic light; the intrinsic fluorescence from a protein is largely a function of tryptophanyl and tyrosine residues.[1] Fluorescence is also useful for characterizing nucleic acids[2]; most studies with nucleic acids use extrinsic probes.[3–5]

In the context of the current discussion, extrinsic fluorescence describes fluorescence from a probe such as dansyl chloride,[6,7] which can be covalently bound to a protein or 5, 5′-bis(8-anilino-1-naphthalenesulfonate),[8] which is noncovalently bound to a protein. These probes are known as extrinsic fluorophors.[9]

Fluorescence is a form of luminescence[*]; phosphorescence and chemiluminescence are also forms of luminescence.[10] Phosphorescence is similar to fluorescence in that phosphorescence is the emission of energy after electromagnetic radiation:

---

[*] Luminescence can be defined as the emission of light from an object such as rock or from a chemical compound. Within the context of the current chapter, luminescence includes intrinsic fluorescence and phosphorescence and extrinsic fluorescence resulting fluorophores which are covalently or noncovalently bound to proteins or other biomacromolecules. See Curie, D., *Luminescence in Crystals*, Methuen, London, U.K., 1963; Goldberg, P. ed., *Luminescence of Inorganic Solids*, Academic Press, New York, 1996; Barenboim, G.M., Domanski, A.N., and Turoverov, K.K., *Luminescence of Biopolymers and Cells*, Plenum Press, New York, 1969; Winefordner, J.D., Schulman, S.G., and O'Haver, T.C., *Luminescence Spectrometry in Analytical Chemistry*, Wiley-Interscience, New York, 1972; Goldberg, M.C., *Luminescence Applications in Biological, Chemical, Environmental, and Hydrological Sciences*, American Chemical Society, Washington, DC, 1989.

phosphorescence differs from fluorescence in having a longer lifetime and the wavelength shift in the emitted light is longer than that seen in fluorescence.[11–15] Both fluorescence and phosphorescence involve the decay of a molecule in its excited state to a ground state. Fluorescence is the emission of energy as an electron returns from the S_1 state (singlet) to a ground state while phosphorescence is the emission of energy from a triplet state to a ground state. The triplet state is derived from the signet state by intersystem crossing; direct passage from ground state to the triplet state is a forbidden transition. The transition from ground state to the singlet state is associated with UV/VIS absorption of light. The reader is directed to several excellent basic discussions of luminescence.[16–19]

Chemiluminescence is luminescence derived from chemical reactions.[20–27] While anecdotal observations of chemiluminescence data back to antiquity,[22] current work on chemiluminescence is considered to have started with the work of Radziszewski[28] on lophine (Figure 9.1). Luminol (5-amino-2,3-dihydro-1,4-phthalazinedione) and luminol derivatives (Figure 9.2) are commonly used chemiluminescent compounds. Lophine derivatives are used as phenolic enhancers for luminol–hydrogen peroxide–horseradish peroxidase chemiluminescence used in immunoassay systems for the detection of horseradish–antibody conjugates.[29] The oxidation of luminol to the activated aminophthalate product, which decays with the emission of light, is accomplished with a variety of oxidizing agents including reactive oxygen species; oxidation is also accomplished with horseradish peroxidase. Lucigenin (10,10′-dimethyl-9,9′-bisacridinium) is another common chemiluminescent compound (Figure 9.3), which is used in detection methods for reactive oxygen species such as superoxide anion,[30] Lucigenin has also been used as conformational probe for pressure-induced conformational change in proteins.[31] Lucigenin is a derivative of acridine and there are other acridine derivatives that are used as chemiluminescent probes (Figure 9.4). Acridinium cation derivatives form a common luminescent derivative, N-methylacridone via the action of oxidizing agents such as hydrogen peroxide in base. Ultrasound has been recently described for the "activation" of an acridinium ester.[32] N-Methylacridinium has been used for the affinity purification of acetylcholine esterase[33]; N-methylacridinium is an inhibitor of cholinesterases.[34,35] Acridinium esters (Figure 9.5) can be converted into derivatives that can be used to label antibodies and used for sensitive immunoassays in microplate and microarray formats.[36–39] The use of chemiluminescent signal can provide detection as low as 10^{-18} to 10^{-23} M.[40–46] Acridine derivatives can bind noncovalently to proteins, which can create some problems for the used of acridine derivatives to covalently label proteins.[47] N-Methylacridinium cation is bound to sites in horse liver alcohol dehydrogenase and human serum albumin with micromolar K_d affinity.[48] Exposure of the protein–acridinium complex to light (xenon lamp) resulted in dismutation of the N-methyl acridinium chloride to N-methylacridone and N-methyl-9,10-dihydroacridine (Figure 9.6); reduction of N-methyl acridinium to N-methyl-9,10-dihydroacridine is also accomplished by NAD(P)H in neutral phosphate buffer.[49] Other acridine derivatives such as proflavin (Figure 9.4) bind to specific sites in proteins and have been used to study protein conformation.[50–54]

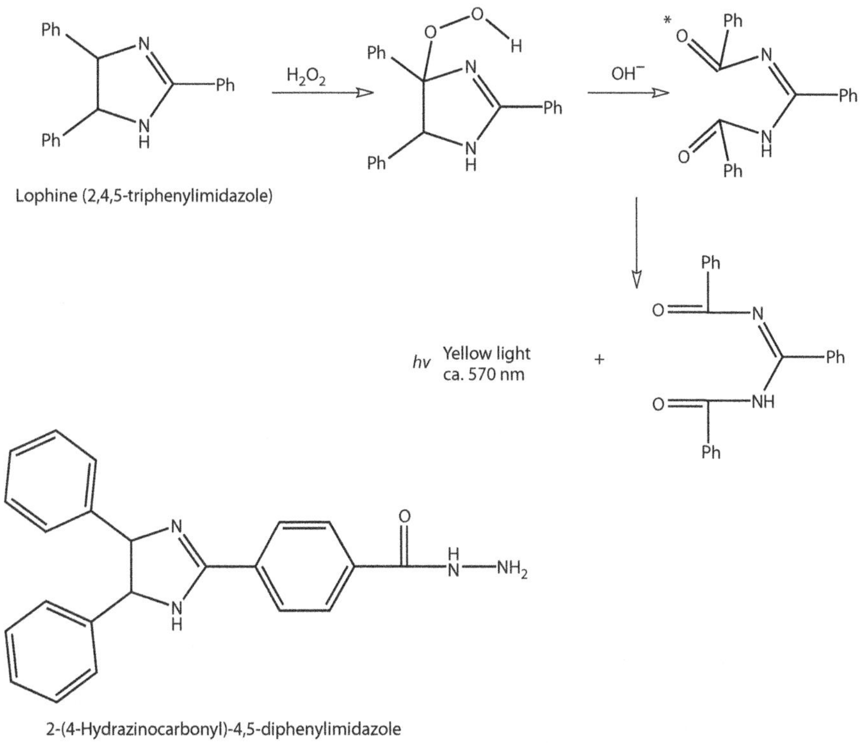

Lophine (2,4,5-triphenylimidazole)

H_2O_2

OH^-

hv Yellow light ca. 570 nm +

2-(4-Hydrazinocarbonyl)-4,5-diphenylimidazole

FIGURE 9.1 The structure of lophine and the excitation pathway for lophine chemilumi-nescence (see Radziszewski, B., Über die phosphorescenz der organishen und organishen Körper, *Leibigs Ann.* 203, 305–336, 1880; White, E.H. and Harding, M.J.C., The chemilu-minescence of lophine and its derivatives, *J. Am. Chem. Soc.* 86, 5686–5687, 1964; Kimura, M., Lu, G.H., NIshigawa, H. et al., Singlet oxygen generation from lophine hydroperox-ides, *Luminescence* 22, 72–76, 2007). Also shown is 2-(4-hydrazinocarbonylphenyl)-4,5-diphenylimidazole, a derivatives of lophine which can be coupled to matrix (see Pontén, E., Appleblad, P., Stigbrand, M. et al, Immobilized 2-(4-hydrazinocarbonylphenyl)-4,5-diphenylimidazole as solid phase luminophore in peroxyoxalate chemiluminescence, *Fresh J. Anal. Chem.* 356, 84–89, 1996).

Bioluminescence[55–57] is a luminescent response caused by a biological agent such an enzyme instead of a chemical such as an oxidizing agent. The mechanism of bioluminescence and chemiluminescence is the same: the relaxation of an excited state by ejection of a photon.[58] The luminescent response of luminol (see above) can be caused by either chemical agents (hydrogen peroxide in base) or by horseradish peroxidase. Thus, there is considerable "crossover" between chemiluminescence and bioluminescence in the literature.[59–61] Bioluminescence is only of indirect interest to the primary thrust of this book and the reader is directed to a selection of work in this area.[62–64] As noted above, horseradish peroxidase-based luminescence is of importance in microplate and microarray technology[65,66]: Luminescence has other

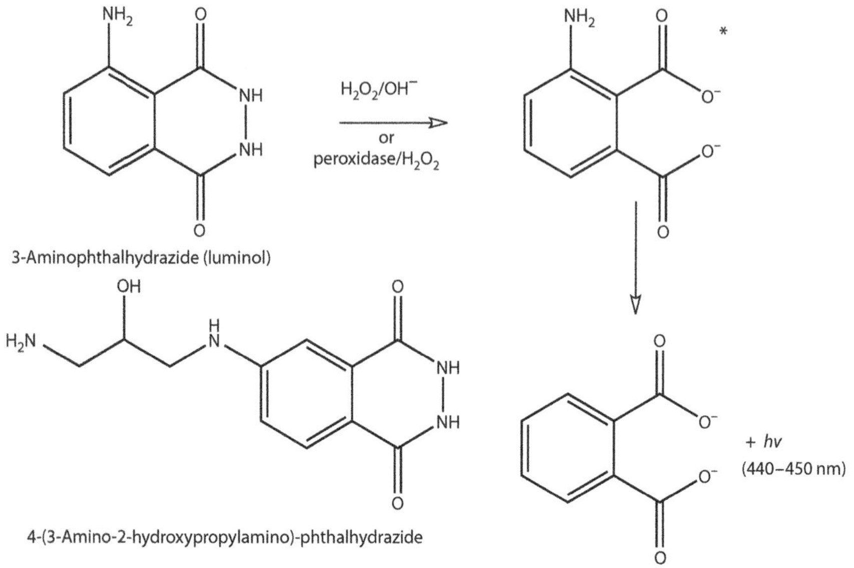

3-Aminophthalhydrazide (luminol)

4-(3-Amino-2-hydroxypropylamino)-phthalhydrazide

FIGURE 9.2 The structure of luminol (5-amino-2,3-dihydro-1,4-phthalazinedione) and a derivative. The modification of the amino group on luminol decreases quantum yield. 4-(3-Amino-2-hydroxypropylamino)-phthalhydrazide provides a derivative for coupling to proteins (see Seitz, W.R., Immunoassay labels based on chemiluminescence and bioluminescence, *Clin. Biochem.* 12, 120–125, 1954; García-Campaña, A.M. and Baeyens, W.R.G. eds., *Chemiluminescence in Analytical Chemistry*, Marcel Dekker, New York, 2001; Marquette, C.A. and Blum, L.J., Applications of the luminol chemiluminescent reaction in analytical chemistry, *Anal. Bioanal. Chem.* 385, 546–554, 2006). Luminol can also serve as substrate for horseradish peroxidase (most often the presence of an enhancer) providing a luminescent label for immunoassay (see Whitehead, T.P., Thorpe, G.H.G., Carter, T.J.N. et al., Enhanced luminescence procedure for sensitive determination of peroxidase-labelled conjugates in immunoassay, *Nature* 305, 158–159, 1983; Kim, B.B., Pisarev, V.V., and Egorov, A.M., A comparative study of peroxidases from horseradish and *Arthromyces ramosus* as labels in luminol-mediated chemiluminescent assays, *Anal. Biochem.* 199, 1–6, 1991; Ii, M., Yoshida, H., Aramaki, H. et al., Improved enzyme immunoassay for human basic fibroblast growth factor using a new enhanced chemiluminescence system, *Biochem. Biophys. Res. Commun.* 193, 540–555, 1993). Carnitinyl maleate luminol has been used as a probe for mitochondrial free radical production (see Lippman, R.D., Chemiluminescent measurement of free radicals and antioxidant molecular protection inside living rat mitochondria, *Exp. Gerontol.* 15, 335–351, 1980; see also Lippman, R.D., Site-specific chemiluminescent probes for measurement of radicals and peroxidation phenomena, in *Bioluminescence and Chemiluminescence: Instruments and Aplications*, Vol. II, ed. K. Van Dyke, CRC Press, Boca Raton, FL, pp. 13–23, 1985).

broad applications in biology and medicine[67–71] and the discussion is incomplete without the mention of the contribution of native fluorescent proteins such as green fluorescent protein to the study of protein conformation.[72–77]

Intrinsic fluorescence of proteins is mostly a product of tryptophanyl residues[16]; however, tyrosine fluorescence can be useful[78–84] and, despite low quantum yield,

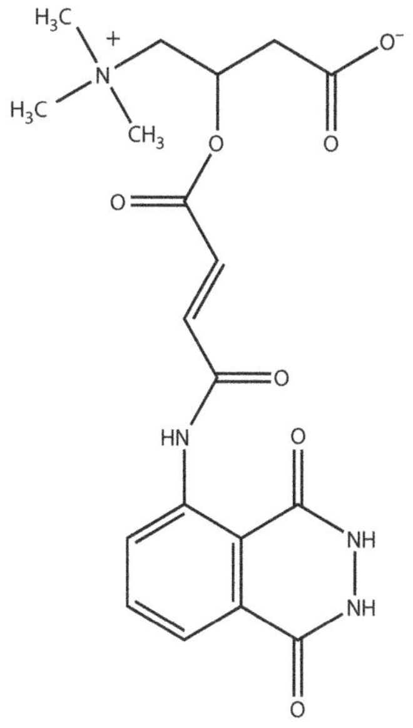

Carnitinyl maleate luminol

FIGURE 9.2 (continued)

there are also examples of the utility of phenylalanine fluorescence.[85–87] The fluorescence of tryptophanyl residues is sensitive to local solvent environment[88–93] and is thus useful for the study of conformational change. Tryptophan absorbs UV light with maxima at 218 and 280 nm (279 nm with a shoulder at 282 nm); the 280 nm absorption (283 nm) is used for excitation in luminescence studies with fluorescence emission occurring at 318 and at 340 nm and somewhat complex phosphorescence at longer wavelengths (400–500 nm) at low temperature (77 K).[13,94]

Phosphorescence usually requires higher sample concentration due to weaker emission and is temperature-sensitive.[95–97] Phosphorescence is due to decay of the triplet state to the ground state and is considered a forbidden transition and hence occurs over a longer period of time. As mentioned above, the triplet state is derived from the S_1 state by intersystem crossover. The temperature sensitivity of protein phosphorescence is not due to the intersystem crossover but more likely a reflection of various quenching effects.[97] Phosphorescence occurs at longer wavelengths than fluorescence and is referred to as being "redshifted." Strambini and colleagues have studied the phosphorescence of indole and indole derivatives (N-acetyltryptophan amide; 1-methyltryptophan) in aqueous solution at 20°C.[98] The phosphorescence lifetime (tau; τ) decreases in the presence of buffer salts such as phosphate acetate; NaCl had little effect while Tris was intermediate between NaCl and phosphate or acetate. The phosphorescence lifetime of the various indole derivatives was markedly less than that observed for the tryptophan in ribonuclease T_1. Tau decreased with increasing pH.

The incorporation of tryptophan analogues such as aza derivatives into proteins can provide derivatives with unique fluorescent emission spectra.[99,100] The quenching of tryptophan fluorescence by substances such as acrylamide is also used for the

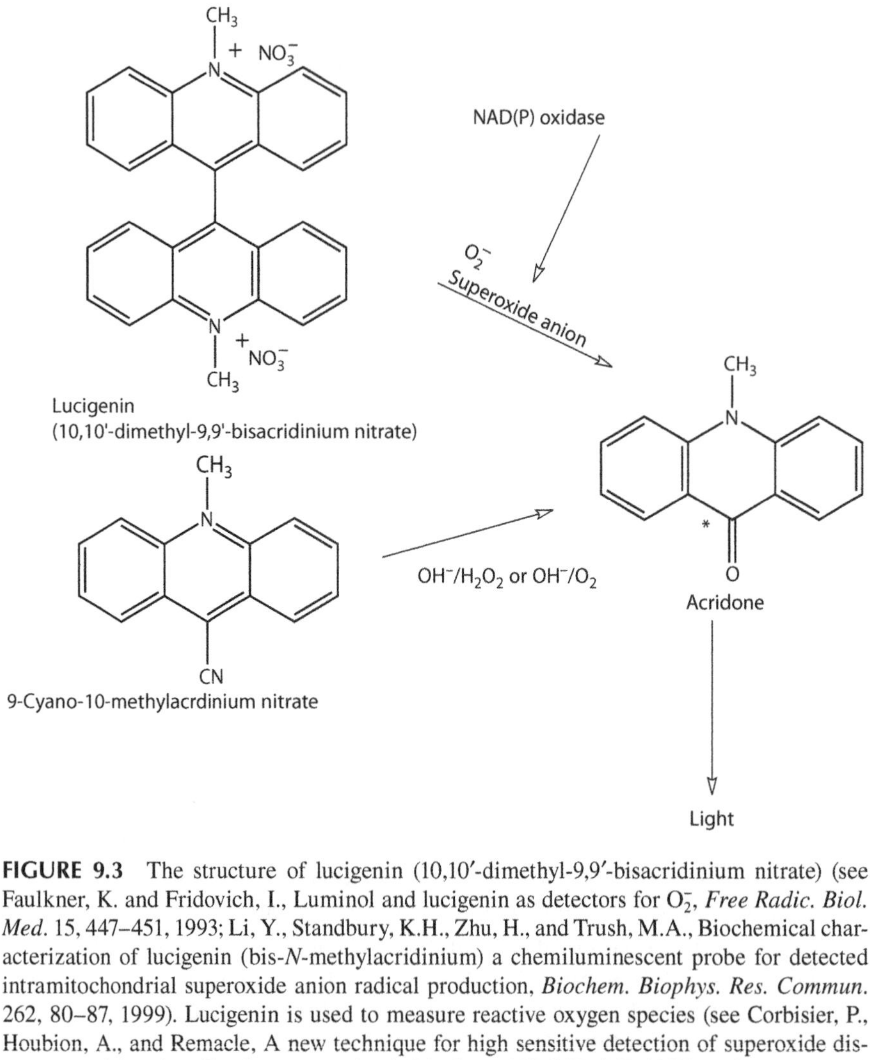

FIGURE 9.3 The structure of lucigenin (10,10′-dimethyl-9,9′-bisacridinium nitrate) (see Faulkner, K. and Fridovich, I., Luminol and lucigenin as detectors for O_2^-, *Free Radic. Biol. Med.* 15, 447–451, 1993; Li, Y., Standbury, K.H., Zhu, H., and Trush, M.A., Biochemical characterization of lucigenin (bis-*N*-methylacridinium) a chemiluminescent probe for detected intramitochondrial superoxide anion radical production, *Biochem. Biophys. Res. Commun.* 262, 80–87, 1999). Lucigenin is used to measure reactive oxygen species (see Corbisier, P., Houbion, A., and Remacle, A new technique for high sensitive detection of superoxide dismutase activity chemiluminescence, *Anal. Biochem.* 164, 240–247, 1987) such as produced with inflammation (Müller-Peddinghaus, R., In vitro determination of phagocyte activity by luminol- and lucigenin-amplified chemiluminescence, *Int. J. Pharmacol.* 6, 455–466, 1984).

study of protein structure.[101–108] Internal quenching in proteins can also be useful by sulfur-containing amino acids[109–111] and by internal fluorescence resonance energy transfer (FRET).[112–114]

There are a large number of studies on the use of intrinsic fluorescence for the study of protein conformation. This chapter focuses on some general concepts and specific applications of luminescence spectroscopy to proteins of pharmaceutical (including diagnostics) interest. Table 9.1 lists some studies on the use of intrinsic fluorescence to characterize proteins developed for biotechnological purposes. Table 9.2 lists some studies on the use of phosphorescence of the study of protein conformation. It can be

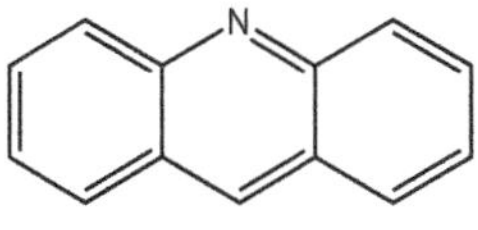

Acridine

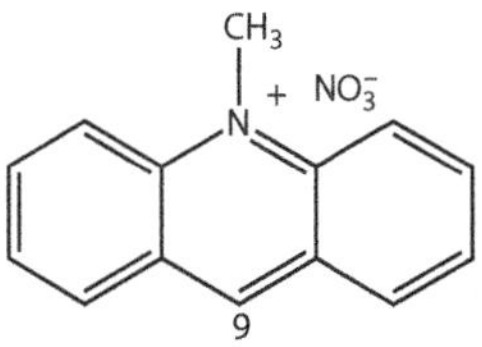

N-Methylacridinium nitrate

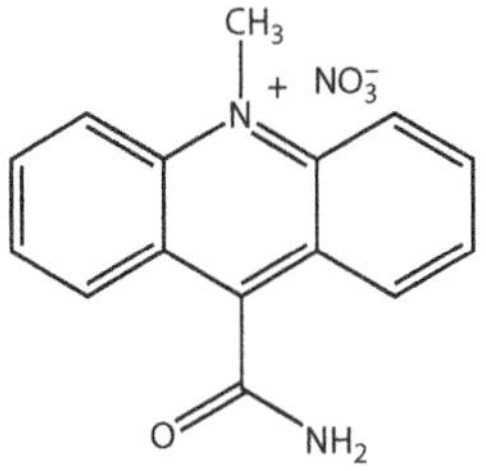

10-Methylacridinium-9-cardoxamide

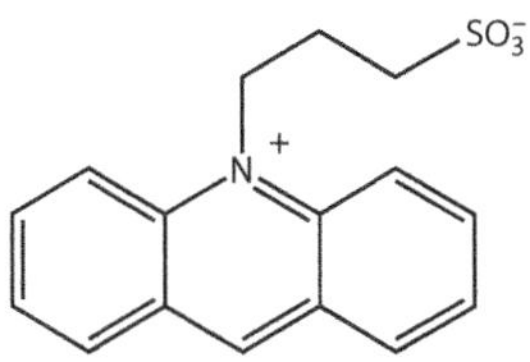

N-Sultopropylacridinium

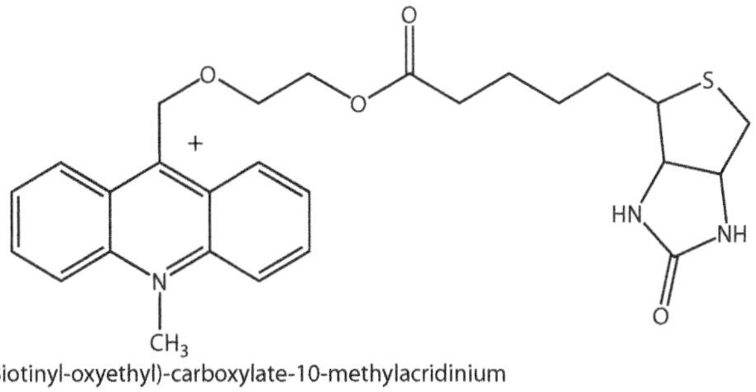

9-(2-Biotinyl-oxyethyl)-carboxylate-10-methylacridinium

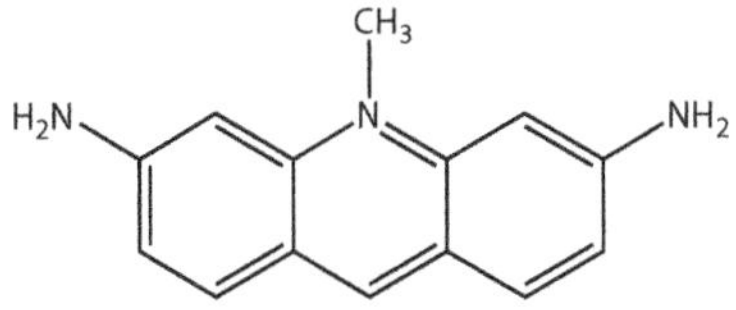

Acriflavine

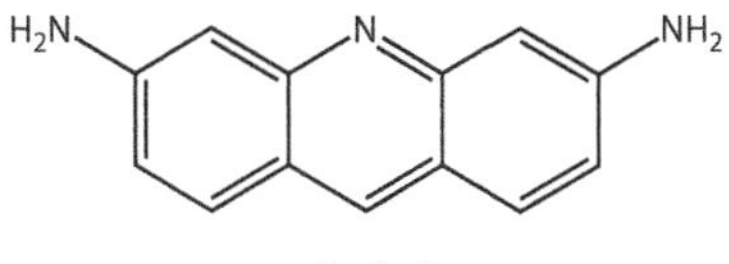

Proflavine

FIGURE 9.4 The structures of acridine and various chemiluminescent derivatives. The parent *N*-methylacridinium nucleus is shown with substituted derivatives. Most of the useful derivatives of the acridinium nucleus involve substitution at the 9 position on the acridine ring. An example is provide by *N*-methylacridinium-9-carboxamide (see Biwersi, J., Tulk, B., and Verkman, A.S., Long wavelength chloride-sensitive fluorescent indicators, *Anal. Biochem.* 219, 139–143, 1994) and by lucignin which also be considered to be a 9-substituted acridinium derivative. Functionally substituted derivatives such as biotinylated forms can be obtained (see also Scorilas, A., Agiamornioti, K., and Papadopoulos, K., Novel biotinylated acridinium derivatives: New reagents for fluorescent immunoassays and proteomics, *Clin. Chim. Acta* 457, 159–167, 2005; see also Wilson, R., Akhavan-Tafti, H., DeSilva, R., and Schaap, A.P., Electrochemiluminescence determination of 2′,6′-difluorophenyl 10-methylacridian-9-carboxylate, *Anal. Chem.* 73, 763–767, 2001).

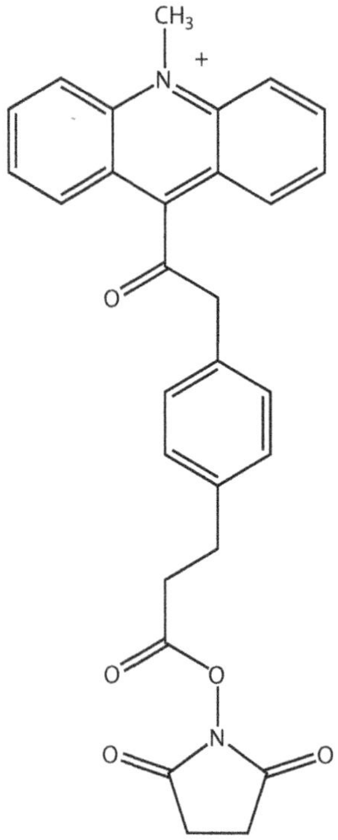

N-Hydroxysuccinimide acridinium ester

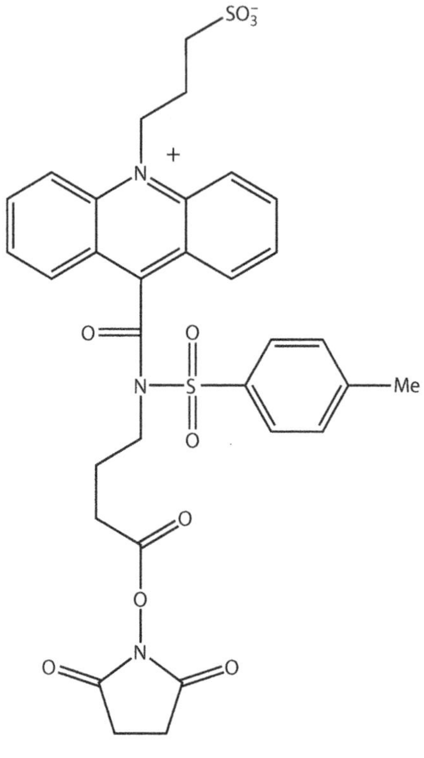

Acridinium active ester

FIGURE 9.5   The structures of acridinium derivatives used for the modification of proteins. These derivatives have been quite useful for sensitive immunoassays including microarrays (Seitz, W.R., Immnoassay labels based on chemiluminescence and bioluminescence, *Clin. Biochem.* 17, 120–125, 1984; Weeks, I., Sturgess, M., Brown, R.C., and Woodhead, J.S., Immunoassays using acridinium esters, *Methods Enzymol.* 133, 366–386, 1986; Goto, S., Takahashi, A., Kamisango, K., and Matsubara, K., Single-nucleotide polymorphism analysis by hybridization protection assay on solid support, *Anal. Biochem.* 307, 25–32, 2002; Roda, A., Pasini, P., Michelini, E., and Guardigli, M., Biotechnological applications of bioluminescence and chemiluminescence, *Trends Biotechnol.* 22, 295–303, 2004; see also Adamczyk, M., Gebler, J.C., Schreder, K., and Wu, J., Quantitative determination of noncovalently bound acridinium in protein conjugates by liquid chromatography/electrospray ion trap mass spectrometry, *Rapid Commun. Mass Spectrom.* 15, 670–674, 2001).

argued that any studies on the use of luminescence for the study of protein conformation are of interest to the biopharmaceutical community. It is not possible to discuss all of these studies within the scope of the current work. The several studies, which will be discussed below, have been selected on the basis of the author's experience with the characterization of biotechnological products. Hillgren and colleagues used

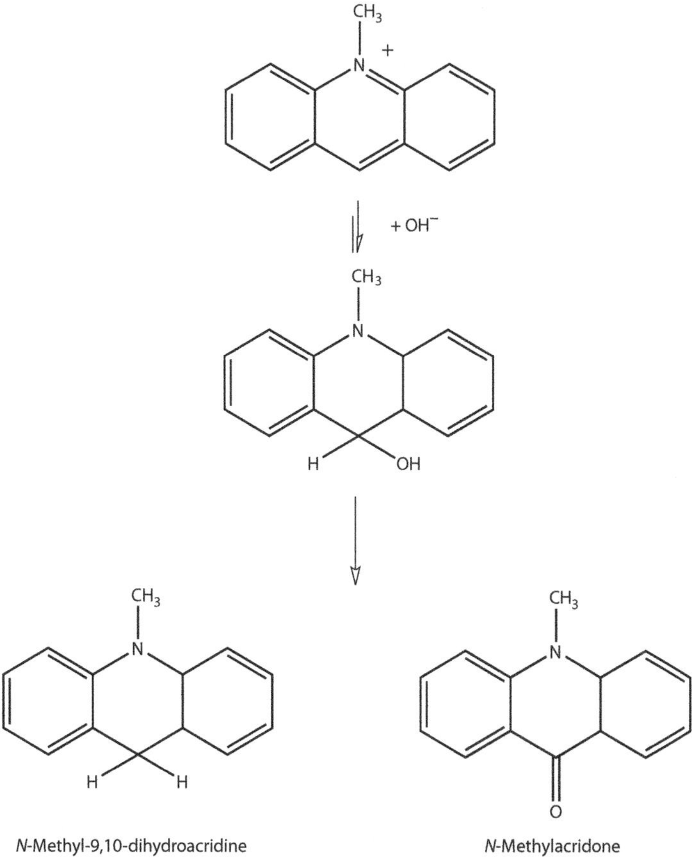

FIGURE 9.6 The dismutation of an acridinium salt by liver alcohol dehydrogenase or human serum albumin in the presence of light. (From Einarsson, R. and Zeppezauer, M., *Eur. J. Biochem.* 59, 295, 1975.)

extrinsic and intrinsic fluorescence to characterize the interaction of Tween 80 (poly-oxyethylene 20 sorbitan monolaureate) and lactate dehydrogenase (LDH).[171] Tween 80 is used in protein formulation[172] and as an aid in clinical assays.[173] Hillgren and colleagues[171] used pyrene as an extrinsic fluorescent probe to demonstrate an interaction between Tween 80 and LDH below the critical micelle concentration for Tween 80. Unlike the interaction of sodium dodecyl sulfate (SDS) with LDH, interaction with Tween 80 does not result in denaturation. Intrinsic fluorescence demonstrated a marked effect of SDS on LDH structure while Tween 80 had little, if any, effect on LDH structure. Luykx and colleagues[174] used high-performance anion-exchange chromatography with fluorescence detection to determine the amounts of recombinant α- and β-erythropoietin in the presence of albumin and other excipients. These investigators also used size exclusion chromatography (SEC) and high-performance liquid chromatography (HPLC) with fluorescence detection to study the conformational changes of interferon α_2b during analysis.[175] Changes in structure occurring

TABLE 9.1

Examples of the Application of Intrinsic Fluorescence for the Characterization of Therapeutic Protein Conformation

Protein	Description of Study	Refs.
β-Lactoglobulin	Use of intrinsic to study change in protein conformation caused by gamma irradiation. Change was observed with irradiation of liquid state material but little change with solid state protein. CD was also used in these studies.	[115]
Epoetium-α	Intrinsic fluorescence used to characterize the product of the oxidation of epoetinum-α with *t*-butylhydroperoxide and hydrogen peroxide (selective oxidation of Met-54 to sulfoxide). This results in a blueshift of the fluorescence emission. Analysis by CD did not show any major change in structure.	[116]
Monoclonal antibody	Use of intrinsic fluorescence for the study of thermal stability during development of a liquid formulation (effect of pH, ionic strength, buffer composition). The results are compared with DSC data.	[117]
Fibroblast growth factor	Use of intrinsic fluorescence to study the pH and thermal stability of FGF. The effect of polyanions on the thermal stability of FGF is also examined by intrinsic fluorescence. The results obtained with intrinsic fluorescence are compared with data obtained with CD, extrinsic fluorescence (ANS), and derivative UV spectroscopy.	[118]
Keratinocyte growth factor-2	Evaluation of polyanions (heparin, sucrose octasulfate, inositol hexaphosphate) as stabilizers of keratinocyte growth factor-2 using intrinsic fluorescence, extrinsic fluorescence (bis-ANS) and DSC. Thermal stability used as marker for formulation stability.	[119]
Insulin	Intrinsic fluorescence used to characterize the interaction of insulin with a triblock copolymer [PEG-(fumaric-sebacic acids)-PEG]. CD and isothermal calorimetry were also used to characterize this interaction.	[83]
Chymotrypsin	Intrinsic fluorescence was used to study the stabilization of α-chymotrypsin by polyamines. Thermal stability and stability to aggregation by 2,2,2-trifluorethanol. Other techniques used included near- and far-UV CD, DSC, and turbidity measurements.	[120]
Lysozyme, myoglobin	Intrinsic fluorescence was used to characterize the effectiveness of sucrose and trehalose as stabilizers for drying of proteins by supercritical fluids as compared to conventional lyophilization. Other analytical techniques include residual water, x-ray power diffraction, and DSC; techniques for evaluation of integrity after reconstitution include (in addition to intrinsic fluorescence) UV/VIS spectroscopy, CD, and SDS-PAGE.	[121]
Lysozyme	Use of intrinsic fluorescence to evaluate the effect of ionic strength on the binding of heparin to lysozyme forming complex coacervates. Other techniques used include infrared spectroscopy, and DSC. The work included a 12-week stability study.	[122]
Thrombin	Intrinsic fluorescence and extrinsic fluorescence (bis-ANS) used to evaluate the effect of pressure and urea on the conformation of human thrombin. Enzyme activity was also measured.	[123]
AGE[c]	Advanced glycation end products derived from proteins.	[124–126]

TABLE 9.1 (continued)
Examples of the Application of Intrinsic Fluorescence for the
Characterization of Therapeutic Protein Conformation

Protein	Description of Study	Refs.
β-Lactoglobulin	Use of intrinsic fluorescence to characterize the effects of hydrostatic pressure on β-lactoglobulin.	[127]
β-Lactoglobulin	Use of intrinsic fluorescence for study of the effect of pH, ionic strength, and temperature on the structure of β-lactoglobulin.	[128]
CPY[b]	Intrinsic fluorescence and extrinsic fluorescence (ANS) were used to study the pressure-denaturation of carboxypeptidase Y. The pressure-induced denaturation was shown to consist of three processes. Deglycosylation of carboxypeptidase Y lowered the pressure required for denaturation.	[129]
Trypsin	Use of intrinsic and extrinsic (ANS) to study the pressure-induced denaturation of trypsin. It was possible to identify a stable intermediate form of trypsin different from that obtained by denaturation with chaotropic agents or by heat.	[130]
β-Lactoglobulin	Intrinsic fluorescence was used to characterize the products obtained from the reaction of β-lactoglobulin with poly(ethylene) glycol. Characterization included studies with FTIR spectroscopy and immunological measurements.	[131]
Ribonuclease	Intrinsic fluorescence was used to characterize the conformational change of RNase in dilute acetic acid (pH 2.6). While there was total loss of enzymatic activity in dilute acid, there was no change in intrinsic fluorescence, there was a small increase in mass as determined by electrospray mass spectrometry.	[132]
Monoclonal antibody	Intrinsic fluorescence use to study microheterogeneity of a monoclonal antibody (h1B4) expressed in mammalian cell culture (CV1P cells). The heterogeneity was demonstrated on isoelectric focusing. Heterogeneity could not be demonstrated on materials isolated on anion-exchange chromatography using intrinsic fluorescence or far-UV CD. Other studies suggested that the observed heterogeneity was due deamidation of the immunoglobulin heavy chain.	[133]
β-Nerve growth factor	Use of intrinsic fluorescence to study the equilibrium denaturation in chaotropic agents (guanidine hydrochloride, urea). CD and fluorescence polarization were also used to characterize the equilibrium denaturation process.	[134]
Human growth hormone	Intrinsic fluorescence is used to study the equilibrium denaturation[a] of human growth hormone (expressed in *Escherichia coli*)	[135]
Antithrombin	Intrinsic fluorescence used to measure the quality of the interaction of heparin with antithrombin to increase anticoagulant activity.	[136,137]
Thrombin	Use of intrinsic fluorescence to study the interaction of thrombin with various sulfated polysaccharides (dextran sulfate, heparin, and chondroitin sulfate A).	[138]
Ricin	Intrinsic fluorescence was used to assess the effect of pH on the conformation of ricin.	[139]

(*continued*)

TABLE 9.1 (continued)
Examples of the Application of Intrinsic Fluorescence for the
Characterization of Therapeutic Protein Conformation

Protein	Description of Study	Refs.
Protein S	Use of intrinsic fluorescence to study the properties of chemically modified (conversion of γ-carboxyglutamic acid to γ-methylglutamic acid) protein S. Calcium ion quenching of intrinsic fluorescence is reduced.	[140]
Human C1 inhibitor and C1-S	Intrinsic fluorescence used to demonstrate the increased stability of the complex between C1-inhibitor and C1-s to conformational change during denaturation by guanidine hydrochloride.	[141]
Human albumin, pigeon liver malic enzyme	Changes in the quenching of intrinsic fluorescence by acrylamide used as measure of conformational change. Denaturation with guanidine hydrochloride resulted in increased quenching of fluorescence.	[142]
Parvalbumin, α-lactalbumin	Intrinsic fluorescence used to study thermal transitions of proteins.	[86]
Lysozyme	Urea decreases the retention of lysozyme on high-performance cation exchange columns. High concentrations of urea or elevated temperature resulting in the denaturation of lysozyme decrease capacity factor; changes in capacity factor correlate with changes in intrinsic fluorescence.	[143]
Thrombin	Intrinsic fluorescence and extrinsic fluorescence (dansyl labeling of active site serine) used to evaluate stability difference between α- and γ-human thrombin. This study also evaluated enzymatic activity and the mobility (ESR) of a spin-label at the enzyme active site.	[144]
Bence-Jones protein	Intrinsic fluorescence used to characterize a Bence-Jones cryoprotein. The native protein has extremely low intrinsic fluorescence (excitation 280–295 nm) which markedly increases with denaturation by heat or chaotropic agents.	[145]
β-lactoglobulin	Intrinsic fluorescence use to study the effect of heat on bovine β-lactoglobulin. Changes in the characteristics of the emission spectra (excitation 280–296 nm) occurring during heat denaturation permitted differentiation of the two tryptophanyl residues.	[146]

[a] Equilibrium denaturation that the denaturation is a reversible event which may or may not have an intermediate. It is recognized that a definition of denaturation is not clear (Lapanje, S., *Physicochemical Aspects of Protein Denaturation*, Wiley-Interscience, New York, 1959) but is considered to involve a major conformational change in a macromolecule without cleavage of a covalent bond. While not absolutely necessary, equilibrium denaturation assumes the existence of an intermediate between the native and denatured forms (see Brems, D.N., Brown, P.L., Heckenlaible, L.A. et al., Equilibrium denaturation of insulin and proinsulin, *Biochemistry* 29, 9289–9293, 1990; Egan, D.A., Logna, T.M., Liang, H. et al., Equilibrium denaturation of recombinant human FK binding protein in urea, *Biochemistry* 32, 1920–1927, 1993; Ogasahara, K., Matsushita, E., and Yutani, K., Further examination of the intermediate state in the denaturation of the tryptophan synthase α subunit. Evidence that the equilibrium denaturation intermediate is a molten globule, *J. Mol. Biol.* 234, 1197–1206, 1993). Equilibrium denaturation was first applied to the modification of DNA with formaldehyde (McGhee, J.D. and von Hippel, P.H., Formaldehyde as a probe of DNA structure. 3. Equilibrium denaturation of DNA and synthetic polynucleotides, *Biochemistry* 16, 3267–3276, 1977).

[b] CP-Y, carboxypeptidase Y.

[c] AGE, advanced glycation end products.

TABLE 9.2
Examples of the Application of Intrinsic Phosphorescence for the Characterization of Protein Conformation

Protein	Description of Study	Refs.
Odorant-binding protein (ODP)	Phosphorescence spectroscopy was used to study the conformation of porcine and bovine odorant-binding proteins. Phosphorescence was detected at 23°C (ambient temperature) and suggested a number of native conformers. The phosphorescence spectrum is consistent with data obtained by crystallographic analysis.	[147]
Recombinant human erythropoietin	The phosphorescence spectrum of recombinant human erythropoietin is a product of separate contributions from Trp 51 and Trp 64; the phosphorescence from the other tryptophan residue, Trp 88 was effectively quenched under all conditions used. Phosphorescence spectra was obtained from the native protein and three derivatives with point mutations where tryptophan was replaced by phenylalanine over temperature range from low temperature glasses (glycerol/buffer, 60/40; 77 K) to 23°C.	[13]
D-Galactose/ D-glucose binding protein (GGBP)	Phosphorescence spectroscopy was used to study the effect of glucose and calcium on stability of D-galactose/D-glucose binding protein (GGBP) from *Escherichia coli*. There were two distinct vibrational bands in the phosphorescence spectrum, suggesting two classes of chromophores. Both glucose and calcium affect the phosphorescence spectrum.	[148]
Alkaline phosphatase	Phosphorescence used to study the conformation of alkaline phosphatase from *E. coli*. The phosphorescence intensity and lifetime of the single tryptophanyl residue decreases with denaturation with guanidine hydrochloride, EDTA, urea, or acid. The results are consistent with the presence of an intermediate between the native and denatured state.	[149]
LADH, alkaline phosphatase, azurin	The effect of encapsulation in silica hydrogels on the intrinsic phosphorescence of liver alcohol dehydrogenase (LADH), alkaline phosphatase, and azurins was evaluated. The effect of encapsulation on phosphorescence lifetime and acrylamide quenching was dependent on the protein. A common feature was the demonstration of conformation heterogeneity.	[150]
Oxytocin	Time-resolved phosphorescence is used to study the tyrosine residue in oxytoxin, oxytocin analogues, tyrosine, and some tyrosine peptides.	[151]
Factor VIII	Study of Tb^{3+} phosphorescence when bound to factor VIII and component polypeptide chains.	[152]
β-Lactoglobulin Ribonuclease T_1 Superoxide dismutase Alkaline phosphatase Liver alcohol dehydrogenase	Phosphorescence lifetime and phosphorescence quenching were used to assess the effect of deuterium oxide (heavy water) on protein structure. D_2O has little effect on the phosphorescence lifetime of *N*-acetyl-tryptophanamide but increases the phosphorescence lifetime in proteins (increase in protein rigidity).	[153]

(continued)

TABLE 9.2 (continued)

Examples of the Application of Intrinsic Phosphorescence for the Characterization of Protein Conformation

Protein	Description of Study	Refs.
Thermostable esterases	The flexibility of a series of evolved thermostable esterase was evaluated used intrinsic tryptophan phosphorescence. Increases in stability were frequently associated with decreases in protein flexibility.	[154]
Apoazurin; alcohol dehydrogenase; alkaline phosphatase	Quenching of intrinsic protein phosphorescence by acrylamide was used to evaluate the effect of pressure on protein structure. Quenching of intrinsic protein phosphorescence by acrylamide was used to measure the internal migration rate.	[155]
Horse liver alcohol dehydrogenase	Stopped-flow analysis of the decay of intrinsic fluorescence was used to study the denaturation of horse liver alcohol dehydrogenase by chaotropic agents. Heterogeneity of the kinetics of denaturation suggest the existence of multiple stable conformations for native horse liver alcohol dehydrogenase.	[156]
Alkaline phosphatase	Room-temperature tryptophan phosphorescence used to measure hydrogen–deuterium exchange in the hydrophobic core of proteins. Phosphorescence lifetime increases following hydrogen–deuterium exchange as shown for other proteins. Time-resolved tryptophan phosphorescence is sensitive to the changes in the environment of the emitting residue.	[157,158]
Various	Pressure increased τ with azurin and increased τ with ribonuclease T_1 and phosphoglycerate kinase. It is suggested that pressure reduces internal cavities and increased hydration.	[159]
Lysozyme	The phosphorescence of lyophilized lysozyme was studied as a function of humidity (protein hydration). Phosphorescence intensity decreased as function of protein hydration. The heterogeneity of changes in phosphorescence decay suggests the existence of multiple conformations in the native state and that hydration decreases rigidity.	[160]
α-Lactalbumin; lysozyme	The effect of temperature on the intrinsic phosphorescence of α-lactalbumin and lysozyme was evaluated in this study. There was little effect of temperature on quantum yield below 30 K but decreased markedly with increasing temperature above this temperature. It is suggested that internal quenching is responsible for the effect on quantum yield.	[161]
RNase T_1; alkaline phosphosphatase; alcohol dehydrogenase	Intrinsic tryptophan phosphorescence was used to evaluate the effect of glycerol	[162]
Azurin; horse liver alcohol dehydrogenase; paralbumin, alkaline phosphatase	The difference in the ability of charged species ($SH^{-1'}$; NO_2^{-1}) versus uncharged species (H_2S; CS_2) to quench the phosphorescence of exposed or buried tryptophanyl residues was evaluated in this study. Uncharged compound were more effective than charged compounds in quenching the phosphorescence of buried tryptophanyl residues; there was not difference in the ability to quench exposed tryptophanyl residues.	[163]

TABLE 9.2 (continued)
Examples of the Application of Intrinsic Phosphorescence for the Characterization of Protein Conformation

Protein	Description of Study	Refs.
Murine myeloma IgA	Tryptophan–disulfide interactions result in an anomalous component in the tryptophan phosphorescence decay. This is suggested to be a sensitive indicator of conformational changes in immunoglobulins.	[164]
Liver alcohol dehydrogenase (LADH)	Comparison of phosphorescence spectrum of LADH in solution and in crystalline state. There is a redshift in the phosphorescence spectrum of an exposed tryptophan residue and no change in the phosphorescence spectrum of a buried tryptophan residue. The observed redshift is likely due to intermolecular contacts and not conformational differences between the solution form and crystalline form. The results are consistent with the identity of the solution structure and the structure determined by crystallographic analysis.	[165]
Human erythropoietin	Erythropoietin produced by recombinant DNA technology in Chinese hamster ovary cells was characterized by CD, UV spectroscopy, fluorescence and phosphorescence spectroscopy, size-exclusion chromatography, and sedimentation equilibrium.	[166]
Ragweed pollen allergen Ra5	The phosphorescence and fluorescence of ragweed pollen allergen Ra5 were studied at room temperature and ethylene glycol–water at low temperature (165°C), The low temperature phosphorescence spectra was heterogeneous, suggesting that the two tryptophanyl residues are located in distinctly different environments.	[167]
Lysozme	Phosphorescence of native and site-specific oxidized[a] lysozyme permitted the differentiation of the two tryptophanyl residues. This study also used optically detected magnetic resonance.	[168]
N/A	Review on the application of phosphorescence spectroscopy to protein structure.	[169]
α-Trypsin	Phosphorescence spectroscopy is used to study the denaturation of α-trypsin by guanidine hydrochloride.	[170]

[a] Imoto, T., Hartdegen, F.J., and Rupley, J.A., Oxidation of lysozyme by iodine: Isolation of an inactive product and its conversion to an oxindolealanine–lysozyme, *J. Mol. Biol.* 80, 637–648, 1973.

during reverse phase HPLC (RP-HPLC) were confirmed by separate observations with fluorescence and circular dichorism (CD) under the same solvent conditions (38%–70% acetonitrile in 0.1% trifluoroacetic acid). Unlike phosphorescence where this is considerable study of solid state materials, fluorescence is a technique usually applied to proteins (and other fluorophores) in solution. Thermoluminescence[176] was reported for proteins in the solid state and in solution by Nummedal and Steen in 1969.[177] These investigators described the x-ray-induced thermoluminescence of trypsin and trypsinogen. The luminescence obtained from the glassy and dry states was described as phosphorescence resulting from tryptophanyl residues. More recently, Ramachander and coworkers[178] reported on the solid state fluorescence of proteins as a tool for the study of protein formulation and stability.

Table 9.2 provides a list of some studies on the use of phosphorescence on protein conformation. Phosphorescence has a long lifetime compared to fluorescence and is subject to quenching by various substances including oxygen such that early workers rarely observed phosphorescence in solution.[179] Saviotti and Galley[180] observed that while the phosphorescence of small aromatic compounds was rapidly quenched by oxygen and other solutes in aqueous solution at 23°C, some proteins demonstrate phosphorescence at this temperature. The variability in phosphorescence lifetime in proteins is considered to be remarkable and likely a reflection of the critical nature of the environment around the various tryptophanyl residues.[12] A further complication is the proposition that solution heterogeneity is responsible for the multiexpontential decay of phosphorescence in single tryptophan proteins.[181] Incorporation of protein into a glass formed from a mixture of sucrose and trehalose increased protein phosphorescence, reflecting immobilization of the protein and a decrease in the diffusion of quenching materials such as oxygen.[182,183] Removal of oxygen is critical.[184] A device has been developed for the reducing oxygen in phosphorescence experiments.[185]

Vaccines and related biologicals such as gene therapy vectors provide particular challenges for characterization. Table 9.3 contains selected studies on the use of

TABLE 9.3
The Use of Fluorescence for the Characterization of Vaccines, Vaccine-Related Biologicals Including Gene Therapy Vectors

Product	Description of Study	Refs.
Human papilloma virus-like particles[a]	Intrinsic fluorescence was used to follow the assembly and maturation of human papilloma virus-like particles and complements light scattering techniques.	[186]
Hepatitis C virus envelope glycoprotein E1	Intrinsic and extrinsic (ANS) fluorescence were used to study a truncated form (Ely) of heparin C virus envelope glycoprotein E1, which was expressed in yeast, Ely forms protein particles, which could form the basis for a vaccine. The various fluorescence studies were used together with CD and dynamic light scattering to characterize the Ely protein particles.	[187]
A trivalent protein vaccine	Intrinsic fluorescence (front face fluorescence) and other techniques including DSC and derivative spectroscopy were used to characterize a protein vaccine adsorbed to an aluminum salt adjuvant.	[188]
Clostridium difficile toxins	Intrinsic and extrinsic (ANS) fluorescence were used to characterize the formaldehyde treatment of *Clostridium difficile* toxins to form toxoids. Other techniques included dynamic light scattering, high-resolution UV spectroscopy, CD, and turbidity measurements. This work is in support of vaccine development.	[189]
Norwalk virus-like particles[a]	Intrinsic fluorescence was used to study methods for the formulation of Norwalk virus-like particles in aqueous suspension. The goal is the preparation of a stable Norwalk virus-like particle vaccines.	[190]

TABLE 9.3 (continued)

The Use of Fluorescence for the Characterization of Vaccines, Vaccine-Related Biologicals Including Gene Therapy Vectors

Product	Description of Study	Refs.
Carbon nanospheres	The intrinsic fluorescence of carbon nanospheres, which are used as intracellular carriers can be used to track the intracellular localization of these materials.	[191]
Conjugate polysaccharide vaccine	Intrinsic fluorescence and NMR were used to characterize a meningococcal C polysaccharide–tetanus toxoid conjugate vaccine.	[192]
Norwalk virus-like particles[a]	Intrinsic fluorescence and extrinsic fluorescence (ANS) were used together with DSC, dynamic light scattering, CD, Second derivative UV-VIS spectrophotometry, and electron microscopy to study the stability (pH and temperature) of Norwalk virus-like particles.	[193]
S. aureus saccharide conjugate vaccine	Intrinsic fluorescence and size-exclusion chromatography were used to establish batch-to-batch consistency in the manufacture of a bivalent conjugate vaccine composed of *Staphylococcus aureus* capsular polysaccharide and a mutant nontoxic recombinant *Pseudomonas aeruginose* exotoxin A.	[194]
Adenovirus type 5	Intrinsic and extrinsic fluorescence (propidium iodide) were used together with second derivative UV-VIS spectrophotometry and dynamic light scattering to evaluated the pH and thermal stability of adenovirus type 5.	[195]
Human respiratory syncytial virus	Intrinsic and extrinsic (ANS) fluorescence together with CD, second derivative UV-VIS spectroscopy, and dynamic light scattering were used to study the thermal and pH stability of human respiratory syncytial virus (human RSV).	[196]
Pertussis toxin vaccine	Intrinsic fluorescence was used to characterize the formaldehyde treatment of pertussis toxin in the preparation of acellular pertussis vaccines. Other technologies used included monoclonal antibody binding, analytical ultracentrifugation, SEC, multiangle laser light scattering, and immunoblotting.	[197]
Classical swine fever virus	Intrinsic fluorescence used to study structural changes in class swine fever virus, resulting from inactivation with hydrostatic pressure or ultraviolet irradiation. It is suggested that pressure or UV irradiation can be used for vaccine preparation.	[198]
Infectious bursal disease virus	Intrinsic fluorescence used to characterize the product from the treatment of infectious bursal disease virus which resulted in viral inactivation. The results suggest that hydrostatic pressure can be used to prepare material for use as a vaccine.	[199]
Bordetella pertussis vaccine	Intrinsic fluorescence was used to characterize the formaldehyde inactivation of *Bordetella pertussis* antigens to form toxoid forms suitable for vaccines. Other techniques included DC, SDS-PAGE, and size-exclusion chromatography. It is suggested that these analytical techniques when combined with immunological and toxicology assays will enable the consistency of vaccine manufacture.	[200]

(continued)

TABLE 9.3 (continued)
The Use of Fluorescence for the Characterization of Vaccines, Vaccine-Related Biologicals Including Gene Therapy Vectors

Product	Description of Study	Refs.
Hepatitis A virus	Intrinsic fluorescence, extrinsic fluorescence (TNS[b] for protein and TOTO-1 for RNA[c]) was used to evaluate structural changes in hepatitis A virus in the process of preparing VAQTA (vaccine prepared by formalin-inactivation and adsorption to aluminum salts).	[201]
Human HIV type 1	RNA and DNA complexes of nucleocapsid protein p7.Zn (NCp7.Zn) of human HIV-1 were studied by phosphorescence and optically detected magnetic resonance, The redshift of the phosphorescence spectrum and the decrease in τ is suggested to reflect the stacking interaction between nucleic acid and NCp.7.Zn.	[202]
A-1 influenza virus (H1N1) and Sendai parainfluenza virus	Intrinsic fluorescence used to evaluate the effect of formalin on viruses in the preparation of vaccines. Fluorescence anisotropy measurements were used in these studies.	[203,204]
Brome mosaic virus	Intrinsic fluorescence and extrinsic fluorescence (bis-ANS) were used to study the pressure-induced dissociation of brome mosaic virus into subunits. Other techniques used included size exclusion chromatography and electron microscopy.	[205]

[a] Virus-like particles can be used as vaccine components (see Roy, P. and Noad, R., Virus-like particles as vaccine delivery systems: myths and facts, *Hum. Vaccin.* 4, 5–12, 2008; Jennings, G.T. and Bachmann, M.F., The coming of age of virus-like particle vaccines, *Biol. Chem.* 389, 521–536, 2008).

[b] TNS, 2-(*p*-toluidinylnapthalene-6-sulfonic acid); see Schlessinger, J. and Steinberg, I.Z., Circular polarization of fluorescence of probes bound to chymotrypsin. Change in asymmetric environment upon electronic excitation, *Proc. Natl. Acad. Sci. USA* 69, 769–772, 1972; Albani, J.R., Fluorescence origin of 6,*p*-toluidinyl-napthalene-2-sulfonate (TNS) bound to proteins, *J. Fluoresc.*, 19, 399, 2009.

[c] TOTO: 1,1'-(4,4,7,7-tetramethyl-4,7-diazaundecamethylene)-bis-4-[3-methyl-2,3-dihydro-(benzo-1,3-thiazole)-2-methylidene]-quinolinium tetraiodide. See Rye, H.S., Debora, J.M., Quesda, M.A. et al., Fluorometric assay using dimeric dyes for double- and single-stranded DNA and RNA with pictogram sensitivity, *Anal. Biochem.* 208, 144–150, 1993.

fluorescence for the characterization of vaccines, vaccine-related biologicals, and gene therapy products.

Albumin is present in blood plasma at a high concentration and was one of the first biological products. Albumin continues to see considerable clinical use for the original indications and is also used in laser glues and as a bridge-to-liver transplant. The ready availability of this material has allowed albumin to be a model protein for chemistry studies including fluorescence. Table 9.4 contains a limited list of studies on the fluorescence of albumin. One of the major functions of albumin is the binding of low-molecular-weight drugs and many of the studies in Table 9.4 describe studies on the changes in intrinsic fluorescence on the binding of these materials.

TABLE 9.4
Studies on the Intrinsic Fluorescence and Phosphorescence of Serum Albumins

Protein	Study Description	Ref.
Human serum albumin[a]	Intrinsic and extrinsic fluorescence (ANS) were used to study the binding of alkanes to human serum albumin. The competitive binding of ANS in the presence of alkanes provided information on all hydrophobic binding sites in human serum albumin while intrinsic fluorescence reported on binding near the single tryptophan residue in human serum albumin.	[206]
Human serum albumin	The effect of oxidation of human serum albumin with hypochlorite on intrinsic fluorescence and extrinsic fluorescence (dansyl sarcosine; dansyl-1-sulfonamide) was determined in this study. Oxidation reduced the number of binding sites but not binding site affinity.	[207]
Bovine serum albumin[a]	Intrinsic fluorescence used to study the role of trehalose in the moisture-induced aggregation of bovine serum albumin.	[208]
Bovine serum albumin	Intrinsic fluorescence was used with isothermal titration calorimetry, CD, and UV-VIS spectroscopy to demonstrate lack of binding of methimazole.	[209]
Human serum albumin	The quenching of intrinsic phosphorescence in human serum albumin was shown to be enantioselective. The reduction in phosphorescence lifetime varied depending on the R and S enantiomers examined.	[210]
Human serum albumin	Intrinsic fluorescence used to study the effect of nonenzymatic glycation on the denaturation of human serum albumin by guanidine hydrochloride	[211]
Human serum albumin	Intrinsic fluorescence used for the assay of advanced glycation end products such as argpyrimidine and pentosidine.	[212]
Human serum albumin	Oxygen is an effective quencher of phosphorescence. This study describes an effective deoxygenation system which has proved useful in the measurement of protein phosphorescence.	[185]
Bovine serum albumin	The relationship of the structure of human serum albumin as obtained from crystallographic analysis and solution structure was studied by phosphorescence depolarization using erythrosine-BSA as model protein.	[213]
Human serum albumin	Recombinant native human serum albumin and several site-specific mutants were prepared in a yeast expression system. Intrinsic fluorescence and thermal denaturation was used to characterize these proteins. Change in intrinsic fluorescence was used to study the binding of warfarin.	[214]
Human serum albumin	Intrinsic fluorescence, far-UV DC, and extrinsic fluorescence (ANS) used to study the stabilization of human serum albumin by inorganic anions during urea denaturation. Perchlorate was the most effective followed by thiocyanate and sulfate. Stabilization involved the prevention of the formation of an intermediate.	[215]
Human serum albumin	The quenching of intrinsic tryptophan fluorescence of human serum albumin was used to measure affinity of drug binding.	[216]

(continued)

TABLE 9.4 (continued)
Studies on the Intrinsic Fluorescence and Phosphorescence of Serum Albumins

Protein	Study Description	Ref.
Bovine serum albumin	Phosphorescence decay was used to study the effect of glycerol on the structure of bovine serum albumin. The complex pattern of phosphorescence decay in the presence of glycerol suggests the existence of multiple native protein conformations.	[217]
Ovalbumin	Intrinsic fluorescence was used to compare results obtained with DSC and limited proteolysis (chymotrypsin) in the study of the thermal denaturation of solid ovalbumin (granules with 11% moisture). The studies also included bovine serum albumin and γ-globulin.	[218]
Human serum albumin; bovine serum albumin	The phosphorescence lifetime of human serum albumin was show to increase on the addition of sodium dodecyl sulfate. The increase was greater with bovine serum albumin reflecting the participation of an additional tryptophan residue. It is suggested that the increase in phosphorescence lifetime reflects an increase in protein rigidity in the presence of SDS.	[219]
Human serum albumin; bovine serum albumin	Intrinsic fluorescence and optically detected magnetic resonance were used to study the binding of fatty acids to HSA and BSA. Oleic acid had an effect on the phosphorescence and the magnetic resonance spectra of BSA which was suggested to be an effect of Trp-134 as there was no effect on HSA which has a single tryptophan residues (Trp-134).	[220]
Human serum albumin	Low-temperature phosphorescence studies were used to determine the relative contribution of tyrosine and tryptophan to the phosphorescence of human serum albumin.	[221]

[a] HSA, human serum albumin; BSA, bovine serum albumin.

Some of the above cited studies use both intrinsic fluorescence and extrinsic fluorescence. Intrinsic fluorescence depends on the intrinsic properties of the protein or other biomacromolecule while extrinsic fluorescence depends on the properties of a compound, which binds covalently or noncovalently to a macromolecule. Extrinsic phosphorescence can also be of value. The phosphorescence spectrum from terbium binding to proteins[222] can be used to understand calcium ion binding sites.[152,223] Most of studies on extrinsic luminescence have focused on fluorescence studies. The reader is referred to Chapter 14 and to a reference text on the chemical modification of proteins[224] for more discussion of extrinsic fluorophores.

REFERENCES

1. Kyte, J., *Structure in Protein Chemistry*, 2nd edn., Garland Science, New York, 2007.
2. Lakowicz, J.R., Radiative decay engineering: Biophysical and biomedical applications, *Anal. Biochem.* 298, 1–24, 2001.

3. Sarkar, D., Das, P., Basak, S., and Chattopadhyay, N., Binding interaction of cationic phenazinium dyes with calf thymus DNA: A comparative study, *J. Phys. Chem. B* 112, 9243–9249, 2008.
4. Wu, M., Wu. W., Gao, X. et al., Synthesis of a novel fluorescent probe based on acridine skeleton used for sensitive determination of DNA, *Talenta* 75, 995–1001, 2008.
5. Ho, H.A., Najari, A., and Leclerc, M., Optical detection of DNA and proteins with cationic polythiophenes, *Acc. Chem. Res.* 41, 168–178, 2008.
6. Takeda, K., Yoshida, I., and Yamamoto, K., Changes of fluorescence lifetime and rotational correlation time of bovine serum albumin labeled with 1-dmethylaminonaph-thalene-5-sulfonyl chloride in guanidine and thermal denaturation, *J. Protein Chem.* 10, 17–23, 1991.
7. Homchaudhuri, L., Kumar, S., and Swaminathan, R., Slow aggregation of lysozyme in alkaline pH monitored in real time employing the fluorescence anisotropy of covalently labelled dansyl probe, *FEBS Lett.* 580, 2097–2101, 2006.
8. Hawe, A., Friess, W., Sutter, M., and Jiskoot, W., Online fluorescent dye detection method for the characterization of immunoglobulin G aggregation by size exclusion chromatography and asymmetrical flow field flow fractionation, *Anal. Biochem.* 379, 115–122, 2008.
9. Lakowicz, J.R., *Principles of Fluorescence Spectroscopy*, 2nd edn., Kluwer/Plenum, New York, 1999.
10. Baenboim, G.M., Domanskii, A.N., and Turoverov, K.K., *Luminescence of Biopolymers and Cells*, Plenum Press, New York, 1969.
11. Kasha, M., Phosphorescence and the role of the triplet state in the electronic excitation of complex molecules, *Chem. Rev.* 41, 401–419, 1947.
12. Papp, S. and Vanderkool, J.M., Tryptophan phosphorescence at room temperature as a tool to study protein structure and dynamics, *Photochem. Photobiol.* 49, 775–784, 1989.
13. Kerwin, B.A., Aoki, K.H., Gonelli, M., and Strambini, G.B., Differentiation of the local structure around tryptophan 51 and 64 in recombinant human erythropoietin by trypto-phan phosphorescence, *Photochem. Photobiol.* 84, 1172–1181, 2008.
14. Schauerte, J.A., Steel, D.G., and Gafni, A., Time-resolved room temperature tryptophan phosphorescence in proteins, *Methods Enyzmol.* 278, 49–71, 1997.
15. Cioni, P. and Strambini, G.B., Tryptophan phosphorescence and pressure effects on protein structure, *Biochim. Biophys. Acta* 1595, 116–130, 2002.
16. Guilbault, G.G. ed., *Practical Fluorescence*, 2nd edn., Marcel Dekker, New York, 1990.
17. Permyakov, E.A., *Luminescent Spectroscopy of Proteins*, CRC Press, Boca Raton, FL, 1993.
18. Strambini, G.B., Kerwin, B.A., Mason, B.D., and Gonnelli, M., The triplet-state lifetime of indole derivatives in aqueous solution, *Photochem. Photobiol.* 80, 462–470, 2004.
19. Dashnau, J.L., Zelent, B., and Vanderkool, J.M., Tryptophan interactions with glycerol/water and trehalose cryosolvents: Infrared and fluorescence spectroscopy and ab initio calculations, *Biophys. Chem.* 114, 71–83, 2005.
20. Gundermann, K.-D. and McCapra, R., *Chemiluminescence in Organic Chemistry*, Springer-Verlag, Berlin, Germany, 1987.
21. McElroy, W.D. and Seliger, H.E., The chemistry of light emission, *Adv. Enzymol. Relat. Areas Mol. Biol.* 25, 119–166, 1963.
22. Garcia-Campaña, A.M. and Bacyens, W.R.G. eds., *Chemiluminescence in Analytical Chemistry*, Marcel Dekker, New York, 2001.
23. Acuña, A.U. and Amat-Guerri, E., Early history of solution fluorescence: The *Lignum nephriticum* of Nicolás Monardes, *Springer Ser. Fluoresc.* 4, 3–20, 2008.
24. Mestre, Y.F., Zamora, L.L., and Calatayud, J.M., Flow-luminescence: A growing modality of pharmaceutical analysis, *Luminescence* 16, 213–235, 2001.

25. Yamaguchi, M., Yoshida, H., and Nohta, H., Luminol-type chemiluminescence derivatization reagents for liquid chromatography and capillary electrophoresis, *J. Chromatogr.* 950, 1–19, 2002.

26. Liu, Y.M. and Chang, J.K., Ultrasensitive chemiluminescence detection in capillary electrophoresis, *J. Chromatogr. A* 959, 1–13, 2002.

27. Francis, P.S., Barnett, N.W., Lewis, S.W., and Lim, K.F., Hypothalites and related oxidants as chemiluminescence reagents: A review, *Luminescence* 19, 95–115, 2004.

28. Radziszewski, B., Über die Phosphorescenz der organischen and organischer Körper, *Liebigs Ann.* 203, 305–336, 1880.

29. Kuroda, N., Shimoda, R., Wada, M., and Nakashimi, K., Lophine derivatives and analogues as new phenolic enhancers for the luminol-hydrogen peroxide-horseradish peroxidase chemiluminescence system, *Anal. Chim. Acta* 403, 131–136, 2000.

30. Faulkner, K. and Fridovich, I., Luminol and lucigenin as detectors for O_2^-, *Free Radic. Biol. Med.* 15, 447–451, 1993.

31. Herron, J.N., Ely, K.R., and Edmundson, A.B., Pressure-induced conformational changes in a Bence-Jones protein (Mcg), *Biochemistry* 24, 3453–3459, 1985.

32. Lai, Y., Qi, Y., Wang, J., and Chen, G., Using acridinium ester as the sonochemiluminescent probe for labeling of protein, *Analyst* 134, 131–137, 2009.

33. Dudai, Y., Silman, I., Shinitzky, M., and Blumberg, S., Purification by affinity chromatography of a the molecular forms of acetylcholinesterase present in fresh electric-organ tissue of electric eel, *Proc. Natl. Acad. Sci. USA.* 69, 2400–2408, 1972.

34. Eriksson, H. and Augustinsson, K.B., A mechanistic model for butyrylcholinesterase, *Biochim. Biophys. Acta* 567, 161–173, 1979.

35. Eastman, J., Wilson, E.J., Cerveñansky, C., and Rosenberry, T.L., Fasiculin 2 binds to the peripheral site on acetylcholinesterase and inhibits substrate hydrolysis by slowing a step involving proton transfer during enzyme acylation, *J. Biol. Chem.* 270, 19694–19701, 1995.

36. Weeks, L., Sturgess, M., Brown, R.C., and Woodhead, J.S., Immunoassays using acridinium esters, *Methods Enzymol.* 133, 366–387, 1986.

37. Brown, R.C., Aston, J.P., St. John, A., and Woodhead, J.S., Comparison of poly- and monoclonal antibodies as labels in a two-site immunochemiluminometric assay for intact parathyroid hormone, *J. Immunol. Methods* 109, 139–144, 1988.

38. Russell, J., Colpitts, T., Hoelts-McCormick, S. et al., Defined protein conjugates as signaling agents in immunoassays, *Clin. Chem.* 50, 1921–1929, 2004.

39. Scorilas, A., Agiamarnioti, K., and Papadopoulos, K., Novel biotinylated acridinium derivatives: New reagents for fluorescence immunoassays and proteomics, *Clin. Chim. Acta* 357, 159–167, 2005.

40. Khalil, O.S., Zurek, T.F., Pepe, C. et al., Detection apparatus for multiple heterologous chemiluminescence immunoassay configurations, *Anal. Biochem.* 196, 61–68, 1991.

41. Batmanghelich, S., Brown, H.C., Woodhead, J.S. et al., Preparation of a chemiluminescent imidoester for the non-radioactive labeling of proteins, *J. Photochem. Photobiol.* 12, 193–201, 1992.

42. Melucci, D., Roda, B., Zattoni, A. et al., Field-flow fractionation of cells with chemiluminescent detection, *J. Chromatogr. A* 1056, 229–236, 2004.

43. Shiesh, S.C., Chou, T.C., Lin, X.Z., and Kao, P.C., Determination of C-reactive protein with an ultra-sensitivity immunochemoluminometric assay, *J. Immunol. Methods* 311, 87–95, 2006.

44. Herranz, S., Ramón-Azcón, J., Benito-Peño, E. et al., Preparation of antibodies and development of a sensitive immunoassay with fluorescence detection for triazine herbicides, *Anal. Bioanal. Chem.* 391, 1801–1812, 2008.

45. Fichorova, R.N., Richardson-Harman, N., Alfano, M. et al., Biological and technical variables affecting immunoassay recovery of cytokines from human serum and simulated vaginal fluid: A multicenter study, *Anal. Chem.* 80, 4741–4751, 2008.

46. Zhu, S., Zhang, Q., and Guo, L.H., Parts-per-trillion level detection of estradiol by competitive fluorescence immunoassay using DNA/dye conjugates as antibody multiplier labels, *Anal. Chim. Acta* 624, 141–146, 2008.

47. Adamczyk, M., Gebier, J.C., Shreder, K., and Wu, J., Quantitative determination of noncovalently bound acrindinium in protein conjugates by liquid chromatography/ electrospray ion trap mass spectrometry, *Rapid Commun. Mass Spectrom.* 15, 670–674, 2001.

48. Einarsson, R. and Zeppezauer, M., Catalysis of the photochemical dismutation of *N*-methylacridinium cation to *N*-methylacridone and *N*-methyl-9,10-dihydroacridine by hydrophobic sites of horse-liver alcohol dehydrogenase and human serum albumin, *Eur. J. Biochem.* 59, 295–304, 1975.

49. Creighton, D.J., Hajda, J., Mooser, G., and Sigman, D.S., Model dehydrogenase reactions. Reduction of *N*-methylacridinium ion by reduced nicotinamide adenine dinucleotide and its derivatives, *J. Am. Chem. Soc.* 95, 6855–6857, 1973.

50. Vandersypen, H.A. and Heremans, K.A., Proflavin–chymotrypsin interaction: High pressure temperature-jump studies, *Arch. Int. Physiol. Biochem.* 82, 205, 1974.

51. Li, E.H., Orton, C., and Feinman, R.D., The interaction of thrombin and heparin: Proflavine dye binding studies, *Biochemistry* 13, 5012–5017, 1974.

52. Stoesz, J. and Lumry, R.W., The effects of chemical modification on the refolding transition of α-chymotrypsin, *Biophys. Chem.* 10, 105–112, 1979.

53. Fojo, A.T., Whitney, P.I., and Awad, W.J., Jr., Effects of acylation and guanidation on alkaline conformation of chymotrypsin, *Arch. Biochem. Biophys.* 224, 636–642, 1983.

54. Banerjee, D. and Pal, S.K., Conformational dynamics at the active site of α-chymotrypsin and enzymatic activity, *Langmuir* 24, 8163–8168, 2008.

55. Harvey, E.N., *Bioluminescence*, Academic Press, New York, 1952.

56. Cormier, M.J. and Tooter, J.R., Bioluminescence, *Annu. Rev. Biochem.* 33, 431–458, 1964.

57. Cormier, M.J., Lee, J., and Wampler, J.E., Bioluminescence: Recent advances, *Annu. Rev. Biochem.* 44, 255–272, 1975.

58. Stanley, P.E., Some brief notes on nomenclature and units and standards used in bioluminescence and chemiluminescence, *Methods Enzymol.* 305, 47–50, 2000.

59. Sadikot, R.T. and Blackwell, T.S., Bioluminescence imaging, *Proc. Am. Thorac. Soc.* 2, 537–540, 2005.

60. Roda, A., Guardiagl, M., Pasini, P. et al., Bio- and chemiluminescence imaging in analytical chemistry, *Anal. Chim. Acta* 541, 25–36, 2005.

61. Kricka, L.J., Application of bioluminescence and chemiluminescence in biomedical sciences, *Methods Enzymol.* 305, 333–345, 2000.

62. DeLuca, M.A. and McElroy, W D. eds., *Bioluminescence and Chemiluminescence: Base Chemistry and Analytical Applications*, Academic Press, New York, 1981.

63. Campbell, A.K., *Chemiluminescence*, VCH, Weinheim, Germany, 1988.

64. Klein, H.A., *Bioluminescence*, Lippincott, Philadelphia, PA, 1965.

65. Marquette, C.A. and Blum, L.J., Applications of the luminol chemiluminescent reaction in analytical chemistry, *Anal. Bioanal. Chem.* 385, 546–554, 2006.

66. Wolter, A., Niessner, R., and Seidel, M., Detection of *Escherichia coli* O157:H7, *Salmonella typhimurium*, and *Legnionella pneumonia* in water using a flow-through chemiluminescence microarray readout system, *Anal. Chem.* 80, 5854–5863, 2008.

67. Phillips, D., Luminescence lifetimes in biological systems, *Analyst* 119, 543–550, 1994.

68. Dickson, E.F., Pollak, A., and Diamandis, E.F., Time-resolved detection of lanthanide luminescence for ultrasensitive bioanalytical assays, *J. Photochem. Photobiol. B* 27, 3–19, 1995.

69. Rangarajan, B., Coons, L.S., and Scranton, A.B., Characterization of hydrogels using luminescence spectroscopy, *Biomaterials* 17, 649–661, 1996.

70. Gaye-Seye, M.D., Aaron, J.J., Párkányi, C., and Motohashi, N., Luminescence and photophysical properties of benzo[*a*]phenothiazines—Therapeutic, physico-chemical, and analytical applications, *Curr. Drug Targets* 7, 1083–1093, 2006.

71. Fraga, H., Firefly luminescence: A historical perspective and recent developments, *Photochem. Photobiol. Sci.* 7, 146–158, 2008.

72. Jeong, E.J., Park, K., Joung, H.A. et al., Detection of glucose-induced conformational change in hexokinase II using fluorescence complementation assay, *Biotechnol. Lett.* 29, 797–802, 2007.

73. Ohashi, T., Gallacy, S.D., Briscoe, G., and Erickson, H.P., An experimental study of GFP-based FRET, with application to intrinsically unstructured proteins, *Protein Sci.* 16, 1429–1438, 2007.

74. Shaner, N.C., Patterson, G.H., and Davidson, M.W., Advances in fluorescent protein technology, *J. Cell Sci.* 120, 4247–4260, 2007.

75. Becker, K., Van Alstine, J., and Bülow, L., Multipurpose peptide tags for protein isolation, *J. Chromatogr. A* 1202, 40–46, 2008.

76. Nienhaus, G.U., The green fluorescent protein: A key tool to study chemical processes, *Angew. Chem. Int. Ed. Engl.* 47, 8992–8994, 2008.

77. Martinez-Alonso, M., Garcia-Fruitós, E., and Villaverde, A., Yield, solubility and conformational quality of soluble proteins are not simultaneously favored in recombinant *Escherichia coli*, *Biotechnol. Bioeng.* 101, 1353–1358, 2008.

78. Pearce, S.F. and Hawrot, E., Intrinsic fluorescence of binding-site fragments of the nicotinic acetylcholine receptor: Perturbations produced upon binding α-bungarotoxin, *Biochemistry* 29, 10649–10659, 1990.

79. Peranteau, A.G., Kuzmig, P., Angell, Y. et al., Increase in fluorescence upon the hydrolysis of tyrosine peptides: Application to proteinase assays, *Anal. Biochem.* 227, 242–245, 1995.

80. Wilson, C.J. and Copeland, R.A., Spectroscopic characterization of arrestin interactions with competitive ligands: Study of heparin and phytic acid binding, *J. Protein Chem.* 16, 755–763, 1997.

81. Gao, J., Yin, D., Yao, Y. et al., Progressive decline in the ability of calmodulin isolated from aged brain to activate the plasma membrane Ca-ATPase, *Biochemistry* 37, 9536–9548, 1998.

82. Stepanenko, O.V., Kuznetsova, I.M., Turoverov, K.K. et al., Conformational change of the dimeric DsbC molecule induced by GdnHCl. A study by intrinsic fluorescence, *Biochemistry* 43, 5296–5303, 2004.

83. Pourhosseini, P.S., Soboury, A.A., Najafi, F., and Sarbolouki, M.N., Interaction of insulin with a triblock copolymer of PEG-(fumaric-sebacic acids)-PEG: Thermodynamic and spectroscopic studies, *Biochim. Biophys. Acta* 1774, 1274–1280, 2007.

84. Vlasova, T.N. and Ugarova, N.N., Quenching of the fluorescence of Tyr and Trp residues of firefly luciferase from *Luciola mingrelica* by the substrates, *Biochemistry* (Mosc) 72, 962–967, 2007.

85. Agro, A.F., Intrinsic fluorescence of a protein devoid of tyrosine and tryptophan: Horse hepatocuprein, *FEBS Lett.* 39, 164–166, 1974.

86. Permyakov, E.A. and Burstein, E.A., Some aspects of studies of thermal transitions in proteins by means of their intrinsic fluorescence, *Biophys. Chem.* 19, 265–271, 1984.

87. VanScyoc, W.S., Sorensen, B.R., and Rusinova, E., Calcium binding to calmodulin mutants monitored by domain-specific intrinsic phenylalanine and tyrosine fluorescence, *Biophys. J.* 83, 2767–2786, 2002.
88. King, L.A. and Miller, J.N., Factors affecting the luminescence of tryptophan at 77 K, *Biochim. Biophys. Acta* 446, 206–213, 1976.
89. Ryu, K. and Dordick, J.S., How do organic solvents affect peroxidase structure and function?, *Biochemistry* 31, 2588–2598, 1992.
90. Mozo-Villarias, A., Second-derivative fluorescence spectroscopy of tryptophan in proteins, *J. Biochem. Biophys. Methods* 50, 163–178, 2002.
91. Nayar, S., Brahma, A., Mukherjee, C., and Bhattacharyya, D., Second derivative fluorescence spectra of indole compounds, *J. Biochem.* 131, 427–435, 2002.
92. Lotte, K., Plessow, R., and Brockhinke, A., Static and time-resolved fluorescence investigations of tryptophan analogues—A solvent study, *Photochem. Photobiol. Sci.* 3, 348–359, 2004.
93. Muiño, P.L. and Callis, P.R., Solvent effects on the fluorescence quenching of tryptophan by amides via electron transfer, experimental and computational studies, *J. Phys. Chem. B* 113, 2572–2577, 2009.
94. Sardar, P.S., Maity, S.S., Das, L., and Ghosh, S., Luminescence studies of perturbation of tryptophan residues of tubulin in the complexes of tubulin with colchicine and colchicine analogues, *Biochemistry* 46, 14544–14456, 2007.
95. Barboy, N. and Feitelson, J., Quenching of tryptophan phosphorescence in alcohol dehydrogenase from horse liver and its temperature dependence, *Photochem. Photobiol.* 41, 9–13, 1985.
96. Strambini, G.B., Cioni, P., and Felicicli, R.A., Characterization of tryptophan environments in glutamate dehydrogenase from temperature-dependent phosphorescence, *Biochemistry* 26, 4968–4975, 1987.
97. Strambini, G.B. and Gabelieri, E., Temperature dependence of tryptophan phosphorescence in proteins, *Photochem. Photobiol.* 51, 643–648, 1990.
98. Strambini, G.B., Kerwin, B.A., Mason, B.D., and Gonnelli, M., The triplet-state lifetime of indole derivatives in aqueous solution, *Photochem. Photobiol.* 80, 462–470, 2004.
99. Budisa, N. and Pal, P.P., Designing novel spectral classes of proteins with a tryptophan-expanded genetic code, *Biol. Chem.* 385, 893–904, 2003.
100. Lepthien, S., Hoezl, M.S., Merkel, L., and Budisa, N., Azatryptophans endow proteins with intrinsic blue fluorescence, *Proc. Natl. Acad. Sci. USA* 105, 16095–16100, 2008.
101. Ray, S., Bhattacharyya, M., and Chakrabarti, A., Conformational study of spectrin in presence of submolar concentrations of denaturants, *J. Fluoresc.* 15, 61–70, 2005.
102. Portugal, C.A., Crespo, J.G., and Lima, J.C., Amalous "unquencing" of the fluorescence decay times of β-lactoglobulin induced by the known quencher, acrylamide, *J. Photochem. Photobiol. B* 82, 117–126, 2006.
103. Mátyus, L., Szöllosi, J., and Jenei, A., Steady-state fluorescence quenching applications for studying protein structure and dynamics, *J. Photochem. Photobiol. B* 83, 223–236, 2006.
104. Fukuma, H., Hakashima, K., Ozaki, Y., and Noda, I., Two-dimensional fluorescence correlation spectroscopy IV: Resolution of fluorescence of tryptophan residues in alcohol dehydrogenase and lysozyme, *Spectrochim. Acta A Mol. Biomol. Spectrosc.* 65, 517–522, 2006.
105. Baghuraman, H. and Chattopadhyay, A., Effect of ionic strength on folding and aggregation of the hemolytic peptide mellitin in solution, *Biopolymers* 83, 111–121, 2006.
106. Gull, N., Kumar, S., Ahmad, B. et al., Influence of urea additives on micellar morphology/protein conformation, *Colloids Surf. B Biointerfaces* 51, 10–15, 2006.

107. Jana, S., Chauduri, T.K., and Deb, J.K., Effects of guanidine hydrochloride on the conformation and enzyme activity of streptomycin adenyltransferase monitored by circular dichroism and fluorescence, *Biochemistry* (Mosc.) 71, 1230–1237, 2006.

108. Fatima, S. and Khan, R.H., Effect of polyethylene glycols on the function and structure of thiol proteases, *J. Biochem.* 142, 65–72, 2007.

109. Hennecke, J., Sillen, A., Huber-Wunderlich, M. et al., Quenching of tryptophan fluorescence by the active-site disulfide bridge in the DebA protein from *Escherichia coli*, *Biochemistry* 36, 6391–6400, 1997.

110. Martinho, J.M., Santoe, A.M., Fedorov, A. et al., Fluorescence of the single tryptophan of cutinase: Temperature and pH effect on protein conformation and dynamics, *Photochem. Photobiol.* 78, 15–22, 2003.

111. Vanhooren, A., Illyen, E., Majer, Z., and Hanssens, I., Fluorescence contributions of the individual Trp residues in goat α-lactalbumin, *Biochim. Biophys. Acta* 1764, 1586–1591, 2006.

112. Chen, J., Toptygin, D., Brand, L., and King, J., Mechanism of the efficient tryptophan fluorescence quenching in human γD-crystallin studied by time-resolved fluorescence, *Biochemistry* 47, 10705–10721, 2008.

113. Shi, K., Duft, D., and Parks, J.H., Fluorescence quenching induced by conformational fluctuations in unsolvated polypeptides, *J. Phys. Chem. B* 112, 12801–12815, 2008.

114. Erker, W., Hübler, R., and Decker, H., Tryptophan quenching as linear sensor for oxygen binding of arthropod hemocyanins, *Biochim. Biophys. Acta* 1780, 1143–1147, 2008.

115. de la Hoz, L. and Netto, F.M., Structural modifications of β-lactoglobulin subjected to gamma irradiation, *Int. Dairy J.* 18, 1126–1132, 2008.

116. Labrenz, S.R., Calmann, M.A., Reavner, G.A., and Tolman, G., The oxidation of methionine-54 of epoetinum α does not affect molecular structure or stability, but does decrease biological activity, *PDA J. Pharm. Sci. Technol.* 62, 211–223, 2008.

117. Garidel, P., Hegyi, M., Bassareb, S., and Weichel, M., A rapid, sensitive and economical assessment of monoclonal antibody conformational stability by intrinsic tryptophan fluorescence spectroscopy, *Biotechnol. J.* 3, 1201–1211, 2008.

118. Fan, H., Vitharana, S.N., Chen, T. et al., Effects of pH and polyanions on the thermal stability of fibroblast growth factor, *Mol. Pharm.* 4, 232–240, 2007.

119. Derrick, T., Grillo, A.O., Vitharana, S.M. et al., Effect of polyanions on the structure and stability of repifermin (keratinocyte growth factor-2), *J. Pharm. Sci.* 95, 761–776. 2007.

120. Rezaei-Ghaleh, N., Ebrahim-Habibi, A., Moosavi-Movahedi, A.A., and Nemat-Gorgani, M., Effect of polyamines on the structure, thermal stability and 2,2,2-trifluoroethanol-induced aggregation of α-chymotrypsin, *Int. J. Biol. Macromol.* 41, 597–604, 2007.

121. Jovanović, N., Bouchard, A., Hofland, G.W. et al., Distinct effects of sucrose and trehalose on protein stability during supercritical fluid drying and freeze-drying, *Eur. J. Pharm. Sci.* 27, 336–345, 2006.

122. van de Weert, M., Anderson, M.B., and Frokjaer, S., Complex coacervation of lysozyme and heparin: Complex characterization and protein stability, *Pharm. Res.* 21, 2354–2359, 2004.

123. Lima, L.M., Zingali, R.B., Foguel, D., and Monteiro, R.Q., New insights into conformational and functional stability of human α-thrombin probed by high hydrostatic pressure, *Eur. J. Biochem.* 271, 3580–3587, 2004.

124. Ahmed, N., Argirov, O.K., Minhas, H.S. et al., Assay of advanced glycation endproducts (AGEs): Surveying AGEs by chromatographic assay with derivatization by 6-aminoquinolyl-*N*-hydroxysuccinimdyl-carbamate and application to N^{ε}-(1-carboxyetyl)lysine modified albumin, *Biochem. J.* 364, 1–14, 2002.

125. Liggins, J. and Furth, A.J., Role of protein-bound carbonyl groups in the formation of advanced glycation endproducts, *Biochim. Biophys. Acta* 1361, 123–130, 1997.

126. Cochrane, S.M. and Furth, A.J., The role of bound lipid and transition metal in the formation of fluorescent advanced glycation endproducts by human serum albumin, *Biochem. Soc. Trans.* 21, 97S, 1993.

127. Valente-Mesquita, V.L., Botelho, M.M., and Ferreira, S.T., Pressure-induced subunit dissociation and unfolding of dimeric β-lactoglobulin, *Biophys. J.* 75, 471–476, 1998.

128. Renard, D., Lefebvre, J., Giffin, M.C., and Griffin, W.G., Effects of pH and salt environment on the association of β-lactoglobulin revealed by intrinsic fluorescence studies, *Int. J. Biol. Macromol.* 22, 41–49, 1998.

129. Dumoulin, M., Ueno, H., Hayashi, R., and Balny, C., Contribution of the carbohydrate moiety to conformational stability of the carboxypeptidase Y high pressure study, *Eur. J. Biochem.* 262, 475–483, 1999.

130. Ruan, K., Lange. R., Bec, N., and Balny, C., A stable partly denatured state of trypsin induced by high hydrostatic pressure, *Biochem. Biophys. Res. Commun.* 239, 150–154, 1997.

131. Nijs, M., Azarkan, M., Smolders, N. et al., Preliminary characterization of bovine β-lactoglobulin after its conjugation to polyethylene glycol, *Biotechnol. Bioeng.* 54, 50–49, 1997.

132. Pan, X.M., Sheng, X.R., Yang, S.M., and Zhou, J.M., Probing subtle acid-induced conformational changes of ribonuclease A by electrospray mass spectrometry, *FEBS Lett.* 402, 25–27, 1997.

133. Tsai, P.K., Bruner, M.W., Irwin, J.I. et al., Origin of the isoelectric heterogeneity of monoclonal immunoglobulin h1B4, *Pharm. Res.* 10, 1580–1586, 1993.

134. Timm, D.E. and Neet, K.E., Equilibrium denaturation studies of mouse β-nerve growth factor, *Protein Sci.* 1, 236–244, 1992.

135. Brems, D.N., Brown, P.L., and Becker, G.W., Equilibrium denaturation of human growth hormone and its cysteine-modified forms, *J. Biol. Chem.* 265, 5504–5511, 1990.

136. Ogamo, A., Metori, A., Uchiyama, H., and Nagasawa, K., Generation of affinity for antithrombin III by supplemental sulfation of heparin species with low affinity for the protein, *J. Biochem.* 108, 588–592, 1990.

137. Uchiyama, S., Metori, A., Ogamo, A., and Nagasawa, K., Contribution of chemical 6-*O*-sulfation of the aminodeoxyhexose residues in whale heparin wit high affinity for antithrombin III to its anticoagulant properties, *J. Biochem.* 107, 377–380, 1990.

138. Oshima, G., Time-dependent conformational change of thrombin molecules induced by sulfated polysaccharides, *Chem. Pharm. Bull.* 37, 1324–1328, 1989.

139. Buchueva, T.L. and Tonevitsky, A.G., The effect to of pH on the conformation and stability of the structure of plant toxin—Ricin, *FEBS Lett.* 215, 155–159, 1987.

140. Walker, F.J., Properties of chemically modified protein S: Effect of the conversion of γ-carboxyglutamic acid to γ-methylglutamic acid on functional properties, *Biochemistry* 25, 6305–6311, 1986.

141. Lennick, M., Brew, S.A., and Ingham, K.C., Changes in protein conformation and stability accompany complex formation between human C1 inhibitor and C1-s, *Biochemistry* 24, 2561–2568, 1985.

142. Chang, G.G. and Lee, H.J., Monitoring protein conformational changes by quenching of intrinsic fluorescence, *J. Biochem. Biophys. Methods* 9, 351–355, 1984.

143. Parente, E.S. and Wetlaufer, D.B., Influence of urea on the high-performance cation-exchange chromatography of hen egg white lysozyme, *J. Chromatogr.* 288, 389–398, 1984.

144. Bauer, R.S., Chang, T.L., and Berlinger, L.J., Stability differences between high coagulant (α) and noncoagulant (γ) human thrombins. Denaturation, *J. Biol. Chem.* 255, 5900–5903, 1980.

145. Finazzi, A.A., Crifo, C., Natali, P.G., and Chersi, A., Differential denaturation of a crystalline Bence-Jones type cryoprecipitate as monitored by fluorescence, *Ital. J. Biochem.* 27, 36–42, 1978.

146. Mills, O.E., Effect of temperature on tryptophan fluorescence of β-lactoglobulin, *Biochim. Biophys. Acta* 434, 36–42, 1976.

147. D'Auria, S., Staiano, M., Varriale, A. et al., The tryptophan phosphorescence of porcine and mutant bovine odorant-binding proteins: A probe for the local protein structure and dynamics, *J. Proteome Res.* 7, 1151–1158, 2008.

148. D'Auria, S., Varriale, A., Gonnelli, M. et al., Tryptophan phosphorescence studies of the D-galactose/D-glucose-binding protein from *Escherichia coli* provide a molecular portrait with structural and dynamics features of the protein, *J. Proteome Res.* 6, 1306–1312, 2007.

149. Zhang, H.R., Guo, S.Y., Li., L., and Cai, M.Y., Study on *Escherichia coli* alkaline phosphatase conformation by phosphorimetry in the presence of denaturant, *Spectrochim. Acta A Mol. Biomol. Spectrosc.* 59, 3185–3191, 2003.

150. Gonnellli, M. and Strambini, G.B., Structure and dynamics of proteins encapsulated in silica hydrogels by Trp phosphorescence, *Biophys. Chem.* 104, 155–169, 2003.

151. Rousslang, K.W., Reid, P.J., Holloway, D.M. et al., Time-resolved phosphorescence of tyrosine, tyrosine analogues, and tyrosyl residues in oxytocin and small peptides, *J. Protein Chem.* 21, 547–555, 2002.

152. Wakabayashi, H., Zhen, Z., Schmidt, K.M., and Fay, P.J., Mn^{2+} binding to factor VIII subunits and its effect on cofactor activity, *Biochemistry* 42, 145–153, 2003.

153. Cioni, P. and Strambini, G.B., Effect of heavy water on protein flexibility, *Biophys. J.* 82, 3246–3253, 2002.

154. Gershenson, A., Schauerte, J.A., Giver, L., and Arnold, F.H., Tryptophan phosphorescence study of enzyme flexibility and unfolding in laboratory-evolved thermostable esterases, *Biochemistry* 39, 4658–4665, 2000.

155. Cioni, P. and Strambini, G.B., Pressure/temperature effects on protein flexibility from acrylamide quenching of protein phosphorescence, *J. Mol. Biol.* 291, 955–964, 1999.

156. Gonnelli, M. and Strambini, G.B., Time-resolved protein phosphorescence in the stopped-flow: Denaturation of horse liver alcohol dehydrogenase by urea and guanidine hydrochloride, *Biochemistry* 36, 16212–16220, 1997.

157. Schlyer, B.D., Steel, D.G., and Gafni, A., Long time-scale probing of the protein globular core using hydrogen-exchange and room temperature phosphorescence, *Biochem. Biophys. Res. Commun.* 223, 670–674, 1996.

158. Fischer, C.J., Schauerte, J.A., Wisser, K.C. et al., Hydrogen exchange at the core of *Escherichia coli* alkaline phosphatase studied by room-temperature tryptophan phosphorescence, *Biochemistry* 39, 1455–1461, 2000.

159. Cioni, P. and Strambini, G.B., Pressure effects on protein flexibility monomeric proteins, *J. Mol. Biol.* 242, 291–301, 1994.

160. Shah, N.K. and Ludescher, R.D., Influence of hydration on the internal dynamics of hen egg white lysozyme in the dry state, *Photochem. Photobiol.* 58, 169–174, 1993.

161. Smith, C.A. and Maki, A.H., Temperature dependence of the phosphorescence quantum yield of various α-lactalbumin and of hen egg-white lysozyme, *Biophys. J.* 64, 1885–1895, 1993.

162. Gonnelli, M. and Strambini, G.B., Glycerol affects on protein flexibility: A tryptophan phosphorescence study, *Biophys. J.* 65, 131–137, 1993.

163. Wright, W.W., Owen, C.S., and Vanderkool, J.M., Penetration of analogues of H_2O and CO_2 in proteins studied by room temperature phosphorescence of tryptophan, *Biochemistry* 31, 6538–6544, 1992.

164. Li., Z. and Galley, W.C., Evidence for ligand-induced conformational changes in proteins from phosphorescence spectroscopy, *Biophys. J.* 56, 353–360, 1989.

165. Gabellieri, ED., Strambini, G.B., and Gualtieri, P., Tryptophan phosphorescence and the conformation of liver alcohol dehydrogenase in solution and in the crystalline state, *Biophys. Chem.* 30, 61–67, 1988.

166. Davis, J.M., Arakawa, T., Strickland, T.W., and Yphantis, D.A., Characterization of recombinant human erythropoietin produced in Chinese hamster ovary cells, *Biochemistry* 26, 2633–2638, 1987.

167. Galley, W.C., Williams, R.E., and Goodfriend, L., Unusual emission properties of the tryptophans at the surface of short ragweed allergen Ra5, *Biochemistry* 21, 378–383, 1982.

168. Rousslang, K.W., Thomasson, J.M., Rose, J.B., and Kwiram, A.L., Triplet state of tryptophan in proteins. 2. Differentiation between tryptophan residues 62 and 108 in lysozyme, *Biochemistry* 18, 2296–2300, 1979.

169. Sokolovsky, M. and Daniel, E., Applications of phosphorescence to the study of proteins, *Methods Enzymol.* 49, 236–249, 1978.

170. Ghiron, C.A., Longworth, J.W., and Ramachandran, N., Triplet-triplet energy transfer in α-trypsin, *Proc. Natl. Acad. Sci. USA* 70, 3703–3706, 1973.

171. Hillgren, A., Evertsson, H., and Aldén, M., Interaction between lactate dehydrogenase and Tween 80 in aqueous solution, *Pharm. Res.* 19, 504–510, 2002.

172. Okumu, F.W., Dao le, N., Fielder, P.J. et al., Sustained delivery of human growth hormone from a novel gel system; SABER, *Biomaterials* 23, 4353–4358, 2002.

173. Gram, J. and Jespersen, J., Improved assay of antithrombin-III. Effects of certain additives on thrombin and on chromogenic peptide substrates, *Clin. Chem.* 31, 1683–1688, 1985.

174. Luykx, D.M., Dingemanse, P.J., Goerdayal, S.S., and Jongen, P.M., High-performance anion-exchange chromatography combined with intrinsic fluorescence detection to determine erythropoietin in pharmaceutical products, *J. Chromatogr. A*, 1078, 113–119, 2005.

175. Luykx, D.M., Goerdayal, S.S., Dingemanse, P.J. et al., HPLC and tandem detection to measure conformational properties of biopharmaceuticals, *J. Chromatogr. B Analyt. Technol. Biomol. Life Sci.* 821, 45–52, 2005.

176. Daniels, F., Boyd, C.A., and Saunders, D.F., Thermoluminescence as a research tool, *Science* 117, 343–349, 1953.

177. Nummedal, D. and Steen, H.B., On the x-ray-induced thermoluminescence of some proteins in the dry state and in solution, *Radiat. Res.* 39, 241–251, 1969.

178. Ramachander, R., Jiang, Y., Li, C. et al., Solid state fluorescence of lyophilized proteins, *Anal. Biochem.* 376, 173–182, 2008.

179. Campbell, I.D. and Dwek, R.A., *Biological Spectroscopy*, Benjamin/Cummings Publishing, Menlo Park, CA, Chapter 5, 1984.

180. Saviotti, M.L. and Galley, W.C., Room temperature phosphorescence and the dynamic aspects of protein structure, *Proc. Natl. Acad. Sci. USA* 71, 4154–4158, 1974.

181. Cioni, P., Gabellieri, E., Gonnelli, M., and Strambini, G.B., Heterogeneity of protein conformation in solution from the lifetime of tryptophan phosphorescence, *Biophys. Chem.* 52, 25–34, 1994.

182. Wright, W.W., Carlos Baez, J., and Vanderkooi, J.M., Mixed trehalose/sucrose glasses used for protein incorporation as studied by infrared and optical spectroscopy, *Anal. Biochem.* 307, 167–172, 2002.

183. Cioni, P., Onuffer, J.J., and Strambini, G.B., Characterization of tryptophan phosphorescence of aspartate aminotransferase from *Escherichia coli, Eur. J. Biochem.* 209, 759–764, 1992.

184. Schlyer, B.D., Schauerte, J.A., Steel, D.G., and Gafni, A., Time-resolved room temperature protein phosphorescence: Nonexpotential decay from single emitting tryptopans, *Biophys. J.* 67, 1192–1202, 1994.

185. Banks, D.D. and Kerwin, B.A., A deoxygenation system for measuring protein phosphorescence, *Anal. Biochem.* 324, 106–114, 2004.
186. Hanslip, S.J., Zaccal, N.R., Middelberg, A.P., and Falconer, R.J., Intrinsic fluorescence as an analytical probe of virus-like particle assembly and saturation, *Biochem. Biophys. Res. Commun.* 375, 351–355, 2008.
187. He, F., Joshi, S.B., Bosman, F. et al., Structural stability of hepatitis C virus envelope glycoprotein E1: Effect of pH and dissociative detergents, *J. Pharm. Sci.* 98, 3340–3357, 2009.
188. Vessely, C., Estey, T., Randolph, T.W. et al., Stability of a trivalent recombinant protein vaccine formulation against botulinum neurotoxin during storage in aqueous solution, *J. Pharm. Sci.* 98, 2970–2993, 2009.
189. Salnikova, M.S., Joshi, S.B., Rytting, J.H. et al., Physical characterization of *Clostridium difficile* toxins and toxoids: Effect of the formaldehyde crosslinking on thermal stability, *J. Pharm. Sci.* 97, 3735–3752, 2008.
190. Kissmann, J., Ausar, S.F., Foubert, T.R. et al., Physical stabilization of Norwalk virus-like particles, *J. Pharm. Sci.* 97, 4208–4218, 2008.
191. Selvi, B.R., Jagadeesan, D., Suma, B.S. et al., Intrinsically fluorescent carbon nano-spheres as a nuclear targeting vector: Delivery of membrane-impermeable molecule to modulate gene expression in vivo, *Nano Lett.* 8, 3182–3188, 2008.
192. Cuello, M., Cabrera, O., Martinez, I. et al., New meningococcal C polysaccharide-tetanus toxoid conjugate: Physico-chemical and immunological characterization, *Vaccine* 25, 1798–1805, 2007.
193. Ausar, S.F., Foubert, T.R., Hudson, M.H. et al., Conformational stability and disassembly of Norwalk virus-like particles. Effect of pH and temperature, *J. Biol. Chem.* 281, 19478–19488, 2006.
194. Ho, M.M., Bolgiano, B., Martino, A. et al., Preclinical laboratory evaluation of a bivalent *Staphylococcus aureus* saccharide–exotoxin A protein conjugate vaccine, *Hum. Vaccin.* 2, 89–98, 2006.
195. Rexroad, J., Evans, R.K., and Middaugh, C.R., Effect of pH and ionic strength of the physical stability of adenovirus type 5, *J. Pharm. Sci.* 95, 237–243, 2006.
196. Ausar, S.F., Rexroad, J., Frolov, V.G. et al., Analysis of the thermal and pH stability of human respiratory syncytial virus, *Mol. Pharm.* 2, 491–499, 2005.
197. Fowler, S., Byron, O., Jumel, K. et al., Novel configurations of high molecular weight species of the pertussis toxin vaccine component, *Vaccine* 21, 2678–2688, 2003.
198. Freitas, T.R., Gaspar, L.P., Caldas, L.A. et al., Inactivation of classical swine fever virus: Association of hydrostatic pressure and ultraviolet irradiation, *J. Virol. Methods* 108, 205–211, 2003.
199. Tian, S.M., Ruan, K.C., Qian, J.F. et al., Effects of high pressure on the structure and biological activity of infectious bursal disease virus, *Eur. J. Biochem.* 267, 4486–4494, 2000.
200. Bolgiano, B., Crane, D.T., Xing, D. et al., Physico-chemical analysis of *Bordetella pertussis* antigens, *Biologicals* 27, 155–162, 1999.
201. Volkin, D.B., Burke, C.J., Marfia, K.E. et al., Size and conformational stability of the hepatitis A virus used to prepare VAQTA, a highly purified inactivated vaccine, *J. Pharm. Sci* 86, 666–673, 1997.
202. Lam, W.C., Maki, A.H., Casas-Finet, J.R. et al., Phosphorescence and optically detected magnetic resonance investigation of the binding of the nucleocapsid protein of the human immunodeficiency virus type 1 and related peptide to RNA, *Biochemistry* 33, 10693–10700, 1994.
203. Hörer, O.L., Fluorescence anisotropy of viruses treated with formaldehyde and dimethyl-sulfoxide, *Revue Roumaine Virologie* 41, 29–35, 1990.

204. Hörer, O.L., Fluorescence anisotropy of UV-irradiated viruses, *Virologie* 50, 115–118, 1989.
205. Silva, J.L. and Weber, G., Pressure-induced dissociation of brome mosaic virus, *J. Mol. Biol.* 199, 149–159, 1988.
206. Takehara, K., Yuki, K., Shirasawa, M. et al., Binding properties of hydrophobic molecules to human serum albumin studied by fluorescence titration, *Anal. Sci.* 25, 115–120, 2009.
207. Lissi, E., Alicia Biasutti, M., Abuin, E., and León, L., A fluorescence assay of human serum albumin binding sites modification by hypochlorite, *J. Photochem. Photobiol. B* 94, 77–81, 2009.
208. Jain, N.K. and Roy, I., Role of trehalose in moisture-induced aggregation of bovine serum albumin, *Eur. J. Pharm. Biopharm.* 69, 824–834, 2008.
209. Singh, S.K. and Kishore, N., Calorimetric and spectroscopic studies on the interaction of methimazole with bovine serum albumin, *J. Pharm. Sci.* 97, 2362–2372, 2008.
210. Wei, Y., Dong, C., Liu, D. et al., Enantioselective quenching of room-temperature phosphorescence lifetimes of proteins: Bovine and human serum albumins, *Biomactromolecules* 8, 761–764, 2007.
211. Mendez, D.L., Jensen, R.A., McElroy, L.A. et al., The effect of non-enzymatic glycation on the unfolding of human serum albumin, *Arch. Biochem. Biophys.* 444, 92–99, 2005.
212. Ahmed, N. and Thornalley, P.J., Chromatographic assay of glycation adducts in human serum albumin glycated in vitro by derivatization with 6-aminoquinolyl-*N*-hydroxysucciinimidyl-carbamate and intrinsic fluorescence, *Biochem. J.* 364, 15–24, 2002.
213. Ferrer, M.L., Duchowicz, R., Carrasco, B. et al., The conformation of serum albumin in solution: A combined phosphorescence depolarization–hydrodynamic modeling study, *Biophys. J.* 80, 2422–2430, 2001.
214. Watanabe, H., Kragh-Hansen, U., Tanase, S. et al., Conformational stability and warfarin-binding properties of human serum albumin studied by recombinant mutants, *Biochem. J.* 357, 269–274, 2001.
215. Muzammil, S., Kumar, Y., and Tayyab, S., Anion-induced stabilization of human serum albumin prevents formation of intermediate during urea denaturation, *Proteins* 40, 29–38, 2000.
216. Epps, D.E., Raub, T.J., Caiolfa, V. et al., Determination of the affinity of drugs toward serum albumin by measurement of the quenching of the intrinsic tryptophan fluorescence of the protein, *J. Pharm. Pharmacol.* 51, 41–48, 1999.
217. Hogiu, S., Enescu, M., and Pascu, M.L., Dynamic and thermodynamic effects of glycerol on bovine serum albumin in aqueous solution: A tryptophan phosphorescence study, *J. Photochem. Photobiol. B* 40, 55–60, 1997.
218. Gorinstein, S., Zemser, M., Friedman, M., and Chang, S.M., Simultaneous differential scanning calorimetry, x-ray diffraction and FTIR spectrometry in studies of ovalbumin denaturation, *Int. J. Pept. Protein Res.* 45, 248–256, 1995.
219. Enescu, M., Ionescu, R., Dumbarveanu, G., and Pascu, M.L., Phosphorescence lifetime studies of interactions between serum albumins and sodium dodecyl sulfate, *Photochem. Photobiol.* 57, 367–370, 1993.
220. Mao, S.Y. and Maki, A.H., Comparative phosphorescence and optically detected magnetic resonance studies of fatty acid binding to serum albumin, *Biochemistry* 26, 3576–3582, 1987.
221. Waldemeyer, J., Korkidis, K., and Geacintov, N.E., Relative contributions of tryptophan and tyrosine to the phosphorescence emission of human serum albumin at low temperatures, *Photochem. Photobiol.* 35, 299–304, 1982.

222. De Jersey, J., Jeffers Morley P., and Martin, R.B., Lanthanide probes in biological systems: Characterization of luminescence excitation spectra of terbium complexes with proteins, *Biophys. Chem.* 13, 233–243, 1981.

223. Schiødt, J., Harrit, N., Christensen, U., and Petersen, L.C., Two different Ca^{2+} ion binding sites in factor VIIa and in des (1–38) factor VIIa, *FEBS Lett.* 306, 265–268, 1992.

224. Lundblad, R.L. *Chemical Reagents for the Protein Modification*, CRC Press, Boca Raton, FL, 2004.

10 Near-Infrared Spectroscopy and Macromolecular Conformation

Infrared spectroscopy is based on the absorption of energy related to the vibration of atoms in a molecule; UV-VIS spectroscopy involves the absorption of energy resulting in transition between electronic states. The spectra is expressed as absorption of radiation as a function of wavenumber, which is the reciprocal of wavelength ($1/\lambda = v/c$) and is shown in cm^{-1} (reciprocal centimeters).[1-3] Infrared spectroscopy is divided into near-infrared, mid-infrared, and far-infrared. This chapter is concerned with near-infrared (NIR) spectroscopy; Chapter 11 is concerned with mid-infrared spectroscopy and Fourier-transform while Chapter 12 is concerned with Raman spectroscopy (also known as feeble infrared fluorescence).

NIR spectroscopy comprises the infrared spectrum overlapping the UV-VIS spectrum ($12,500-4,000\,cm^{-1}$; $780-2500\,nm$) and can be used for the measurement of overtones of the fundamentals for the stretching modes of C–H, N–H, and O–H[4,5] (Table 10.1). The ability to separate sample measuring position and the spectrophotometer by some distance using light fibers (see below) permits facile use of NIR spectroscopy in process monitoring[6-10] as well as for in vivo applications.[11-14]

The emphasis of this chapter is on the application of NIR spectroscopy in the study of the conformation of proteins and other biological polymers as such macromolecules are developed for therapeutic use. Other applications of NIR spectroscopy in biotechnology are also worth a mention. These are the use of NIR spectroscopy for moisture determination, determination of protein concentration, and process monitoring.

Determination of moisture in a final drug product[15-20] and food product[21-24] is critical for product quality.[25-33] The traditional method is the Karl Fischer titration method.[34-43] While this method is accurate, it is destructive in that drug product is consumed in the analysis. NIR spectroscopy has been shown to be useful for the determination of water in solid protein preparations.[44] Another study showed that NIR spectroscopy had the same precision as the Karl Fischer method and noted that there can be an effect of excipient (disaccharide), which must be considered.[45] Moisture can be determined by in-line technology, permitting control of wet granulation processes,[46] which are used in neutraceutical manufacture.[47] NIR spectroscopy can be used of the in situ analysis of product in vials[48] and has been used to monitor freeze-drying processes.[49,50] There has been recent progress in improving the accuracy of the

TABLE 10.1
Positions of Near-Infrared Bands

Functional Group	Wavelength (nm)	Wavenumber (cm⁻¹)
	300	33,333
	400	25,000
	500	20,000
	600	16,667
	700	14,286
	800	12,500
	900	11,111
	1000	10,000
	1200	8,000
	1500	6,667
–OH 3600 cm⁻¹	2000	5,000
–NH₂ 3400 cm⁻¹		
–C–H 3300 cm⁻¹		
–SH 2580 cm⁻¹	3000	3,333

technology for the measurement of moisture content in solid biopharmaceuticals.[51] It is noted that NIR spectroscopy is used for the determination of water in petroleum-based products[52] and for the presence of organic solvents in water.[53]

It is possible to estimate protein concentration by NIR spectroscopy. One indirect protein assay involves the measurement of relative concentrations of oxy- and deoxy-hemoglobin as a measure of oxygen saturation in whole blood.[54–57] It is possible to directly estimate total protein concentration in solution using NIR spectroscopy. Harthun and colleagues[58] developed a method for the use of NIR spectroscopy in the determination of protein concentration in cell culture; this technology is applicable for process analytical technology (PAT).[59] Other examples of NIR spectroscopy for protein analysis include the measurement of lysozyme concentration,[60] glucose isomerase,[61] and the simultaneous determination of albumin, IgG, and glucose.[62] Other applications of NIR spectroscopy for protein analysis include the quantitative analysis of fat, protein, and lactose in raw milk[63] and absorbed protein concentration in aluminum hydroxide suspensions.[64] NIR spectroscopy is not as sensitive as other methods and requires a sample concentration in the range of 0.3–1 mg/mL.[65,66] Attenuated total reflectance (ATR) is used for solid-state thin film and protein adsorption studies, permitting the studies of protein solutions with lower protein concentration.[66]

NIR spectroscopy permits the noninvasive monitoring of biotechnology manufacturing processes.[67–70] There is considerable experience in the use of NIR spectroscopy in the chemical industry.[71–74] NIR spectroscopy has advantages over the use of infrared spectroscopy for in-process monitoring. One distinct advantage is the ability to use glass cell windows instead of salt windows,[52] which provides the ability to use glass fiber optic probes.[75]

Technology has been developed for the determination of various critical process parameters during cell culture and fermentations.[76–89] The reader is directed to several sources, which consider the available instrumentation.[90,91] There is also considerable interest in the use of NIR spectroscopy for in vivo monitoring of metabolites.[92,93] There has been limited application of NIR spectroscopy to nucleic acids.[94–98]

NIR spectroscopy can be used for the study of a variety of polymers.[99–105] Workman[106–108] has reviewed the application of NIR spectroscopy to the study of synthetic polymers and rubbers. The significant NIR bands for use in protein conformation are overtones of fundamentals occurring in the mid-infrared and far-infrared region (see Table 10.1). NIR spectroscopy, as with infrared spectroscopy, is used to determine secondary structure in structure in proteins.[109] The advantages of NIR spectroscopy as compared to infrared spectroscopy have been described above but it should be emphasized that the two techniques reflect the same fundamental vibrational event. The fundamental or first harmonic represents a transition from the ground vibrational state to the first excited level; transitions to higher energy levels, as discussed above, are referred to as overtones or secondary harmonics.[3]

The major use of NIR spectroscopy for the study of proteins uses absorption bands (see Table 10.1) reflecting the secondary structure of proteins.[110–114] There is good correlation between infrared spectroscopy and other analytical techniques such as circular dichorism (CD) and x-ray crystallography for the estimation of secondary structure configuration (α-helix; β-sheet)[115] and it is reasonable to assume that such correlations can be extended to NIR spectroscopy. As noted above, NIR spectroscopy does require relative high protein concentrations (mg/mL) for solutions studies; however, studies can be performed with protein adsorbed to surfaces (ATR). Of greater importance to our current discussion is the ability to use NIR spectroscopy to study protein conformation (and moisture as discussed above) in lyophilized materials. Fourier-transfer infrared spectroscopy (FTIR) has been used to study lyophilized proteins.[116–123] While this has been useful, infrared spectroscopy suffers from some disadvantages in that the technique is sensitive to the presence of moisture, presenting issues with respect to sample preparation. Aggregation has been observed in the preparation of protein in KBr pellets for infrared spectroscopy.[124–127] NIR spectroscopy can be used for the direct examination of drug product in a glass vial,[128] permitting the noninvasive analysis of protein conformation. This is important in the development of formulation as perseveration of native structure during lyophilization is desirable from quality viewpoint of the final product.[120,129,130] Jovanović and coworkers[131] used near-infrared imaging with principal component analysis to study the homogeneity of lyophilized protein preparations; these investigators also compared lyophilization with supercritical fluid drying. Other investigators[132] used near-infrared imaging to study the effect of excipient composition on heterogeneity and stability of dried protein products. There are several studies on the use of near-infrared imaging for the study of the uniformity of solid dosage forms.[133,134]

There are several studies that are of particular interest to the overall thrust of this book in that such studies are concerned with the use of NIR spectroscopy to study protein denaturation. Cho and coworkers observed that a combination of heat and pressure caused a shift in the spectra of lysozyme at 2180 nm (ca. 4590 cm^{-1}).[135,136] These early studies suggested the feasibility of using NIR spectroscopy for the study of conformational changes associated with protein denaturation. The development of two-dimensional infrared spectroscopy[137] permitted the analysis of complex infrared spectrum obtained during protein denaturation.[138] Sefara and coworkers[138] used two-dimensional Fourier transform mid-infrared and near-infrared spectroscopy to study the conformational change of β-lactoglobulin in the presence of 2-bromoethanol. Changes in bands at 4290, 4360, and $4420\,cm^{-1}$ were assigned to C–H stretching and deformation on amino acid side chains while the changes in a band at $4610\,cm^{-1}$ were assigned to a combination of amide B (N–H stretching) and amide II (N–H bending). Correlation spectroscopy suggested that these changes reflected a transition from β-sheet to α-helix. Wang and coworkers[139] used two-dimensional Fourier transfer NIR spectroscopy to study the heat denaturation of ovalbumin in aqueous solution. The reader is referred to several review articles for the application of NIR spectroscopy to study of protein denaturation.[140–142] It is recognized that NIR spectroscopy requires relatively high protein concentration (mg/mL).

REFERENCES

1. Stuart, B.H., *Infrared Spectroscopy: Fundamentals and Applications*, John Wiley & Sons, Chichester, U.K., 2004.
2. Stuart, B., *Biological Applications of Infrared Spectroscopy*, ACOL Series, Chichester, U.K., 1997.
3. Campbell, I.D. and Dwek, R.A., *Biological Spectroscopy*, Benjamin/Cummings, Menlo Park, CA, 1984.
4. Shaw, R.A. and Mantsch, H.H., Near-IR spectrophotometers, in *Encyclopedia of Spectroscopy and Spectrometry*, eds. J.C. Linden, G.E. Trauter, and J.L. Holmes, Academic Press, New York, 2000.
5. Siesler, H.W., Ozaki, Y., Kawato, S., and Heise, H.M. eds., *Near-Infrared Spectroscopy. Principles, Instruments, Applications*, Wiley-VCH, Weinheim, Germany, 2003.
6. White, J.G., On-line moisture detection for a microwave vacuum dryer, *Pharm. Res.* 11, 728–732, 1995.
7. Hailey, P.A., Doherty, P., Tapsell, P. et al., Automated system for the on-line monitoring of powder blending processes using near-infrared spectroscopy. Part I. System development and control, *J. Pharm. Biomed. Anal.* 14, 551–559, 1996.
8. Bouversesse, E., Casolino, C., and de la Pezuela, C., Application of standardization methods to correct the spectral differences induced by a fibre optic probe used for the near-infrared analysis of pharmaceutical tablets, *J. Pharm. Biomed. Anal.* 18, 35–42, 1998.
9. Tosi, S., Rossi, M., Tamburini, E. et al., Assessment of in-line near-infrared spectroscopy for continuous monitoring of fermentation processes, *Biotechnol. Prog.* 19, 1816–1821, 2003.
10. Roychoudhury, P., O'Kennedy, R., McNeil, B., and Harvey, L.M., Multiplexing fibre optic near infrared (NIR) spectroscopy as an emerging technology to monitor industrial bioprocesses, *Anal. Chim. Acta* 590, 110–117, 2007.
11. Huang, Z., Zeng, H., Hamzavi, I. et al., Cutaneous melanin exhibiting fluorescence emission under near-infrared light excitation, *J. Biomed. Opt.* 11(3): 34010, 2006.
12. Cerussi, A.E., Berger, A.J., Bevilacqua, F. et al., Sources of absorption and scattering contrast for near-infrared optical mammography, *Acad. Radiol.* 8, 211–218, 2001.
13. Piao, D., Xie, H., Zhang, W. et al., Endoscopic, rapid near-infrared optical tomography, *Opt. Lett.* 31, 2876–2878, 2006.

14. Alencar, H., Funovics, M.A., Figureiredo, J. et al., Colonic adenocarcinomas: Near-infrared microcatheter imaging of smart probes for early detection—Study in mice, *Radiology* 244, 232–238, 2007.

15. Hsu, C.C., Ward, C.A., Pearlman, R. et al., Determining the optimum residual moisture in lyophilized protein pharmaceuticals, *Dev. Biol. Stand.* 74, 255–270, 1992.

16. May, J.C., Wheeler, R.N., Etz, N., and Del Grasso, A., Measurement of final container residual moisture in freeze-dried biological products, *Dev. Biol. Stand.* 74, 153–164, 1992.

17. Dawson, P.J., Effect of formulation and freeze-drying on the long-term stability of rDNA-derived cytokines, *Dev. Biol. Stand.* 74, 273–282, 1992.

18. Skrabanja, A.T., de Meere, A.L., de Ruiter, A.A., and van den Oetelaar, P.J., Lyophilization of biotechnology products, *PDA J. Pharm. Sci. Technol.* 48, 311–317, 1994.

19. Ma, X., Wang, D.Q., Bouffard, A., and MacKenzie, A., Characterization of murine monoclonal antibody to tumor necrosis factor (TNF-mab) formulation for freeze-drying cycle development, *Pharm. Res.* 18, 196–201, 2001.

20. Donovan, P.D., Corvari, V., Burton, M.D., and Rajagopalan, N., Effect of stopper processing conditions on moisture content and ramifications for lyophilized products. Comparison of "low" and "high" moisture uptake stoppers, *PDA J. Pharm. Sci. Technol.* 61, 51–58, 2007.

21. Nagel, F.J., Tramper, J., Bakker, M.S., and Rinzema, A., Model for on-line moisture-content control during solid-state fermentation, *Biotechnol. Bioeng.* 72, 231–243, 2001.

22. Nelson, S.O. and Trabelsi, S., Principles for microwave moisture and density measurement in grain and seed, *J. Microw. Power Electromagn. Energy* 39, 107–117, 2006.

23. McKenna, D., Measuring moisture in cheese by near infrared absorption spectroscopy, *J. AOAC Int.* 84, 672–677, 2006.

24. Baik, O.D., Marcotte, M., Sablani, S.S., and Castaigne, F., Thermal and physical properties of bakery products, *Crit. Rev. Food Sci. Nutr.* 41, 321–352, 2001.

25. Kahn, K.D., New desiccant keeps quality in, moisture out, *Food Ind.* 20, 18, 1948.

26. Miller, G.A., Food quality—A matter of time, temperature, and moisture control, *Hospitals* 42(13), July, 85–86, 1968.

27. Giovanelli, G. and Paradiso, A., Stability of dried and intermediate moisture tomato pulp during storage, *J. Agric. Food Chem.* 50, 7277–7281, 2006.

28. Tang, X., De Rooij, M., and De Jong, L., Volume change measurements rice by environmental scanning electron microscopy and stereoscopy, *Scanning* 29, 197–205, 2007.

29. Barbaree, J.M., Methods for determining favorable conditions for freeze-drying biological reference materials, *Dev. Biol. Stand.* 36, 335–341, 1975.

30. Towns, J.K., Moisture content in proteins: Its effects and measurement, *J. Chromatogr. A* 705, 115–127, 1995.

31. Yoshioka, S., Aso, Y., and Kojima, S., Determination of molecular mobility of lyophilized bovine serum albumin and gamma-globulin by solid state ^{1}H NMR and relation to aggregation-susceptibility, *Pharm. Res.* 13, 926–930, 1996.

32. Savage, M., Torres, J., Franks, L. et al., Determination of adequate moisture content for efficient dry-heat viral inactivation in lyophilized factor VIII by loss on drying and by near infrared spectroscopy, *Biologicals* 26, 119–124, 1998.

33. Chang, L.L., Shepherd, D., Sun, J. et al., Effect of sorbitol and residual moisture on the stability of lyophilized antibodies: Implications for the mechanism of protein stabilization in the solid state, *J. Pharm. Sci.* 94, 1445–1455, 2005.

34. Fischer, K., Neues Verfahren zur maszanalytischen Bestimmung des Wassrgelhaltes von Flüssigkeiten und festen Körpern, *Angew. Chem.* 45, 394–396, 1935.

35. Rao, P.B. and Bryan, W.P., Measurement of strongly held water of lysozyme, *J. Mol. Biol.* 97, 119–122, 1975.

36. Bryan, W.P. and Rao, P.B., Comparison of standards in the Karl Fischer method for water determination, *Anal. Chim. Acta* 84, 149–155, 1976.
37. Strickley, R.G. and Andereson, B.D., Solid-state stability of human insulin. I. Mechanism and effect of water on kinetics of degradation n lyophiles from pH 2–5, *Pharm. Res.* 1142–1153, 1996.
38. Margolis, S.A., Sources of systemic bias in the measurement of water by the coulometric and volumetric Karl Fischer methods, *Anal. Chem.* 69, 4864–4871, 1997.
39. Costantino, H.R., Curley, J.G., and Hsu, C.C., Determining the water sorption monolayer of lyophilized pharmaceutical proteins, *J. Pharm. Sci.* 86, 1390–1393, 1997.
40. Margolis, S.A. and Angelo, J.B., Interlaboratory assessment of measurement precision and bias in the coulometric Karl Fischer determination of water, *Anal. Bioanal. Chem.* 374, 505–512, 2002.
41. Hawe, A. and Friess, W., Physico-chemical lyophilization behavior of mannitol, human serum albumin formulations, *Eur. J. Pharm. Sci.* 28, 224–232, 2006.
42. http://www.radiometer-analytical.com/en_product_details.asp?pid=257
43. http://www.scientificgear.com/Karl-Fisher-titration
44. Stokvold, A., Drystad, K., and Libnau, F.O., Sensitive NIRS measurement of increased moisture in stored hydroscopic freeze dried product, *J. Pharm. Biomed. Anal.* 28, 867–873, 2002.
45. Lin, T.P. and Hsu, C.C., Determination of residual moisture in lyophilized protein pharmaceuticals using a rapid and non-invasive method: Near infrared spectroscopy, *PDA J. Pharm. Sci. Technol.* 56, 196–205, 2002.
46. Rantanen, J., Räsänen, E., Tenhunen, J. et al., In-line moisture measurement during granulation with a four-wavelength near infrared sensor: An evaluation of particle size and binder effects, *Eur. J. Pharm. Biopharm.* 50, 271–276, 2000.
47. Etube, N.K., Mark, W., and Hahm, H., Preformulation studies and characterization of proposed chondroprotective agents: Glucosamine and chondroitin sulfate, *Pharm. Dev. Technol.* 7, 457–469, 2002.
48. Birrer, G.A., Liu, J., Halas, J.M., and Nucera, G.G., Evaluation of a container closure integrity test model using visual inspection with confirmation by near infra-red spectroscopic analysis, *PDA J. Pharm. Sci. Technol.* 54, 373–382, 2000.
49. Liu, J., Physical characterization of pharmaceutical formulations in frozen and freeze-dried solid states: Techniques and applications in freeze-drying development, *Pharm. Dev. Technol.* 11, 3–28, 2006.
50. De Beer, T.A., Vercruysse, P., Burggraeve, A. et al., In-line and real-time process monitoring of a freeze drying process using Raman and NIR spectroscopy as complementary process analytical technology (PAT) tools, *J. Pharm. Sci.* 98, 3430–3446, 2009.
51. Brülls, M., Folestad, S., Sparén, A. et al., Applying spectral peak area analysis in near-infrared spectroscopy moisture assays, *J. Pharm. Biomed. Anal.* 44, 127–136, 2007.
52. Baughman, E., Near-infrared spectroscopy in analysis of crudes and transportation fuels, in *Encyclopedia of Analytical Chemistry*, Vol. 8, ed. R.A. Meyers, John Wiley & Sons, Chichester, U.K., pp. 6842–6849, 2000.
53. Ding, Q., Boyd, B.L., and Small, G.W., Determination of organic contaminants in aqueous samples by near-infrared spectroscopy, *Appl. Spectrosc.* 54, 1047–1054, 2000.
54. Ferrari, M., Wilson, D.A., Hanley, D.F. et al., Noninvasive determination of hemoglobin saturation in dogs by derivative near-infrared spectroscopy, *Am. J. Physiol.* 256, H1493–H1499, 1989.
55. Edwards, A.D., Richardson, C., van der Zee, P. et al., Measurement of hemoglobin flow and blood flow by near-infrared spectroscopy, *J. Appl. Physiol.* 75, 1884–1889, 1993.
56. Tamaki, T., Uchiyama, S., Tamura, T., and Nakano, S., Changes in muscle oxygenation during weight-lifting exercise, *Eur. J. Appl. Physiol.* 68, 465–469, 1994.

57. Sato, H., Kiguchi, M., Kawaguchi, F., and Maki, A., Practicality of wavelength selection to improve signal-to-noise ratio in near-infrared spectroscopy, *Neuroimage* 21, 1554–1562, 2004.

58. Harthun, S., Matischak, K., and Fried, P., Determination of recombinant protein in animal cell culture supernatant by near-infrared spectroscopy, *Anal. Biochem.* 251, 73–78, 1997.

59. Rathore, A.S., Yu, M., Yebaoh, S., and Sharma, A., Case study and application of process analytical technology (PAT) towards bioprocessing: Use of on-line high-performance liquid chromatography (HPLC) for making real-time pooling decisions for process chromatography, *Biotechnol. Bioeng.* 100, 306–316, 2008.

60. Hu, S.Y., Arnold, M.A., and Wiencek, J.M., Temperature-independent near-infrared analysis of lysozyme aqueous solutions, *Anal. Chem.* 72, 698–702, 2000.

61. Olesberg, J.T., Arnold, M.A., Hu, S.Y., and Wiencek, J.M., Temperature-insensitive near-infrared method for determination of protein concentration during protein crystal growth, *Anal. Chem.* 72, 4985–4990, 2000.

62. Kasemsumran, S., Du, Y.P., Murayama, K. et al., Simultaneous determination of human serum albumin, gamma-globulin, and glucose in a phosphate buffer solution by near-infrared spectroscopy with moving window partial least-squares regression, *Analyst* 128, 1471–1477, 2003.

63. Sasić, S. and Ozaki, Y., Short-wave near-infrared spectroscopy of biological fluids. 1. Quantitative analysis of fat, protein, and lactose in raw milk by partial least-squares regression and band assignment, *Anal. Chem.* 73, 64–71, 2001.

64. Lai, X., Zheng, Y., Jacobsen, S. et al., Determination of adsorbed protein concentration in aluminum hydroxide suspensions by near-infrared transmission spectroscopy, *Appl. Spectrosc.* 62, 784–790, 2008.

65. Shaw, R.A., Kotowich, S., Mantsch, H.H., and Leroux, M., Quantitation of protein, creatinine, and urea in urine by near-infrared spectroscopy, *Clin. Biochem.* 29, 11–19, 1996.

66. Singh, B.R., Basic aspects of the technique and application of infrared spectroscopy of peptides and proteins, in *Infrared Analysis of Peptides and Proteins*, ed. B.R. Singh, American Chemical Society, Washington, DC, 2000.

67. Suehara, K. and Yano, T., Bioprocess monitoring using near-infrared spectroscopy, *Adv. Biochem. Eng. Biotechnol.* 90, 173–198, 2004.

68. Reich, G., Near-infrared spectroscopy and imaging: Basic principles and pharmaceutical applications, *Adv. Drug Deliv. Rev.* 57, 1109–1143, 2005.

69. Becker, T., Hitzmann, B., Muffler, K. et al., Future aspects of bioprocess monitoring, *Adv. Biochem. Eng. Biotechnol.* 105, 249–293, 2007.

70. Roggo, Y., Chalus, P., Mauer, L. et al., A review of near infrared spectroscopy and chemometrics in technologies, *J. Pharm. Biomed. Anal.* 44, 683–700, 2007.

71. Aaljoki, K., Hukkanen, H., and Jokinen, P., Application of spectroscopy, especially near-infrared, for process and pilot monitoring, *Process Control Qual.* 6, 125–131, 1994.

72. Buttner, G., The use of NIR analysis for refineries, *Process Control Qual.* 9, 197–203, 1997.

73. Pasquini, C. and Bueno, A.F., Characterization of petroleum using near-infrared spectroscopy: Quantitative modeling for the true boiling point curve and specific gravity, *Fuel* 86, 1927–1934, 2007.

74. Araujo, A.M., Santos, L.M., Fortuny, M. et al., Evaluation of water content and average droplet size in water-in-crude oil emulsions by means of near-infrared spectroscopy, *Energy Fuels* 22, 3450–3458, 2008.

75. Goldman, D.S., Near-infrared spectroscopy in process analysis, in *Encyclopedia of Analytical Chemistry*, Vol. 9, ed. R.A. Meyers, John Wiley & Sons, Chichester, U.K., pp. 8256–8264, 2000.

76. Yeung, K.S., Hoare, M., Thornhill, N.F. et al., Near-infrared spectroscopy for bioprocess monitoring and control, *Biotechnol. Bioeng.* 63, 684–693, 1999.

77. Riley, M.R., Monitoring of animal cell culture by near-infrared spectroscopy, *Recent Res. Dev. Biotechnol. Bioeng.* 3, 143–166, 2000.

78. Vaidyanathan, S., Arnold, S.A., Matheson, L. et al., Assessment of near-infrared spectral information for rapid monitoring of bioprocess quality, *Biotechnol. Bioeng.* 74, 376–388, 2001.

79. Riley, M.R., Crider, H.M., Nite, M.E. et al., Simultaneous measurement of 19 components in serum-containing animal cell culture media by Fourier transfer infrared spectroscopy, *Biotechnol. Prog.* 17, 376–378, 2001.

80. Arnold, S.A., Gaensakoo, R., Harvey, L.M., and McNeil, B., Use of at-line and in-situ near-infrared spectroscopy to monitor biomass in an industrial fed-batch *Escherichia coli* process, *Biotechnol. Bioeng.* 80, 405–413, 2002.

81. Jung, B., Lee, S., Yang, I.-H. et al., Automated on-line noninvasive optical glucose monitoring in a cell culture system, *Appl. Spectrosc.* 56, 51–57, 2002.

82. Arnold, S.A., Crowley, J., Woods, N. et al., In-situ near infrared spectroscopy to monitor key analytes in mammalian cell culture, *Biotechnol. Bioeng.* 84, 13–19, 2003.

83. Tarumi, M., Shimada, M., Murakami, T. et al., Simulation study of *in vitro* glucose measurement by NIR spectroscopy and a method of error reduction, *Phys. Med. Biol.* 48, 2373–2380, 2003.

84. Brülls, M., Folestad, S., Sparén, A., and Rasmuson, A., In-situ near-infrared spectroscopy monitoring of the lyophilization process, *Pharm. Res.* 20, 494–499, 2003.

85. Cimander, C. and Mandenius, C.F., Bioprocess control from a multivariate process trajectory, *Bioprocess Biosyst. Eng.* 26, 401–411, 2004.

86. Rodriques, L.O., Alves, T.P., Cardosa, J.P., and Menezes, J.C., Improving drug manufacturing with process analytical technology, *Drugs* 9, 44–48, 2006.

87. Nordon, A., Littlejohn, D., Dann, A.S. et al., In situ monitoring of the seed stage of a fermentation process using non-invasive NIR spectrometry, *Analyst* 133, 660–666, 2008.

88. Holm-Nielsen, J.B., Lomborg, C.J., Oleskowicz-Popiel, P., and Esbensen, K.H., On-line near infrared monitoring of glycerol-boosted anaerobic digestion processes: Evaluation of process analytical technologies, *Biotechnol. Bioeng.* 99, 302–313, 2008.

89. Rodrigues, L.O., Vieira, L, Cardosa, J.P., and Menezes, J.C., The use of NIR as a multiparametric in situ monitoring technique in filamentous fermentation systems, *Talanta* 75, 1356–1361, 2008.

90. http://www.pharmamanufacturing.com/articles/2008/172.html

91. http://www.pharmamanufacturing.com/experts/nir_spectroscopy_pat.html

92. Pickup, J.C., Hussain, F., Evans, N.D., and Sachedina, N., In vivo glucose monitoring: The clinical reality and the promise, *Biosens. Bioelectron.* 20, 1897–1902, 2005.

93. Amerov, A.K., Chen, J., and Arnold, M.A., Molar absorptivities of glucose and other biological molecules in aqueous solutions over the first overtone and combination regions of the near infrared spectrum, *Appl. Spectrosc.* 58, 1195–1204, 2004.

94. Symons, M.C., Spectroscopy of aqueous solutions: Protein and DNA interactions with water, *Cell. Mol. Life Sci.* 57, 999–1007, 2000.

95. Jung, D.H., Kim, B.H., Ko, Y.K. et al., Covalent attachment and hybridization of DNA oligonucleotides on patterned single-walled carbon nanotube films, *Langmuir* 20, 8886–8891, 2004.

96. Zhu, C.Q., Wu, Y.Q., Zheng, H. et al., Determination of nucleic acids by near-infrared fluorescence quenching of hydrophobic thiacyanine dye in the presence of Triton X-100, *Anal. Sci.* 20, 945–949, 2004.

97. Fang, F., Zheng, H., Li, L. et al., Determination of nucleic acids with a near infrared cyanine dye using resonance light scattering technique, *Spectrochim. Acta A Mol. Biomol. Spectrosc.* 64, 698–702, 2006.

98. Cathcart, H., Nicolosi, V., Hughes, J.M. et al., Ordered DNA wrapping switches on luminescence in single-walled nanotube dispersions, *J. Am. Chem. Soc.* 130, 12734–12744, 2008.

99. Schirm, B., Benend, H., and Wätzig, H., Improvements in pentosan polysulfate sodium quality assurance using fingerprint electrophoretograms, *Electrophoresis* 22, 1150–1162, 2001.

100. Cervera, M.F., Karjalainen, W., Airaksinen, S. et al., Physical stability and moisture sorption of aqueous chitosan-amylose starch films plasticized with polyols, *Eur. J. Pharm. Biopharm.* 58, 69–76, 2004.

101. Floyd, C.J. and Dickens, S.H., Network structure of bis-GMA- and UDMA-based resin systems, *Dent. Mater.* 22, 1143–1149, 2006.

102. Sun, B., Lin, Y., and Wu, F., Structure analysis of poly(*N*-isopropylacrylamide) using near-infrared spectroscopy and generalized two-dimensional correlation infrared spectroscopy, *Appl. Spectrosc.* 61, 765–771, 2007.

103. Watanabe, A., Morita, S., and Ozaki, Y., Temperature-dependent changes in hydrogen bonds in cellulose Iα studied by infrared spectroscopy in combination with perturbation-correlation moving-window correlation spectroscopy: Comparison with cellulose Iβ, *Biomacromolecules* 8, 2969–2975, 2007.

104. Pomerantz, Z., Levi, M.D., Salitra, G. et al., UV-VIS-NIR spectroelectrochemical and *in situ* conductance studies of unusual stability of n- and p-doped poly(dimethyldioctylquaterthiophene-ala-oxadiazole) under high cathodic and anodic polarizations, *Phys. Chem. Chem. Phys.* 10, 1032–1042, 2008.

105. Bista, R.K. and Bruch, R.F., Near-infrared spectroscopy of newly developed PEGylated lipids, *Spectrochim. Acta A Mol. Biomol. Spectrosc.* 71, 410–416, 2008.

106. Workman, J.J. Jr., Near-infrared spectroscopy of polymers and rubbers, in *Encyclopedia of Analytical Chemistry*, Vol. 8, ed. R.A. Meyers, John Wiley & Sons, Chichester, U.K., pp. 7829–7856, 2000.

107. Workman, J.J., Interpretive spectroscopy for near infrared, *Appl. Spectrosc. Rev.* 31, 251–320, 1996.

108. Workman, J.J., Review of process and non-invasive and infrared spectroscopy: 1993–1999, *Appl. Spectrosc. Rev.* 34, 1–89, 1999.

109. Haris, P.I., Fourier transform infrared spectroscopic studies of peptides: Potentials and pitfalls, in *Infrared Analysis of Peptides and Proteins*, ed. B.R. Singh, American Chemical Society, Washington, DC, pp. 54–95, Chapter 3m, 2000.

110. Kim, Y., Rose, C.A., Liu, Y. et al., FT-IR and near-infrared FT-Raman studies of the secondary structure of insulinotropin in the solid state: α-helix to β-sheet conversion induced by phenol and/or by high shear force, *J. Pharm. Sci.* 83, 1175–1180, 1994.

111. Debelle, L., Alix, A.J., Wei, S.M. et al., The secondary structure and architecture of human elastin, *Eur. J. Biochem.* 258, 533–539, 1998.

112. Chehin, R., Iloro, I., Marcos, M.J. et al., Thermal and pH-induced conformational changes of β-sheet protein monitored by infrared spectroscopy, *Biochemistry* 38, 1525–1530, 1999.

113. Manning, M.C., Use of infrared spectroscopy to monitor protein structure and stability, *Expert Rev. Proteomics* 2, 731–743, 2005.

114. Wu, D., Feng, S., and He, Y., Short-wave near-infrared spectroscopy of milk powder for brand identification and component analysis, *J. Dairy Sci.* 91, 939–949, 2008.

115. Dong, A., Huang, P., and Caughey, W.S., Protein secondary structures in water from second-derivative amide I infrared spectra, *Biochemistry* 29, 3303–3308, 1990.

116. Prestrelski, S.J., Arakawa, T., and Carpenter, J.F., Separation of freezing- and drying-induced denaturation of lyophilized proteins using stress-specific stabilization. II. Structural studies using infrared spectroscopy, *Arch. Biochem. Biophys.* 303, 465–473, 1993.

117. Levy, M.C., Lefebvre, S., Andry, M.C. et al., Fourier-transform infrared spectroscopic studies of cross-linked human serum albumin microcapsules. 2. Influence of reaction time on spectra and correlation with microcapsule morphology and size, *J. Pharm. Sci.* 83, 419–422, 1994.

118. Prestrelski, S.J., Pikal, K.A., and Arakawa, T., Optimization of lyophilization conditions for recombinant human interleukin-2 by dried-state conformational analysis using Fourier-transform infrared spectroscopy, *Pharm. Res.* 12, 1250–1259, 1995.

119. Constantino, H.R., Grisbenow, K., Mishra, P. et al., Fourier-transform infrared spectroscopic investigation of protein stability in the lyophilized form, *Biochim. Biophys. Acta* 1253, 69–74, 1995.

120. Chang, B.S., Beauvais, R.M., Dong, A., and Carpenter, J.F., Physical factors affecting the storage stability of freeze-dried interleukin-1 receptor antagonist: Glass transition and protein conformation, *Arch. Biochem. Biophys.* 331, 249–258, 1996.

121. Remmela, R.L. Jr., Stushnoff, C., and Carpenter, J.F., Real-time in situ monitoring of lysozyme during lyophilization using infrared spectroscopy: Dehydration stress in the presence of sucrose, *Pharm. Res.* 14, 1548–1555, 1997.

122. Carpenter, J.F., Prestrelski, S.J., and Dong, A., Application of infrared spectroscopy to development of stable lyophilized protein formulations, *Eur. J. Pharm. Biopharm.* 45, 231–238, 1998.

123. Breen, E.D., Curley, J.G., Overcashier, D.E. et al., Effect of moisture on the stability of a lyophilized humanized monoclonal antibody formulation, *Pharm. Res.* 18, 1345–1353, 2001.

124. Chan, H.K., Ongpipattanakul, B., and Au-Yeung, J., Aggregation of rhDNase occurred during the compression of KBr pellets used for FITR spectroscopy, *Pharm. Res.* 13, 238–242, 1996.

125. Meyer, J.D., Manning, M.C., and Carpenter, J.F., Effects of potassium bromide disk formation on the infrared spectra of dried model proteins, *J. Pharm. Sci.* 93, 496–506, 2004.

126. Wolkers, W.F. and Oldenhof, H., *In situ* FTIR assessment of dried *Lactobacillus bulgaricus*; KBr disk formation effects physical properties, *Spectroscopy* 19, 89–99, 2005.

127. Luthra, S., Kalonia, D.S., and Pikal, M.J, Effect of hydration on the secondary of lyophilized proteins as measured by Fourier transform infrared spectroscopy, *J. Pharm. Sci.* 96, 2910–2921, 2007.

128. Li, B., O'Meara, M.H., Lubach, J.W. et al., Effects of sucrose and mannitol on asparagine deamidation rates of model peptides in solution and in the solid state, *J. Pharm. Sci.* 94, 1723–1735, 2005.

129. Carpenter, J.F., Chang, B.S., Garzon-Rodriguez, W., and Randolph, T.W., Rational design of stable lyophilized protein formulations, *Pharm. Biotechnol.* 13, 109–133, 2003.

130. Izutsu, K., Funkimaki, Y., Kuwabara, A. et al., Near-infrared analysis of protein secondary structure in aqueous solutions and freeze-dried solids, *J. Pharm. Sci.* 95, 781–789, 2006.

131. Jovanović, N., Gerich, A., Bouchard, A., and Jiskoot, W., Near-infrared imaging for studying homogeneity of protein–sugar mixtures, *Pharm. Res.* 23, 2002–2013, 2006.

132. Ragoonanan, V. and Aksan, A., Heterogeneity in desiccated solutions: Implications for biostabilization, *Biophys. J.* 94, 2212–2227, 2008.

133. Gendrin, C., Roggo, Y., and Collet, C., Content uniformity of pharmaceutical solid dosage forms by near infrared hyperspectral imaging: A feasibility study, *Talanta* 73, 733–741, 2007.

134. Amigo, J.M., Cruz, J., Bautista, M. et al., Study of pharmaceutical samples by NIR chemical-image ad multivariate analysis, *TRAC—Trends Anal. Chem.* 27, 696–713, 2008.

135. Cho, R.K., Ozaki, Y., Ahn, J.J. et al., The applicability of near infrared reflectance spectroscopy for monitoring changes in secondary structure of denatured proteins, in *Near Infra-Red Spectroscopy*, Horwood, Chichester, U.K., pp. 333–338, 1992.
136. Cho, R.K., Lee, J.H., Ahn, J.J. et al., The applicability of near-infrared reflectance spectroscopy for determining solubility and digestibility of heated protein under high pressure, *J. Near Infrared Spectrosc.* 3, 73–79, 1996.
137. Noda, I., Two-dimensional infrared spectroscopy, *J. Am. Chem. Soc.* 111, 8116–8118, 1989.
138. Sefara, N.L., Magtoto, N.P., and Richardson, H.H., Structural characterization of β-lactoglobulin in solution using two-dimensional FT mid-infrared and FT near-infrared correlation spectroscopy, *Appl. Spectrosc.* 51, 536–540, 1997.
139. Wang, Y., Murayama, K., Myojo, Y. et al., Two-dimensional Fourier transform near-infrared spectroscopy study of heat denaturation of ovalbumin in aqueous solution, *J. Phys. Chem. B* 102, 6655–6662, 1998.
140. Ozaki, Y., Murayama, K., and Wang, Y., Application of two-dimensional near-infrared correlation spectroscopy to protein research, *Vibr. Spectrosc.* 20, 127–132, 1999.
141. Buchet, R., Wu, Y., Lachenal, G. et al., Selecting two-dimensional cross-correlational functions to enhance interpretation of near-infrared spectra of proteins, *Appl. Spectrosc.* 55, 155–162, 2001.
142. Jiang, Y. and Wu, P.Y., Application of near-infrared spectroscopy in the study of protein and polymers with amide groups, *Prog. Chem.* 20, 2021–2033, 2008.

11 Use of Mid-Infrared and Fourier Transform Infrared Spectroscopy to Study Conformation of Biomacromolecules

Near-infrared (NIR) spectroscopy was discussed in Chapter 10; this chapter is concerned with mid-infrared (200–4000 cm^{-1}) and the far-infrared (10–200 cm^{-1}). Most of the discussion in this chapter will focus on the region of 1400–4000 cm^{-1}, which is referred to as the common infrared region.[1] Fundamental infrared absorbance bands of significance for the study of biological macromolecules are shown in Table 11.1. As will be discussed below, most of the interest in the use of infrared spectroscopy for the study of proteins has focused on the amide I bond (Tables 11.2 and 11.3).

There has been limited application of far-infrared (terahertz) spectroscopy for the study of proteins and other biological macromolecules. Hydration has been demonstrated to increase the far-infrared absorbance of lysozyme[5] and myoglobin.[6] The studies on lysozyme, which examined the range from 15 to 45 cm^{-1} showed a band at 19 cm^{-1} at low hydration; at higher hydration, the spectrum from 20 to 45 cm^{-1} closely resembles the spectra of pure water. The later study on myoglobin[6] showed that absorbance by protein was increased by bound water. The myoglobin samples ranged from hydrated powders to highly concentrated (70%) protein solutions. Other far-infrared studies on proteins have focused on metal–protein interactions.[7–9] Far-infrared spectroscopy has been used to study the conformation of DNA,[10,11] to determine the configuration of oligosaccharides,[12] and for the determination of the state of water in solid drug dosage forms.[13] The rest of the discussion in this chapter will concern the use of mid-infrared spectroscopy, which will be referred to as infrared spectroscopy.

Infrared spectroscopy may be the most utilized technique in biological spectroscopy. A single search on PUBMED yielded more that 44,000 citations; there are more than 11,000 obtained when the search combined protein and infrared spectroscopy. Early studies used solid protein films or solutions in deuterium oxide (D_2O)[2] for the study of secondary structure of proteins.[14,15] This is illustrative of the problems of using infrared spectroscopy in the presence of water. The use of Fourier transform infrared spectroscopy (FTIR spectroscopy)[16] enables the correction of the protein

TABLE 11.1

Fundamental Infrared Absorbance Bands for the Study of Biological Macromolecules[a,b]

Band (cm^{-1})[c]	Functional Groups
971	Dianionic phosphate (monophosphate esters; nucleic acids)
1024 1025	Glycogen; other carbohydrate
1030, 1031, 1034	Collagen
1299	Amide III
1450	Valine CH_3 asymmetric bending
	Tyrosine ring vibrations
1460	$-CH_3$
1465	Alanine CH_2 bending
1567	Amide II
1600	Tyrosine ring vibrations
1650	Asparagine C=O stretching
1600–1700	Amide I
1720	Aspartic C=O stretching
1730	$-C=O$
2580	$-S-H$
3300	C–H
3400	$-NH_2$
3600	$-O-H$

[a] The information in this table is derived from the following sources
  - Naumann, D., Infrared spectroscopy in microbiology, in *Encyclopedia of Analytical Chemistry*, Vol. 1, ed. R.A. Meyers, Wiley, Chichester, U.K., 2000.
  - Stuart, B.H., Infrared spectroscopy of biological applications, in *Encyclopedia of Analytical Chemistry*, Vol. 1, ed. R.A. Meyers, Wiley, Chichester, U.K., 2000.
  - Gremlich, H.-U., Infrared and Raman spectroscopy, in *Handbook of Analytical Techniques*, eds. H. Günzler and A. Williams, Wiley-VCH, Weinheim, Germany, pp. 465–507, Chapter 17, 2001.
  - Movasaghi, Z., Rehman, S., and ur Rehman, I., Fourier transform infrared spectroscopy of biological tissues, *Appl. Spectrosc. Rev.* 43, 134–179, 2008.

[b] This is a partial list. The reader is directed to the following citation for a more exhaustive listing of infrared absorbance bands for biological molecules: Movasaghi, Z., Rehman, S., and ur Rehman, I., Fourier transform infrared (FTIR) spectroscopy of biological tissues, *Appl. Spectrosc. Rev.* 41, 134–179, 2008.

[c] It is recognized that the assignment of bands to a functional group is approximate. The reader is directed to the several databases available for infrared spectroscopy:
  - Hashimoto, A. and Kameoka, T., Applications of infrared spectroscopy to biochemical, food, and agricultural processes, *Appl. Spectrosc. Rev.* 43, 416–451, 2008.
  - Benito, M.T.J., Ojeda, C.B., and Rojas, E.S., Process analytical chemistry: Applications of near-infrared spectrophotometry in environmental and food analysis: An overview, *Appl. Spectrosc. Rev.* 43, 452–484, 2008.

TABLE 11.1 (continued)
Fundamental Infrared Absorbance Bands for the Study of Biological Macromolecules[a,b]

- Mellet, Y.L., Coomans, D.H., and de Vel, O., Spectral data; modern classification methods, in *Encyclopedia of Analytical Chemistry*, Vol. 12, ed. R.A. Meyers, Wiley, Chichester, U.K., pp. 10909–10928, 2000.
- Debska, B.I. and Guzowska-Swider, B., Spectral databases, infrared, in *Encyclopedia of Analytical Chemistry*, Vol. 12, ed. R.A. Meyers, Wiley, Chichester, U.K., pp. 10928–10953, 2000.
- Robinson, J.W. ed., *CRC Handbook of Spectroscopy*, Vol. II, CRC Press, Cleveland, OH, 1974.

TABLE 11.2
Infrared Absorbencies of the Amide Group[a]

Designation	Wavenumber (cm^{-1})	Description
Amide A	3300	N–H stretch
Amide B	3110	Overtone (2 × amide II)
I[b]	1653	Carbonyl stretch, C–N stretch, N–H bend
II	1567	N–H bend; C–N stretch
III	1299	C–N stretch

[a] Adapted from Stuart, B.H., Infrared spectroscopy of biological applications, in *Encyclopedia of Analytical Chemistry*, Vol. 1, ed. R.A. Meyers, Wiley, Chichester, U.K., 2000.

[b] The amide I band is considered the most significant absorbency band for protein conformation structure.

- Haris, P.I. and Chapman, D., The conformational analysis of peptides using Fourier transform IR spectroscopy, *Biopolymers* 37, 251–263, 1995.
- Cooper, E.A. and Knutson, K., Fourier transform infrared spectroscopy investigations of protein structure, *Pharm. Biotechnol.* 7, 101–143, 1995.
- Vogel, R. and Siebert, F., Vibrational spectroscopy as a tool for probing protein function, *Curr. Opin. Chem. Biol.* 4, 518–523, 2000.
- Barth, A. and Zscherp, C., What vibrations tell us about proteins, *Q. Rev. Biophys.* 35, 369–430, 2002.
- Dev, S.B., Keller, J.T., and Rha, C.K., Secondary structure of 11 S globulin in aqueous solution investigated by FT-IR derivative spectroscopy, *Biochim. Biophys. Acta* 957, 272–280, 1988.
- Middaugh, C.R., Mach, H., Ryan, J.A. et al., Infrared spectroscopy, *Methods Mol. Biol.* 40, 137–156, 1995.
- Ganin, Z., Chung, H.S., Smith, A.W. et al., Amide I two-dimensional infrared spectroscopy of proteins, *Acc. Chem. Res.* 41, 432–441, 2008.

TABLE 11.3

α-Helix Amide I Bands in Selected Proteins[a]

Protein	Wavenumber (cm⁻¹)	Ref.
α-Chymotrypsin	1654	[2]
Elastase	1657	[2]
IgG	1651	[2]
Lysozyme	1654	[2]
Trypsin	1654	[2]
α-Lactalbumin	1653	[3]
Myoglobin	1657[b]	[4]

[a] A mean of 1654 cm⁻¹ was obtained for 16 proteins [2].
[b] Spectra from protein adsorbed on silanized silicon wafer.

spectrum for the absorbance of water[17]; FTIR also permits resolution of the complex spectra for protein in the infrared region.[18–22] Studies of proteins and peptides in D_2O continue to provide useful information.[23–29] It is noted that the conformation of proteins differs between D_2O and H_2O.[30–33] Studies with β-lactoglobulin[34] and pepsin[35] suggest that D_2O increases the thermal stability of proteins as evaluated by differential scanning calorimetry (DSC).

The estimation of protein secondary structural elements (α-helix, β-sheet, turns, etc.) by FTIR is in agreement with estimations derived from circular dichroism (CD) spectroscopy and x-ray crystallography.[36–38]

Studies that used infrared spectroscopy to evaluate structural changes in proteins secondary to denaturation are used as models for possible changes in proteins, which may occur during stressing-associated accelerated stability studies. We are of the opinion that infrared spectroscopy would be of greatest value in evaluating changes during stability studies or active pharmaceutical ingredient or final drug product. Is denaturation a good model? Are the changes in secondary structure (infrared spectroscopy studies) of relevance to denaturation? Protein denaturation was briefly mentioned in Chapter 1 as part of a larger discussion of the importance of protein conformation in biopharmaceutical quality. Protein denaturation can be defined as a major change in protein conformation without covalent changes.[39] Denaturation may or may not be a reversible process and reversibility may depend on the denaturing agent.[40–42] While the pathways for protein denaturation are not clearly understood, changes in secondary structure as measured by infrared spectroscopy are part of the process; the key question is—how do such changes relate to the critical quality attributes of the biomacromolecule? The same question can be raised about any of the physicochemical tests described in this book. Thus, the key to understanding is not a single assay but a collection of assays, which address product quality and, the relationship of results obtained to the in vivo performance of the material under investigation.

An example is provided by the effect of denaturation on epitopes (see Chapter 20). Tipton and colleagues[43] prepared nine monoclonal antibodies against horseradish peroxidase, which could be classified into three different groups. Two groups (I and III) recognized epitopes in different antigenic regions on the molecule while the other group, II, recognized overlapping epitopes. Antibodies in groups II and III recognized heme-associated epitopes at the active site. Only one antibody in groups II and III is bound to both the apo-protein and denatured protein. Another group[44] prepared 10 monoclonal antibodies against a nonstructural protein from an avian retrovirus. The antibodies could be divided into three groups (A, B, and C) based on epitope specificity. Denaturation of the antigen (boiling in SDS with 2-mercaptoethanol) did not affect epitopes recognized by groups B and C while reactivity of group A was totally abolished. This suggests that epitope A is conformationally dependent while epitopes B and C are independent of antigen conformation. The studies provide examples of substantial conformation change with retention of specific biological activity.

The above comment on the relationship of epitopes and conformation is meant to be cautionary and should not be interpreted as disparaging of studies on conformation. Indeed, spectral techniques provide a noninvasive and potentially nondestructive approach to the study of biopharmaceutical quality. There are few studies that have directly examined the relationship between biological activity and secondary structure as measured by infrared spectroscopy. Hiddinga and coworkers[45] evaluated the conformation of several proenkephalin peptides with CD and FTIR. The three peptides had similar random solutions structures but possess unique structural potential to assume secondary structures in less polar environments at the effector binding site. Purcell and coworkers[46] showed that there are major changes in the secondary structure of RNase at high concentration of some vanadyl compounds where this is loss of enzyme activity. There is a decrease in α-helix and an increase β-helix as determined by FTIR spectroscopy. This group later showed that the binding of 3'-azido-3'-deoxythymidine (AZT)[47] and aspirin[48] to RNase was associated with changes in secondary structure as shown by FTIR spectroscopy and CD measurements.[49] The inhibition of trypsin by Pb(II) ions is associated with a change in secondary structure.[50] There are a large number of reports on changes in secondary structure associated with protein denaturation.[34,35,51-60] Some studies that have direct relevance to biopharmaceuticals are presented in Table 11.4. The use of infrared spectroscopy for the study of bovine pancreatic ribonuclease a (RNAse a) is shown in Table 11.5. Table 11.6 shows some artifactual bands.

There are several reports on the relationship of FTIR spectroscopy and biological activity and/or tertiary structure, which should be discussed in more detail. Tetenbaum and Miller[78] studied the reduction of soybean trypsin inhibitor (SBTI) by tris(2-carboxyethyl)-phosphine (TCEP) using x-ray absorption spectroscopy, CD, and FTIR spectroscopy. X-ray absorption spectroscopy showed a rapid, noncooperative reduction of the two disulfide bonds in SBTI; while biological activity was not measured, it is assumed that biological activity is lost as fully reduced SBTI is inactive. There are some early changes in FTIR and CD spectra but then changes occur more, slowly lagging behind the reduction of the disulfide bonds; it is suggested that the noncooperative reduction of the disulphide bonds results in an initial collapse of

TABLE 11.4
Selected Applications of FTIR Spectroscopy to Proteins of Commercial Biotechnology Interest

Protein	Observations	Refs.
Porcine plasma proteins subjected to pH or thermal denaturation	FTIR 2D correlation spectroscopy showed a decrease at 1653 cm^{-1} (α-helix), an increase at 1683 cm^{-1} (aggregation) which occurs after the change at 1653 cm^{-1}). There is also an increase at 1618 cm^{-1} (intermolecular β-sheet associated with aggregation) also following decrease at 1653 cm^{-1}	[61]
Aggregation of recombinant human interferon γ	Increase in intermolecular β-sheet (1620 cm^{-1}) and decrease in α-helix (1656 cm^{-1})	[62]
Surface-linked monolayers of engineered myoglobin[a]	FTIR was used to characterize the protein bound to the silicon substrate (silicon wafers, transparent from 4000–400 cm^{-1}), native structure was maintained on the protein monolayer. The native structure was preserved under denaturing conditions including extreme thermal stress (150°C)	[4]
Highly concentrated MAB solutions (100 mg/mL)	FTIR spectroscopy showed geometry-dependent differences (ATR versus transmittance) while other analytical techniques (CD, DSC) did not demonstrate any differences with the highly concentrated solutions	[63]
Thermal hydrogels as cell scaffolds	Conversion of α-helix to β-sheet (intermolecular) on heating	[64]
Recombinant human granulocyte-macrophage colony stimulating factor	FTIR spectra evaluated in D$_2$O and H$_2$O. Increase at 1639 and 1673 cm^{-1} suggest intermolecular β-sheet (aggregation); decrease at 1658 and 1649 cm^{-1} in H$_2$O indicates loss of α-helix; in D$_2$O, decrease is at 1656 cm^{-1}	[65]
Tumor necrosis factor-α (TNF-α)	Conversion of β-sheet protein (1637 cm^{-1}) to α-helical form (1653 cm^{-1}) on denaturation with acid, heat, or trifluoroethanol. This change in structure was also shown by CD spectroscopy	[66,67]
Pegylated β-lactoglobulin	Absence of 1620 cm^{-1} used to demonstrate lack of aggregation in the starting material	[68]
Pegylated bone-derived neurotrophic factor (BDNF)	Infrared spectroscopy showed that pegylation of BDNF resulted in a decrease in β-sheet (1638 cm^{-1}) and an increase in random structure (1644 cm^{-1}). There was a major change from apparent β-sheet absorbance at 1637–1638 cm^{-1} to random (1644 cm^{-1}) but at 1688 cm^{-1}; these results are interpreted as due to dehydration and not secondary structure changes	[69]

[a] The myoglobin is engineered to replace an alanine with a cysteine (A126C) which is then coupled to a glass substrate modified with (3-iodopropyl)trimethyloxysilane) producing an oriented protein monolayer. See Stayton, P.K., Olinger, J.M., Jiang, M. et al., Genetic engineering of surface attachment sites yields oriented protein monolayers, *J. Am. Chem. Soc.* 114, 9298–9299, 1992.

TABLE 11.5

The Use of Infrared Spectroscopy of the Study of Bovine Pancreatic Ribonuclease A (RNase A)

Study Description	Ref.
Second derivative infrared spectroscopy of RNase A in D_2O; tentative assignment of secondary structure bands.	[15]
Modification with formaldehyde did not influence the observed secondary structure of RNase A as judged by infrared spectra.	[70]
Thermally denatured RNase retains some secondary structure; there is a difference in the thermally denatured protein and the chemically denatured protein.	[71]
Thermally denatured RNase retains some secondary structure (α-helix and β-sheet).	[72]
Comparison of pressure-induced and thermal-induced denaturation of RNase A demonstrates difference between the two processes by FTIR. Denaturation is reversible although difference with pressure-denatured materials suggesting partially unfolded intermediate during refolding	[73]
Use of FTIR spectroscopy to study the refolding of denatured (urea, thermal) RNase A	[74]
Use of FTIR spectroscopy to compare the pressure-induced and thermal-induced denaturation (unfolding) and refolding of RNase A; pressure-denatured form has more secondary structure than thermally denatured protein	[75]
Use of FTIR spectroscopy to compare the thermal-induced and pressure-induced unfolding (denaturation) of RNase A and 10 variant proteins (C-terminal engineered variants). Some secondary structure was retained by the denatured proteins although there were qualitative difference in the spectra indicating differences in the structures of the denatured proteins	[75]
Pressure-denaturation of RNase A in the presence and absence of reducing agents. FTIR did not show large difference between the pressure-denatured native and reduced protein. Upon decompression under reducing conditions, amorphous aggregates are formed which are not found with the native protein. It is suggested that pressure produces aggregation-prone conformations	[76]
FTIR spectroscopy is used to study the thermal-induced unfolding of RNase in solution and when embedded in spherical polyelectrolyte brushes. Unfolding (denaturation) of RNase in solution is reversible; the unfolding of RNase bound to the polyelectrolyte occurs at a lower temperature and is irreversible.	[77]

structure rather than unfolding. Liu and coworkers[79] used CD, FTIR spectroscopy, and fluorescence to study the structural changes in hemoglobin induced by dimethylsulfoxide (DMSO). Tertiary structure is disrupted at 50% DMSO as judged by loss of the Soret CD spectrum at 400 nm, which is characteristic of heme–protein interaction.[80] Exposure to 52% DMSO in D_2O did not result in a change in α-helical content as judged by any change in a band at $1652\,cm^{-1}$ of the FTIR spectrum; an increase in DMSO concentration to 80% did result in the loss of secondary structure. The biological function of hemoglobin was lost at a DMSO concentration as low as 20%. Intrinsic fluorescence was "redshifted" as DMSO concentration increased. These two studies suggest that changes in FTIR spectrum can occur later than changes in tertiary structure or biological activity.

It is also useful to consider the effect of different denaturing agents (chaotropic agent, heat, pH, etc.)[39,81] on FTIR spectroscopy of various proteins. Meersman

TABLE 11.6
Spurious Bands in Infrared Spectroscopy[a]

Wavenumber (cm⁻¹)	Wavenlength (μ)	Components
3700	2.70	H_2O in solvent
3450	2.90	Hydrogen-bonded water in KBr disks
2330	4.3	CO_2 from dry ice
1720	5.8	Phthalate from tubing
1270	7.9	Silicone grease
1000–1110	9–10	Silicones (SiOSi)

[a] Derived from Robinson, J.W. ed., *CRC Handbook of Spectroscopy*, CRC Press, Cleveland, OH, 1974.

and colleagues[58] compared the heat denaturation of myoglobin with pressure unfolding and cold unfolding. Heat denaturation provided a spectral pattern consistent with transition from α-helix to β-sheet while cold or pressure denaturation yields spectra consistent with an early-folding intermediate. This same group[82] compared pressure- and temperature-induced unfolding of interferon-γ using infrared spectroscopy. High-pressure transformed α-helix to a disordered structure; heat treatment yielded spectral changes consistent with conversion from α-helix to antiparallel β-sheet. Boye and coworkers[83] used FTIR spectroscopy to study the thermal stability of β-lactoglobulins A and B in the presence of SDS, urea, cysteine, or *N*-ethylmaleimide. Thermal stability was increased in the presence 10 mM SDS and decreased in the presence of 50 mM SDS. Pedona and colleagues[84] have used FTIR to examine the effect of SDS, pD, and temperature on the structure of a protein disulfide oxidoreductase from *Pyrococcus furiosus*. This is a very useful study. First, it compares the FTIR spectra of the protein in H_2O and D_2O; absorbance, deconvoluted spectra, and second derivative spectra are reported. There is a difference in the effect of pD on the thermostability of α-helices and β-sheets. SDS has a significant effect on structure at concentrations above 0.5%; α-helices are more sensitive than β-sheets; at 4.0% SDS, the structure is unfolded. It is noted that these experiments performed at high protein concentration (30–40 mg mL⁻¹) in small volumes (ca. 40 μL) required 1.5 mg protein with several processing steps. While there are some differences in the quality of changes in secondary structure, a change from α-helix to antiparallel β-sheet is a consistent feature with protein denaturation. Argument does continue with respect to the denaturation pathways and the characteristics of the denatured protein as demonstrated by studies on RNase.[71,73,75–77,85–89] While there may be differences in FTIR spectroscopy data, it does provide evidence about temporal relationship of secondary structure and tertiary structure in protein folding.[74] It is of interest that modification of proteins with formaldehyde appears to "lock-in" secondary structure conformation as judged by FTIR spectroscopy[70]; DSC shows the absence of a thermal transition. There are also some questions as to whether a denatured protein is ever a random coil,[90] making it a bit difficult to know whether there is an "end game" in the change of secondary structure during denaturation. There are multiple changes in the FTIR spectra of stressed proteins,[55] demonstrating changes in various secondary structure elements (see Table 11.4) providing a bit of a challenge for unequivocal interpretation. A consideration of the various studies in this area suggests that major structural change is associated with a decrease in α-helix with an increase in β-helix; thus there is a decrease in absorbance at $1650-1656\,cm^{-1}$ [61,64,91–94] and an increase in absorbance in the region of $1670-1690\,cm^{-1}$.[61,93] Intermolecular β-sheet interaction, resulting in aggregation is detected at $1616-1625\,cm^{-1}$,[92,93] while intramolecular β-sheet formation can also be observed at $1638\,cm^{-1}$.[94]

The use of FTIR spectroscopy for the study of biopharmaceutical proteins in solution is complicated by requirement of high sample concentration. With the exception of products such as albumin, which is used in gram quantities, biopharmaceutical proteins are formulated in final drug product and used therapeutically in microgram to milligram quantities. The cost of producing biopharmaceutical proteins makes solution FTIR spectroscopy unfeasible for a quality control assay for product release. In addition, other techniques such as CD can provide information on solution secondary structure at lower proteins concentrations (see Chapter 14) can provide information at lower protein concentration.[95] FTIR spectroscopy does have the great advantage, as with NIR spectroscopy and x-ray scattering,[96–98] of obtaining spectral data from dried or adsorbed protein. Notwithstanding the challenges in the application of infrared spectroscopy to the study of proteins, there is a large literature in this area. Literature considered directly relevant to the production of biopharmaceutical products has been presented in Table 11.4.

There are a number of reports on the application of FTIR spectroscopy to lyophilized or dried protein, which are relevant to our emphasis on biopharmaceutical products. van de Weert and colleagues[99] reviewed the application of FTIR spectroscopy to proteins including various sampling techniques. These investigators suggest that differences in FTIR spectra observed between lyophilized and solution structure might reflect actual conformational change or might reflect spectral changes due to the removal of water resulting in a change in the environment of the peptide bond. Other investigators[100] suggest that the change in secondary structure on lyophilization does represent an actual change on drying of the protein. The reader is directed to several recent reports of the application of infrared spectroscopy to the study of lyophilized proteins.[101–108] While this has been useful, infrared spectroscopy suffers from some disadvantages in that the technique is sensitive to the presence of moisture presenting issues with respect to sample preparation. Aggregation has been observed in the preparation of protein in KBr pellets for infrared spectroscopy.[100,109–111]

REFERENCES

1. Campbell, I.D. and Dwek, R.A., *Biological Spectroscopy*, Benjamin Cummings, Menlo Park, CA, 1984.
2. Stuart, B.H., Infrared spectroscopy of biological tissues, in *Encyclopedia of Analytical Chemistry*, Vol. 1, ed. R.A. Meyers, Chichester, U.K., 2000.
3. Dzwolak, W., Kato, M., Shimizu, A., and Taniguchi, Y., FTIR study on heat-induced and pressure-assisted cold-induced changes in structure of bovine α-lactalbumin: Stabilizing role of calcium ions, *Biopolymers* 62, 29–39, 2001.

4. Jiang, M., Nölting, B., Stayton, P.S., and Sliger, S.G., Surface linked molecular mono-layers of an engineered myoglobin: Structure, stability, and function, *Langmuir* 12, 1278–1283, 1996.

5. Moeller, K.D., Williams, G.P., Steinhauser, S. et al., Hydration-dependent far-infrared absorption in lysozyme detected using synchrotron radiation, *Biophys. J.* 61, 276–280, 1992.

6. Zhang, C. and Durgin, S.M. Hydration-induced far-infrared absorption increase in myoglobin, *J. Phys. Chem. B* 110, 23607–23613, 2006.

7. Berthomieu, C., Marboutin, L., Dupeyrat, F., and Bouyer, P., Electrochemically induced FTIR difference spectroscopy in the mid- to far infrared (200 micron) domain: A new setup for the analysis of metal-ligand interactions in redox proteins, *Biopolymers* 82, 363–367, 2006.

8. Dörr, S., Schade, U., Hellwig, F., and Ortolani, M., Characterization of temperature-dependent iron-imidazole vibrational modes in far infrared, *J. Phys. Chem. B* 111, 14418–14422, 2007.

9. Bao, X.L., Lv, Y., Yang, B.C. et al., A study of the soluble complexes formed during calcium binding soybean protein hydrolysates, *J. Food Sci.* 73, C117–C121, 2008.

10. Fischer, B.M., Walther, M., and Uhd Jepsen, P., Far-infrared vibrational modes of DNA components studies by tetrahertz time-domain spectroscopy, *Phys. Med. Biol.* 47, 3807–3814, 2002.

11. Woods, K.N., Lee, S.A., Holman, H.Y., and Wiedemann, H., The effect of solvent dynamics on the low frequency collective motions of DNA in solution and unoriented films, *J. Chem. Phys.* 124, 224706, 2006.

12. Tul'chinsky, V.M., Zurabyan, S.E., Asankozhoev, K.A. et al., Study of the infrared spectra of oligosaccharides in the region $1000–40\,cm^{-1}$, *Carbohydr. Res.* 51, 1–8, 1976.

13. Zeitler, J.A., Kogermann, K., Rantanen, J. et al., Drug hydrate systems and dehydration processes studied by terahertz pulsed spectroscopy, *Int. J. Pharm.* 334, 78–84, 2007.

14. Timasheff, S.N. and Susi, H., Infrared investigation of the secondary structure of β-lactoglobulins, *J. Biol. Chem.* 241, 249–251, 1966.

15. Susi, H. and Byler, D.M., Protein structure by Fourier transform infrared spectroscopy: Second derivative spectra, *Biochem. Biophys. Res. Commun.* 115, 391–397, 1983.

16. Susi, H. and Byler, D.M., Resolution-enhanced Fourier transform infrared spectroscopy of enzymes, *Methods Enzymol.* 130, 290–311, 1986.

17. Singh, B.R. ed., *Infrared Analysis of Peptides and Proteins*, American Chemical Society, Washington, DC, 2000.

18. George, A. and Veis, A., FTIRS in H_2O demonstrates that collagen monomers undergo a conformational transition prior to thermal self-assembly in vitro, *Biochemistry* 30, 2372–2377, 1991.

19. Zuber, G., Prestresski, S.J., and Benedek, K., Application of Fourier transform infrared spectroscopy to studies of protein solutions, *Anal. Biochem.* 207, 150–156, 1992.

20. McPhie, P., Estimation of secondary/tertiary structure, *Dev. Biol. Stand.* 96, 29–36, 1998.

21. Secundo, F. and Guerrieri, N., ATR-FR/IR study on the interactions between gliadins and dextrin and their effect on protein secondary structure, *J. Agric. Food Chem.* 53, 1757–1764, 2006.

22. Navea, S., Tauler, R., and de Juan, A., Application of the local regression method interval partial least squares to the elucidation of protein secondary structure, *Anal. Biochem.* 336, 231–242, 2005.

23. Dong, A., Kendrick, B., Kreilgârd, L. et al., Spectroscopic study of secondary structure and thermal denaturation of recombinant human factor XIII in aqueous solution, *Arch. Biochem. Biophys.* 347, 213–220, 1997.

24. Allison, S.D., Chang, B., Randolph, T.W., and Carpenter, J.F., Hydrogen bonding between sugar and protein is responsible for inhibition of dehydration-induced protein unfolding, *Arch. Biochem. Biophys.* 365, 289–298, 1999.

25. Barth, A., The infrared absorption of amino acid side chains, *Prog. Biophys. Mol. Biol.* 74, 141–173, 2000.

26. Schweitzer-Stenner, R., Eker, F., Huang, Q., and Griebenow, K., Dihedral angles of trialanine in D_2O determined by combining FTIR and polarized visible Raman spectroscopy, *J. Am. Chem. Soc.* 123, 9628–9633, 2001.

27. Militello, V., Casarino, C., Emauele, A. et al., Aggregation kinetics of bovine serum albumin studied by FTIR spectroscopy and light scattering, *Biophys. Chem.* 107, 175–187, 2004.

28. Kim, Y.S. and Hochstrasser, R.M., Dynamics of amide-I modes of the alanine dipeptide in D_2O, *J. Phys. Chem. B* 109, 6884–6891, 2005.

29. Yuan, B., Murayama, K., and Yan, H., Study of thermal dynamics of defatted bovine serum albumin in D_2O solution by Fourier transform infrared spectra and evolving factor analysis, *Appl. Spectrosc.* 61, 921–927, 2007.

30. Cioni, P. and Strambini, G.B., Effect of heavy water of protein flexibility, *Biophys. J.* 82, 3246–3253, 2002.

31. Chellgren, B.W. and Creamer, T.P., Effects of H_2O and D_2O on polyproline II helical structure, *J. Am. Chem. Soc.* 126, 14734–14735, 2004.

32. Goodard, Y.A., Korb, J.P., and Bryant, R.G., Structural and dynamical examination of the low-temperature glass transition in serum albumin, *Biophys. J.* 91, 3841–3847, 2006.

33. Das, A., Sinha, S., Acharya, B.R. et al., Deuterium oxide stabilizes conformation of tubulin: A biophysical and biochemical study, *BMB Rep.* 41, 62–67, 2008.

34. Verheul, M., Roefs, S.P., and de Kruif, K.G., Aggregation of β-lactoglobulin and influence of D_2O, *FEBS Lett.* 421, 273–276, 1998.

35. Dee, D., Pencer, J., Nieh, M.P. et al., Comparison of solution structures and stabilities of native, partially unfolded and partially refolded pepsin, *Biochemistry* 45, 13982–13992, 2006.

36. Byler, D.M. and Susi, H., Examination of the secondary structure of proteins by deconvoluted FTIR spectra, *Biopolymers* 25, 469–487, 1986.

37. Dong, A., Huang, P., and Caughey, W.S., Protein secondary structures in water from second-derivative amide I infrared spectra, *Biochemistry* 29, 3303–3308, 1990.

38. Sarver, J.W. Jr. and Krueger, W.C., An infrared and circular dichroic combined approach to the analysis of protein secondary structure, *Anal. Biochem.* 199, 61–67, 1991.

39. Lapanje, S., *Physicochemical Aspects of Protein Denaturation*, Wiley-Interscience, New York, 1978.

40. Rodriguez-Larrea, D., Ibarra-Molero, B., de Maria, L. et al., Beyond Lumry-Eyring: An unexpected pattern of operational reversibility/irreversibility in protein denaturation, *Proteins* 70, 19–24, 2008.

41. Chodankar, S., Aswal, V.K., Kohlbrecher, J. et al., Structural evolution during protein denaturation as induced by different methods, *Phys. Rev. E Stat. Nonlin. Soft Matter Phys.* 77, 031901, 2008.

42. Elkordy, A.A., Forbes. R.T., and Barry, B.W., Study of protein conformational stability and integrity using calorimetry and FT-Raman spectroscopy correlated with enzymatic activity, *Eur. J. Pharm. Sci.* 33, 177–190, 2008.

43. Tipton, D.A., Walker, W.S., and Schonbaum, G.R., Epitope mapping of horseradish peroxidase with use of monoclonal antibodies, *Hybridoma* 9, 319–330, 1990.

44. Hou, H.S., Su, Y.P., Shieh, H.K., and Lee, L.H., Monoclonal antibodies against different epitopes of nonstructural protein ςNS of avian retrovirus S1133, *Virology* 282, 168–175, 2001.

45. Hiddinga, H.J., Katzenstein, G.E., Middaugh, C.R., and Lewis, R.V., Secondary structure characteristics of proenkephalin peptides E, B, and F, *Neurochem. Res.* 15, 393–399, 1990.

46. Purcell, M., Novetta-Delen, A., Arakawa, H. et al., Interaction of RNase A with VO^{3-} and VO^{2+} ions. Metal ion binding mode and protein secondary structure, *J. Biomol. Struct. Dyn.* 17, 473–480, 1999.

47. Gaudreau, S., Novetta-Dellen, A., Neault, J.F. et al., 3′-azido-3′-deoxythymidine binding to ribonuclease: Model for drug-protein interaction, *Biopolymers* 72, 435–551, 2003.

48. Neault, J.F., Ragi, C., Novetta-Dellen, A. et al., Aspirin interaction with ribonuclease A, *Cell Biochem. Biophys.* 46, 27–33, 2006.

49. Neault, J.F., Diamantoglou, S., Beuregard, M. et al., Protein unfolding in drug–RNase complexes, *J. Biomol. Struct. Dyn.* 25, 387–394, 2008.

50. Yang, L., Gao, Z., Cao, Y. et al., Effect of Pb(II) on the secondary structure and biological activity of trypsin, *Chembiochem* 6, 1191–1195, 2005.

51. Wang, P.T. and Heremans, K., Pressure effects on protein secondary structure and hydrogen deuterium exchange in chymotrypsinogen: A Fourier transform infrared spectroscopic study, *Biochim. Biophys. Acta* 956, 1–9, 1988.

52. van Stokkum, I.H., Linadell, H., Hadden, J.M. et al., Temperature-induced changes in protein structures studied by Fourier transform infrared spectroscopy and global analysis, *Biochemistry* 34, 10508–10518, 1995.

53. Cai, S. and Singh, B.R., Identification of β-turn and random coil amide III infrared bands for secondary structure estimation of proteins, *Biophys. Chem.* 80, 7–20, 1999.

54. Arrondo, J.L. and Goñi, F.M., Structure and dynamics of membrane proteins as studied by infrared spectroscopy, *Prog. Biophys. Mol. Biol.* 72, 367–405, 1999.

55. Ma, C.Y., Rout, M.K., and Mock, W.Y., Study of oat globulin conformation by Fourier transform infrared spectroscopy, *J. Agric. Food Chem.* 49, 3328–3334, 2001.

56. Hoekstra, F.A., Golovina, E.A., Tetteroo, F.A., and Wolkers, W.F., Inductions of desiccation tolerance in plant somatic embryos: How exclusive is the protective role of sugars?, *Cryobiology* 43, 140–150, 2001.

57. Barth, A. and Zscherp, C., What vibrations tell us about proteins, *Q. Rev. Biophys.* 35, 369–430, 2002.

58. Meersman, F., Smeller, L., and Heremans, K., Comparative Fourier transform infrared spectroscopy study of cold-, pressure-, and heat-induced unfolding and aggregation of myoglobin, *Biophys. J.* 82, 2635–2644, 2002.

59. Lefèvre, T., Arseneault, K., and Pézolet, M., Study of protein aggregation using two-dimensional correlation infrared spectroscopy and spectral stimulations, *Biopolymers* 73, 705–715, 2004.

60. Takekiyo, T., Imai, T., Kato, M., and Taniguchi, Y., Understanding high pressure stability of helical conformation of oligopeptides and helix bundle protein: High pressure FT-IR and RISM theoretical studies, *Biochim. Biophys. Acta* 1764, 355–363, 2006.

61. Saguer, E., Alvarez, P., Sedman, J. et al., Heat-induced gel formation of plasma proteins: New insights by FTIR 2D correlation spectroscopy, *Food Hydrocoll.* 23, 874–879, 2009.

62. Kenovick, B.S., Cleland, J.L., Lam, Y. et al., Aggregation of recombinant human interferon gamma: Kinetics at structural transition, *J. Pharm. Sci.* 87, 1069–1076, 1998.

63. Harn, N., Allan, C., Oliver, C., and Middaugh, C.R., Highly concentrated monoclonal antibody solutions: Direct analysis of physical structure and thermal stability, *J. Pharm. Sci.* 96, 532–546, 2007.

64. Yan, H., Nykanen, A., Ruokalainen, J. et al., Thermo-reversible protein fibrillar hydrogels as cell scaffolds, *Faraday Discuss.* 139, 71–84, 2008.

65. Jiang, H., Song, Z., Ling, M. et al., FTIR studies of recombinant human granulocyte-macrophage colony-stimulating factor in aqueous solution: Secondary structure, disulfide reduction and thermal behavior, *Biochim. Biophys. Acta* 1294, 121–128, 1996.

66. Narhi, L.O., Philo, J.S., Li, T. et al., Induction of α-helix in the β-sheet protein tumor necrosis factor-α: Thermal- and trifluoroethanol-induced denaturation at neutral pH, *Biochemistry* 35, 11447–11453, 1996.

67. Narhi, L.O., Philo, J.S., Li, T. et al., Induction of α-helix in the β-sheet protein tumor necrosis factor-α: Acid-induced denaturation, *Biochemistry* 35, 11454–11460, 1996.

68. Nijs, M., Azarkan, M., Smolders, N. et al., Preliminary characterization of bovine beta-lactoglobulin after its conjugation to polyethylene glycol, *Biotechnol. Bioeng.* 54, 40–49, 1997.

69. Heller, M.C., Carpenter, J.F., and Randolph, T.W., Conformational stability of lyophilized PEGylated proteins in a phase-separating system, *J. Pharm. Sci.* 88, 58–64, 1999.

70. Mason, J.T. and O'Leary, T.J., Effects of formaldehyde fixation on protein secondary structure: A calorimetric and infrared spectroscopic investigation, *J. Histochem. Cytochem.* 39, 225–229, 1991.

71. Sosnick, T.R. and Trewhella, J., Denatured states of ribonuclease A have compact dimensions and residual secondary structure, *Biochemistry* 31, 8329–8335, 1992.

72. Seshadri, S., Oberg, K.A., and Fink, A.L., Thermally denatured ribonuclease A retains secondary structure as shown by FTIR, *Biochemistry* 33, 1351–1355, 1994.

73. Takeda, N., Kato, M., and Taniguchi, Y., Pressure- and thermally-induced reversible changes in the secondary structure of ribonuclease A studied by FT-IR spectroscopy, *Biochemistry* 34, 5980–5987, 1995.

74. Reinstädler, D., Fabian, R., Backmann, J., and Naumann, D., Refolding of thermally and urea-denatured ribonuclease monitored by time-resolved FTIR spectroscopy, *Biochemistry* 35, 15822–15830, 1996.

75. Panick, G. and Winter, R., Pressure-induced unfolding/refolding of ribonuclease A: Static and kinetic Fourier transform infrared spectroscopy study, *Biochemistry* 39, 1862–1869, 2000.

76. Meersman, F. and Heremans, K., High pressure induced the formation of aggregation-prone states of proteins under reducing conditions, *Biophys. Chem.* 104, 297–304, 2003.

77. Wittemann, A. and Ballauff, M., Temperature-induced unfolding of ribonuclease A embedded in spherical polyelectrolyte brushes, *Macromol. Biosci.* 5, 13–20, 2005.

78. Tetenbaum, J. and Miller, L.M., A new spectroscopic approach to examining the role of disulfide bonds in the structure and unfolding of soybean trypsin inhibitor, *Biochemistry* 40, 12215–12219, 2001.

79. Liu, C, Bo, A., Cheng, G. et al., Characterization of the structural and functional changes of hemoglobin in dimethyl sulfoxide by spectroscopic techniques, *Biochim. Biophys. Acta* 1385, 53–60, 1998.

80. Chattopadhyay, K. and Mazumdar, S., Structural and conformational stability of horseradish peroxidase: Effect of temperature and pH, *Biochemistry* 39, 263–270, 2000.

81. Haschemeyer, R.H. and Haschemeyer, A.E.V., *Proteins. A Guide to Study by Physical and Chemical Methods*, Wiley-Interscience, New York, 1973.

82. Goossens, K., Haelewyn, J., Meersman, F. et al., Pressure- and temperature-induced unfolding and aggregation of recombinant human interferon-γ: A Fourier transform infrared spectroscopy study, *Biochem. J.* 370, 529–535, 2003.

83. Boye, J.I., Ma, C.Y., and Ismail, A., Thermal stability of β-lactoglobulins A and B: Effect of SDS, urea, cysteine, and *N*-ethylmaleimide, *J. Dairy Res.* 71, 207–215, 2004.

84. Pedona, E., Saviano, M., Bartolucci, S. et al., Temperature-, SDS-, and pD-induced conformational changes in protein disulfide oxidoreductase from the archeon *Pyrococcus furiosus*: A dynamic simulation and Fourier transform infrared spectroscopic study, *J. Proteome Res.* 4, 1972–1980, 2005.

85. Seshadri, S., Oberg, K.A., and Fink, A.L., Thermally denatured ribonuclease A retains secondary of ribonuclease A studied by FT-IR spectroscopy, *Biochemistry* 34, 5980–5987, 1994.

86. Fabian, H. and Mantsch, H.H., Ribonuclease A revisited: Infrared spectroscopic evidence for lack of native-like secondary structure in the thermally denatured state, *Biochemistry* 34, 13651–13655, 1995.

87. Stelea, S.D., Pancoska, P., Benight, A.S., and Keiderling, T.A., Thermal unfolding of ribonuclease A in phosphate at neutral pH: Deviations from the two-state model, *Protein Sci.* 10, 970–978, 2001.

88. Fabian, H. and Nauman, D., Methods to study protein folding by stopped-flow FT-IR, *Methods* 34, 28–40, 2004.

89. Zhang, J. and Yan, Y.E., Oligomerization and aggregation of bovine pancreatic ribonuclease A: Backbone hydration probed by infrared band-shift, *Protein Pept. Lett.* 15, 650–657, 2008.

90. Baldwin, R.I. and Zimm, B.H., Are denatured proteins ever random coils? *Proc. Nat. Acad. Sci. USA* 97, 12391–12392, 2000.

91. Scirè, A., Marabotti, A., Aurilia, V. et al., Molecular strategies for protein stabilization: The case of a trehalose/maltose-binding protein from *Thermus thermophilis*, *Proteins* 73, 839–850, 2008.

92. Herman, P., Straiano, M., Marabotti, A. et al., D-Trehalose/D-maltose-binding protein from the hyperthermophilic archeon *Thermococcus literalis*: The binding of trehalose and maltose results in different conformational states, *Proteins* 63, 754–767, 2006.

93. Pikel, M.J., Rigsbee, D., and Roy, M.L., Solid state stability of protein III. Calorimetric (DSC) and spectroscopic (FTIR) characterization of thermal denaturation in freeze dried human growth hormone (hGH), *J. Pharm. Sci.* 97, 5122–5131, 2008.

94. Lu, J., Wang, X.-J., Liu, Y.-X., and Chiang, C.-B., Thermal and FTIR investigation of freeze-dried protein-excipient mixtures, *J. Therm. Anal. Calorimetry* 89, 913–919, 2007.

95. Singh, B.R., Basic aspects of the technique and applications of infrared spectroscopy of peptides and proteins, in *Infrared Analysis of Peptides and Proteins*, ed. B.R. Singh, American Chemical Society, Washington, DC, 2000.

96. Castellano, A.C., Barteri, M., Bianconi, A. et al., X-ray small angle scattering of the human transferring protein aggregates. A fractal study, *Biophys. J.* 64, 520–524, 1993.

97. Elshemey, W.M., Desouky, O.S., and Ashour, A.H., Low-angle x-ray scattering from lyophilized blood constituents, *Phys. Med. Biol.* 46, 531–539, 2001.

98. Desouky, O.S., Elshemey, W.M., Selim, N.S., and Ashour, A.H., Analysis of low-angle x-ray scattering peaks from lyophilized biological samples, *Phys. Med. Biol.* 46, 2099–2106, 2001.

99. van de Weert, M., Haris, P.I., Hennink, W.E., and Crommelin, D.J.A., Fourier transform infrared spectrometric analysis of protein conformation: Effect of sampling method and stress factors, *Anal. Biochem.* 297, 160–169, 2001.

100. Luthra, S., Kalonia, D.S., and Pikal, M.J., Effect of hydration on the secondary structure of lyophilized proteins as measured by Fourier transform infrared (FTIR), *J. Pharm. Sci.* 96, 2910–2921, 2007.

101. Prestrelski, S.J., Arakawa, T., and Carpenter, J.F., Separation of freezing- and drying-induced denaturation of lyophilized proteins using stress-specific stabilization. II. Structural studies using infrared spectroscopy, *Arch. Biochem. Biophys.* 303, 465–473, 1993.

102. Levy, M.C., Lefebvre, S., Andry, M.C. et al., Fourier-transform infrared spectroscopic studies of cross-linked human serum albumin microcapsules. 2. Influence of reaction time on spectra and correlation with microcapsule morphology and size, *J. Pharm. Sci.* 83, 419–422, 1994.

103. Prestrelski, S.J., Pikal, K.A., and Arakawa, T., Optimization of lyophilization conditions for recombinant human interleukin-2 by dried-state conformational analysis using Fourier-transform infrared spectroscopy, *Pharm. Res.* 12, 1250–1259, 1995.

104. Constantino, H.R., Grisbenow, K., Mishra, P. et al., Fourier-transform infrared spectroscopic investigation of protein stability in the lyophilized form, *Biochim. Biophys. Acta* 1253, 69–74, 1995.

105. Chang, B.S., Beauvais, R.M., Dong, A., and Carpenter, J.F., Physical factors affecting the storage stability of freeze-dried interleukin-1 receptor antagonist: Glass transition and protein conformation, *Arch. Biochem. Biophys.* 331, 249–258, 1996.

106. Remmela, R.L. Jr., Stushnoff, C., and Carpenter, J.F., Real-time in situ monitoring of lysozyme during lyophilization using infrared spectroscopy: Dehydration stress in the presence of sucrose, *Pharm. Res.* 14, 1548–1555, 1997.

107. Carpenter, J.F., Prestrelski, S.J., and Dong, A., Application of infrared spectroscopy to development of stable lyophilized protein formulations, *Eur. J. Pharm. Biopharm.* 45, 231–238, 1998.

108. Breen, E.D., Curley, J.G., Overcashier, D.E. et al., Effect of moisture on the stability of a lyophilized humanized monoclonal antibody formulation, *Pharm. Res.* 18, 1345–1353, 2001.

109. Chan, H.K., Ongpipattanakul, B., and Au-Yeung, J., Aggregation of rhDNase occurred during the compression of KBr pellets used for FITR spectroscopy, *Pharm. Res.* 13, 238–242, 1996.

110. Meyer, J.D., Manning, M.C., and Carpenter, J.F., Effects of potassium bromide disk formation on the infrared spectra of dried model proteins, *J. Pharm. Sci.* 93, 496–506, 2004.

111. Wolkers, W.F. and Oldenhof, H., *In situ* FTIR assessment of dried *Lactobacillus bulgaricus*; KBr disk formation effects physical properties, *Spectroscopy* 19, 89–99, 2005.

12 Use of Raman Spectroscopy to Evaluate Biopharmaceutical Conformation

Raman spectroscopy, not to be confused with Ramen noodles, a staple of biomedical support, is a useful technique for the study of biological macromolecule conformation.[1-4] Raman spectroscopy is actually a type of light scattering and has been described as "feeble fluorescence." Raman scattering is different from Rayleigh scattering; Rayleigh scattering is elastic light scattering, which does not involve the transfer of energy. Raman scattering is inelastic scattering, resulting from energy transfer and the wavelength of the scattered light differs from the incident light. The Raman effect usually results in Stokes scattering where the scattered light has a lower frequency/higher wavelength (red-shifted) as a result of energy loss or anti-Stokes scattering where the frequency of scattered light is of higher frequency/lower wavelength (blue-shifted) as a result of energy gain. Implicit in anti-Stokes is either emission from a level higher than ground state or ending at a state lower than ground state. The amount of anti-Stokes scattering is usually much less than Stokes scattering. Furthermore, it should be emphasized that the relative amount of photon scattered is quite low, requiring considerable energy in the incident radiation provided by lasers. Raman spectra data (scattering data) can be presented as an absolute band in analogy with infrared spectroscopy or presented as a Raman shift (the difference between the wavelength of the incident radiation and the emitted radiation); this number would be positive with Stoke radiation and negative with anti-Stokes radiation. A Raman shift is conceptually related to the Stokes shift, which is a term used in fluorescence to describe the shift in wavelength between the exciting beam and the emission.[5]

Raman spectroscopy was discovered in 1928 by Raman and Krishnan.[6,7] This technique had limited use until 20 years ago and since then there has been marked increase in its use as more sophisticated technologies have become available.[8]

There is extensive use of Raman spectroscopy in art and archeology.[9-14] The reader is directed to the collection edited by Edwards and Chalmers.[9] The chapter by Long[11] is most useful as an introduction to Raman scattering. The chapter by Edwards and Chalmers[12] provides a useful comparison between Raman scattering and other physical techniques such as infrared microscopy, thermal analysis, and surface scattering. The reader is also directed to a collection edited by Creagh and Bradley,[13] which discusses the use of radiation in art and archeometry; archeometry is the application of modern methods of physical and chemical analysis to the study of archeology artifacts.

Raman spectroscopy can provide useful information about protein conformation but, because of weak signal produced by proteins, usually requires high sample concentration. As discussed in Chapter 11 on infrared spectroscopy, this requirement for high sample concentration makes the application of Raman spectroscopy to biopharmaceutical proteins difficult. Sensitivity is improved with resonance Raman spectroscopy and surface-enhanced resonance Raman scattering but utility is decreased. Raman spectroscopy is considered complementary to infrared spectroscopy but has the advantage of performing analysis in the presence of water. As a result of the low yield of scattered photons, Raman spectroscopy studies require high input energy using lasers for excitation.[15–17]

There is considerable interest in the application of the Raman effect to the study of proteins (slightly more than 4000 citations were obtained from a PUBMED search). However, most studies do not have direct application to biotechnology programs. For example, resonance Raman scattering is only useful for the study of proteins with prosthetic groups. Most work with Raman scattering occurs with complex, but small, organic chemicals.

There are a few studies using Raman scattering, which are of value for the biopharmaceutical industry. The phosphorylation and conformational effects in casein can be observed with a $4 \mu M$ solution (ca. 0.1 mg/mL).[18] The samples were dried on a surface before near-infrared Raman spectroscopy. Hédoux and coworkers[19] used low-frequency (10–350 cm^{-1}) Raman scattering to show that the thermal denaturation of lysozyme was a two-stage process with changes in tertiary structure preceding changes in secondary structure. High concentration of protein (14.58 M) was required for these studies. The requirement for high concentrations of sample makes the application of Raman scattering to biopharmaceuticals problematic due to reasons discussed in Chapter 11. Thus, the studies on proteins of therapeutic interest have been quite limited. Yu and workers[20] use Raman scattering to study the denaturation of insulin; protein concentrations of 10–100 mg/mL were used in this study.

As with near-infrared (Chapter 10) and infrared (Chapter 11) spectroscopy, Raman scattering can be applied to solid samples such as those obtained from lyophilization.[22–25] Raman scattering, either as such or as Raman microscopy, is also applied for the identification of particulates in drug product[26] and for the characterization of colloidal systems[27] including hydrogels.[28]

Anti-Stokes Raman scattering is less common than Stokes Raman scattering. Yamamoto and coworkers[29] used time-resolved Raman scattering (excitation 532 nm) to study the thermal unfolding of RNase A (7 mM). The temperature rise was determined by the ratio of anti-Stokes scattering to Stokes scattering. Coherent anti-Stokes Raman scattering (CARS) has received attention for the study of proteins.[30] CARS involves high-intensity laser pulses focused on a small volume. The CARS signal is blue-shifted (anti-Stokes), permitting detection in the presence of a fluorescent background. Chikishev and coworkers[31] used polarization sensitive coherent anti-Stokes scattering spectroscopy of the amide I band (1640–1680 cm^{-1}) to study the conformation of lysozyme and other proteins; protein concentration of

50 mg/mL were used in D_2O to avoid complications with water. Resonance CARS was used to study the several fluorescent centers within red fluorescent protein.[21]

The anisotropy of the polarization of the scattering can be measured (Raman optical activity [ROA]) and provides information on the conformation of the molecule.[32–34] The wavelength of incident radiation falls within the near-infrared and mid-infrared spectrum. ROA requires higher sample concentrations and long data acquisition times.[35] Solution structures of several β peptides by Raman backscattering or ROA in methanol required concentrations from 16 to 50 mg/mL.[36] There are several excellent reviews on the use of ROA.[37,38]

Resonance Raman spectra (RRS)[39–42] occurs when the wavelength of the incident light is within the absorption band of a chromophore resulting in enhancement of the scattered radiation. Resonance Raman spectroscopy can be used to study specific sites in proteins which contain porphyrin and related molecules.[43–47] Time-resolved resonance Raman spectroscopy can be used on protein solutions at a concentration of 10 μM.[45] Resonance Raman scattering is of only limited value in the study of biopharmaceuticals because of the requirement for prosthetic group. Juszczak and colleagues did report at UV resonance Raman study on poly(ethylene glycol)-modified hemoglobin.[48]

Raman microscopy[49–52] and, more recently, Raman crystallography[53,54] are applications of Raman spectroscopy. Raman microscopy can be described as chemical imaging by molecular vibrations.[55] As cited, Raman microscopy has applications in the study of gels and surface-bound macromolecules of biopharmaceutical interest. There is also potential for application of the Raman Effect to the study of viruses[56–59] and vaccines.[60,61]

Surface-enhanced Raman resonance scattering (SERRS) is resonance Raman spectroscopy of molecules adsorbed to a surface, usually silver.[62–64]

Raman scattering[65] and Raman microscopy[66] both have potential for diagnostic application. Raman spectroscopy has also been used to study biocompatibility of implanted biomaterials such as ceramic materials, carbon fibers, and polymers as well as ocular implant materials and contact lenses.[67–69] It is noted that Raman spectroscopy is also applied to glycosaminoglycans and proteoglycans.[70]

REFERENCES

1. Spiro, T.G. and Gaber, B.P., Laser Raman scattering as a probe of protein structure, *Annu. Rev. Biochem.* 46, 553–572, 1977.
2. Ma, C.Y., Rout, M.K., Chan, W.M., and Phillips, D.L., Raman spectroscopic study of oat globulin conformation, *J. Agric. Food Chem.* 48, 1542–1547, 2000.
3. Wen, Z.Q., Cao, X., and Vance, A., Conformation and side chains environments of recombinant human interleukin-1 receptor antagonist (rh-IL-1ra) probed by Raman, Raman optical activity, and UV-resonance Raman spectroscopy, *J. Pharm. Sci.* 97, 2228–2241, 2008.
4. Balakrishnan, G., Weeks, C.L., Ibrahim, M. et al., Protein dynamics from time-resolved UV Raman spectroscopy, *Curr. Opin. Struct. Biol.* 18, 623–629, 2008.
5. Stokes, G.G., On the change of refrangibility of light, *Philos. Trans. R. Soc. Lond.* 142, 463–562, 1852.
6. Raman, C.V. and Krishnan, K.S., A new type of secondary radiation, *Nature* 121, 501, 1928.

7. Krishnan, R.S. and Shankar, R.K., Raman effect—History of the discovery, *J. Raman Spectrosc.* 10, 1–8, 1981.

8. Vandenbeele, P. and Moens, L., Introducing students to Raman spectroscopy, *Anal. Bioanal. Chem.* 385, 209–211, 2006.

9. Edwards, H.G., Probing history with Raman spectroscopy, *Analyst* 129, 870–879, 2004.

10. Edwards, H.G.M. and Chalmers, J.M. eds., *Raman Spectroscopy in Archeology and Art History*, Royal Society of Chemistry, Cambridge, U.K., 2005.

11. Long, D.A., Introduction to Raman spectroscopy, in *Raman Spectroscopy in Archeology and Art History*, eds. H.G.M. Edwards and J.M. Chalmers, Royal Society of Chemistry, Cambridge, U.K., pp. 17–40, Chapter 2, 2005.

12. Edwards, H.G.M. and Chalmers, J.M., Practical Raman spectroscopy and complementary techniques, in *Raman Spectroscopy in Archeology and Art History*, eds. H.G.M. Edwards and J.M. Chalmers, Royal Society of Chemistry, Cambridge, U.K., pp. 41–67, Chapter 3, 2005.

13. Creagh, D.C. and Bradley, D.A. eds., *Radiation in Art and Archeometry*, Elsevier, Amsterdam, the Netherlands, 2000.

14. Osticioli, I., Wolf, M., and Anglos, D., An optimization of parameters for application of a laser-induced breakdown spectroscopy microprobe for the analysis of works of art, *Appl. Spectrosc.* 62, 1242–1249, 2008.

15. Thermo Scientific; http://www.thermo.com

16. Perkin-Elmer; http://las.perkinelmer.com/

17. InPhotonics; http://www.inphotonics.com/raman.htm

18. Jarvis, R.M., Blanch, E.W., Golovanov, A.P. et al., Quantification of casein phosphorylation with conformational interpretation using Raman spectroscopy, *Analyst* 132, 1053–1060, 2007.

19. Hédous, A., Ionov, R., Willart, J.F. et al., Evidence of two-stage thermal denaturation process in lysozyme: A Raman scattering and differential scanning calorimetry investigation, *J. Chem. Phys.* 124, 14703, 2006.

20. Yu, N.T., Liu, C.S., and O'Shea, D.C., Laser Raman spectroscopy and the conformation of insulin and proinsulin, *J. Mol. Biol.* 70, 117–132, 1972.

21. Kruglik, S.G., Subramaniam, V., Greve, J., and Otto, C., Resonance CARS study of the structure of "green" and "red" chromophores within the red fluorescent protein DsRed, *J. Am. Chem. Soc.* 124, 10902–10993, 2002.

22. Koenig, J.L. and Frushour, B.G., Raman scattering of chymotrypsinogen A, ribonuclease, and ovalbumin in the aqueous solution and solid state, *Biopolymers* 11, 2505–2520, 1972.

23. Yt, N.T. and Jo, B.H., Comparison of protein structure in crystals, in lyophilized state, and in solution by laser Raman scattering. I. Lysozyme, *Arch. Biochem. Biophys.* 156, 469–474, 1973.

24. Yu, N.T. and Jo, B.H., Comparison of protein structure in crystals and in solution by laser Raman scattering. II. Ribonuclease and carboxypeptidase A, *J. Am. Chem. Soc.* 95, 5033–5037, 1973.

25. Yu, N.T., Comparison of protein structure in crystals, in lyophilized state, and in solution by laser Raman scattering, 3. α-Lactalbumin, *J. Am. Chem. Soc.* 96, 4664–4668, 1973.

26. Cao, X., Wen, Z., Vance, A. et al., Raman microscopic application in biopharmaceuticals: Unique in situ identification of particulates in drug, *Microsc. Microanal.* 14(Suppl 2), 1608–1609, 2008.

27. Wartewig, S. and Neubert, R.H.H., Pharmaceutical applications of mid-IR and Raman spectroscopy, *Adv. Drug Deliv. Rev.* 57, 1144–1170, 2005.

28. van Manen, H.-J., van Apeldoam, A.A., Ruud, V. et al., Intracellular degradation of microspheres based on cross-linked dextran hydrogels or amphiphilic block copolymers: A comparative Raman microscopy study, *Int. J. Nanomed.* 2, 241–252, 2007.

29. Yamamoto, K., Mizutani, Y., and Kitagawa, T., Nanosecond temperature jump and time-resolved Raman study of thermal unfolding on of ribonuclease A, *Biophys. J.* 79, 485–495, 2000.

30. Baena, J.R. and Lendi, B., Raman spectroscopy in chemical bioanalysis, *Curr. Opin. Chem. Biol.* 8, 534–539, 2004.

31. Chikishev, A.Y., Lucassen, G.W., Koroteev, N.I. et al., Polarization sensitive coherent anti-Stokes Raman scattering spectroscopy of the amide I band of proteins in solutions, *Biophys. J.* 63, 976–985, 1992.

32. Nafie, L.A. and Freedman, T.B., Raman optical activity, *Methods Enzymol.* 226, 470–482, 1993.

33. Barron, L.D., Hecht, L., Blanch, E.W., and Bell, A.F., Solution structure and dynamics of biomolecules from Raman crystal optical activity, *Prog. Biophys. Mol. Biol.* 73, 1–49, 2000.

34. Barron, L.D., Blanch, E.W., and Hecht, L., Unfolded proteins studied by Raman optical activity, *Adv. Protein Chem.* 62, 51–90, 2002.

35. Hug, W. and Hangartner, G., A novel high-throughput Raman spectrometer for polarization difference measurements, *J. Raman Spectrosc.* 30, 841–852, 1999.

36. Kapitán, J., Zhu, F., Hecht, L. et al., Solution structures of β peptides from Raman optical activity, *Angew. Chem. Int. Ed. Engl.* 47, 6392–6394, 2008.

37. Blanch, E.W., Hecht, L., and Barron, L.D., Vibrational optical activity of proteins, nucleic acids, and viruses, *Methods* 29, 196–209, 2003.

38. Zhu, F., Issacs, NW., Hecht, L. et al., Raman optical activity of proteins, carbohydrates, and glycoproteins, *Chirality* 18, 103–115, 2006.

39. Callender, R., Resonance Raman scattering of visual pigments, *Annu. Rev. Biophys. Bioeng.* 6, 33–55, 1977.

40. Schick, G.A. and Bocian, D.F., Resonance Raman studies of hydroporphyrins and chlorophylls, *Biochim. Biophys. Acta* 895, 127–165, 1987.

41. Desbois, A., Resonance Raman spectroscopy of c-type cytochromes, *Biochimie* 76, 693–707, 1994.

42. Efremov, E.V., Ariese, F., and Gooijer, C., Achievements in resonance Raman spectroscopy review of a technique with a distinct analytical chemistry potential, *Anal. Chim. Acta* 606, 119–134, 2008.

43. Ikemura, K., Mukai, M., Shimada, H. et al., Red-excitation resonance Raman analysis of the nu(Fe = O) mode of ferryl-oxo hemoproteins, *J. Am. Chem. Soc.* 130, 14384–14385, 2008.

44. Mak, P.J. and Kincaid, J.R., Resonance Raman spectroscopic studies of hydroperoxo derivatives of cobalt-substituted myoglobin, *J. Inorg. Biochem.* 102, 1952–1957, 2008.

45. Singh, U.P., Obayashi, E., Takahashi, S. et al., The effects of heme modification on reactivity, ligand binding properties and iron-coordination structures of cytochrome P_{450}nor, *Biochim. Biophys. Acta* 1384, 103–111, 1998.

46. Anzenbacher, P. and Hudeĉeik U., Differences in flexibility of active sites of cytochrome P_{450} probed by resonance Raman and UV-VIX absorption spectroscopy, *J. Inorg. Biochem.* 87, 209–213, 2001.

47. Youn, H.D., Yim, Y.I., Kim, K. et al., Spectral characterization and chemical modification of catalase-peroxidase from *Streptomyces* sp., *J. Biol. Chem.* 270, 13740–13747, 1995.

48. Juszczak, L.J., Manjula, B., Bonaventura, C. et al., UV resonance Raman study of β93-modified hemoglobin A: Chemical modifier-specific effects and added influence of attached poly(ethylene glycol) chains, *Biochemistry* 41, 376–385, 2002.

49. Barry, B. and Mathies, R., Resonance Raman microscopy of rod and cone receptors, *J. Cell. Biol.* 94, 479–482, 1982.

50. Appel, R., Xu, W., Zerda, T.W., and Hu, Z., Direct observations of polymer network structures in macroporous *N*-isopropylacrylamide gel by Raman microscopy, *Macromolecules* 31, 5071–5074, 1998.

51. Balss, K.M., Llanos, G., Papandreou, G., and Maryanoff, C.A., Quantitative spatial distribution of sirolimus and polymers in drug-eluting stents using confocal Raman microscopy, *J. Biomed. Mater. Res. A* 85, 258–270, 2008.

52. Ivleva, N.P., Wagner, M., Horn, H. et al., Towards a nondestructive chemical characterization of biofilm matrix by Raman microscopy, *Anal. Bioanal. Chem.* 393, 197–206, 2009.

53. Carey, P.R., Raman crystallography and other biochemical applications of Raman microscopy, *Annu. Rev. Phys. Chem.* 57, 527–554, 2006.

54. Gong, B., Chen, J.H., Chase, E. et al., Direct measurement of a pK(a) near neutrality for the catalytic cytosine in the genomic HDV ribozyme using Raman crystallography, *J. Am. Chem. Soc.* 129, 13335–13342, 2007.

55. Cheng, J.-X. and Xie, X.S., Coherent anti-Stokes Raman scattering microscopy: Instrumentation, theory, and applications, *J. Phys. Chem. B* 108, 827–840, 2004.

56. Thomas, G.J. Jr., Raman spectroscopy and virus research, *Appl. Spectrosc.* 30, 483–494, 1976.

57. Shie, M., Dobrov, E.N., and Tichonenko, T.I., A comparative study of the structures of tobacco mosaic virus and cucumber virus 4 by laser Raman spectroscopy, *Biochem. Biophys. Res. Commun.* 81, 907–914, 1978.

58. Thomas, G.J. Jr., Li, Y., Fuller, M.T., and King, J., Structural studies of P22 phage, precursor particles, and proteins by laser-Raman spectroscopy, *Biochemistry* 21, 3866–3878, 1982.

59. Alexander, T.A., Surface-enhanced Raman spectroscopy: A new approach to rapid identification of intact viruses, *Spectroscopy* 36, 38–42, 2008.

60. Batenjany, M.M., Boni, L.T., Guo, Y. et al., The effect of cholesterol in a liposomal Muc1 vaccine, *Biochim. Biophys. Acta* 1514, 280–290, 2001.

61. Gordon, S., Saupe, A., McBurney, W. et al., Comparison of chitosan nanoparticles and chitosan hydrogels for vaccine delivery, *J. Pharm. Pharmacol.* 60, 1591–1600, 2008.

62. Smith, W.F., Surface-enhanced resonance Raman scattering, *Methods Enzymol.* 226, 482–495, 1993.

63. Faulds, K., Smith, W.E., and Graham, D., DNA detection by surface enhanced resonance Raman scattering (SERRS), *Analyst* 130, 1125–1131, 2005.

64 Smith, W.E., Practical understanding and use of surface enhanced Raman scattering/ surface-enhanced resonance Raman scattering in chemical and biological analysis, *Chem. Soc. Rev.* 37, 955–964, 2008.

65. Filik, J. and Stone, N., Drop coating deposition Raman spectroscopy of protein mixtures, *Analyst* 132, 544–550, 2007.

66. Dong, J., Atwood, C.S., Anderson, V.E. et al. Metal binding and oxidation of amyloid-β with isolated senile place cores: Raman microscopic validation, *Biochemistry* 42, 2768–2777, 2003.

67. Bertoluzza, A., Fagnano, C., Monti, P. et al., Raman spectroscopy in the study of biocompatibility, *Clin. Mater.* 9, 49–68, 1992.

68. Bertoluzza, A., Bottura, G., Taddei, D. et al., Vibrational spectra of controlled-structure hydroxyapatite coatings obtained by the polymeric route, *J. Raman Spectrosc.* 27. 759–764, 1996.

69. Taddei, P., Tinti, A., and Fini, G., Vibrational spectroscopy of polymeric biomaterials, *J. Raman Spectrosc.* 32, 619–629, 2001.

70. Ellis, R., Green, E., and Winlove, C.P., Structural analysis of glycosaminoglycans and proteoglycans by means of Raman microspectrometry, *Connect. Tissue Res.* 50, 29–36, 2009.

13 Use of UV-VIS Spectrophotometry for the Characterization of Biopharmaceutical Products

Ultraviolet–visible (UV-VIS) spectrophotometry is considered to include the spectral range from 180 to 750 nm (UV, 200–400 nm; VIS, 400–750 nm).[1] The visible range overlaps slightly with the near-infrared region of the spectrum. Near-UV circular dichroism (CD; 180–240 nm) and far-UV CD spectra (250–300 nm) reside in the UV spectral range.[2] The same separation is made infrequently with UV-VIS spectrometry. The relationship between UV-VIS spectrophotometry, CD, and fluorescence prompted the development of a spectrophotometer, which could be used for both UV spectroscopy and CD spectroscopy (and also fluorescence).[3,4] Absorbance in the visible range is confined (mostly) to the measurement of chromophoric groups such as heme, measurement of chemical modification such as the nitration of tyrosine, and the aggregation of proteins (light scattering). Light scattering has been discussed in Chapter 8. This chapter is focused on the ultraviolet spectral region; the absorbing groups (chromophoric groups) are the aromatic amino acids, tryptophan, tyrosine, and phenylalanine, the disulfide bond, and the peptide bond. The aromatic amino acids absorb in the range of 260–290 nm, cysteine at 240–250 nm, and the peptide bond absorbs at wavelengths less than 230 nm.[2] Tryptophan is the major chromophore in the ultraviolet region with an extinction coefficient of approximately 5600 M^{-1} cm^{-1} (280 nm). Tyrosine has an extinction coefficient of approximately 1150 M^{-1} cm^{-1} (280 cm) at neutral pH, which increases to approximately 1500 M^{-1} cm^{-1} in 0.1 M KOH.[5] The extinction coefficient for phenylalanine is much less (approximately 200 M^{-1} cm^{-1} at 260 nm; approximately 1 M^{-1} cm^{-1} at 280 nm).[5] Acetylation of the tyrosine hydroxyl group reduces the extinction coefficient of tyrosine to less than 100.[6] The absorbance of ultraviolet light by cysteine and the peptide bond are of lesser importance and it has been recognized that the spectra of proteins is, for all practical purposes, a function of the content of tryptophan, tyrosine, and phenylalanine.[7] The early observation of the quantitative difference between the ultraviolet spectrum of a protein and constituent amino acids/small peptides suggested that structure of a protein influenced such spectrum.[8–11]

The absorbance of ultraviolet light by tryptophan, tyrosine, and, to a much less extent, phenylalanine, provides the basis for the estimation of protein concentration by absorbance at 280 nm.[12,13] This method provides a useful approximation of protein concentration and it is assumed that an absorbance of 1.0 AU at 280 nm is obtained for a protein concentration of 1 mg mL^{-1}. A precise determination of protein concentration requires knowledge of the extinction coefficient and correction for light scattering.[14–16] Absorbance in the 270–290 nm range has also been used for the determination of tryptophan and tyrosine content in proteins.[17–22]

The denaturation of a protein results in change in the UV-VIS spectra.[8,10,23–26] In one of the earlier studies,[27] Donavan showed that the spectral changes observed on the denaturation of protein were larger than would be expected for the transfer of the responsible chromophores from 20% ethylene glycol into water. In the 230 nm range, the transfer of an indole chromophore from ethylene glycol to water produced a much larger change than that produced by a phenolic chromophore. The transfer of cysteine or phenylalanine produced only small changes. Selected examples of the use of ultraviolet absorbance to measure protein denaturation are shown in Table 13.1.

The denaturation (unfolding) of nucleic acids also results in a hyperchromic response,[39] which is used for the characterization of polynucleotides.[40–42] Examples of biopharmaceutical interest include aptatmers[43,44] and other potential nucleic acid-based therapeutics.[45,46] UV spectroscopy was used to study the interaction of siRNA and a cell-penetrating peptide, which could be a carrier for the siRNA.[46a]

Difference spectroscopy[47–49] is a technique uses a sample cuvette and a reference cuvette. The difference in spectrum between the two cuvettes is measured.[50] This has been used mostly in studies on solvent perturbation as discussed below. UV difference spectroscopy was used to determine the number of haptens coupled to bovine serum albumin.[51] UV difference spectroscopy was also used to characterize the product of the reaction of barstar and 5,5'-dithiobis (2-nitrobenzoic acid),[52] and the binding of gadolinium ions to apoovotransferrin.[53] While UV spectrophotometry is used frequently for the determination of biopharmaceutical concentration and for the characterization of biopharmaceutical proteins, difference spectroscopy as such is rarely used. Solvent perturbation is a form of difference spectroscopy, which is used somewhat more frequently and is discussed below. Difference spectroscopy is used more frequently in other modalities such a Raman[54–56] or Fourier transform infrared spectroscopy (FTIR) spectroscopy.[57–59]

Solvent perturbation spectrophotometry[60–62] is a form of difference spectroscopy[63] where a modifier such as ethylene glycol is added to the sample and the spectral change measured relative to a control.[64–75] One unique application investigated the use of poly(ethylene) glycol as a perturbant.[23,63] Custom double cells were the first cells developed for this purpose.[60]

Derivative spectroscopy is likely the most useful approach in UV-VIS spectroscopy for the study of protein conformation. Derivative spectroscopy is the representation of basic absorbance data (A, absorbance) as a derivative function with respect

TABLE 13.1

Use of Changes in Ultraviolet Absorbance to Measure Protein Denaturation

Protein	Study Description	Ref.
RNase A	UV absorbance is used to measure the effect of methanol on the thermal denaturation of RNase A. Denaturation (unfolding) is measured by change in absorbance at 286 nm (tyrosine exposure). As the concentration of methanol increased, denaturation temperature decreased. In 50% methanol, increasing the pH from 2 to 6 increased the transition temperature. Aggregation was assessed by light scattering.	[28]
Bovine growth hormone	UV absorbance is used to study conformational change/denaturation in bovine growth hormone. Changes in absorbance at 290 nm in the presence of urea. UV spectral changes correlated with change in intrinsic fluorescence.	[29,30]
Lactate dehydrogenase	Cold denaturation of lactate dehydrogenase measured by change in absorbance at 240 nm.	[31]
Human insulin-like growth factor I	Absorbance at 280 and 320 nm used to evaluate solubility of human insulin-like growth factor I in various excipients with thermal challenge.	[32]
Chymotrypsin, cutinase, model compounds	Study of the thermal unfolding of proteins studied by UV absorbance. Measurement at 250 and 278 nm permitted determination of the state of ionization of tyrosine residues in proteins.	[33]
Basic fibroblast growth factor	Use of UV absorbance to determine aggregation in pharmaceutical protein formulations.	[34]
α-Chymotrypsinogen A	Use of UV spectrophotometry to study denaturation of α-chymotrypsinogen A by pH and temperature.	[35]
β-Trypsin	Use of UV absorbance (286 nm) to follow thermal denaturation of β-trypsin at acid pH (pH 2.8) with correlation with CD.	[36]
Porcine growth hormone	Use of UV spectroscopy to study pH-dependent structural transitions in the presence and absence of 4.0 M urea. Correlation with CD and fluorescence studies.	[37]
Kunitz soybean trypsin inhibitor	UV absorbance used to study reversible denaturation of Kunitz soybean trypsin inhibitor.	[38]

to wavelength.[76,77] Derivative spectroscopy permits the precise assignment of local absorbance maxima, permitting the identification of difference between reference sample and experimental sample. Second-derivative and fourth-derivative spectroscopy are commonly used for the study of low-molecular-weight drugs including multicomponent analysis.[78–80] Derivative UV-VIS spectroscopy has been used for the characterization of proteins[81–85]; some studies of biopharmaceutical interest are shown in Table 13.2.

TABLE 13.2

Examples of the Use of UV-VIS Spectrophotometry for the Characterization of Biopharmaceuticals

Protein	Study	Ref.
Ovine growth hormone	Ovine growth hormone is used as model protein for the evaluation of 2-hydroxypropyl-β-cyclodextrin as a solubilizing/stabilization excipient. The use of fourth derivative ultraviolet spectroscopy suggested no conformational change in the presence of this excipient.	[86]
Human FK-binding protein	Second derivative UV spectroscopy used with CD, intrinsic fluorescence, and NMR to characterize structural changes during urea-induced denaturation.	[87]
Recombinant human growth hormone	This study is a continuation of efforts to understand factors influencing the retention of proteins and peptides in reverse phase chromatography. The effect of 1-propanol and acetonitrile on the conformation of recombinant human growth hormone was examined by second derivative spectroscopy and intrinsic fluorescence. Concentrations of 1-propanol above 10% (v/v) resulted in conformational change with completion of process at 30% 1-propanol (change was dependent on temperature).	[88]
Chemically modified bovine serum albumin	Fourth derivative spectroscopy was used with CD to evaluate conformational change in bovine serum albumin caused by modification with glycine methyl ester/carbodiimide. This was part of a study on the effect of protein charge on distribution in an aqueous two-phase system.	[89]
IgG	Second derivative spectroscopy used to study the effect of guanidine on the conformation of a monoclonal IgG antibody. Spectroscopy and intrinsic fluorescence suggested that changes in antibody affinity occur prior to observable conformational change in either the Fc or Fv domains.	[90]
Botulinum neurotoxin	Second derivative spectroscopy (with CD, FTIR) used to determine conformational change in botulinum neurotoxin occurring on removal of bound Zn^{2+} by EDTA.	[91]
Pepstatin-insensitive carboxyl proteases and pepstatin-sensitive carboxyl proteases	Fourth derivative spectroscopy was used to measure the effect of pressure on the conformation of carboxyl proteases. There was an effect of pressure on peptide hydrolysis (greater effect with the pepstatin-insensitive proteases) but not the hydrolysis of acid-denatured myoglobin. Pressure did appear to stabilize the proteases against autolysis. There appears to be a relationship between sensitivity to pepstatin and response to pressure.	[92]
Recombinant human growth hormone	Second derivative UV spectroscopy and FTIR spectroscopy used to change denatured, aggregated forms of recombinant human growth hormone.	[93]
Not applicable	Review on the use of UV-VIS derivative spectroscopy under high pressure.	[94]
RNAse, lysozyme, bovine serum albumin, concanavalin A	The effect of silicone oil on the conformation of model proteins was evaluated by CD and second derivative UV spectroscopy; little conformational change was observed.	[95]

TABLE 13.2 (continued)
Examples of the Use of UV-VIS Spectrophotometry for the Characterization of Biopharmaceuticals

Protein	Study	Ref.
Trigger factor	Second derivative spectroscopy (with intrinsic fluorescence, far-UV CD, biological activity, and ANS binding) used to study denaturation of tissue factor by guanidine hydrochloride.	[96]
Interferon-β-1a	Fourth derivative spectroscopy used with CD and fluorescence to construct a phase diagram for interferon-β-1a, which can be used for formulation studies.	[97]
Norwalk virus-like particles	Fourth derivative spectroscopy used with CD, fluorescence, DSC, DLS, and transmission electron microscopy used to evaluate the pH and thermal stability of Norwalk virus-like particles.	[98]
Fibroblast growth factor 20	Second derivative spectroscopy used with CD, intrinsic fluorescence, and ANS binding to construct a conformation phase diagram (pH and temperature) for fibroblast growth factor 20.	[99]
Model proteins (bovine serum albumin, human fibroblast growth factor 20, bovine CGSF) and gelatins (recombinant and porcine)	Second derivative UV spectroscopy used with CD and fluorescence to evaluate the stabilizing effect of gelatins on model proteins.	[100]
Various proteins	Second derivative spectroscopy used to determine the binding of SDS to tryptophanyl residues in proteins. This study also evaluated the use of UV photoacoustic spectroscopy to study protein–detergent complexes.	[101]
Monoclonal antibody	Second derivative spectroscopy and DSC used to evaluate the effect of excipients on monoclonal antibody conformation and stability when challenged with freeze–thaw cycles.	[102]

REFERENCES

1. Campbell, I.D. and Dwek, R.A., *Biological Spectroscopy*, Benjamin-Cummings, Menlo Park, CA, 1986.
2. Colón, W., Analysis of protein structure by solution optical spectroscopy, *Methods Enzymol.* 309, 605–632, 1999.
3. Ramsay, G. and Eftink, M.R., A multidimensional spectrophotometer for monitoring thermal unfolding transitions of macromolecules, *Biophys. J.* 66, 519–523, 1994.
4. Olis Instruments: http://olisweb.com/products/rsm/db620.php
5. Fasman, G.D. ed., *Practical Handbook of Biochemistry and Molecular Biology*, CRC Press, Boca Raton, FL, 1989.
6. Myers II, B. and Glazer, A.N., Spectroscopic studies of the exposure of tyrosine residues in proteins with special reference to the subtilisins, *J. Biol. Chem.* 246, 412–419, 1971.
7. Smith, F.C., Ultraviolet absorption spectra of certain aromatic amino acids and of the serum proteins, *Proc. R. Soc. B. Lond.* 104, 198–205, 1929.

8. Doty, P. and Geiduschek, E.P., Optical properties of proteins, in *The Proteins*, eds. H. Neurath and K. Bailey, Academic Press, New York, Chapter 5, 1953.

9. Weber, G. and Teale, F.W.J., Interaction of proteins with radiation, in *The Proteins*, Vol. III, eds. H. Neurath, Academic Press, New York, Chapter 17, 1965.

10. Yanari, S. and Bovey, F.A., Interpretation of the ultraviolet changes of proteins, *J. Biol. Chem.* 235, 2818–2826, 1961.

11. Harrap, B.S., Gratzer, W.B., and Doty, P., The structure of proteins, *Annu. Rev. Biochem.* 30, 269–292, 1961.

12. Taylor, J.F., The purification of proteins, in *The Proteins*, 2nd edn., Vol. 1, Part A, eds. H. Neurath and K. Bailey, Academic Press, New York, Chapter 1, 1953.

13. Grimsley, G.R. and Pace, C.N., Spectrophotometric determination of protein concentration, *Curr. Protoc. Protein Sci.*, November, 2004, Chapter 3, Unit 3.1, 2004.

14. Fridovich, I., Farkas, W., Schwert, G.W. Jr., and Handler, P., Instrumental artifacts in the determination of difference spectra, *Science* 125, 1141–1142, 1957.

15. Leach, S.J. and Scheraga, H.A., Effect of light scattering on ultraviolet difference spectra, *J. Am. Chem. Soc.* 82, 4790–4792, 1960.

16. Pace, C.N., Vajdos, F., Fee, L. et al., How to measure and predict the molar coefficient of a protein, *Protein Sci.* 4, 2411–2423, 1995.

17. Edelhoch, H., Spectroscopic determination of tryptophan and tyrosine in proteins, *Biochemistry* 6, 1945–1954, 1967.

18. Servillo, L., Colonna, G., Balestrieri, C. et al., Simultaneous determination of tyrosine and tryptophan residues in proteins by second-derivative spectroscopy, *Anal. Biochem.* 126, 251–257, 1982.

19. Nozaki, Y., Determination of tryptophan, tyrosine, and phenylalanine by second derivative spectrophotometry, *Arch. Biochem. Biophys.* 277, 324–333, 1990.

20. Botsoglou, N.A., Fletouris, D.J., Papageogiou, G.E., and Mantis, A.J., Derivative spectrophotometric method for the analysis of tyrosine in unhydrolyzed protein, food, and feedstuff samples, *J. Agric. Food Chem.* 41, 1635–1639, 1993.

21. Bray, M.R., Carriere, A.D., and Clarke, A.J., Quantitation of tryptophan and tyrosine residues in proteins by 4th-derivative spectroscopy, *Anal. Biochem.* 221, 278–284, 1994.

22. Gatellier, P., Kondjoyan, A., Portanguen, S. et al., Determination of aromatic amino acid content in cooked meat by derivative spectrophotometry: Implications for nutritional quality of meat, *Food Chem.* 114, 1074–1078, 2009.

23. Farruggia, B., Garcia, G., D'Angelo, C., and Picó, G., Destabilization of human serum albumin by polyethylene glycol studied by thermodynamic equilibrium and kinetic approaches, *Int. J. Biol. Macromol.* 20, 43–51, 1997.

24. Nikolovski, Z., Buzón, V., Ribó, M. et al., Thermal unfolding of eosinophili cationic protein/ribonuclease 3: A nonreversible process, *Protein Sci.* 15, 2816–2927, 2006.

25. Dar, T.A., Singh, L.R., Islam, A. et al., Guanidinium chloride and urea denaturations of β-lactoglobulin at pH 2.0 and 25°C: The equilibrium intermediate contains non-native structures (helix, tryptophan and hydrophobic patches), *Biophys. Chem.* 127, 140–148, 2007.

26. Ghosh, S., Interaction of trypsin with sodium dodecyl sulfate in aqueous medium: A conformational view, *Colloids Surf. B. Biointerfaces* 66, 178–186, 2008.

27. Donavan, J.W., Changes in ultraviolet absorption produces by alteration of protein conformation, *J. Biol. Chem.* 244, 1961–1967, 1969.

28. Fink, A.L. and Painter, B., Characterization of the unfolding of ribonuclease A in aqueous methanol solvents, *Biochemistry* 26, 1665–1671, 1987.

29. Holzman, T.F., Brems, D.N., and Dougherty, J.J. Jr., Reoxidation of reduced bovine growth hormone from a stable secondary structure, *Biochemistry* 25, 6907–6917, 1986.

30. Holzman, T.F., Dougherty, J.J. Jr., Brems, D.N., and MacKenzie, N.E., pH-induced conformational states of bovine growth hormone, *Biochemistry* 29, 1255–1261, 1990.

31. Hatley, R.H.M. and Franks, F., The cold-induced denaturation of lactate dehydrogenase at sub-zero temperatures in the absence of perturbants, *FEBS Lett.* 257, 171–173, 1989.

32. Fransson, J., Hallén, D., and Florin-Robertsson, E., Solvent effects on the solubility and physical stability of human insulin-like growth factor I, *Pharm. Res.* 14, 606–612, 1997.

33. Melo, E.F., Aires-Barros, M.R., Costa, S.M., and Cabral, J.M., Thermal unfolding of proteins at high pH range studied by UV absorbance, *J. Biochem. Biophys. Methods* 34, 45–59, 1997.

34. Shahrokh, Z., Stratton, P.R., Eberlein, G.A., and Wang, Y.J., Approaches to analysis of aggregates and demonstrating mass balance in pharmaceutical protein (basic fibroblast growth factor) formulations, *J. Pharm. Sci.* 83, 1645–1650, 1994.

35. Chalikian, T.V., Völker, J., Anafi, D., and Breslauer, K.J., The native and the heat-induced denatured states of α-chymotrypsinogen A: Thermodynamic and spectroscopic studies, *J. Mol. Biol.* 274, 237–252, 1997.

36. Burmano, M.H., Rogana, E., and Swaisgood, H.E., Thermodynamics of unfolding of β-trypsin at pH 2.8, *Arch. Biochem. Biophys.* 382, 57–62, 2000.

37. Parkinson, E.J., Morris, M.B., and Bastiras, S., Acid denaturation of recombinant porcine growth hormone: Formation and self-association of folding intermediates, *Biochemistry* 39, 12345–12354, 2000.

38. Roychaudhuri, R., Sarath, G., Zeece, M., and Markwell, J., Reversible denaturation of the soybean Kunitz trypsin inhibitor, *Arch. Biochem. Biophys.* 412, 20–26, 2003.

39. Thomas, R., The denaturation of DNA, *Gene* 135, 77–79, 1993.

40. Cox, R.A., The secondary structure of ribosomal ribonucleic acid in solution, *Biochem. J.* 98, 841–957, 1966.

41. Diaz-Ruiz, J.R. and Kaper, J.M., Cucumber mosaic virus-associated RNA 5. VI. Characterization and denaturation–renaturation behavior of the double-stranded form, *Biochim. Biophys. Acta* 564, 275–288, 1979.

42. Shea, R.G., Ng., P., and Bischofberger, N., Thermal denaturation profiles and gel mobility shift analysis of oligonucleotide triplexes, *Nucleic Acids Res.* 18, 4850–4866, 1990.

43. Smirnov, I. and Shafer, R.H., Effect of loop sequence and size on DNA aptamers stability, *Biochemistry* 39, 1462–1468, 2000.

44. Bozza, M., Sheardy, R.D., Dilone, E. et al., Characterization of the secondary structure and stability of an RNA aptamers that binds vascular endothelial growth factor, *Biochemistry* 45, 7639–7643, 2006.

45. Saccà, B., Lacroix, L., and Mergny, J.L., The effect of chemical modifications on the thermal stability of different G-quadruplex-forming oligonucleotides, *Nucleic Acid Res.* 33, 1182–1192, 2005.

46. Roy, S., Tanious, F.A., Wilson, W.D. et al., High-affinity homologous peptide nucleic acid probes for targeting a quadruplex-forming sequence from a MYC promoter element, *Biochemistry* 46, 10433–10443, 2007; (a) Law, M., Jarfari, M., and Chen, P., Physicochemical characterization of siRNA-peptide complexes, *Biotechnol. Prog.* 24, 957–963. 2008.

47. Holloway, M.R. and White, H.A., A double-beam rapid-scanning stopped-flow spectrophotometer, *Biochem. J.* 149, 221–231, 1975.

48. Cross, D.G., Hydrogen exchange in nucleosides and nucleotides. Measurement of hydrogen exchange by stopped-flow and ultraviolet difference spectroscopy, *Biochemistry* 14, 357–362, 1975.

49. Loontiens, F.G. and Dhollander, G., Temperature-induced ultraviolet difference absorption spectrometry for determination of enthalpy changes. Binding of 4-methylumbelliferyl glycosides to four lectins, *FEBS Lett.* 175, 249–254, 1984.

50. Salahuddin, P., Studies on the acid-induced unfolding of human serum albumin, *Protein Pept. Lett.* 16, 324–332, 2009.

51. Adamczyk, M., Buko, A., Chen, Y.Y. et al., Characterization of protein–hapten conjugates. 1. Matrix-assisted laser desorption ionization mass spectrometry of immuno BSA–hapten conjugates and comparison with other characterization methods, *Bioconjug. Chem.* 5, 631–635, 1994.

52. Ramachandran, S. and Udgaonkar, J.B., Stabilization of barstar by chemical modification of the buried cysteines, *Biochemistry* 35, 8776–8785, 1996.

53. Wang, J.L., Li, Y.Q., and Yang, B.S., Spectral study on the binding of gadolinium ions with apoovotransferrin, *Spectrochim. Acta A Mol. Biomol. Spectrosc.* 67, 1101–1105, 2007.

54. Rousseau, D.L., Shelnutt, J.A., Henry, E.R., and Simon, S.R., Raman difference spectroscopy of tertiary and quaternary structure changes in methaemoglobins, *Nature* 285, 49–51, 1980.

55. Yue, K.T., Lee, M., Zheng, J., and Callender, R., The determination of the pKa of histidine residues in proteins by Raman difference spectroscopy, *Biochim. Biophys. Acta* 1078, 296–302, 1991.

56. Callender, R. and Deng, H., Nonresonance Raman difference spectroscopy: A general probe of protein structure, ligand binding, enzymatic catalysis, and the structures of other biomacromolecules, *Annu. Rev. Biophys. Biomol. Struct.* 23, 215–245, 1994.

57. Baenziger, J.E., Miller, K.W., McCarthy, M.P., and Rotheschild, K.J., Probing conformational changes in the nicotine acetylcholine receptor by Fourier transform infrared different spectroscopy, *Biophys. J.* 62, 64–66, 1992.

58. Ryan, S.F., Demers, C.N., Chew, J.P., and Baenziger, J.E., Structural effects of neutral and anionic lipids on the nicotinic acetylcholine receptor. An infrared difference spectroscopy study, *J. Biol. Chem.* 271, 24590–24597, 1996.

59. Zscherp, C. and Barth, A., Reaction-induced difference spectroscopy for the study of protein reaction mechanisms, *Biochemistry* 40, 1875–1883, 2001.

60. Herskovtiz, T,T. and Laskowski, M. Jr., Location of chromophoric residues in proteins by solvent perturbation. I. Tyrosyls in serum albumin, *J. Biol. Chem.* 237, 2481–2492, 1962.

61. Solli, N.J. and Hersokvits, T.T., Solvent perturbation studies and analysis of proteins and model compound data in denaturing solvents, *Anal. Biochem.* 34, 70–78, 1973.

62. Georgievea, D.N., Genow, N., Ramashanker, K.R. et al., Spectroscopic investigation of phenolic groups ionization in the vipoxin neurotoxic phospholipase A2: Comparison with the x-ray structure in the region of the tyrosine residues, *Spectrochim. Acta A Mol. Biomol. Spectrosc.* 55A, 239–244, 1999.

63. Bailey, J.E., Beaven. J.H., Chignell, D.A., and Gratzer, W.H., An analysis of perturbation in the ultraviolet absorption spectra of proteins and model compounds, *Eur. J. Biochem.* 7, 5–14, 1968.

64. Swaney, J.H., An alternative technique for difference spectroscopy, *Anal. Biochem.* 43, 388–393, 1972.

65. Quast, U., Engel, J., Steffen, E. et al., The effect of cleaving the reactive-site peptide bond Lys-15–Ala-16 on the conformation of bovine trypsin–kallikrein inhibitor (Kunitz) as revealed by solvent-perturbation spectra, circular dichroism and fluorescence, *Eur. J. Biochem.* 52, 511–514, 1975.

66. Furied, B. and Furie, B.C., Spectra changes in bovine factor X associated with activation by the venom coagulant protein of *Vipera russelli*, *J. Biol. Chem.* 251, 6807–6814, 1975.

67. Demchenko, A.P., On the effect of temperature on the ultraviolet spectra of protein chromophores, *Biophys. Chem.* 9, 393–396, 1975.

68. Larsson, L.J., Lindahl, P., Hallén-Sandgren, C., and Björk, I., The conformational changes of α$_2$-macroglobulin induced by methylamine or trypsin. Characterization by extrinsic and intrinsic spectroscopic probes, *Biochem. J.* 243, 47–54, 1987.

69. Villanueva, G.B., Perret, V., and Fenton II, J.W., Conformational integrity of human α-thrombin, *Thromb. Res.* 36, 377–387, 1984.

70. Rock, C.O., Environment of the aromatic chromophores of acyl carrier protein, *Arch. Biochem. Biophys.* 225, 122–129, 1983.

71. Callahan, H.J., Liberti, P.A., and Maurer, P.H., Solvent perturbation difference spectra of a purified sheep antibody, *Immunochemistry* 11, 197–202, 1974.

72. Sogami, M. and Ogura, S., Structural transitions of bovine plasma albumin. Location of tyrosyl and tryptophanyl residues by solvent perturbation difference spectra, *J. Biochem.* 73, 323–334, 1973.

73. Farruggia, B., Nerli, B., Di Nuci, H. et al., Thermal features of the bovine serum albumin unfolding by polyethylene glycols, *Int. J. Biol. Macromol.* 26, 23–33, 1999.

74. Isenman, D.E., Ellerson, J.R., Painter, R.H., and Dorrington, K.J., Correlation between the exposure of aromatic chromophores at the surface of the Fc domains of immunoglobulin G and their ability to bind complement, *Biochemistry* 16, 233–240, 1977.

75. Björk, I. and Larsson, K., Exposure to solvent of tyrosyl and tryptophanyl residues of bovine antithrombin in the absence and presence of high-affinity and low-affinity antithrombin, *Biochim. Biophys. Acta* 621, 273–282, 1980.

76. Talsky, G., Mayring, L., and Kreuzer, H., Higher order derivative spectroscopy for fine resolution of UV/VIS spectra, *Angew. Chem. Int. Ed.* 17, 532–533, 1978.

77. Perkampus, H.-H., *Encyclopedia of Spectroscopy*, VCH, Weinheim, pp. 166–118, 1995.

78. Vetuschi, C., Giannandrea, A., Carluccin, G. et al., Determination of hydrochlorothiazide and irbesartan in pharmaceuticals by fourth-order UV derivative spectrophotometry, *Farmaco* 60, 665–670, 2005.

79. Koba, M., Koba, K., and Przyborowski, L., Application of UV-derivative spectrophotometry for determination of some bisphosphonate drugs in pharmaceutical formulations, *Acta Pol. Pharm.* 65, 289–294, 2008.

80. Karpińska, J., Sokól, A., and Skoczhlas, M., An application of UV-derivative spectrophotometry and bivariate calibration algorithm for study of photostability of levomepromazine hydrochloride, *Spectrochim. Acta A Mol. Biomol. Spectrosc.* 71, 1562–1564, 2008.

81. Matsushima, A., Inoue, Y., and Shibata, K., Derivative absorption spectrophotometry of native proteins, *Anal. Biochem.* 65, 362–368, 1975.

82. Butler, W.L., Fourth derivative spectra, *Methods Enzymol.* 56, 501–515, 1979.

83. Havel, H.A., Derivative near-ultraviolet absorption techniques for investigating protein structure, in *Spectroscopic Methods for Determining Protein Structure in Solution*, ed. H.A. Havel, VCH, New York, Chapter 4, 1996.

84. Lange, R. and Balny, C., UV-visible derivative spectroscopy under high pressure, *Biochim. Biophys. Acta* 1595, 80–93, 2002.

85. Lucas, L.H., Ersoy, B.A., Kueltzo, L.A. et al., Probing protein structure and dynamics by second-derivative ultraviolet absorption analysis of cation-π interactions, *Protein Sci.* 15, 2228–2243, 2006.

86. Brewster, M.E., Hora, M.E., Simpkins, J.W., and Bodor, N., Use of 2-hydroxypropyl-β-cyclodextrin as a solubilizing and stabilizing excipient for protein drugs, *Pharm. Res.* 8, 792–795, 1991.

87. Egan, D.A., Logan, T.M., Liang, H. et al., Equilibrium denaturation of recombinant human FK binding protein in urea, *Biochemistry* 32, 1920–1927, 1993.

88. Wicar, S., Mulkerrin, M.B., Bathory, G. et al., Conformational changes in the reversed phase liquid chromatography of recombinant human growth hormone as a function of organic solvent: The molten globule state, *Anal. Chem.* 66, 3908–3915, 1994.

89. Obludziner, A., Camperi, S.A., and Cascone, O., Effect of the surface charge on partitioning of chemically modified bovine serum albumin in aqueous two-phase systems, *Bioseparation* 5, 369–374, 1995.

90. Wang, X.D., Luo, J., and Guo, Z.Q., Perturbation of the antigen-binding site and staphylococcal A-binding site of IgG before significant changes in global concentration during denaturation: An equilibrium study, *Biochem. J.* 325, 707–710, 1997.

91. Fu, F.N., Lomneth, R.B., Cai, S., and Singh, B.R., Role of zinc in the structure and toxic activity of botulinum neurotoxin, *Biochemistry* 37, 5267–5278, 1998.

92. Fujiwara, S., Kunugi, S., Oyama, H., and Oda, K., Effects of pressure on the activity and spectroscopic properties of carboxyl proteinases. Apparent correlation of pepstatin-insensitivity and pressure response, *Eur. J. Biochem.* 268, 645–655, 2001.

93. St. John, R.J., Carpenter, J.F., Balny, C., and Randolph, T.W., High pressure refolding of recombinant human growth hormone from insoluble aggregates. Structural transformations, kinetic barriers, and energetics, *J. Biol. Chem.* 276, 46856–46863, 2001.

94. Lange, R. and Balny, C., UV-visible derivative spectroscopy under high pressure, *Biochim. Biophys. Acta* 1595, 80–93, 2002.

95. Jones, L.S., Kaufmann, A., and Middaugh, C.R., Silicone oil induced aggregation of proteins, *J. Pharm. Sci.* 94, 918–927, 2005.

96. Liu, C.P., Li, Z.Y., Huang, G.C. et al., Two distinct intermediates of trigger factor and populated during guanidine denaturation, *Biochimie* 87, 1023–1031, 2005.

97. Fan, H., Ralston, J., Dibiase, M. et al., Solution behavior of IFN-β-1a: An empirical phase diagram based approach, *J. Pharm. Sci.* 94, 1893–1911, 2005.

98. Ausar, S.F., Foubert, T.R., Hudson, M.H. et al., Conformational stability and disassembly of Norwalk virus-like particles. Effect of pH and temperature, *J. Biol. Chem.* 281, 19478–19488, 2006.

99. Fan, H., Vitharana, S.N., Chen, T. et al., Effects of pH and polyanions on the thermal stability of fibroblast growth factor 20, *Mol. Pharm.* 4, 232–240, 2007.

100. Thyagarajapuram, N., Olsen, D., and Middaugh, C.R., Stabilization of proteins by recombinant human gelatins, *J. Pharm. Sci.* 96, 3304–3315, 2007.

101. Bugs, M.R., Bortoleto-Bugs, R.K., and Cornélio, M.L., Photoacoustic spectroscopy of aromatic amino acids in proteins, *Eur. Biophys. J.* 37, 205–212, 2008.

102. Kueltzo, L.A., Wang, W., Randolph, T.W., and Carpenter, J.F., Effects of solution conditions, processing parameters, and container materials on aggregation of a monoclonal antibody during freeze-thawing, *J. Pharm. Sci.* 97, 1801–1812, 2008.

14 Use of Optical Rotatory Dispersion and Circular Dichroism to Study Therapeutic Biomacromolecule Conformation

Plane polarized light is composed of two equal, opposite (left and right) vectors, which cancel each other. When plane polarized light passes though a medium containing one or more asymmetric centers, the two vectors move at different rates. This differential rate, which derives from differences in the respective refractive indices, results in a rotation in the plane of polarization (optical rotation), which can be either left-handed or right-handed. Optical rotatory dispersion (ORD)[1] measures the change in rotation as a function of the wavelength of incident light. If there is also unequal absorption of light, the light is elliptically polarized, resulting in circular dichroism (CD). The reader is directed to two excellent early works[2,3] for a description of the development of ORD and CD. The optical activity of a polypeptide,[1,4–6] oligonucleotide,[7–9] oligosaccharide,[10–12] or conjugate[13–15] is both a configurational and a conformational property of individual asymmetric centers in the backbone of the polymer and the superimposition of the several diastereoisomers (secondary and tertiary structure).[16]

Optical rotatory dispersion has been mostly used to study the secondary structure of proteins. Most of the studies with ORD occurred more than 20 years ago but there are some recent studies of interest. Current studies with time-resolved ORD use sample concentrations of 15–70 µM.[17–19] Time-resolved optical rotatory dispersion[20,21] has provided an approach to looking at very fast reactions such as binding of carbon monoxide by hemoglobin. The reader is referred to early work on the application of ORD to protein structure.[22–28] Data obtained from ORD can be combined with other data obtained, for example, from small-angle x-ray diffraction, to prepare protein structure models but this does not appear to be a dominant source of information.[29]

There is greater use of CD than ORD for the study of macromolecular conformation although a number of studies use both techniques.[30–35] CD studies can be separated into near UV and far UV studies. The far UV range (230–180 nm) measures peptide bond absorbance while the near UV range (330–260 nm) measures the absorbance aromatic amino acids residues and disulfide bonds.[36–38] The independent contribution of disulfide bonds to the near-UV CD is demonstrated by the studies of Yoshida and workers[39] on the trypsin inhibitor from Adzuki beans. This is a small protein (8 kDa), which contains six cysteine residues (12 cysteine residues).[40] This protein contains no tyrosine or tryptophan and only one phenylalanine. The high degree of disulfide cross-linkage, which is also seen in lima bean trypsin inhibitor appears to provide stability.[41]

The reader is directed to the excellent review by Kelly and Price[36] for a consideration of instrumentation, solvent, and sample preparation issues critical for successful CD studies. Buffer selection can be a major consideration for far UV studies as many common buffer materials and other solvents absorb in this region.[42,43] Since the near UV response is dependent on composition, sample size can vary. Depending on cell volume, the sample can vary from 10 to 100 μg[36] corresponding to a concentration of 0.2–1.0 mg/mL; in the Adzuki bean trypsin inhibitor cited above, the absence of aromatic chromophores required a concentration of 4.5 mg/mL. CD spectra in

---

TABLE 14.1
Some Examples of the Application of Circular Dichroism for the Study of Monoclonal Antibodies

Antibody	Study Description	Ref.
Human IgM monoclonal antibody	Use of CD together with oligosaccharide analysis, isoelectric focusing, and biological activity to demonstrate product equivalence during process optimization	[50]
Lambda light-chain Mcg	Use of CD and immunoreactivity to demonstrate similarity to lambda light-chain Mcg to T-cell receptor	[51]
IgG	CD and calorimetry used to study the thermal stability and pH stability of IgG domains. Strong correlation between calorimetry and changes in secondary structure	[52]
PEG-modified monoclonal antibodies	CD was used to evaluation monoclonal antibodies modified with poly(ethylene)glycol (PEG) at lysine residues. CD evaluation was combined with ¹H-NMR and titration with trinitrobenzene sulfonic acid	[53]
Monoclonal antibody	CD used to study the effect of arginine of protein structure (arginine is used to suppress protein aggregation during refolding and purification)	[54]
Monoclonal antibody	CD, DSC, and analytical ultracentrifugation used to assess the effect of acid conditions (0.1 M citrate, pH 2.7–3.9) used during affinity purification on MAB structure	[55]
Monoclonal antibody	Use of CD, DSC, and x-ray crystallography used to rationalize the structural differences responsible for aspirate isomerization in the CDR region in monoclonal antibodies	[56]
Monoclonal antibody	CD spectra used to evaluate the effect of various osmolytes (arginine, proline, sorbitol) on MAB structure. β-Structure of native antibody decreases on storage at 4°C (15 mM sodium phosphate, pH 7.2); the presence of osmolytes reduced the loss of ordered structure	[57]

the far UV region provide information on the secondary structure[37,38] of a protein while the near UV region can provide information about the environment of the aromatic amino acid residues. CD has provided insight in to protein folding issues, membrane protein structure, and the "molten globule," but it has not been extensively used in the characterization of therapeutic biological macromolecules. Major problems in the use of CD are experimental design, operator expertise, and data interpretation. The reader is directed to the several pragmatic papers on the use of CD for the study of proteins[36,44]; these considerations can be used for the study of other biological macromolecules.

Notwithstanding the above discussion, CD and ORD are used in the development and characterization of biological therapeutics.[45] It is important to consider the objective before applying the technique. The following studies are presented as examples to provide guidance for use of ORD and CD in the study of biopharmaceuticals.

Circular dichroism has found limited use in the characterization of polyclonal immunoglobulin preparations.[46–49] Early studies by Litman and colleagues used CD to study intact and fragmented immunoglobulins and showed the presence of β-sheet conformation.[46] Szenczi and colleagues used CD to study the formulation of therapeutic polyclonal IgG preparations.[47] Hartman and colleague used CD with other technologies to evaluate the effect of heating on the conformation of polyclonal IgG.[48] Liu and colleagues used CD to evaluate the conformation of the Fc domain after methionine oxidation.[49] There has been somewhat more use of CD in the study of monoclonal antibodies as described in Table 14.1. Other CD studies of interest for the development of biological therapeutics are shown in Table 14.2.

Vibration circular dichroism[71] has been used to evaluate protein secondary structure.[72–75]

TABLE 14.2

Selected Application of the use of Circular Dichroism to Study Proteins of Therapeutic Interest

Protein	Study Description	Ref.
5S ribonucleic acid	Denaturation (unfolding) of 5S RNA evaluated by CD, UV absorption spectroscopy, analytical ultracentrifugation	[58]
Lysozyme modified with diethylpyrocarbonate	CD was used to study hen egg while lysozyme (HEWL) which was modified with diethylpyrocarbonate (CD difference spectra in the 240–260 nm region). Reduction and carboxymethylation abolished the difference spectra observed with monoethoxyformylated lysozyme with decarbethoxylation	[59]
Human IL-6	CD used with immunological reactivity and biological activity to assess the role of disulfide bonds in the function of human IL-6. Derivatives were prepared by reduction and carboxymethylation and by site-directed mutagenesis	[60]
Recombinant chloroperoxidase	CD was used to characterize the refolding of recombinant chloroperoxidase expressed in *Escherichia coli*	[61]

(continued)

TABLE 14.2 (continued)
Selected Application of the use of Circular Dichroism to Study Proteins of Therapeutic Interest

Protein	Study Description	Ref.
Cardiotoxin III	CD, 2D NMR, hydrogen-deuterium exchange, and ANS binding used to study the conformational changes in the precipitation of proteins with trichloroacetic acid (2,2,2-trichloroacetic acid); cardiotoxin III is used as a model protein	[61]
Human α_1-antitrypsin	CD, intrinsic fluorescence, and biological activity used to evaluate the effect of heparin and glucose on the stability of human α_1-antitrypsin	[62]
Recombinant TGF β3	CD was used to study the conformation of recombinant transforming growth factor β3. Analytical ultracentrifugation was used to assess product aggregation; CD analysis showed that conformational change is associated with the aggregation process	[63]
Human antithrombin	CD is used to measure the conformational change in antithrombin occurring on oxidation with hydrogen peroxide	[64]
Human serum albumin	CD and intrinsic fluorescence were used to study the conformation of human serum albumin after oxidation with chloramine-T. This study is a validation of the use of chloramine-T oxidation as model for in vivo oxidative damage	[65]
Bovine serum albumin	CD was used to measure the conformational change in BSA with decreasing pH; volume and adiabatic compressibility are measured with ultrasound. The volume and adiabatic compressibility decrease in going from pH 7 to pH 2; CD measures a decrease in α-helix	[66]
Concentrated monoclonal antibody	CD, fluorescence, UV absorbance, FTIR, and DSC are used to characterize highly monoclonal antibody preparations. The collective results suggest that the decrease in structural stability observed on heating at the high protein concentration is due to aggregation/self-association and not a decrease in intrinsic stability	[67]
Human CD81	CD used with analytical ultracentrifugation to characterize a full-length human CD81 produced in *Pichia pastoris* and solubilized with detergent	[68]
Pegylated albumin	Human serum albumin was modified with poly(ethylene glycol)-phenylisothiocyanate. CD analysis of the modified albumin showed little change in secondary structure	[69]

REFERENCES

1. Jirgensons, B., *Optical Rotatory Dispersion of Proteins and Other Macromolecules*, Springer-Verlag, Berlin, Germany, 1969.
2. Djerassi, C., *Optical Rotatory Dispersion. Application to Organic Chemistry*, McGraw-Hill, New York, 1960.
3. Crabbe, P., *Optical Rotatory Dispersion and Circular Dichroism in Organic Chemistry*, Holden-Day, San Francisco, CA, 1965.

4. Usui, K., Tomizaki, K.Y., Ohyama, T. et al., A novel peptide microarray for protein detection and analysis utilizing a dry peptide array system, *Mol. Biosyst.* 2, 113–121, 2006.

5. Shanmugam, G. and Polavarapu, P.L., Vibrational circular dichroism of protein films, *J. Am. Chem. Soc.* 126, 10292–10295, 2004.

6. Van Mierle, C.P.M., de Jongh, H.H., and Visser, A.J.W.G., Circular dichroism of protein in solution and at interfaces, in *Physical Chemistry at Biological Interfaces*, ed. A. Baszkin, Marcel Dekker, New York, Chapter 20, 2008.

7. Inoue, Y. and Satoh, K., Oligonucleotide studies. Optical rotatory dispersion of five homodinucleotides, *Biochem. J.* 113, 843–851, 1969.

8. Kresse, J., Nagpal, K.I., Nagyvary, J., and Uchic, J.I., The use of s-2-cyanoethyl phosphorothioate in the preparation of oligo 5′-deoxy-5′-thiothymidylates, *Nucleic Acids Res.* 2, 1–9, 1975.

9. Sacca, B., Meyer, R., and Niemeyer, C.M., Temperature-dependent FRET spectroscopy for the high-throughput analysis of self-assembled DNA nanostructures in real time, *Nat. Protoc.* 4, 271–285, 2009.

10. Pysh, E.S., Optical activity in the vacuum ultraviolet, *Annu. Rev. Biophys. Bioeng.* 5, 63–75, 1975.

11. Cowman, M.K. and Matsuoka, S., Experimental approaches to hyaluronan structure, *Carbohydr. Res.* 340, 791–809, 2005.

12. Petrovic, A.G., Bose, P.K., and Polavarapu, P.L., Vibrational circular dichroism of carbohydrate films formed from aqueous solutions, *Carbohydr. Res.* 339, 2713–2720, 2004.

13. Oda, Y., Yanagisawa, H., Maruyama, M. et al., Design, synthesis and evaluation of D-galactose-β-cyclodextrin conjugates as drug-carrying molecules, *Bioorg. Med. Chem.* 16, 8830–8840, 2008.

14. Emami, J., Hamishehkear, H., Najafabadi, A.R. et al., A novel approach to prepare insulin-loaded poly(lactic-*co*-glycolic acid) microcapsules and the protein stability study, *J. Pharm. Sci.* 98, 1712–1731, 2009.

15. Dhalluin, C., Ross, A., Leuthold, L.A. et al., Structural and biophysical characterization of the 40 kDA PEG-interferon-α2a and its individual positional isomers, *Bioconjug. Chem.* 16, 504–517, 2005.

16. Weber, G. and Teale, F.W.J., Interaction of protein with radiation, in *The Proteins*, 2nd edn., ed. H. Neurath, Academic Press, New York, Chapter 17, 1965.

17. Chen, E., Swartz, T.E., Bogomolni, R.A., and Kliger, D.S., A LOV story: The signaling state of the phot1 LOV2 photocycle involves chromophore-triggered protein structure relaxation, as probed by far-UV time-resolved optical rotatory dispersion spectroscopy, *Biochemistry* 46, 4619–4624, 2007.

18. Chen, E., Goldbeck, R.A., and Kliger, D.S., The earliest events in protein folding: A structural requirement for ultrafast folding in cytochrome C, *J. Am. Chem. Soc.* 126, 11175–11181, 2004.

19. Bohr, H. and Bohr, J., Microwave enhanced kinetics observed in ORD studies of a protein, *Bioelectromagnetics* 21, 68–72, 2000.

20. Shapiro, D.B., Goldbeck, R.A., Che, D. et al., Nanosecond optical rotatory dispersion spectroscopy: Application to photolyzed hemoglobin-CO kinetics, *Biophys. J.* 68, 326–334, 1995.

21. Chen, E., Genson, T., Gross, A.B. et al., Dynamics of protein and chromophore structural changes in the photocycle of photoactive yellow protein monitored by time-resolved optical rotatory dispersion, *Biochemistry* 42, 2062–2071, 2003.

22. Jirgensons, B., Optical rotatory dispersion and conformation of various globular proteins, *J. Biol. Chem.* 238, 2716–2722, 1963.

23. Herskovits, T.T. and Mscanti, L., Conformation of protein and polypeptides. II. Optical rotatory dispersion and conformation of the milk proteins and other proteins in organic solvents, *J. Biol. Chem.* 240, 639–644, 1965.

24. Steinhardt, J., Krijin, J., and Leidy, J.G., Differences between bovine and human serum albumins: Binding isotherms, optical rotatory dispersion, viscosity, hydrogen ion titration, and fluorescence efforts, *Biochemistry* 10, 4005–4015, 1971.

25. Yaron, A., Katchalski, E., and Berger, A., The chain length dependence of the conformation for oliogomers of L-lysine in aqueous solution: Optical rotatory dispersion, *Biopolymers* 10, 1107–1120, 1971.

26. Liberti, P.A., Maurer, P.H., and Clark, L.G., Antigenicity of polypeptides (poly-α-amino acids). Physicochemical studies of a calcium dependent antigen–antibody reaction, *Biochemistry* 10, 1632–1639, 1971.

27. Chen, Y.-H., Yang, J.T., and Martinez, H.M., Determination of the secondary structures of proteins by circular dichroism and optical rotatory dispersion, *Biochemistry* 11, 4120–4131, 1972.

28. Funding, L., Jacobsen, C., Steersgaard, J. et al., Properties and immunochemical reactivities of carboxy-modified human serum albumin, *Int. J. Pept. Protein Res.* 7, 245–250, 1975.

29. Momany, F.A., Sessa, D.J., Lawton, J.W. et al., Structural characterization of α-zein, *J. Agric. Food Chem.* 54, 543–547, 2006.

30. Dobrov, E.N., Kust, S.V., Yakovleva, O.A., and Tikchonenko, T.L., Structure of single-stranded virus RNA *in situ.* II. Optical activity of five tobacco mosaic-like viruses and their components, *Biochim. Biophys. Acta* 475, 623–637, 1977.

31. Inaki, Y., Ishikawa, T., and Takemoto, K., Synthesis and interactions of poly-L-lysine containing nucleic acid bases, *Nucleic Acids Symp. Ser.* 8, s137–s140, 1980.

32. Raghavendra, K. and Ananthanarayanan, V.S., β-Structure of polypeptides in non-aqueous solutions. I. Spectral characteristics of the polypeptide backbone, *Int. J. Pept. Protein Res.* 17, 412–419, 1981.

33. Egelandsdal, B., Conformation and structure of mildly heat-treated ovalbumin in dilute solutions and gel formation at higher protein concentrations, *Int. J. Pept. Protein Res.* 28, 560–568, 1986.

34. McKenzie, H.A. and Frier, R.D., The behavior of R-ovalbumin and its individual components, A1, A2, and A3 in urea solution: Kinetics and equilibria, *J. Protein Chem.* 22, 207–214, 2003.

35. Ikeda, M., Hasegawa, T., Numata, M. et al., Instantaneous inclusion of a polynucleotide and hydrophobic guest molecules into a helical core of cationic β-1,3-glucan polysaccharide, *J. Am. Chem. Soc.* 129, 3979–3988, 2007.

36. Kelly, S.M. and Price, N.C., The use of circular dichroism in the investigation of protein structure and function, *Curr. Pept. Protein Sci.* 1, 349–384, 2000.

37. Oakley, M.T., Bulheller, B.M., and Hunt, J.D., First principles of protein circular dichroism in the far-ultraviolet and beyond, *Chirality* 18, 18341–18347, 2006.

38. Johnson, W.C. Jr., Protein secondary structure and circular dichroism: A practical guide, *Protein: Struct. Func. Genet.* 7, 205–214, 1990.

39. Yoshida, C., Yoshikawa, M., and Takagi, T., Near UV circular dichroism of trypsin inhibitor of Adzuki beans attributable to disulfide groups, *J. Biochem.* 80, 448–451, 1975.

40. Yoshida, O. and Yoshikawa, M., Purification and characterization of proteinase inhibitors from Adzuki beans (*Phaseolus angularis*), *J. Biochem.* 78, 935–945, 1975.

41. Krogdahl, A. and Holm, H., Soybean proteinase inhibitors and human proteolytic enzymes: Selective inactivation of inhibitors by treatment with human gastric juice, *J. Nutr.* 111, 2045–2051, 1981.

42. Lundblad, R.L., *Biochemistry and Molecular Biology Compendium*, CRC Press, Boca Raton, FL, 2007.

43. Martin, S.G., Circular dichroism, in *Protein Labfax*, ed. N.C. Price, Bios/Academic Press, Oxford, U.K., Chapter 18, 1996.
44. Kelly, S.M., Jess, T.J., and Price, N.C., How to study proteins by circular dichroism, *Biochim. Biophys. Acta* 1751, 119–139, 2005.
45. Bierau, H., Tranter, G.E., LePevelen, D.D. et al., Higher-order structure comparison of proteins derived from different clones or processes. Unbiased assessment of spectra by quantitative CD, *Bioprocess Int.*, 6(8), 52–59, 2008.
46. Litman, G.W., Good, R.A., Frommel, D., and Rosenberg, A., Conformational significance of the intrachain disulfide linkages in immunoglobulins, *Proc. Natl. Acad. Sci. USA* 67, 1085–1092, 1970.
47. Szenczi, A., Kardos, J., Medgyesi, G.A., and Zavodszky, P., The effect of solvent environment on the conformation and stability of human polyclonal IgG in solution, *Biologicals* 34, 5–14, 2006.
48. Hartmann, W.K., Saptahrishi, N., Yang, X.Y. et al., Characterization and analysis of thermal denaturation of antibodies by size-exclusion high-performance liquid chromatography, *Anal. Biochem.* 325, 227–239, 2004.
49. Liu, D., Ren, D., Huang, H. et al., Structure and stability changes of human IgG1 Fc as a consequence of methionine oxidation, *Biochemistry* 47, 5088–5100, 2008.
50. Maiorella, G.L., Ferris, R., Thomson, J. et al., Evaluation of product equivalence during process optimization for manufacture of a human IgM monoclonal antibody, *Biologicals* 21, 197–205, 1993.
51. Lake, D.F., Helgerson, S., Landsperger, W.J., and Marchalonis, J.J., Physical and epitope analysis of a recombinant human T-cell receptor $V\alpha/V\beta$ construct support the similarity to immunoglobulin, *J. Protein Chem.* 16, 309–320, 1997.
52. Vermeer, A.W. and Nords, W., The thermal stability of immunoglobulin unfolding and aggregation of a multi-chain protein, *Biophys. J.* 78, 394–404, 2000.
53. Larson, R.S., Menard, V., Jacobs, H., and Kim, S.W., Physicochemical characterization of poly(ethylene glycol)-modified anti-GAD antibodies, *Bioconjug. Chem.* 12, 861–869, 2001.
54. Arakawa, T., Kita, Y., Ejima, D. et al., Aggregation suppression of proteins by arginine during thermal unfolding, *Protein Pept. Lett.* 13, 921–927, 2006.
55. Ejima, D., Tsumoto, K., Fukada, H. et al., Effects of acid exposure on the conformation, stability, and aggregation of monoclonal antibodies, *Proteins* 66, 954–962, 2007.
56. Wakankar, A.A., Borchardt, R.T., Eigenbrot, C. et al., Aspartate isomerization in the complementary-determining regions of two closely related monoclonal antibodies, *Biochemistry* 46, 1534–1544, 2007.
57. Kashanian, S., Pakneiad, M., Ghobadi, S. et al., Effect of osmolytes on the conformational stability of mouse monoclonal antidigoxin antibody in long-term storage, *Hybridoma* 27, 99–106, 2008.
58. Fox, J.W. and Wong, K.P., Acquisition of native conformation of ribosomal 5S ribonucleic acid from *Escherichia coli*. Hydrodynamic and spectroscopic studies on the unfolding and refolding of ribonucleic acid, *Biochemistry* 21, 2096–2102, 1982.
59. Li, C., Moore, D.S., and Rosenberg, R.C., Circular dichroism studies of diethyl pyrocarbonate-modified histidine in hen egg white lysozyme, *J. Biol. Chem.* 268, 11090–11096, 1993.
60. Rock, F.L., Li, X., Chong, P. et al., Role of disulfide bonds in recombinant human interleukin 6 conformation, *Biochemistry* 33, 5146–5154, 1994.
61. Zong, Q., Osmulski, P.A., and Hager, L.P., High-pressure-assisted reconstitution of recombinant chloroperoxidase, *Biochemistry* 34, 12420–12425, 1995.
62. Sivaraman, T., Kumar, T.K.S., Jayaraman, G., and Yu, C., The mechanism of 2,2,2-trichloroacetic acid-induced protein precipitation, *J. Protein Chem.* 16, 291–297, 1997.

63. Finotti, P. and Polverino de Laureto, P., Differential effects of heparin and glucose on structural conformation of human α_1 antitrypsin: Evidence for a heparin-induced cleaved form of the inhibitor, *Arch. Biochem. Biophys.* 347, 19–29, 1997.

64. Pellaud, J., Schote, U., Arvinte, T., and Seelig, J., Conformation and self-association of human recombinant transforming growth factor-β3 in aqueous solution, *J. Biol. Chem.* 274, 7699–7704, 1999.

65. Van Patten, S.M., Hanson, E., Bernasconi, R. et al., Oxidation of methionine residues in antithrombin. Effects on biological activity and heparin binding, *J. Biol. Chem.* 274, 10268–10276, 1999.

66. Anraku, M., Kragh-Hansen, U., Kawai, K. et al., Validation of the chloramine-T induced oxidation of human serum albumin as a model for oxidative damage in vivo, *Pharmaceut. Res.* 20, 684–692, 2003.

67. El Kadi, N., Taulier, N., Le Huérou, J.Y. et al., Unfolding and refolding of bovine serum albumin at acid pH: Ultrasound and structural studies, *Biophys. J.* 91, 3397–3404, 2006.

68. Harn, N., Allan, C., Oliver, C., and Middaugh, C.B., Highly concentrated monoclonal antibody solutions: Direct analysis of physical structure and thermal stability, *J. Pharm. Sci.* 96, 532–546, 2007.

69. Jamshad, M., Rajesh, S., Stamataki, Z. et al., Structural characterization of recombinant human CD81 produced in *Pichia pastoris*, *Protein Expr. Purif.* 57, 206–216, 2008.

70. Meng, F., Manjula, H.N., Smith, P.K., and Acharya, S.A., PEGylation of human serum albumin: Reaction of PEG-phenyl-isothiocyanate with protein, *Bioconjug. Chem.* 19, 1352–1360, 2008.

71. Baello, B.I., Pancoska, P., and Keiderling, T.A., Vibrational circular dichroism spectra of proteins in the amide III region: Measurement and correlation of bandshape to secondary structure, *Anal. Biochem.* 250, 212–221, 1997.

72. Keiderling, T.A., Silva, R.A., Yoder, G., and Dukor, R.K., Vibrational circular dichroism spectroscopy of selected oligopeptides conformations, *Bioorg. Med. Chem.* 7, 133–141, 1999.

73. Polavarapu, P.I. and Zhao, C., Vibrational circular dichroism: A new spectroscopic tool for biomolecular structural determination, *Fresenius J. Anal. Chem.* 366, 727–734, 2000.

74. Keiderling, T.A., Protein and peptide secondary structure and conformational determination with vibrational circular dichroism, *Curr. Opin. Chem. Biol.* 6, 682–688, 2002.

75. Taniguchi, T., Miura, N., Nishimura, S., and Monde, K., Vibrational circular dichroism: Chiroptical analysis of biomolecules, *Mol. Nutr. Food Res.* 48, 246–254, 2004.

15 Use of Nuclear Magnetic Resonance for the Characterization of Biotherapeutic Products

Nuclear magnetic resonance (NMR) spectroscopy studies the response of atomic nuclei with spin and magnetic moment in a strong, static magnetic field.[1–6] Spin is a poorly understood property of some nuclei.[7,8] For example, ^1H has a nuclear spin of ½, ^2H has a spin of 1, ^{13}C has a spin of ½, ^{15}N has a spin of ½, and ^{31}P has a spin of ½[7]; these atoms are the most studied nuclei in biological systems.[8] Differences in the resonant frequencies allow differentiation of the various atomic species with a complex molecule such as a nucleic acid or protein (Table 15.1). Measurement with NMR is accomplished by the absorption and emission of energy by the sample in a strong, static magnetic field in response to radiofrequency from an applied radiation field (10^8 to 10^{10} Hz). As the sample is scanned by the applied radiofrequency/ microwave radiation, there is nuclear spin reorientation and the resonant frequencies are measured and such transitions are sensitive to the electronic environment of thee atom.[9] Data is expressed as chemical shift (δ),[10] with respect to an internal standard, spin–spin coupling (J), relaxation time (T_1, T_2), and signal intensity.[1–6] The signal is unique to the atomic nucleus at a specific radiofrequency and the quality of the signal reflects the electronic environment of the atom. Thus, it was possible to differentiate a methyl hydrogen from a methylene hydrogen,[11] which changed the fundamental basis of organic analytical chemistry.[12–15] The author was a research associate in the laboratories of Stanford Moore and William Stein at the Rockefeller University in the late 1960s. Moore and Stein were organic chemists who were interested in amino acids and proteins. Ted Bella, an analytical chemist, was an important part of the laboratory and provided us with the elemental analyses critical for our work. The determination of carbon, hydrogen, and nitrogen together with absorption spectroscopy and chromatography formed the basis for analysis in the laboratory. Today, it would be difficult to find similar information in a current article; NMR and mass spectrometry are the basic analytical tools. Dawson[16] reviewed the application of NMR for pharmaceutical analysis and listed some of advantages:

- NMR is a nondestructive. The sample material can be considered for another analytical procedure provided care is taken to avoid contamination. The

TABLE 15.1

Characteristics of the Atomic Nuclei Important for the Study of Therapeutic Biologics

Z^a	Isotope	Abundance (%)	Spin (I)	N^b
1	^1H	99.99	1/2	42.6
1	^2H	0.01	1	6.5
1	^3H	—[c]	1/2	45.4
6	^{13}C	1.11	1/2	10.7
7	^{15}N	0.37	1/2	4.3
8	^{17}O	0.04	5/2	5.8
9	^{19}F	100	1/2	40.1
15	^{31}P	100	1/2	17.3
16	^{33}S	0.08	3/2	3.3

Source: Adapted from Lide, D. ed., *CRC Handbook of Chemistry and Physics*, 89th edn., CRC Press, Boca Raton, FL, 2008–2009.

[a] Z, atomic number.

[b] ν, resonant frequency, MHz at an applied field of 1 T.

[c] Radioactive species with extremely low occurrence (Argonne National Laboratory, TVS, http://www.ead.anl.gov/pub/doc/tritium.pdf).

See also Evans, N.S.E., Nuclear magnetic resonance of biomolecules, in *Encyclopedia of Analytical Chemistry*, Vol. 1., ed. R.A. Meyers, Wiley, Chichester, U.K., pp. 585–623, 2000; Craik, D.J. and Scanlon, M.J., Two-, three-, and four-dimensional nuclear magnetic resonance of biomolecules, in *Encyclopedia of Analytical Chemistry*, ed. R.A. Meyers, Wiley, Chichester, U.K., 2000.

amount of sample is more important for biologicals where material is usually quite limited. It is noted that NMR can be used for process analysis.[17,18]

- NMR is the single most powerful tool for organic structure analysis.
- NMR is a robust technology.

Dawson[16] also considers some disadvantages for NMR:

- NMR is a relatively insensitive technique and requires a large sample (frequently milligram quantities).[19–23]
- The instrumentation is complicated and expensive; instruments with high magnetic fields are used (>500 mHz and more likely 900 mHz).[8,24–26]
- Considerable expertise is required for the entire analytical process; consideration of the various operational requirements makes the use of a contract analytical laboratory essential.

Although not mentioned by Dawson, molecular size can be an issue although improvement in instrumentation and computer software is making size somewhat less of a problem. Protein and polysaccharide NMR spectra can be obtained in ^2H_2O[27,28] or $H_2^{17}O$.[23,29–31] The use of oxygen isotope permits the study of bound water while the use of heavy water simplified proton NMR studies by increasing the intensity of the protein proton NMR signals.[8] However, D_2O has an independent effect on protein conformation,[32–39] which should be considered in interpretation of studies performed in the presence of heavy water. ^{15}N and ^{13}C can be used for protein and polysaccharide NMR but the natural abundance of these isotopes is low. This requires the use of recombinant DNA technology to express samples containing ^{15}N or ^{13}C.[8,40–42] ^{31}P NMR[43,44] can be used for the study of nucleotides and polynucleotides.[45–48]

Nuclear magnetic resonance imaging (MRI) makes use of the above nuclear spin properties of atoms for the study of in vivo structures and, as such, is one of major modalities of medical imaging.[9,49] An extensive discussion of MRI is outside of the scope of the current work and the reader is referred to several excellent books on this subject.[49–51]

NMR is used to study molecular crowding[52–54] and can be use for in cell measurements.[55,56] Direct application of the study of biopharmaceuticals is not clear but it is possible that the measurement of translational diffusion[57] of a suitable analyte in albumin could be a surrogate measurement of the bulk effects of this protein.[58,59]

NMR is a powerful tool for the determination of protein structure. However, at the current time, it is unlikely that NMR can be considered of value for the study of the conformation of biological therapeutics within the context of the current work. Expenditure and sample size are the primary issues, which need to be solved for the routine application of NMR for the characterization of biopharmaceutical products. There is no question that the structural data obtained for proteins with NMR is excellent and correlates well with other analytical data.[60–62] However, the degree of detail may not be relevant to product quality as evaluated by safety and efficacy. A change in structure may not be causative of a change in function, immunogenicity, or therapeutic index; indeed the therapeutic index of a biopharmaceutical can be improved with a change in structure.[63–65] Applications of NMR of interest to therapeutic biotechnology are presented in Table 15.2.

TABLE 15.2

Some Selected Applications of NMR to the Study of Biopharmaceuticals

Biological Macromolecules	Study Detail	Ref.
Lysozyme	Proton NMR was used to follow the thermal denaturation of lysozyme. 20% poly(ethylene glycol) increased denaturation as did an increase in lysozyme concentration.	[66]
Trisulfide[a] variant of recombinant human growth hormone	Two-dimensional NMR was used to characterize the native and variant peptide. The variant protein is more hydrophobic than native using hydrophobic interaction chromatography. Mass spectrometry was also used to characterize the variant protein.	[67]

(continued)

TABLE 15.2 (continued)
Some Selected Applications of NMR to the Study of Biopharmaceuticals

Biological Macromolecules	Study Detail	Ref.
Recombinant transforming growth factor β3	NMR used to control batch-to-batch consistency of disulfide location in protein drug products product by recombinant DNA technology.	[68]
Serum γ-globulin (BGG)	Stability of IgG in poly(vinyl alcohol) formulation or dextran evaluated by proton NMR.	[69]
C-Mannosylated human interleukin-12	NMR used to demonstrate C-mannosylation of a tryptophan residue in recombinant human interleukin-12. Mass spectrometry was also used for structural analysis.	[70]
Polysaccharide–protein conjugate vaccine	Use of NMR and other physicochemical techniques to characterize pneumococcal conjugate vaccine and a meningococcal conjugate vaccine.	[71]
Meningococcal group C conjugates vaccines	Use of NMR, CD and fluorescence to evaluate the stability of several oligosaccharide-protein vaccines.	[72]
Therapeutic proteins including factor VII, IgG, insulin, and α_1-antitrypsin	Use of NMR and x-ray crystallography to design ligands for affinity chromatography.	[73]
PEGylated antibody	Protein NMR, reactivity with trinitrobenzene-sulfonic acid, and biological activity used to evaluate the effect of modification of monoclonal antibodies against glutamic acid decarboxylase with poly(ethylene glycol).	[74]
PEGylated human growth hormone-releasing factor	NMR was used to determine if change in conformation was associated with a reduction in the biological activity of human growth hormone-releasing factor associated with pegylation. NMR studies when combined with hydrogen-exchange suggest that the decrease in activity is caused by steric hindrance and not conformational change.	[75]
Hyaluronan	^{13}C-NMR spectroscopy was used to characterize hyaluronan to be used for the formulation of dermatology products.	[76]
Granulocyte colony-stimulating factor (G-CSF)	2D-NMR NOESY spectra was used to compare the structures of various G-CSF isolates from *Escherichia coli* expression systems.	[77]
Peptide hormones	NMR spectroscopy, SDS-PAGE, and mass spectrometry were used to establish the quality of site-specific modification of protein disulfide bonds with sulfone derivatives of poly(ethylene glycol).	[78]
Viscotoxins	NMR was used to study possible conformation change is a fusion proteins containing viscotoxins. The study sample was uniformly labeled with ^{15}N.	[79]
Bovine pancreatic trypsin inhibitor (BPTI)	BPTI was used as a model protein for the solid-state spectroscopy of peptide and protein pharmaceuticals.	[80]

[a] Trisulfide linkages are, in general, unusual but are a constituent of garlic thought to be important for health (Seki, T., Hosono, T., Hosono-Fukao, T. et al., Anticancer effects of diallyl trisulfide derived from garlic, *Asia Pac. J. Clin. Nutr.* 17 (Suppl 1), 249–252, 2008).

REFERENCES

1. Jones, R., Nuclear magnetic resonance, in *An Introduction to Biological Spectroscopy*, ed. S.B. Brown, Academic Press, New York, pp. 236–278, Chapter 6, 1980.
2. Campbell, I.D. and Dwek, R.A., *Biological Spectroscopy*, Benjamin-Cummings, Menlo Park, CA, Chapter 6, 1984.
3. Derome, A.E., *Modern NMR Techniques for Chemistry Research*, Pergamon Press, Oxford, U.K., 1987.
4. Martin, G.E. and Zektzer, A.S., *Two-Dimensional NMR Methods for Establishing Molecular Connectivity: A Chemist's Guide to Experimental Selection, Performance, and Interpretation*, VCH Publishers, New York, 1988.
5. Bruch, M., *NMR Spectroscopy*, 2nd edn., Taylor & Francis, Boca Raton, FL, 1996.
6. Pochapsky, T.C. and Pochapsky, S., *NMR for Physical and Biological Scientists*, Taylor & Francis, Boca Raton, FL, 2002.
7. Lide, D., *CRC Handbook of Chemistry and Physics*, 86th edn., CRC Press/Taylor & Francis, Boca Raton, FL, pp. 9–92, 2007.
8. Craik, D.J. and Scanlon, M.J., Two, three, and four-dimensional nuclear magnetic resonance, in *Encyclopedia of Analytical Chemistry*, ed. R.A. Meyers, pp. 12490–12411, John Wiley & Sons, Chichester, U.K., 2000.
9. Sueten, P., *Fundamentals of Medical Imaging*, Cambridge University Press, Cambridge, U.K., Chapter 6, 2002.
10. von Bramer, S.E., Chemical shifts in nuclear magnetic resonance, in *Encyclopedia of Analytical Chemistry*, ed. R.A. Meyer, John Wiley & Sons, Chichester, U.K., pp. 12023–12040, 2000.
11. Bothner-By, A. and Naar-Coli, C., Organic analytical applications of nuclear magnetic resonance, *Ann. N.Y. Acad. Sci.* 70, 833–840, 1958.
12. Fitch, W.L., Analytical methods for quality control of combinatorial libraries, *Mol. Divers.* 4, 39–45, 1998–1999.
13. Bross-Walch, N., Kühn, T., Moskau, D., and Zerbe, O., Strategies and tools for structure determination of natural products using modern methods of NMR spectroscopy, *Chem. Biodivers.* 2, 147–177, 2005.
14. Paslo, D.J. and Johnson, C.B., *Organic Structure Determination*, Prentice-Hall, Englewood Cliffs, NJ, 1969.
15. Wood, G.W., *Analytical Instrumentation Methods*, Marcel Dekker, New York, 1997.
16. Dawson, B., Nuclear magnetic resonance spectroscopy in pharmaceutical analysis, in *Encyclopedia of Analytical Chemistry*, ed. R.A. Meyers, pp. 7229–7242, John Wiley & Sons, Chichester, U.K., 2000.
17. McCarthy, M.J. and Bobroff, S., Nuclear magnetic resonance and magnetic resonance imaging for process analysis, in *Encyclopedia of Analytical Chemistry*, ed. R.A. Meyers, John Wiley & Sons, Chichester, U.K., pp. 8266–8281, 2000.
18. Kumar, S., Wittmann, C., and Heinzle, E., Minibioreactors, *Biotechnol. Lett.* 26, 1–10, 2004.
19. Wagner, G., Prospects for NMR of large proteins, *J. Biomol. NMR* 3, 375–385, 1993.
20. Page, R.C., Moore, J.D., Nguyen, H.B. et al., Comprehensive evaluation of solution nuclear magnetic resonance spectroscopy sample preparation for helical integral membrane proteins, *J. Struct. Funct. Genomics* 7, 51–64, 2006.
21. Wijesinha-Bettoni, R., Gao, C., Jenkins, J.A. et al., Heat treatment of bovine α-lactalbumin results in partially folded disulfide bond shuffled states with enhanced surface activity, *Biochemistry* 46, 9774–9784, 2007.
22. Zhou, D.H., Shah, G., Cornos, M. et al., Proton-detected solid-state NMR spectroscopy of fully protonated proteins at 40 kHZ magic-angle spinning, *J. Am. Chem. Soc.* 129, 11791–11801, 2007.

23. Persson, E. and Halle, B., Nanosecond to microsecond protein dynamics probed by magnetic relaxation dispersion of buried water molecules, *J. Am. Chem. Soc.* 130, 1774–1787, 2008.

24. Eisenreich, W. and Bacher, A., Advances of high-resolution NMR techniques in the structural and metabolic analysis of plant biochemistry, *Phytochemistry* 68, 2799–2825, 2007.

25. Tate, S., Anisotropic nuclear spin interactions for the morphology analysis of proteins in solution by NMR spectroscopy, *Anal. Sci.* 24, 39–50, 2008.

26. Skinner, A.L. and Lawrence, J.S., High-field solution NMR spectroscopy as a tool for assessing protein interactions with small molecule ligands, *J. Pharm. Sci.* 97, 4670–4695, 2008.

27. Kumosinski, T.F., Pessen, H., Prestrelski, S.J., and Farrell, H.M. Jr., Water interactions with varying molecular states of bovine casein: ^{2}H NMR, *Arch. Biochem. Biophys.* 257, 259–268, 1987.

28. Sklenár, V., Selective excitation techniques for water suppression in one- and two-dimensional NMR spectroscopy, *Basic Life Sci.* 56, 63–84, 1990.

29. Halle, B. and Denison, V.P., Magnetic relaxation dispersion studies of biomolecular solutions, *Methods Enzymol.* 338, 178–201, 2001.

30. Halle, B., Protein hydration dynamics in solution: A critical survey, *Philos. Trans. R. Soc. Lond. B Biol. Sci.* 359, 1207–1223, 2004.

31. Lemaltre, V., Smith, M.E., and Watts, A., A review of oxygen-17 solid-state NMR of organic materials—Toward biological applications, *Solid State Nucl. Magn. Reson.* 26, 215–235, 2004.

32. Smith, S., The effect of ^{2}H$_2$O on the quaternary structure of the fatty acid synthetase multienzyme complex, *Biochim. Biophys. Acta* 251, 477–481, 1971.

33. Krysteva, M.A., Mazuriedr, J., and Spik, G., Ultraviolet difference spectral studies of human serotransferrin and lactotransferrin, *Biochim. Biophys. Acta* 453, 484–493, 1976.

34. Lin, X.Q. and Sano, Y., Effect of Na$^+$ and K$^+$ on the initial crystallization process of lysozyme in the presence of D$_2$O and H$_2$O, *J. Protein Chem.* 17, 479–484, 1998.

35. Cioni, P. and Strambini, G.B., Effect of heavy water on protein flexibility, *Biophys. J.* 82, 3246–3253, 2002.

36. Eker, F., Griebenow, K., and Schweitzer-Stenner, R., Stable conformations of tripeptides in aqueous solution studied by UV circular dichroism spectroscopy, *J. Am. Chem. Soc.* 125, 8178–8185, 2003.

37. Chellgren, B.W. and Creamer, T.P., Effects of H$_2$O and D$_2$O on polyproline II helical structure, *J. Am. Chem. Soc.* 126, 1473–1475, 2004.

38. Goddard, Y.A., Korb, J.P., and Bryant, R.G., Structural and dynamical examination of the low-temperature glass transition in serum albumin, *Biophys. J.* 91, 3841–3847, 2006.

39. Das, A., Sinha, S., Acharya, B.R. et al., Deuterium oxide stabilizes conformation of tubulin: A biophysical and biochemical study, *BMB Rep.* 41, 62–67, 2008.

40. Akasaka, K., Li, H., Yamada, H. et al., Pressure response of protein backbone structure. Pressure-induced amide ^{15}N chemical shifts in BPTI, *Protein Sci.* 8, 1946–1953, 1999.

41. Koharudin, L.M.I., Bonvin, A.M.J.J., Kaptein, R., and Boetens, R., Use of very long-distance NOEs in a fully deuterated protein: An approach for rapid protein fold determination, *J. Magn. Res.* 163, 228–235, 2003.

42. Lindhout, D.A., Boyko, R.F., Carson, D.C. et al., The role of electrostatics in the interaction of the inhibitory region of troponin I with troponin C, *Biochemistry* 44, 14750–14759, 2006.

43. Gornestein, D.G., ^{31}P NMR of DNA, *Methods Enzymol.* 211, 254–286, 1992.

44. Hirschbein, B.L. and Fearon, K.L., ^{31}P NMR spectroscopy in oligonucleotide research and development, *Antisense Nucleic Acid Drug Dev.* 7, 55–61, 1997.

45. Gaus, H., Olsen, P., Sooy, K.V. et al., Trichloroacetaldehyde modified oligonucleotides, *Bioorg. Med. Chem. Lett.* 15, 4118–4124, 2005.

46. Cieślak, J., Ausín, C., Chmielewski, M.K. et al., ^{31}P NMR study of the desulfurization of oligonucleoside phosphorothioates effected by "aged" trichloroacetic acid solutions, *J. Org. Chem.* 70, 3303–3306, 2005.

47. Abdelkafi, M., Ghomi, M., Turpin, P.Y. et al., Common structural features UUCG and UACG tetraloops in very short hairpins determined by UV absorption, Raman, IR, and NMR spectroscopies, *J. Biomol. Struct. Dyn.* 14, 579–593, 1997.

48. Macdonald, P.M, Damha, M.J., Ganeshan, K. et al., Phosphorus 31 solid state NMR characterization of oligonucleotides covalently bound to a solid support, *Nucleic Acids Res.* 24, 2868–2876, 1996.

49. Iturralde. M.P., *Dictionary and Handbook of Nuclear Medicine and Chemical Imaging*, CRC Press, Boca Raton, FL, 1990.

50. Brown, M.A. and Semelka, R.P., *MRI Basic Principles and Applications*, Wiley-Liss, New York, 1995.

51. Buxton, R.B., *Introduction to Functional Magnetic Resonance Imaging*, Cambridge University Press, Cambridge, U.K., 2002.

52. Ryan, T.A., Myers, J., Holowka, D. et al., Molecular crowding on the cell surface, *Science* 239, 61–64, 1988.

53. Minton, A.P., Molecular crowding: An analysis of effects of high concentration of inert cosolutes on biochemical equilibria and rates in terms of volume exclusion, *Methods Enzymol.* 295, 127–149, 1998.

54. Rösgen, J., Molecular crowding and solvation: Direct and indirect impact on protein reactions, *Methods Mol. Biol.* 490, 195–225, 2009.

55. Bernadó, P., Garcia de la Torre, J., and Pons, M., Macromolecular crowding in biological systems: Hydrodynamics and NMR methods, *J. Mol. Recognit.* 17, 397–407, 2004.

56. Latham, M.P., Brown, D.J., McCallum, S.A., and Pardi, A., NMR methods for studying the structure and dynamics of RNA, *Chembiochem* 6, 1492–1505, 2005.

57. García-Pérez, A.I., López-Beltrán, E.A., Klüner, P. et al., Molecular crowding and viscosity as determinants of translational diffusion of metabolites in subcellular organelles, *Arch. Biochem. Biophys.* 362, 329–338, 1999.

58. Belayev, L., Zhao, W., Pattany, P.M. et al., Diffusion-weighted magnetic resonance imaging confirms marked neuroprotective efficacy of albumin therapy in focal cerebral ischemia, *Stroke* 29, 2587–2899, 1998.

59. Laurent, S., Elst, L.V., and Muller, R.N., Comparative study of the physicochemical properties of six clinical low molecular weight gadolinium contrast agents, *Contrast Media Mol. Imaging* 1, 128–137, 2006.

60. Yee, A.A., Savchenko, A., Ignachenko, A. et al., NMR and x-ray crystallography, complementary tools in structural proteomics of small proteins, *J. Am. Chem. Soc.* 127, 16512–16517, 2005.

61. Doerr, A., Together at last: Crystallography and NMR, *Nat. Methods* 3, 6, 2006.

62. Damm, K.L. and Carlson, H.A., Exploring experimental sources of multiple protein conformations in structure-based drug design, *J. Am. Chem. Soc.* 129, 8225–8235, 2007.

63. Yan, Z., Zhao, N., Wang, Z. et al., A mutated human tumor necrosis factor-α improves the therapeutic index in vitro and in vivo, *Cytotherapy* 8, 415–423, 2006.

64. McDonagh, C.F., Kim, K.M., Turcott, E. et al., Engineered anti-CD70 antibody-drug conjugate with increased therapeutic index, *Mol. Cancer Ther.* 7, 2913–2923, 2008.

65. Brunetti-Pierri, N., Grove, N.C., Zuo, Y. et al., Bioengineered factor IX molecules with increased catalytic activity improve the therapeutic index of gene therapy vectors for hemophilia B, *Hum. Gene Ther.* 20, 479–485, 2009.

66. Hancock, T.J. and Hsu, J.T., Thermal denaturation of lysozyme in a cosolvent studied by NMR, *Biotechnol. Prog.* 12, 494–502, 1996.

67. Andersson, C., Edlund, P.O., Gellerfors, P. et al., Isolation and characterization of a trisulfide variant of recombinant human growth hormone formed by during expression in *Escherichia coli*, *Int. J. Pept. Protein Res.* 47, 311–321, 1996.

68. Blommers, M.J.J. and Cerletti, N., High resolution NMR, a useful tool to control batch-to-batch consistency of disulfide bonds in biopharmaceuticals: Application to transforming growth factor-β3, *Pharm. Sci.* 3, 29–36, 1997.

69. Yoshioka, S., Aso, Y., Nakai, Y., and Kojima, S., Effect of high molecular mobility of poly (vinyl alcohol) on protein stability of lyophilized gamma-globulin formulation, *J. Pharm. Sci.* 87, 147–151, 1998.

70. Doucey, M.-A., Hess, D., Blommers, M.J.J., and Hofsteenge, J., Recombinant human interleukin-12 in the second example of C-mannosylated protein, *Glycobiology* 9, 435–441, 1999.

71. Hsieh, C.L., Characterization of saccharide-CRM197 conjugate vaccines, *Dev. Biol.* 103, 93–104, 2000.

72. Ho, M.M., Lemercinier, X., Bolgiano, B., and Corbel, M.J., Monitoring stability of meningococcal group C conjugate vaccines: Correlation of physico-chemical methods and immunogenicity assays, *Dev. Biol.* 103, 139–150, 2000.

73. Lowe, C.R., Lowe, A.R., and Gupta, G., New developments in affinity chromatography with potential application in the production of biopharmaceuticals, *J. Biochem. Biophys. Methods* 49, 561–576, 2001.

74. Larson, R.S., Menard, V., Jacobs, H., and Kim, S.W., Physicochemical characterization of poly(ethylene glycol)-modified anti-GAD antibodies, *Bioconjug. Chem.* 12, 861–869, 2001.

75. Digilio, G., Barbero, L., Bracco, C. et al., NMR structure of two novel polyethylene glycol conjugates of the human growth hormone-releasing factor, hGRF(1–29)-NH$_2$, *J. Am. Chem. Soc.* 125, 3458–3470, 2003.

76. Lago, G., Oruña, L., Cremata, J.A. et al., Isolation, purification and characterization of hyaluronan from human umbilical cord residues, *Carbohydr. Polymers* 62, 321–326, 2005.

77. Župerl, Š., Pristovšek, P., Menart, V. et al., Chemometric approach in quantification of structural identity/similarity of proteins in biopharmaceuticals, *J. Chem. Inf. Model.* 47, 737–743, 2007.

78. Balan, S., Choi, J.-W., Godwin, A. et al., Site-specific PEGylation of protein disulfide bonds using a three-carbon bridge, *Bioconjug. Chem.* 18, 61–76, 2007.

79. Bogomolovas, J., Simon, B., Sattler, M., and Stier, G., Screening of fusion partners for high yield expression and purification of bioactive viscotoxins, *Protein Expr. Purif.* 64, 16–23, 2009.

80. Zhou, D.H., Shah, G., Mullen, C. et al., Proton-detected solid-state NMR spectroscopy of natural-abundance peptide and protein pharmaceuticals, *Angew. Chem. Int. Ed.* 48, 1253–1256, 2009.

16 Use of Chemical Probes for the Study of Protein Conformation

This chapter will discuss the use of noncovalent fluorescent probes for the study of biopolymer conformation. The use of dansyl chloride (1-dimethylamino-naphthalene-5-sulfonyl chloride; DNS-Cl) (Figure 16.1) for the modification of albumin appears to the first use of an introduced chemical as a conformational probe.[1] This paper was one of several by Gregorio Weber and associates[1-3] concerning the development of fluorescent probes from for the sturdy of protein conformation. Chen[4] also proposed the use of dansylated amino acids as a noncovalent fluorescent probe (Figure 16.1) as such a derivative is structurally similar to the naphthalene sulfonates, 8-anilino-1-naphthalene sulfonate (ANS), and 2-toluidinyl-6-naphthalene sulfonate (TNS) described later (Figure 16.2). The fluorescence of dansyl amino acids such as dansyl-DL-tryptophan is markedly increased in organic solvents (0.54 in ethyl acetate) compared to 0.068 in water with a blueshift in the optimum wavelength for emission. Fluorescence was also enhanced with bovine serum albumin (relative fluorescence intensity with dansyl-DL-tryptophan of 55.5) but not with lysozyme (0.2) or trypsin (0.0). It is of interest that the driving force for fluorescence enhancement was the dansyl function as the relative fluorescence intensity was 61.0 for methionine, 43.4 for valine, 41.4 for glutamate, 68.0 for glycine, and 71.0 for proline.

Reporter groups are chemical probes which are designed to reflect local environments (most in proteins) and have been used for a variety of purposes[5-26] (Figures 16.3 through 16.7). The current work includes fluorescent probes as reporter groups (see also Chapters 18, 19, and 23). The current chapter will focus on noncovalent fluorophore probes; there will be a brief mention of covalent probes but there will be more discussion in Chapter 19.

Information regarding Molecular Probes should necessarily be introduced at the start of a discussion of fluorophore reagents. Molecular Probes is now part of Invitrogen[27] but was for many years an independent company located in Eugene, Oregon. Dr. Richard Haugland and his associates made major contributions to the development and application of fluorescent probes (Figure 16.8) for over more than 40 years.[28,29] Fluorescent probes can be separated into those which binding to biopolymers via noncovalent interaction and those that bind via covalent linkage to a functional group on the biopolymer. The use of fluorescent probes is enhanced by changes in properties when bound to a polymer as well as by changes in their polarization properties as elucidated by Weber.[2] Fluorescein was used early as an in

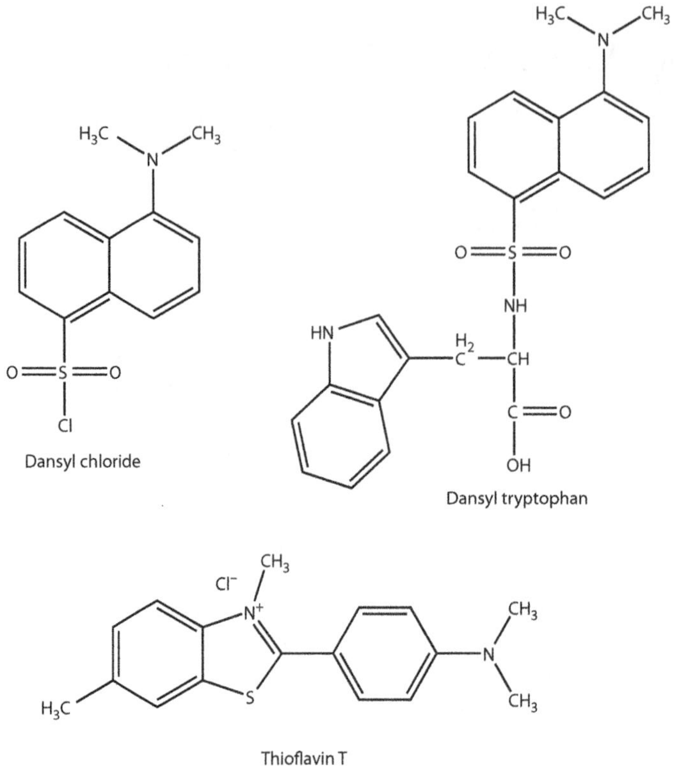

FIGURE 16.1 The structure of dansyl chloride, dansyl tryptophan, and thioflavin T. Thioflavin T is used for the staining of amyloid fibrils (see Hawe, A., Sutter, M., and Jiskoot, W., Extrinsic fluorescent dyes as tools for protein characterization, *Pharmaceut. Res.* 35, 1487–1499, 2008; Chen, R.F., Fluorescence of dansyl amino acids in organic solvents and protein solutions, *Arch. Biochem. Biophys.* 120, 609–620, 1967; Goto, Y., Yagi, H., Yamaguchi, K. et al., Structure, formation and propagation of amyloid fibrils, *Curr. Pharm. Des.* 14, 3205–3218, 2008).

vivo indicator of tissue perfusion[30] and subsequently for the preparation of fluorescent antibodies for immunocytochemistry.[31] Extensive use of fluorescent antibodies continues in immunochemistry.[32–35]

Noncovalent fluorophore probes bind to proteins (and other biological macromolecules) with considerable specificity and the binding is usually associated with a marked increase in fluorescence intensity. This property enables the application of noncovalent fluorophore probes to micromolar concentrations of test materials. Covalent fluorophore probes have the same enhancement of intensity but use is complicated by the addition of the chemical reaction between the dye molecule and the protein or other macromolecule. Covalent fluorophores are used more frequently as antibody-conjugates for signals in solid-phase assays[36,37] and for measurement of biomolecular interactions.[38–48]

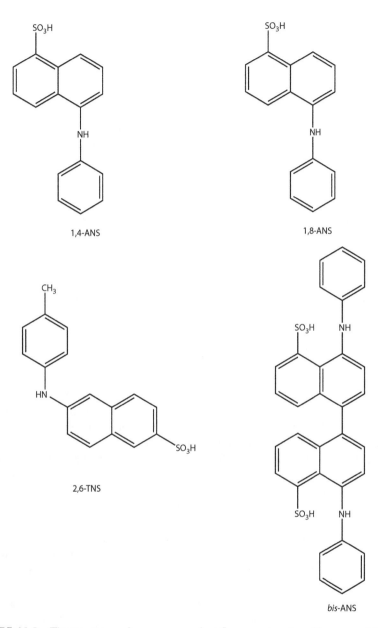

1,4-ANS

1,8-ANS

2,6-TNS

bis-ANS

FIGURE 16.2 The structures of some noncovalent fluorescent probes. Shown are 1,4-ANS (1-anilinonaphthalene-4-sulfonate); 1,8-ANS (1-anilinonaphthalene-8-sulfonate); and 2,6-TNS (2-*p*-toluidinylnaphthalene-6-sulfonate) (see Edelman, G.M. and McClure, W.O, Fluorescent probes and the conformation of proteins, *Acc. Chem. Res.* 1, 65–69, 1968; Cody, V. and Hazel, J, Molecular conformation of ammonium 8-anilino-1-naphthalenesulfonate hemihydrate. A fluorescent probe for thyroxine binding to thyroxine binding globulin, *J. Med. Chem.* 20, 12–17, 1977; Hawe, A., Sutter, M., and Jiskoot, W., Extrinsic fluorescent dyes as tools for protein characterization, *Pharm. Res.* 25, 1487–1499, 2008; Albani, J.-R., Fluorescence origin of 6,P-toluidinyl-naphthalene-2-sulfonate (TNS) bound to proteins, *J. Fluoresc.*, in press, 2009).

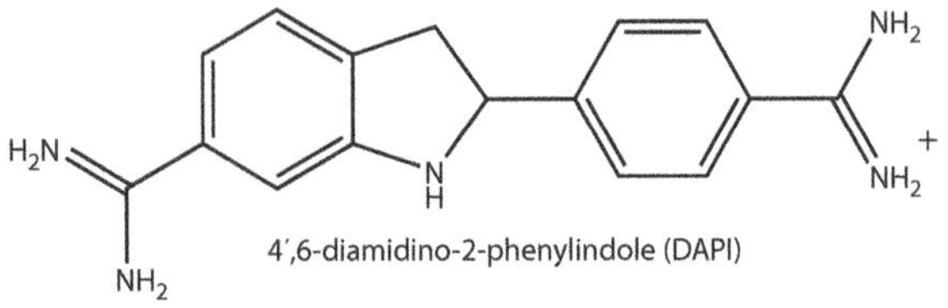

4′,6-diamidino-2-phenylindole (DAPI)

FIGURE 16.3   The structure of 4′,6-diamidino-2-phenylindole (DAPI). DAPI is used as a DNA probe and is also a probe of pepsin conformation (see Mazzini, A., Incerti, M., and Favilla, R., Interaction of DAPI with pepsin as a function of pH and ionic strength, *Biophys. Chem.* 67, 65–74, 1997).

There has been sustained interest in the use of fluorescent probes to study protein in membranes.[49–52] However, in keeping with the somewhat narrow focus of the current work, the current discussion is limited to the consideration of the application of well-known reagents to the study of biopharmaceutical products. Noncovalent fluorophore probes[53,54] are valuable for the study of protein biopharmaceuticals for two major reasons: first, these probes bind to proteins (and other macromolecules) with considerable specificity; second, the binding of the probe changes spectral characteristics.[55] As a general concept, the fluorophore probes of interest, ANS and TNS (Figure 16.2) are poorly soluble in water and bind to hydrophobic regions of proteins with an increase in fluorescence[56,57]; fluorescence also increases in hydrophobic solvents.

TNS is used less than ANS for the study of biopolymer conformation. Albani has reviewed the mechanism of fluorescence increase observed on the binding of TNS to proteins.[57] Albani compared the fluorescence lifetimes of TNS (425 nm) in water, ethanol, and bound to different proteins. Mean fluorescence lifetime was similar in water and ethanol but increased when bound to proteins. Fluorescence intensity is markedly increased in either ethanol or when bound to BSA as compared to water. The results from these studies suggest that the structure of TNS is different when bound to protein as compared to solution in ethanol. McClure and Edelman[58] have also studied the fluorescence of TNS in organic solvents.

Wang and Edelman used TNS to study the conformation change(s) associated with the conversion of pepsinogen to pepsin.[59] They observed an increase in TNS fluorescence during the activation of pepsinogen, which correlated with the appearance of proteolytic activity. Horowitz and colleagues used TNS to document the thermal transitions of human complement component C1q.[60] There is an decrease in fluorescence intensity as temperature increases from 0°C to 50°C; above 50°C, there is marked decrease in activity concomitant with an increase in fluorescence. These investigators also noted changes in intrinsic fluorescence supporting a temperature-dependent conformational change. Steiner and colleagues used changes in TNS fluorescence (increase) to measure the association between plasmin and α_2-macroglobulin.[61] These investigators subsequently reported on a similar change in TNS fluorescence on the reaction of thrombin with α_2-macroglobulin.[62] Wasylewski and colleagues used TNS to show that conformation changes occurred in yeast hexokinase during the transition from 0°C to 40°C (thermal transition).[63]

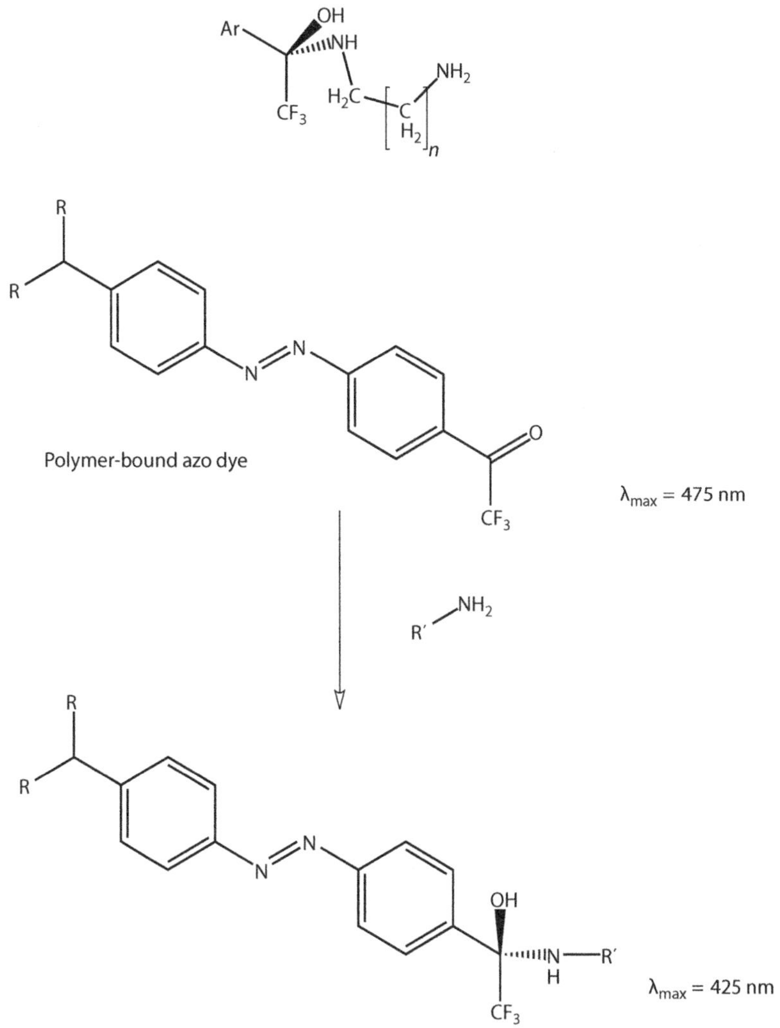

FIGURE 16.4 The structure of a polymer-bound amine reporter group. Reaction with a primary amine is shown; reaction also occurs with secondary and tertiary amine. The formation of the carbinolamine is associated with a 50 nm blueshift in absorbance (see Mohr, G.J., Demuth, C., and Spichiger-Keller, U.E., Applications of chromogenic and fluorogenic reactants in the optical sensing of dissolved aliphatic amines, *Anal. Chem.* 70, 3868–3873, 1998; Mertz, E., Bell, J.B., and Zimmerman, S.C., Kinetics and thermodynamics of amine and diamine signaling by a trifluoroacetyl azobenzene reporter group, *Org. Lett.* 5, 3127–3130, 2003).

There was a decrease in fluorescence intensity of TNS with a blueshift (438 → 420 nm) at 35°C; there was a change in intrinsic fluorescence (decrease intensity with a redshift from 330 to 340 nm), which was poorly correlated with the change in extrinsic fluorescence. It is noted that a blueshift is associated with an increase in

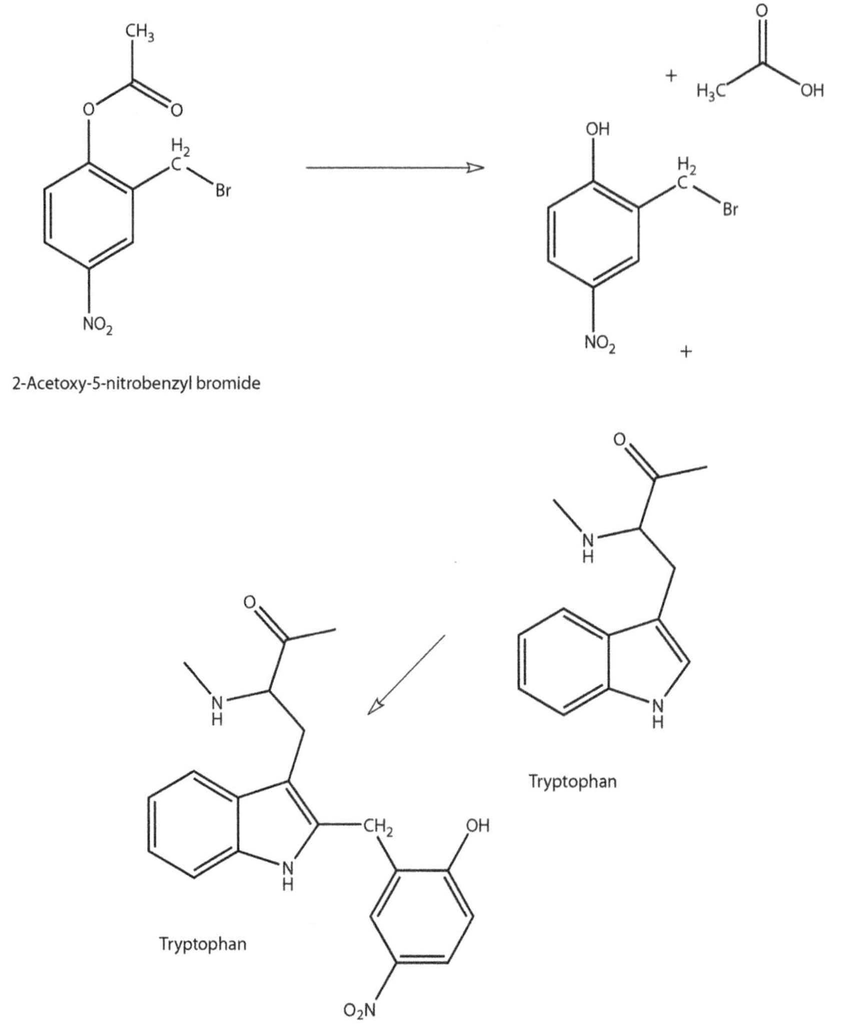

2-Acetoxy-5-nitrobenzyl bromide

Tryptophan

Tryptophan

FIGURE 16.5   The structure of 2-acetoxy-5-nitrobenzyl bromide. This reporter group is generated by the enzymatic hydrolysis of the acetyl function which generates the active reagent (see Uhteg, L.C. and Lundblad, R.L., The modification of tryptophan in bovine thrombin, *Biochim. Biophys. Acta* 491, 551–557, 1977).

hydrophobicity, while a redshift is associated with a decrease in hydrophobicity.[64] It is useful at this juncture to note that the binding of probes such as TNS and ANS to proteins is at least in part electrostatic in nature.[65–67] It is noted that the binding site on calmodulin for another hydrophobic probe, auramine O, a cationic dye (Figure 16.9) is different from that for TNS.[68] The binding of auramine O to proteins has been studied in some detail by Chen.[69] As with TNS and ANS, auramine O is weakly fluorescent in water (quantum yield $= 4 \times 10^{-5}$) but demonstrates increased

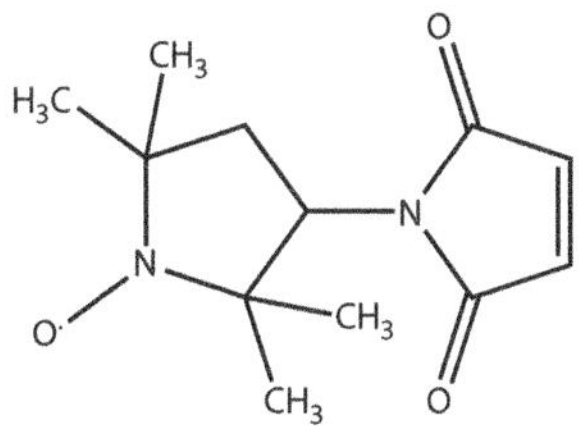

3-Maleimido-2,2,5,5-tetramethyl-1-pyrrolidinyloxyl

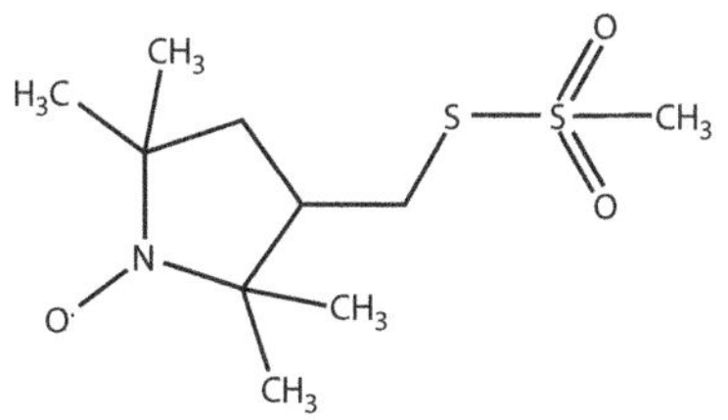

1-Oxyl-2,2,5,5-tetramethylpyrolidin-3-yl methyl methanethiosulfonate

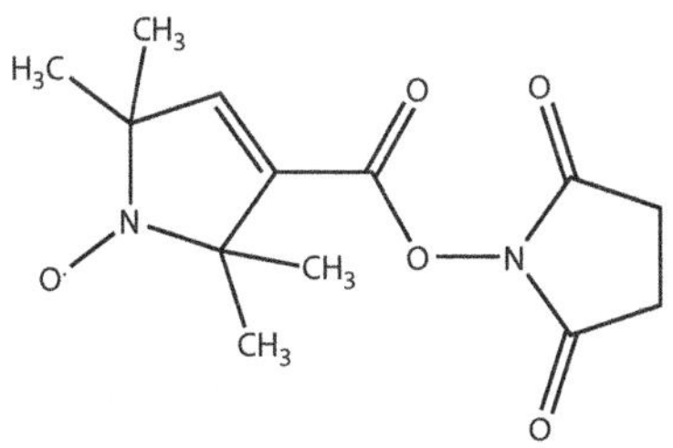

2,2,5,5-Tetramethyl-3-pyrrolin-1-oxyl-3-carboxylic acid   N-hydroxysuccinimide

FIGURE 16.6 Some spin-label reagents for protein (see Song, Y., Means, G.E., Wan, X., and Berliner, L.J., A spin label study of immobilized enzyme spectral subpopulations, *Biotechnol. Bioeng.* 40, 306–312, 1992; Chaudhuri, D., Narayan, M., and Berliner, L.J., Conformation-dependent interaction of α-lactalbumin with model and biological membranes: A spin-label ESR study, *Protein J.* 23, 95–101, 2004; Berliner, L.J., *Spin Labeling*, Kluwer Academic, Boston, MA, 2002).

fluorescence in nonpolar solvents (5.1×10^{-4} in *n*-butanol) and more so when bound to protein (2.5×10^{-2} with human serum albumin). The binding of auramine O has been used to study the conformation of glyceraldehyde 3-phosphate dehydrogenase,[70] actin,[71] and citrate binding to calmodulin.[72] Auramine O also improves the sensitivity of the Lowry protein assay.[73] More recent studies have used TNS in fluorescence resonance energy transfer (FRET) studies on the conformation of human albumin.[74]

The use of ANS to study protein conformation can be traced to the study by Weber and Young on the acid-induced change in the conformation of bovine serum albumin.[75] These investigators first demonstrated that the polarization of

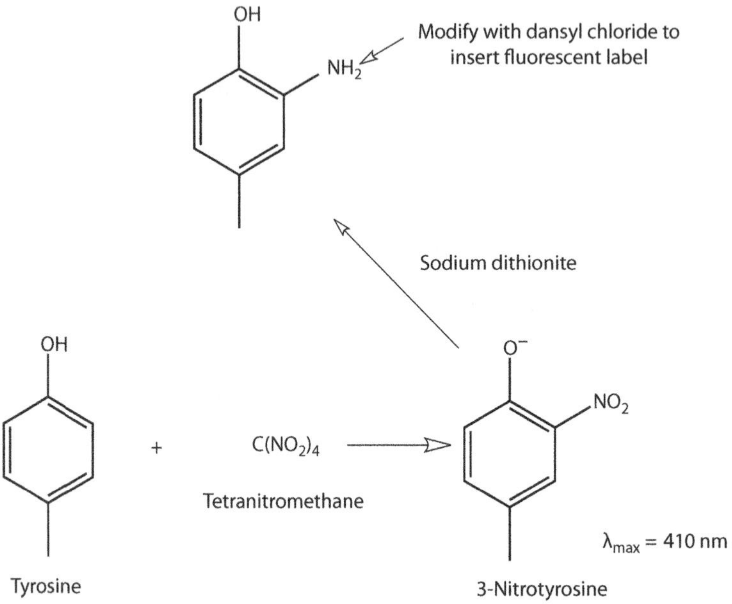

FIGURE 16.7   The reaction of tetranitromethane with tyrosine in proteins. The nitrated tyrosine residue has an absorption maximum at 410 nm which is pH-dependent (maximal absorbance is at alkaline pH) (see Di Bello, C. and Griffin, J.H., Circular dichroism and absorbance properties of nitrotyrosyl chromophores in staphylococcal nuclease and in a model diketopiperazine, *J. Biol. Chem.* 250, 1445–1450, 1975; Scherrer, P. and Stoeckenius, W., Selective nitration of tyrosines-26 and -64 in bacteriorhodopsin with tetranitromethane, *Biochemistry* 23, 6195–6202, 1984; Lee, S.B., Inouye, K., and Tonomura, B., The states of tyrosyl residues in thermolysin as examined by nitration and pH-dependent ionization, *J. Biochem.* 121, 231–237, 1997). Nitration of tyrosyl residues with peroxynitrite results in the same chemical modification (see Cassina, A.M., Hodara, R., Souza, J.M. et al., Cytochrome c nitration by peroxynitrite, *J. Biol. Chem.* 275, 21409–21415, 2000).

the fluorescence of dansyl-albumin decreased on digestion with pepsin. Other experiments showed that the polarization increased with a pH increase from 2 to 5 with the intact protein and not with the pH 3 pepsin digest. These investigators also reported that the quantum yield for ANS fluorescence (excitation at 365 nm) markedly increased on binding to albumin (0.004 in water; 0.75 with albumin). Chance and coworkers subsequently used ANS to demonstrate functional integrity of mitochondrial membranes.[76–78] ANS and *bis*-ANS continue to be used for the study of membrane and artificial membranes[79–83] as a probe of hydrophobicity.[84,85]

The mode of binding of ANS (and TNS) to proteins is complex[55] and likely involves electrostatic interactions as well hydrophobic interactions.[85] Emphasis

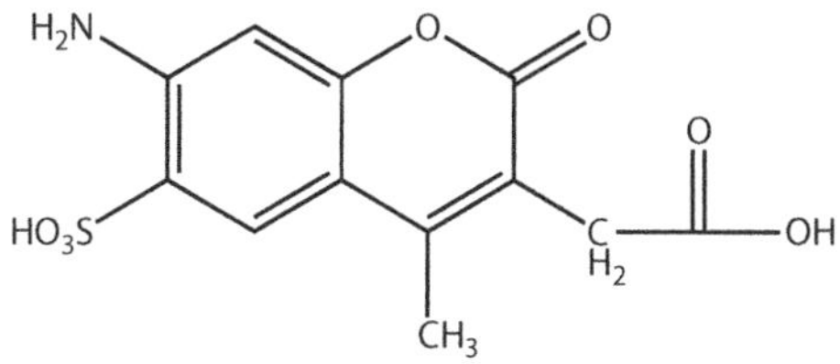

7-Amino-4-methyl-6-sulfocoumarin-3-acetic acid (AMCA-S), Alexa™ 350

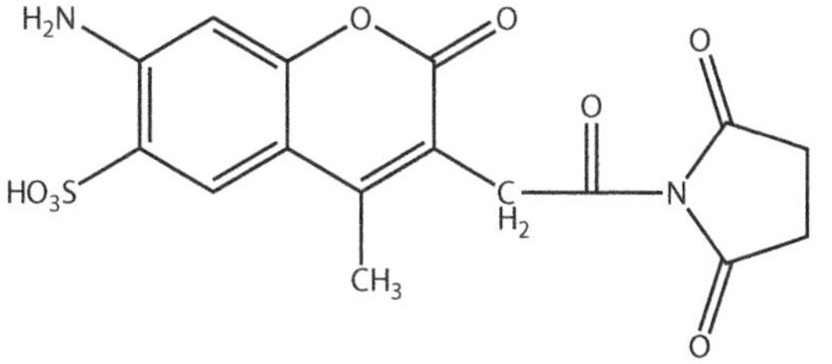

7-Amino-4-methyl-6-sulfocoumarin-3-acetic acid, succinimidyl ester

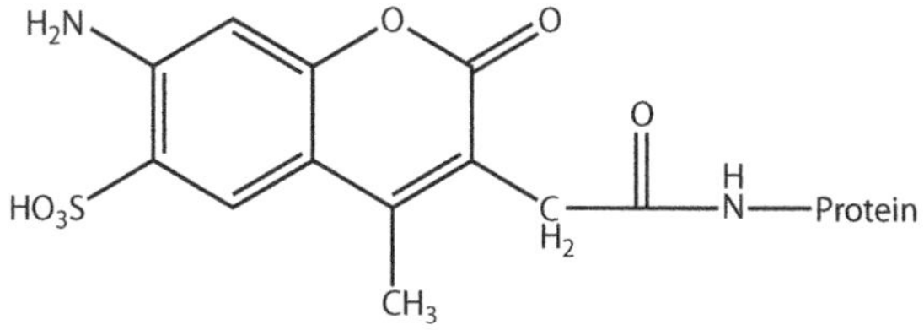

FIGURE 16.8   The structure of Alexa™ 350, a fluorescent reporter group for protein modification. Shown is the parent coumarin derivative, the *N*-hydroxysuccinimide derivative and the product of the reaction with the amino group of a protein. The succinimide derivative will react with any unprotonated amine (see Leung, W.-Y., Trobridge, P.A., Haugland, R.P. et al., 7-Amino-4-methyl-6-sulfocoumarin-3-acetic acid: A novel blue fluorescent dye for protein labeling, *Bioorg. Med. Chem.* 9, 2229–2232, 1999; Lee, S.B., Hassan, M., Fisher, R. et al., Affibody molecules for in vivo characterization of HER2-positive tumors by near infra-red imaging, *Clin. Cancer Res.* 14, 3840–3849, 2008; Harikumar, K.G., Gao, F., Pinon, D.I., and Miller, L.J., Use of multidimensional fluorescence resonance energy transfer to establish the orientation of cholecystokinin docked at the type A cholecystokin receptor, *Biochemistry* 47, 9574–9581, 2008).

is placed on the hydrophobic aspect of ANS binding as it is this property that is responsible for the generation of signal. ANS does bind to lysozyme[65,86] as well as polylysine, polyhistidine, and polyarginine.[86] The binding is not associated with the same quality of fluorescence increase as observed for β-lactoglobulin.[87] Ichimura and Zama[88] did observe an increase in ANS fluorescence with polyarginine, which

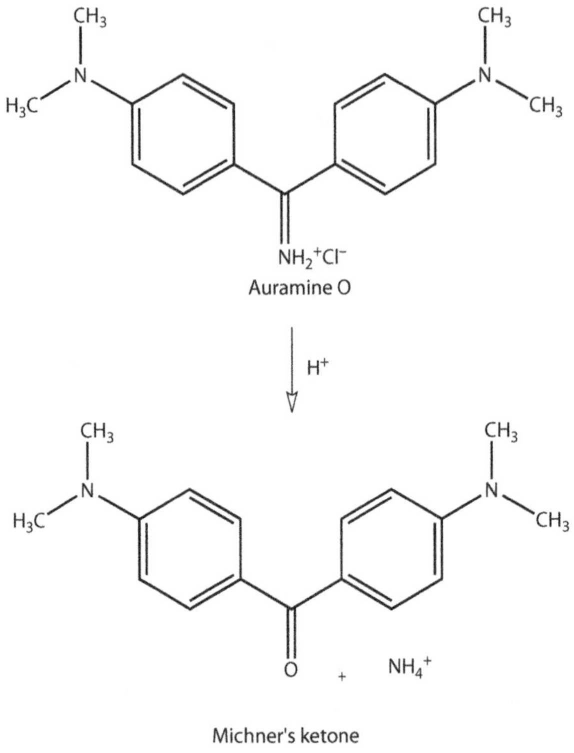

FIGURE 16.9 The structure of auramine O and the acid-catalyzed hydrolysis forming Michner's ketone. Auramine O is a cationic dye which forms fluorescent complexes with proteins (see Conrad, R.H., Heitz, J.R., and Brand, L., Characterization of a fluorescent complex between auramine O and horse liver alcohol dehydrogenase, *Biochemistry* 9, 1540–1546, 1970; Chen, R.F., Fluorescence of free and protein-bound auramine O, *Arch. Biochem. Biophys.* 173, 672–681, 1977). The hydrolysis to Michner's ketone is shown; this reaction has been demonstrated to be catalyzed by heparin (Band, P. and Lukton, A., Catalytic influence of heparin on auramine O hydrolysis: A basis for differentiating heparin from other glycosaminoglycans based on its properties as a polyelectrolyte, *Biopolymers* 23, 2223–2241, 1984).

they suggested was due to a conformational change in the polyarginine driven by the binding of ANS. Modification of lysine residues (carbodiimide-mediated coupling of anthraquinone 2-carboxylic acid) decreased ANS fluorescence in horse radish peroxidase.[89] Gasymov and Glasgow have shown that ANS binding to arginine or lysine via ion part with fluorescent enhancement and hypsochromic (blue) shift in the emission spectra.[90]

The majority of studies using ANS are directed at the study of protein folding.[52,91–99] There has been less interest in the use of ANS for protein characterization although the necessary basic information is available.[100–104] Table 16.1 contains a selection of studies, which could be of value for the characterization of biopharmaceutical materials.

TABLE 16.1
Selected Studies on the Use of 1-Anilino-8-Naphthelenesulfonate (ANS) for Characterization of Biopharmaceuticals

Biopolymer	Study	Ref.
Human serum albumin	Use of ANS fluorescence, far-UV CD, and intrinsic fluorescence to study the urea denaturation of human serum albumin	[105]
A$_1$-Antichymotrypsin	ANS fluorescence and CD spectroscopy were used to study the denaturation of α_1-antichymotrypsin by guanidine hydrochloride	[106]
Human blood coagulation factor VIII	Use of *bis*-ANS fluorescence, CD spectroscopy, and DSC to study the aggregation of human blood coagulation factor VIII	[107]
Bovine serum albumin and fibrinogen	Use of ANS fluorescence and intrinsic fluorescence was used to the conformational changes of proteins adsorbed on siloxanes. There was a change in fibrinogen conformation in the presence of both short-chain and linear siloxanes but change was not observed with albumin	[108,109]
Ovalbumin, ovotransferrin, lysozyme	ANS fluorescence, intrinsic fluorescence, and CD spectroscopy were used to study conformational changes of proteins at air–water interfaces	[110]
Soybean Kunitz trypsin inhibitor	ANS fluorescence, CD spectroscopy, and limited proteolysis were used to study the denaturation of soybean Kunitz trypsin inhibitor in the presence of acid and thermal denaturation	[111]
Human serum albumin	ANS fluorescence and domain-specific ligands were used to study the guanidine hydrochloride denaturation of human serum albumin	[112]
Insulin	ANS fluorescence was used to identify hydrophobic regions in insulin protofibrils	[113]
Interferon-β-1a	ANS fluorescence, FTIR spectroscopy, far-UV CD spectroscopy, and derivative UV-VIS spectroscopy were used to characterize solution stability of interferon-β-1a for formulation studies	[114]
Human insulin	ANS fluorescence was used to characterize the fibrillation of human insulin	[115]
Fibroblast growth factor (FGF)	ANS fluorescence, intrinsic fluorescence, far-UV CD spectroscopy and derivative UV-VIS spectroscopy were used to evaluate the effect of pH and polyanions on structural stability of FGF under thermal stress	[116]
Hemoglobin	Increase in hydrophobicity of hemoglobin on glycation with fructose was measured with ANS fluorescence and far-UV CD demonstration of increased β-structure	[117]
Bovine serum albumin	ANS fluorescence, intrinsic fluorescence, turbidity and Rayleigh scattering were used to study the thermal aggregation of bovine serum albumin; comparison is made with human serum albumin	[118]

(*continued*)

TABLE 16.1 (continued)
Selected Studies on the Use of 1-Anilino-8-Naphthelenesulfonate (ANS) for Characterization of Biopharmaceuticals

Biopolymer	Study	Ref.
A-Chymotrypsin	ANS fluorescence was used to demonstrate the appearance of hydrophobic patches on the thermal denaturation of α-chymotrypsin. There was an increase in fluorescence intensity and a blueshift. Other techniques used including turbidity and changes in intrinsic fluorescence	[119]
Human IgG	*bis*-ANS fluorescence was used as an online fluorescent detection method (HPLC-SEC and asymmetrical flow field flow fractionation) for the analysis of human IgG subjected to thermal stress (heat-stressed formulations)	[120]
Human IgG	ANS fluorescence was used to study IgG thermal stress with isothermal calorimetry	[121]

REFERENCES

1. Weber, G., Polarization of the fluorescence of macromolecules. 2. Fluorescent conjugates of ovalbumin and bovine serum albumin, *Biochem. J.* 51, 155–167, 1952.
2. Weber, G., Polarization of the fluorescence of macromolecules. 1. Theory and experimental method, *Biochem. J.* 51, 145–155, 1952.
3. Laurence, D.J.R., A study of the adsorption of dyes on bovine serum albumin by the method of polarization of fluorescence, *Biochem. J.* 51, 168–180, 1952.
4. Chen, R.F., Fluorescence of dansyl amino acids in organic solvents and protein solutions, *Arch. Biochem. Biophys.* 120, 609–620, 1967.
5. Kirtley, M.E. and Koshland, D.E. Jr., The introduction of a "reporter" group at the active site of glyceraldehyde-3-phosphate dehydrogenase, *Biochem. Biophys. Res. Commun.* 23, 810–815, 1966.
6. Loudon, G.M. and Koshland, D.E. Jr., The chemistry of a reporter group: 2-Hydroxy-5-nitrobenzyl bromide, *J. Biol. Chem.* 245, 2247–2254, 1970.
7. Stallcup, W.B. and Koshland, D.E. Jr., Reactive lysines of yeast glyceraldehyde 3-phosphate dehydrogenase. Attachment of a reporter group to a specific non-essential residue, *J. Mol. Biol.* 80, 63–75, 1973.
8. Zukin, R.S., Hartig, P.R., and Koshland, D.E. Jr., Use of a distant reporter group as evidence for a conformational change in a sensory receptor, *Proc. Natl. Acad. Sci. USA* 74, 1932–1936, 1977.
9. White, D.D., Stewart, S., and Wood, G.C., The use of reporter group circular dichroism in the study of conformational transitions in bovine serum albumin, *FEBS Lett.* 33, 305–310, 1973.
10. Yagisawa, S., Mercuri-nitrophenol as a reporter for the conformational change of hemoglobin, *J. Biochem.* 77, 595–604, 1975.
11. Reisfeld, A., Rothenberg, J.M., Bayer, E.A., and Wilchek, M., Nonradioactive hybridization probes prepared by the reaction of biotin hydrazide with DNA, *Biochem. Biophys. Res. Commun.* 142, 519–526, 1987.

12. Conway, N.E., Fidanza, J., and McLaughlin, L.W., The introduction of reporter groups at multiple and/or specific sites in DNA containing phosphorothioate diesters, *Nucleic Acids Symp. Ser.* (21), 43–44, 1989.
13. Agrawal, S. and Zamecnik, P.C., Site specific functionalization of oligonucleotides for attaching two different reporter groups, *Nucleic Acids Res.* 18, 5419–5423, 1990.
14. Uziel, M. and Houck, K., Direct labeling of DNA-adducts formed from carcinogenic diol-epoxides with a fluorescent reporter compound specific for the *cis vic*-diol group, *Biochem. Biophys. Res. Commun.* 180, 1233–1240, 1991.
15. Skawinski, W.J., Adebodun, F., Cheng, J.T. et al., Labeling of tyrosines in proteins with [^{15}N]tetranitromethane, a new NMR reporter for nitrotyrosines, *Biochim. Biophys. Acta* 1162, 297–308, 1993.
16. Sloop, F.V., Brown, G.M., Foote, R.S. et al., Synthesis of 3-(triethylstannyl)propanoic acid: An organotin mass label for DNA, *Bioconjug. Chem.* 4, 406–409, 1993.
17. Wang, R., Sun, S., Bekos, E.J., and Bright, F.V., Dynamics surrounding Cys-34 in native, chemically denatured, and silica-adsorbed bovine serum albumin, *Anal. Chem.* 67, 145–159, 1995.
18. Jordan, J.D., Dunbar, R.A., and Bright, F.V., Dynamics of acrylodan-labeled bovine and human serum albumin entrapped in a sol-gel-derived biogel, *Anal. Chem.* 67, 2436–2443, 1995.
19. Favilla, R., Parisola, A., and Mazzini, A., Alkaline denaturation and partial refolding of peptide investigated with DAPI as an extrinsic probe, *Biophys. Chem.* 67, 75–83, 1997.
20. Owen, R.M., Gestwicki, J.E., Young, T., and Kiessling, L.L., Synthesis and applications of end-labeled neoglycopolymers, *Org. Lett.* 4, 2293–2296, 2002.
21. Mertz, E. and Zimmerman, S.C., Cross-linked dendrimer hosts containing reporter groups for amine guests, *J. Am. Chem. Soc.* 125, 3424–3425, 2003.
22. Mertz, E., Beil, J.B., and Zimmerman, S.C., Kinetics and thermodynamics of amine and diamine signaling by a trifluoroacetyl azobenzene reporter group, *Org. Lett.* 5, 3127–3130, 2003.
23. O'Neil, L.L. and Wiest, O., A selective, noncovalent assay for base flipping in DNA, *J. Am. Chem. Soc.* 127, 16800–16801, 2005.
24. Grossmann, T.N. and Seitz, O., DNA-catalyzed transfer of a reporter group, *J. Am. Chem. Soc.* 128, 15596–15597, 2006.
25. Samuel, A.P., Moore, E.G., Melchior, M. et al., Water-soluble 2-hydroxyisophthalamides for sensitization of lanthanide luminescence, *Inorg. Chem.* 47, 7535–7544, 2008.
26. Steed, J.W., Coordination and organometallic compounds as anion receptors and sensors, *Chem. Soc. Rev.* 38, 506–519, 2009.
27. Molecular Probes at Invitrogen http://www.invitrogen.com/site/us/en/home/References/Molecular-Probes-The-Handbook.html
28. Stryer, L. and Haugland, R.P., Energy transfer: A spectroscopic ruler, *Proc. Natl. Acad. Sci. USA* 58, 719–726, 1967.
29. Berlier, J.E., Rothe, A., Buller, G. et al., Quantitative comparison of long-wavelength Alexa Fluor dyes to Cy dyes: Fluorescence of the dyes and their bioconjugates, *J. Histochem. Cytochem.* 51, 1699–1712, 2003.
30. Crismon, J.M. and Fuhrman, F.A., Studies on gangrene following cold injury. IV. The use of fluorescein as an indicator of local blood flow: Distribution of fluorescein in body fluids after intravenous injection, *J. Clin. Invest.* 26, 259–267, 1947.
31. Emmart, E.W., Observations on the absorption spectra of fluorescein, fluorescein derivatives and conjugates, *Arch. Biochem. Biophys.* 73, 1–8, 1958.
32. Hemmilä, I.A., *Applications of Fluorescence in Immunoassays*, Wiley, New York, 1991.

33. Bruchez, M.P. and Holz, C.Z., *Quantum Dots: Applications in Biology*, Humana Press, Totowa, NJ, 2007.
34. Hermanson, G.T., *Bioconjugate Techniques*, Elsevier Academic Press, Amsterdam, the Netherlands, 2008.
35. Predki, P.F., *Functional Protein Microarray in Drug Discovery*, CRC Press, Boca Raton, FL, 2007.
36. Nargessi, R.D., Shine, B., and Landon, J., Immunoassays for serum C-reactive protein employing fluorophore-reactants, *J. Immunol. Methods* 71, 17–24, 1984.
37. Koskinen, J.O., Vaarno, J., Meltola, N.J. et al., Fluorescent nanoparticles as labels for immunometric assay of C-reactive protein using two-photon excitation assay technology, *Anal. Biochem.* 328, 210–218, 2004.
38. Krisko, A., Piantanida, I., Kveder, M. et al., The effect of heparin on structural and functional properties of low density lipoproteins, *Biophys. Chem.* 119, 234–239, 2006.
39. Neuenschwander, P.F., Exosite occupation by heparin enhances the reactivity of blood coagulation factor IXa, *Biochemistry* 43, 2978–2986, 2004.
40. Hasegawa, G., Kikuchi, M., Kobayashi, Y., and Saito, Y., Synthesis and characterization of a novel reagent containing dansyl group which specifically alkylated sulfhydryl group: An example of application for chemistry, *J. Biochem. Biophys. Methods* 63, 33–42, 2005.
41. Shi, Q., Wang, X., and Ren, J., Biophysical characterization of the interaction of p21 with calmodulin: A mechanistic study, *Biophys. Chem.* 138, 138–143, 2008.
42. Yegneswaran, S., Hackeng, T.M., Dawson, P.E., and Griffin, J.H., The thrombin-sensitive region of protein S mediates phospholipid-dependent interaction with factor Xa, *J. Biol. Chem.* 283, 33046–33052, 2008.
43. Holmes-Farley, S.R. and Whitesides, G.M., Fluorescence properties of dansyl groups covalently bonded to the surface of oxidatively functionalized low density polyethylene film, *Langmuir* 2, 266–281, 1986.
44. Khan, M.I., Swamy, M.J., Sastry, M.V.K. et al., Saccharide binding to 3Gal/GalNAC specific lectins—Fluorescence, spectroscopic and stopped-flow kinetic studies, *Glycoconj. J.* 5, 75–84, 1988.
45. Kallmayer, H.J. and Schwarz, P., Dansylations of ethylenediamines I. Analytical reactions of ethylenediames, *Pharmazie* 44, 119–123, 1989.
46. Shea, K.J., Sasaki, D.Y., and Stoddard, G.J., Fluorescence probes for evaluating chain solvation in network polymers—An analysis of the solvatochromic shift of the dansyl probe in macroporous styrene divinylbenzene and styrene diisopropenylbenzene copolymers, *Macromolecules* 22, 1722–1730, 1989.
47. Ikeda, H., Nakamura, M., Ise, N. et al., Fluorescent cyclodextrins for molecule sensing: Fluorescent properties, NMR characterization and inclusion phenomena of *N*-dansylleucine-modified cyclodextrins, *J. Am. Chem. Soc.* 119, 10980–10988, 1996.
48. Stubbs, H.J., Lih, J.J., Gustafson, T.L., and Rice, K.G., Influence of core fucosylation on the flexibility of a biantennary N-linked oligosaccharide, *Biochemistry* 35, 937–947, 1996.
49. Nairn, R.C. and Rolland, J.M., Fluorescent probes to detect lymphocyte activation, *Clin. Exp. Immunol.* 39, 1–13, 1980.
50. Zhang, J., Campbell, R.E., Ting, A.Y., and Tsien, R.Y., Creating new fluorescent probes for cell biology, *Nat. Rev. Mol. Cell. Biol.* 3, 906–918, 2002.
51. Giermans, B.N., Adams, S.R., Ellisman, M.H., and Tsien, R.Y., The fluorescent toolbox for assessing protein location and function, *Science* 312, 217–224, 2006.
52. Munishkina, L.A. and Fink, A.L., Fluorescence as a method to reveal structures and membrane-interactions of amyloidogenic proteins, *Biochim. Biophys. Acta* 1768, 1862–1885, 2007.

53. Lakowicz, J.R., *Principles of Fluorescence Spectroscopy*, Springer-Verlag, Berlin, Germany, 2006.

54. Resch-Genger, U. ed., *Standardization and Quality Assurance in Fluorescence Measurements I. Techniques*, Springer-Verlag, Berlin, Germany, 2008.

55. Edelman, G.M. and McClure, W.O., Fluorescent probes and the conformation of proteins, *Acc. Chem. Res.* 1, 65–70, 1968.

56. Cody, V. and Hazel, J., Molecular conformation of ammonium 8-anilino-1-naphthalene-sulfonate hemihydrate. A fluorescent probe for thyroxine binding to thyroxine binding globulin, *J. Med. Chem.* 20, 12–17, 1997.

57. Albani, J.-R., Fluorescence origin of 6,P-toluidinyl-naphthalene-2-sulfonate (TNS) bound to proteins, *J. Fluoresc.* 19, 399–408, 2009.

58. McClure, W.O. and Edelman, G.M., Fluorescent probes for conformational states of proteins. I. Mechanisms of fluorescence of 2-p-toluidinylnaphthalene-6-sulfonate, a hydrophobic probe, *Biochemistry* 5, 1908–1919, 1966.

59. Wang, J.L. and Edelman, G.M., Fluorescent probes for the conformational states of proteins. IV. The pepsinogen-pepsin conversion, *J. Biol. Chem.* 246, 1185–1191, 1971.

60. Horowitz, P.M., Waylewski, Z., and Kolb, W.R., Fluorometric detection of low temperature thermal transitions in the C1Q component of human complement, *Biochem. Biophys. Res. Commun.* 96, 382–387, 1980.

61. Steiner, J.P., Migliorini, M., and Strickland, D.K., Characterization of the reaction of plasmin with α-2-macroglobulin, *Biochemistry* 26, 8487–8495, 1987.

62. Steiner, J.P., Bhattocharya, P., and Strickland, D.K., Thrombin-induced conformational changes of human α-2-macroglobulin: Evidence for two functional domains, *Biochemistry* 24, 2993–3001, 1985.

63. Wasylewski, Z., Criscimagna, N.L., and Horowitz, P.M., A fluorescence study of thermally induced conformational changes in yeast hexokinase, *Biochim. Biophys. Acta* 831, 201–206, 1985.

64. King, L., Effects of denaturant and pressure on the intrinsic fluorescence of titan, *Arch. Biochem. Biophys.* 311, 251–257, 1994.

65. Santambrogio, C. and Grandori, R., Monitoring the Tanford transition in β-lactoglobulin by 8-anilino-1-naphthalene sulfonate and mass spectrometry, *Rapid Commun. Mass Spectrom.* 22, 4049–4054, 2008.

66. Gasymov, B.J., Abduraginov, A.P., and Glasgow, B.J., Evidence for internal and external binding sites on human tear lipocalin, *Arch. Biochem. Biophys.* 468, 15–21, 2007.

67. Gasynov, O.K., Abduragimov, A.R., and Glasgow, B.J., Ligand binding site of lipocalin: Contributions of a trigonal cluster of charged residues probed by 8-anilino-1-naphthalenesufonic acid, *Biochemistry* 47, 1414–1424, 2008.

68. Steiner, R.E., Albaugh, S., Neontas, E., and Norris, L., The interaction of auramine O with calmodulin: Location of the binding site on the connecting strand, *Biopolymers* 32, 73–83, 1992.

69. Chen, R.F., Fluorescence of free and protein-bound auramine O, *Arch. Biochem. Biophys.* 179, 672–681, 1977.

70. Invano, M.V., Klichko, V.T., Nikulin, I.R. et al., Auroamine O as a conformational probe to study glyceraldehyde 3-phosphate dehydrogenase, *Eur. J. Biochem.* 125, 291–297, 1982.

71. Tellam, R.L. and Turner, J.A., The binding of Ca^{2+} to actin monomer is monitored by the fluorescence of actin-bound auramine O, *Biochem. J.* 224, 269–274, 1984.

72. Neufeld, T., Eisenstein, M., Muszkat, K.A., and Fleminger, G., A citrate-binding site in calmodulin, *J. Mol. Recognit.* 11, 20–24, 1998.

73. Sargeant, M.G., Fiftyfold amplification of the Lowry protein assay, *Anal. Biochem.* 163, 476–481, 1987.

74. Shaw, A.K. and Pal, S.K., Resonance energy transfer and ligand binding studies in pH-induced folded states of human serum albumin, *J. Photochem. Photobiol. B* 90, 187–197, 2008.

75. Weber, G. and Young, L.B., Fragmentation of bovine serum albumin by pepsin. 1. The origin of the acid expansion of the albumin molecule, *J. Biol. Chem.* 239, 1415–1423, 1964.

76. Azzi, A., Chance, B., Radda, G.K., and Lee, C.P., A fluorescence probe of energy-dependent structural changes in fragmented membranes, *Proc. Natl. Acad. Sci. USA* 62, 612–619, 1969.

77. Chance, B., Wilson, D.F., Dutton, F.L., and Erecińska, M., Energy-coupling mechanisms in mitochondria: Kinetic, spectroscopic, and thermodynamic properties of an energy-transducing form of cytochrome b, *Proc. Natl. Acad. Sci. USA* 66, 1175–1182, 1970.

78. Chance, B., Fluorescent probe environment and the structural and charge changes in energy coupling of mitochondrial membranes, *Proc. Natl. Acad. Sci. USA* 67, 560–571.

79. Zschörnig, O., Paasche, G., Thieme, C. et al., Modulation of lysozyme charge influences interaction with phospholipid vesicles, *Colloids Surf. B. Biointerfaces* 42, 69–78, 2005.

80. Reig, F., Juvé, A., Ortiz, A. et al., Effect of a laminin amphiphatic sequence on DPPC ordered bilayers, *Luminescence* 20, 326–330, 2005.

81. Kamlekar, R.K. and Swamy, M.J., Studies on the critical micellar concentration and phase transitions of stearoylcarnitine, *Biosci. Rep.* 26, 387–398, 2006.

82. Domènech, O., Redondo, L., Picas, L. et al., Atomic force microscopy characterization of supported planar bilayers that mimic the mitochondrial inner membrane, *J. Mol. Recognit.* 20, 546–553, 2007.

83. Dhanikula, A.B. and Panchagnula, R., Flourescence anisotropy, FT-IR and 31-P NMR studies on the interaction of paclitaxel with lipid bilayers, *Lipids* 43, 569–579, 2008.

84. Lichtenberger, L.M., Ahmed, T.N., Barreto, J.C. et al., Use of fluorescent hydrophobic dyes in establishing the presence of lipids in the gastric mucus gel layer, *J. Clin. Gastroenterol.* 14 (Suppl 1), S82–S87, 1992.

85. Matulis, D., Baumann, C.G., Bloomfield, V.A., and Lovrien, R.E., 1-Anilino-8-naphthalene sulfonate as a protein conformational tightening agent, *Biopolymers* 49, 451–458, 1999.

86. Matulis, D. and Lovrien, R., 1-Anilino-8-naphthalene sulfonate anion-protein binding depends on primarily on ion pair formation, *Biophys. J.* 74, 422–429, 1998.

87. Versee, V. and Berel, A.O., Interaction of 1-anilinonaphthelene-9-sulphonate (ANS) with egg-white lysozyme and bovine α-lactoalbumin. Fluorescence study, *Bull. Soc. Chim. Belg.* 85, 585–594, 1976.

88. Ichimura, S. and Zama, M., The interaction of 8-anilino-1-naphthalenesulfonate with polylysine and polyarginine, *Biopolymers* 16, 1449–1464, 1977.

89. Mogharrab, N., Ghourchian, H., and Amininosab, M., Structural stabilization and functional improvement of horse radish peroxidase upon modification of accessible lysine—Experiments and simulation, *Biophys. J.* 92, 1192–1203, 2007.

90. Gasymov, O.K. and Glasgow, B.J., ANS fluorescence: Potential to augment the identification of the external binding site of proteins, *Biochim. Biophys. Acta* 1774, 403–411, 2007.

91. Jeganathan, S., von Bergen, M., Mandelkow, E.M., and Mandelkow, E., The natively unfolded character of tau and its aggregation to Alzheimer-like paired helical filaments, *Biochemistry* 47, 10526–10539, 2008.

92. Thoppil, A.A. and Kishore, N., Equimolar mixture of 2,2,2-trifluoroethanol and 4-chloro-1-butanol is a stronger induced of molten globule state: Isothermal titration calorimetric and spectroscopic studies, *Protein J.* 26, 507–516, 2007.

93. Fukunaga, Y., Nishimoto, E., Otosu, T. et al., The unfolding of α-momorcharin proceeds through the compact folded intermediate, *J. Biochem.* 144, 457–466, 2008.

94. Sen, P., Ahmad, B., and Kahn. R.H., Formation of a molten globule like state in bovine serum albumin at alkaline pH, *Eur. Biophys. J.* 37, 1303–1308, 2008.
95. Lavery, D.N. and McEwan, J.J., Structural characterization of the native NH$_2$-terminal transactivation domain of the human androgen receptor: A collapsed disordered conformation underlies structural plasticity and protein-induced folding, *Biochemistry* 47, 3360–3369, 2008.
96. Mendu, D.R., Dasari, V.R., Cai, M., and Kim, K.S., Protein folding intermediates of invasion protein IbeA from *Escherichia coli*, *FEBS J.* 275, 458–469, 2008.
97. Naseem, F. and Khan, R.H., Structural intermediates of acid unfolded Con-A in different co-solvents: Fluoroalcohols and polyethylene glycols, *Int. J. Biol. Macromol.* 42, 158–165, 2008.
98. Graziano, J.J., Liu, W., Perera, R. et al., Selecting folded proteins from a library of secondary structural elements, *J. Am. Chem. Soc.* 130, 176–185, 2008.
99. Mixcoha-Hernández, E., Moreno-Vargas, L.M., Rojo-Domínguez, A., and Benítez-Cardoza, C.G., Thermal-unfolding reaction of triosephosphase isomerase from *Trypanosoma cruzi*, *Protein J.* 26, 491–498, 2007.
100. Gabellieri, E. and Strambini, G.B., Perturbation of protein tertiary structure in frozen solutions revealed by 1-anilino-8-naphthalene sulfonate fluorescence, *Biophys. J.* 85, 3214–3220, 2003.
101. Gabrellieri, E. and Strambini, G.B., ANS fluorescence detects widespread perturbations of protein tertiary structure in ice, *Biophys. J.* 90, 3239–3245, 2006.
102. Royer, C.A., Fluorescence spectroscopy, *Methods Mol. Biol.* 40, 65–89, 1995.
103. Hawe, A., Sutter, M., and Jiskoot, W., Extrinsic fluorescent dyes as tools for protein characterization, *Pharm. Res.* 35, 1487–1499, 2008.
104. Cimmpeman, P., Baaranauskiene, L., Jachimoviciūte, S. et al., A quantitative model of thermal stabilization and destabilization of proteins by ligands, *Biophys. J.* 95, 3222–3231, 2008.
105. Muzammil, S., Kumar, Y., and Tayyab, S., Anion-induced stabilization of human serum albumin prevents the formation of intermediate during urea denaturation, *Proteins* 40, 29–38, 2000.
106. Pearce, M.C., Rubin, H., and Bottomley, S.P., Conformational changes and intermediates in the unfolding of α$_1$-antichymotrypsin, *J. Biol. Chem.* 275, 28513–28518, 2000.
107. Grillo, A.O., Edwards, K.L., Kashi, R.S. et al., Conformational origin of the aggregation of recombinant human factor VIII, *Biochemistry* 40, 586–595, 2001.
108. Prokopwicz, M., Banecki, B., Lukasiak, J., and Przyjazny, A., The measurement of conformational stability of proteins adsorbed on siloxanes, *J. Biomater. Sci. Polym. Ed.* 14, 103–118, 2003.
109. Prokopowicz, M., Lukasiak, J., Banecki, B., and Przyjazny, A., In vitro measurement of conformational stability of fibrinogen adsorbed on siloxane, *Biomacromolecules* 6, 39–45, 2005.
110. Lechevalier, V., Corguennec, T., Pezennec, S. et al., Ovalbumin, ovotransferrin, lysozyme: Three model proteins for structural modifications at the air-water interface, *J. Agric. Food Chem.* 51, 6354–6361, 2003.
111. Roychauduri, R., Sarath, G., Zeece, M., and Markwell, J., Stability of the allergenic soybean Kunitz trypsin inhibitor, *Biochim. Biophys. Acta* 699, 207–212, 2004.
112. Ahmad, B., Ahmed, M.E., Haq, S.K., and Khan, R.H., Guanidine hydrochloride denaturation of human serum albumin originates by local unfolding of some stable loops in domain III, *Biochim. Biophys. Acta* 1750, 93–102, 2005.
113. Murali, J. and Jaykumar, R., Spectroscopic studies on native and protofibrillar insulin, *J. Struct. Biol.* 150, 180–189, 2005.
114. Fan, H., Ralston, J., Divase, M. et al., Solution behavior of IFN-β-1a: An empirical phase diagram based approach, *J. Pharm. Sci.* 94, 1893–1911, 2005.

115. Hong, D.P., Ahmad, A., and Fink, A.L., Fibrillation of human insulin A and B chains, *Biochemistry* 45, 9342–9353, 2006.
116. Fan, H., Vitharana, S.N., Chan, T. et al., Effects of pH and polyanions on the thermal stability of fibroblast growth factor 20, *Mol. Pharm.* 4, 232–240, 2007.
117. Bakhti, M., Habibi-Reszaei, M., Moosavi-Movahedi, A.A., and Khazaei, M.R., Consequential alterations in haemoglobin structure upon glycation with fructose: Prevention by acetylsalicylic acid, *J. Biochem.* 141, 827–833, 2007.
118. Vetri, V., Librizzi, F., Leone, M., and Militello, V., Thermal aggregation of bovine serum albumin at different pH: Comparison with human serum albumin, *Eur. Biophys. J.* 36, 717–725, 2007.
119. Rezaei-Ghaleh, N., Ramshini, H., Ebrahim-Habibi, A. et al., Thermal aggregation of α-chymotrypsin: Role of hydrophobic and electrostatic interactions, *Biophys. Chem.* 132, 23–32, 2008.
120. Hawe, A., Friess, W., Sutter, M., and Jiskoot, W., Online fluorescent dye detection method for the characterization of immunoglobulin G aggregation by size exclusion chromatography and asymmetrical flow field flow fractionation, *Anal. Biochem.* 378, 115–122, 2008.
121. Rispens, T., Lakemond, C.M., Derkson, N.I., and Aalberse, R.C., Detection of conformational changes in immunoglobulin G using isothermal titration calorimetry with low-molecular-weight probes, *Anal. Biochem.* 380, 303–309, 2008.

17 Use of Hydrogen Exchange in the Study of Biopharmaceutical Conformation

The study of hydrogen atom exchange as a measure of macromolecule conformation dates to work in Linderstrøm-Lang's laboratory.[1-4] The first study by Hivdt and Linderstrøm-Lang[1] studied the exchange of hydrogen atoms in insulin with deuterium atoms. The technique was quite laborious but yielded excellent data. Approximately half of the hydrogen atoms were exchanged in studies with native proteins. The concept of freely exchangeable hydrogen atoms and firmly bound hydrogen atoms was derived from this result. Oxidation of insulin increased the number of exchangeable hydrogen atoms, establishing the importance of tertiary structure for the stability of internal hydrogen bonds. It is of no small importance to note that these concepts were established in two pages of print.

Most of the interest in hydrogen exchange has been devoted toward the study of protein conformation; there are only a few studies on the use of NMR for the study of nucleic acid conformation.[5]

There has been consistent use of hydrogen–deuterium exchange (H/D exchange) for the study of protein conformation.[6-23] While the work of Linderstrøm-Lang and associates was seminal, Englander and associates at the University of Pennsylvania were critical in the subsequent development of this technology.[3,12,13,17,20] The exchangeable hydrogen atoms (labile protons) include peptide bond hydrogen atoms and the side chain hydrogen atoms of polar amino acids such as lysine, histidine, arginine, glutamic acid, and the hydrogen on the indole ring of tryptophan aspartic acid and the hydrogen atoms of hydroxyl groups such as tyrosine.[3] Deuterium atom replacement of hydrogen (ion-exchange) or hydrogen atom replacement of deuterium was first assessed by infrared spectroscopy of changes in the amide II band (the amide II band decreases as deuterium replaces hydrogen while the amide I band remains constant serving as an internal standard).[10,24] Infrared spectroscopy has continued to be of use in the measurement of H/D exchange.[16,25-31] Infrared spectroscopy only measures the amide hydrogen of the peptide bond backbone. More recent work has used NMR[13,15,32-38] and mass spectrometry has also become a popular method for the study of H/D exchange[18,21,22,39-54] as technology became available for the study of large proteins.

Hydrogen exchange can be measured as "in" or "out."[6,7,10] With "in" exchange, the analyte polymer in H_2O is transferred to D_2O and exchange begins immediately such that a "zero time" value is, for all practical purposes, difficult to obtain.

A similar problem exists with "out" exchange. "Out" exchange measures the hydrogen exchange out after "saturation" of the protein with deuterium or tritium. This presents challenges to the study of exchange rates (kinetics). Exchange during analysis due to scrambling is also an issue in obtaining a reliable sample.[55]

However, it is accepted that there are labile protons that undergo isotope exchange more slowly when "buried" in a native conformation than when exposed in an unfolded or denatured conformation.[11] While this was clear from the early work on insulin cited above, Woodward and coworkers[11] suggested that the exchange of buried protons effectively changes as the protein structure fluctuates. The fluctuation of native structure is a concept that has received considerable attention in recent years.[56–60] This of course means that a value obtained, for example, with infrared spectroscopy or tritium exchange represents an average value while a technique such as mass spectrometry could measure individual "hot spots." On the other hand, such "hot spots" likely exchange a rate to fast to be measured with current technology. The number of "slowly" exchanging protons and "rapidly" exchanging protons under a given set of conditions will be a collective value within a broad distribution (trace labeling of lysine residues as described below provides an experimental approach for measuring specific residue contribution). None of this detracts from the value of hydrogen exchange in measuring protein conformation.

Woodward and associates were also the first to define solvent accessibility as measured by hydrogen exchange[61]; the concept of solvent accessibility had been used earlier by Pettersson[62] to describe solvent perturbation of tyrosine residues by spectral perturbation. The 1975 study from the University of Minnesota[61] described hydrogen exchange in trypsin. The experiments involved the "out" exchange of tritium from trypsin at pH 2.0. The protein could be "completely" labeled with tritium in the reversibly unfolded state or with tritium in the native protein where only the most labile hydrogen atoms would be exchanged. The rate of "out" exchange of tritium from "completely" labeled trypsin markedly increased with increasing temperature. Urea also had an effect on "out" exchange rate. Approximately 20% of the tritium was "out-exchanged" after 30 h at pH 2 at 5°C; the presence of 1.0 M urea resulted in 90% + "out-exchange" over the same time period. Increasing the urea concentration to 3.0 M urea resulted in almost 100% "out-exchange" of tritium in less that 10 h. Brumano and de Almedia Oliveira[63] studied the denaturation of trypsin with 2.0 M urea using intrinsic fluorescence and bis-(8-anilino-1-naphthalene sulfonate) (bis-ANS) binding and obtained data consistent with the hydrogen exchange. Ethanol also increased the rate of "out-exchange" from tritiated trypsin. These investigators concluded that the rate of exchange depended on solvent accessibility and hydrogen bonding.[64–68] These investigators had published two earlier studies of interest for the discussion of conformational effects on hydrogen exchange. Woodward and Rosenberg[69] used oxidized RNase as a model for the random conformational state and for proteins in general. While there are similarities to an amino acid homopolymer, oxidized RNase does demonstrate a distribution of first-order rates; sites that have common properties do not necessarily share a common first-order rate constant. In a subsequent study,[70] these investigators studied the hydrogen exchange in the temperature-induced transition in RNase. Tritiated RNase was prepared by incubation of RNase at 62°C for 15 min in the presence of tritiated water; the zero-time sample was obtained by the gel filtration on a P-2 column. Longer time samples were obtained by stopped the column development at about 50% elution for the desired time and then development continued at the desired time point.[71] The distribution range of the exchange rate for the native protein is 75-fold larger than the random conformation. There are two classes of sites with different activation energies. The work in these studies is not directly applicable to studies on the conformation of biopharmaceuticals but the information provides a strong basis for the use of H/D exchange for protein conformation.

Our focus is directed at technology, which can address comparability of biotherapeutic polymers and, as such, there is considerable interest in studies on the use of hydrogen exchange in protein denaturation.[72–79] As far as end point is concerned, denaturation/unfolding of a protein increases the number of exchanged hydrogen atoms. Fitzgerald and colleagues at Duke University have developed a facile method using H/D exchange to study protein stability. SUPREX (stability of unpurified proteins from rates of H/D exchange)[80–83] to measure protein stability using picomole quantities of material. SUPREX was used most recently to evaluate the stability of a variant hyperoxaluria-associated human alanine:glyoxylate aminotransferase[83]; the minor allele (AGT_{mi}) was less stable than the major allele (AGT_{ma}).

Table 17.1 contains some selected examples of the application of H/D exchange to the study of biotherapeutic molecules. There are several interesting studies that

TABLE 17.1
Selected Examples of the Application of H/D Exchange to the Study of Biopharmaceutical Conformation

Macromolecule	Study	Ref.
Insulin	The adsorption-induced conformational changes of insulin are studied with H/D exchange and mass spectrometry. Insulin adsorbed to a hydrophobic surface showed increased exchange consistent with reduction in stability.	[84]
Thrombin	The epitope for an MAB raised against human thrombin was identified by H/D exchange. Initial studies using proteolysis suggested that epitope was discontinuous. The epitope was identified by "out" exchange of deuterium from a complex between the MAB (bound to a protein G bead) and deuterated thrombin.	[85]
Interferon-γ	H/D exchange and mass spectrometry were used to study the aggregation of interferon-γ induced by guanidine hydrochloride and potassium thiocyanate.	[86]
Factor XIII	The conformational changes occurring on the activation of recombinant factor XIII were studied by H/D exchange.	[87]
Recombinant human macrophage colony-stimulating factor-β (rhM-CSF-β)	The solution structure of rhM-CSFβ was studied with H/D exchange. Solvent accessibility determined with H/D exchange correlates with x-ray crystallographic data.	[88]

(*continued*)

TABLE 17.1 (continued)
Selected Examples of the Application of H/D Exchange to the Study of Biopharmaceutical Conformation

Macromolecule	Study	Ref.
Lysozyme and endopolygalacturonase-II	H/D exchange is used to evaluate protein–carbohydrate interaction. Studies with lysozyme showed that carbohydrate can provide sufficient stable protection to the amide hydrogen atoms from deuterium exchange.	[89]
N/A	H/D exchange and mass spectrometry is used for epitope mapping.	[90]
DNA and RNA tetranucleotides	H/D exchange was performed on a series of tetranucleotides which were then examined by MALDI mass spectrometry to determine the fragmentation pattern of singly protonated nucleic acids.	[91]
Amyloid fibrils and protofibrils	H/D exchange and mass spectrometry were used to determine the regions of peptide involved in the β-sheet network.	[92]
Factor VIIa	H/D exchange was used to determine the conformational changes in blood coagulation factor VIIa, which occur on association with tissue factor.	[93]
Calmodulin	Calmodulin was used as model protein to study the application of H/D exchange to proteins in a solid matrix (lyophilized protein). H/D exchange was affected by relative humidity, calcium, and trehalose.	[94]
Calmodulin	H/D exchange was used to evaluate the effects of excipients (sucrose, trehalose, mannitol, dextran) on the structure of calmodulin in an amorphous matrix.	[95]
Complement factors C3 and C3b	H/D exchange and mass spectrometry were used to study the solution conformation of complement factor C3 and complement factor C3b.	[96]
Interferon-β	Interferon-β was modified with N-ethylmaleimde. Conformational changes occurring secondary to modification were evaluated with H/D exchange.	[97]
α_1-Antitrypsin	α_1-Antitrypsin polymers formed by incubation at moderate temperatures are used as a model of serpin polymerization. H/D exchange and mass spectrometry were used to study the structure of polymers.	[98]
Ribonuclease, myoglobin, lysozyme, β-lactoglobulin	The effect of excipients (trehalose, raffinose, dextran 5000, polyvinyl alcohol, polyvinyl pyrrolidine) on lyophilized proteins. The amorphous powders were exposed to D_2O vapors and evaluated with FTIR and mass spectrometry for conformational change.	[27]
Cytochrome c	Epitope mapping for a monoclonal antibody with H/D exchange. An "in" exchange approach was used in the presence of antibody; regions that gained less deuterium in the presence of antibody than absence were defined as the epitope.	[99]
IgG1	H/D exchange and mass spectrometry were used to study the solution structure of a recombinant monoclonal IgG1 antibody. Exchange properties were changed on deglycosylation.	[100]

merit mention in the text. Dürr and Bosshard[101] used H/D exchange to show that the binding of a monoclonal antibody induced the opening of a coiled coil; the "in" exchange of deuterium into a 29 residues random-coil peptide was far more rapid in the presence of a monoclonal antibody. Other investigators subsequently suggested that this was another example of the "receptor" selection of a specific peptide conformer.[102] Fogle and Fernandez[103,104] used H/D exchange to demonstrate conformational change occurring on the binding of protein to hydrophobic affinity matrices as method for assessing stability during chromatography. Increased binding of protein (α-lactalbumin) to the matrix (phenyl-Sepharose®) increased the "in" exchange of deuterium. Ueberbacher and colleagues[105] also reported conformational changes occurring on the binding of protein to hydrophobic affinity matrices using attenuated total reflectance Fourier transform infrared (ATR-FTIR) spectroscopy.

REFERENCES

1. Hvidt, A. and Linderstrøm-Lang, K., Exchange of hydrogen atoms in insulin with deuterium atoms in aqueous solutions, *Biochim. Biophys. Acta* 14, 574–575, 1954.
2. Hvidt, A. and Linderstrøm-Lang, K., The pH dependence of the deuterium exchange of insulin, *Biochim. Biophys. Acta* 18, 308–312, 1955.
3. Englander, S.W., Mayne, L., Bal, Y., and Sosnick, T.R., Hydrogen exchange: The modern legacy of Linderstrøm-Lang, *Protein Sci.* 6, 1101–1109, 1997.
4. Linderstrom-Lang, K., Deuterium exchange and protein structure, in *Symposium on Protein Structure*, ed. A. Neuberger, Methuen & Company, Ltd., New York, pp. 23–34, Chapter 2, 1958.
5. Mirau, P.A. and Kearns, D.R., Effects of environment, conformation, sequence, and base substituents on the imino proton exchange rates in guanine and inosine containing DNA, RNA, and DNA-RNA complexes, *J. Mol. Biol.* 177, 207–227, 1984.
6. Ottesen, M., Methods for measurement of hydrogen isotope exchange in globular proteins, *Methods Biochem. Anal.* 20, 135–168, 1971.
7. Englander, S.W., Downer, N.W., and Teitelbaum, H., Hydrogen exchange, *Annu. Rev. Biochem.* 41, 903–924, 1972.
8. Woodward, C.K. and Hilton, B.D., Hydrogen exchange kinetics and internal motions in proteins and nucleic acids, *Annu. Rev. Biophys. Bioeng.* 8, 99–127, 1979.
9. Pfeil, W., The problem of the stability of globular proteins, *Mol. Cell. Biochem.* 40, 3–28, 1981.
10. Barksdale, A.D. and Rosenberg, A., Acquisition and interpretation of hydrogen exchange data from peptides, polymers, and proteins, *Methods Biochem. Anal.* 28, 1–113, 1982.
11. Woodward, C., Simon, I., and Tuchsen, E., Hydrogen exchange and the dynamic structure of proteins, *Mol. Cell. Biochem.* 48, 135–160, 1982.
12. Englander, S.W. and Kallenbach, N.R., Hydrogen exchange and structural dynamics of proteins and nucleic acids, *Q. Rev. Biophys.* 16, 521–655, 1983.
13. Englander, S.W. and Mayne, L., Protein folding studied using hydrogen-exchange and two-dimensional NMR, *Annu. Rev. Biophys. Biomol. Struct.* 21, 243–265, 1992.
14. Liepinsh, E., Otting, G., and Wuthrich, K., NMR spectroscopy of hydroxyl protons in aqueous solutions of peptides and proteins, *J. Biomol. NMR* 2, 447–465, 1992.
15. Marion, D., Nuclear magnetic resonance studies of cytochrome c in solution, *Biochimie* 76, 631–640, 1994.
16. Haris, P.I. and Chapman, D., The conformational analysis of peptides using Fourier transform IR spectroscopy, *Biopolymers* 37, 251–263, 1995.
17. Englander, S.W., Sosnick, T.R., Englander, J.J., and Mayne, L., Mechanisms and uses of hydrogen exchange, *Curr. Opin. Struct. Biol.* 6, 18–23, 1996.

18. Miranker, A., Robinson, C.V., Radford, S.E., and Dobson, C.M., Investigation of protein folding by mass spectrometry, *FASEB J.* 10, 93–101, 1996.

19. Raschke, T.M. and Marqusee, S., Hydrogen exchange studies of protein structure, *Curr. Opin. Biotechnol.* 9, 80–86, 1998.

20. Huyghues-Despointes, B.M., Pace, C.N., Englander, S.W., and Scholtz, J.M., Measuring the conformational stability of a protein by hydrogen exchange, *Methods Mol. Biol.* 168, 69–92, 2001.

21. Hoofnagle, A.N., Resing, K.A., and Ahn, N.G., Protein analysis by hydrogen exchange mass spectrometry, *Annu. Rev. Biophys. Biomol. Struct.* 32, 1–25, 2003.

22. Yan, X., Watson, J., Ho, P.S., and Deinzer, M.L., Mass spectrophotometric approaches using electrospray ionization charge states and hydrogen–deuterium exchange for determining protein structures and their conformational changes, *Mol. Cell. Proteomics* 3, 10–23, 2004.

23. Niimura, N., Arai, S., Kurihara, K. et al., Recent results on hydrogen and hydration in biology studied by neutron macromolecular crystallography, *Cell. Mol. Life Sci.* 63, 285–300, 2006.

24. Johansen, J.T., A modified method for determining hydrogen–deuterium exchange in proteins, *Biochim. Biophys. Acta* 214, 551–553, 1970.

25. Downer, N.W., Bruchman, T.J., and Hazzard, J.H., Infrared spectroscopic study of photoreceptor membrane and purple membrane. Protein secondary structure and hydrogen deuterium exchange, *J. Biol. Chem.* 261, 3640–3647, 1986.

26. Závodsky, P., Kardos, J., Svingor, A., and Petsko, G.A., Adjustment of conformational flexibility is a key event in the thermal adaptation of proteins, *Proc. Natl. Acad. Sci. USA* 95, 7406–7411, 1998.

27. Sinha, S., Li, Y., Williams, T.D., and Topp, E.M., Protein conformation in amorphous solids by FTIR and by hydrogen/deuterium exchange with mass spectrometry, *Biophys. J.* 95, 5951–5961, 2008.

28. Iloro, I., Narváez, D., Guillén, N. et al., The kinetics of the hydrogen/deuterium exchange of epidermal growth factor receptor ligands, *Biophys. J.* 94, 4041–4055, 2008.

29. Cannona, P., Rodríguez-Casado, A., and Molina, M., Improving real-time measurement of H/D exchange using a FTIR biospectroscopic probe, *Anal. Bioanal. Chem.* 393, 1289–1295, 2009.

30. Shah, N.B., Wolkers, W.F., Morrissey, M. et al., Fourier transform infrared spectroscopy investigation of native tissue matrix modifications using a gamma irradiation process, *Tissue Eng. Part A* 15, 33–40, 2009.

31. Clarke, A.R. and Waltho, J.P., Protein folding and intermediates, *Curr. Opin. Biotechnol.* 8, 400–410, 1997.

32. Sadqi, M., Casares, S., Abril, M.A. et al., The native state conformational ensemble of the SH3 domain from α-spectrin, *Biochemistry* 38, 8899–8906, 1999.

33. Jonasson, P., Kjellsson, A., Sethson, I., and Jonsson, B.H., Denatured states of human carbonic anhydrase II: An NMR study of hydrogen/deuterium exchange at tryptophan-indole-H(N) sites, *FEBS Lett.* 445, 361–365, 1999.

34. Tagashira, M., Iijima, H., and Toma, K., An NMR study of *O*-glycosylation induced structural changes in the α-helix of calcitonin, *Glycoconj. J.* 19, 43–52, 2002.

35. Digiglio, G., Barbero, L., Bracco, C. et al., NMR structure of two novel polyethylene glycol conjugates of the human growth hormone-releasing factor, hGRF (1–29)-NH₂, *J. Am. Chem. Soc.* 125, 3458–3470, 2003.

36. Polshakov, V.I., Birdsall, B., and Feeney, J., Effects of co-operative ligand binding of protein amide NH hydrogen exchange, *J. Mol. Biol.* 356, 886–903, 2006.

37. Fasoli, E., Ferrer, A., and Barletta, G.L., Hydrogen/deuterium exchange study of subtilisin Carlsberg during prolonged exposure to organic solvents, *Biotechnol. Bioeng.* 102, 1025–1032, 2009.

38. Katta, V. and Chait, B.T., Conformational changes in proteins probed by hydrogen-exchange electrospray-ionization mass spectrometry, *Rapid Commun. Mass Spectrom.* 5, 214–217, 1991.

39. Ehring, H., Hydrogen exchange/electrospray ionization mass spectrometry studies of structural features of proteins and protein/protein interactions, *Anal. Biochem.* 267, 252–259, 1999.

40. Goshe, M.B., Chen, Y.H., and Anderson, V.E., Identification of the sites of hydroxyl radical reaction with peptides by hydrogen/deuterium exchange: Prevalence of reactions with the side chains, *Biochemistry* 39, 1761–1770, 2000.

41. Bienvenut, W.V., Hoogland, C., Greco, A. et al., Hydrogen/deuterium exchange for higher specificity of protein identification by peptide mass fingerprinting, *Rapid Commun. Mass Spectrom.* 16, 616–626, 2002.

42. Kim, M.Y., Maier, C.S., Reed, D.J., and Deinzer, M.L., Conformational changes in chemically modified *Escherichia coli* thioredoxin monitored by H/D exchange and electrospray ionization mass spectrometry, *Protein Sci.* 11, 1320–1329, 2002.

43. Wang, L., Pan, H., and Smith, D.L., Hydrogen exchange-mass spectrometry: Optimization of digestion conditions, *Mol. Cell Proteomics* 1, 132–138, 2002.

44. Wang, L. and Smith, D.L., Downsizing improves sensitivity 100-fold for hydrogen exchange-mass spectrometry, *Anal. Biochem.* 314, 46–53, 2003.

45. Hamuro, Y., Coales, S.J., Southern, M.R. et al., Rapid analysis of protein structure and dynamics by hydrogen/deuterium exchange mass spectrometry, *J. Biomol. Technol.* 14, 171–182, 2003.

46. Cravello, L., Lascoux, D., and Forest, E., Use of different proteases working in acidic conditions to improve sequence coverage and resolution in hydrogen/deuterium exchange of large proteins, *Rapid Commun. Mass Spectrom.* 17, 2387–2393, 2003.

47. Li, X., Chou, Y.T., Husain, R., and Watson, J.T., Integration of hydrogen/deuterium exchange and cyanylation-based methodology for conformational studies of cystinyl proteins, *Anal. Biochem.* 331, 130–137, 2004.

48. Hamuro, Y., Weber, P.C., and Griffin, P.R., High-throughput analysis of protein structure by hydrogen/deuterium exchange mass spectrometry, *Methods Biochem. Anal.* 45, 131–157, 2005.

49. Novak, P. and Giannakopulos, A.E., Chemical cross-linking and mass spectrometry as structure determination tools, *Eur. J. Mass Spectrom.* 13, 105–113, 2007.

50. Tsutsui, Y. and Wintrobe, P.L., Hydrogen/deuterium exchange-mass spectrometry: A powerful tool for probing protein structure, dynamics and interactions, *Curr. Med. Chem.* 14, 2344–2358, 2007.

51. Coales, S.J., Tomasso, J.C., and Hamuro, Y., Effect of electrospray capillary temperature on amide hydrogen exchange, *Rapid Commun. Mass Spectrom.* 22, 1367–1371, 2008.

52. Konenrmann, L., Tong, X., and Pan, Y., Protein structure and dynamics studies by mass spectrometry: H/D exchange, hydroxyl radical labeling, and related approaches, *J. Mass Spectrom.* 43, 1021–1036, 2008.

53. Burkitt, W. and O'Conner, G., Assessment of the repeatability and reproducibility of hydrogen/deuterium exchange mass spectrometry measurements, *Rapid Commun. Mass Spectrom.* 22, 3893–3901, 2008.

54. Yan, X. and Maier, C.S., Hydrogen/deuterium exchange mass spectrometry, *Methods Mol. Biol.* 492, 255–271, 2009.

55. Ferguson, P.L., Pan, J., Wilson, D.J. et al., Hydrogen/deuterium scrambling during quadrupole time-of-flight MS/MS analysis of a zinc-binding protein domain, *Anal. Chem.* 79, 153–160, 2007.

56. Teague, S.J., Implications of protein flexibility for drug discovery, *Nat. Rev. Drug Discov.* 2, 527–541, 2003.

57. Gerstein, M. and Echols, N., Exploring the range of protein flexibility, from a structural proteomics perspective, *Curr. Opin. Chem. Biol.* 8, 14–19, 2006.

58. Dodson, G. and Verma, C.S., Protein flexibility: Its role in structure and mechanism revealed by molecular simulations, *Cell. Mol. Life Sci.* 63, 207–219, 2006.

59. Okura, K., Exploring unique structures: Flexibility is a significant factor in biologic activity, *Biol. Pharm. Bull.* 30, 1025–1036, 2007.

60. Kamerzell, T.J. and Middaugh, C.P., The complex inter-relationships between protein flexibility and stability, *J. Pharm. Sci.* 97, 3494–3517, 2008.

61. Woodward, C., Ellis, L., and Rosenberg, A., Solvent accessibility in folded proteins. Studies of hydrogen exchange in trypsin, *J. Biol. Chem.* 250, 432–439, 1975.

62. Pettersson, G., Structure and function of a cellulose from *Penicillium notatum* as studied by chemical modification and solvent accessibility, *Arch. Biochem. Biophys.* 126, 776–784, 1968.

63. Burmano, M.H.N. and de Almedia Oliveira, M.G., Urea-induced denaturation of β-trypsin: An evidence for a molten globule state, *Protein Pept. Lett.* 11, 133–140, 2004.

64. Rozman, M., The gas-phase H/D exchange mechanism of protonated amino acids, *J. Am. Soc. Mass Spectrom.* 16, 1846–1852, 2005.

65. Grdadolnik, J. and Maréchal, Y., Hydrogen-deuterium exchange in bovine serum albumin protein monitored by Fourier transform infrared spectroscopy. Part II: Kinetics studies, *Appl. Spectrosc.* 59, 1357–1364, 2005.

66. Kovacević, B., Rozman, M., Klasinc, L. et al, Gas-phase structure of protonated histidine and histidine methyl ester: Combined experimental mass spectrometry and theoretical ab initio study, *J. Phys. Chem. A* 109, 8329–8335, 2005.

67. Hoshino, M., Katou, H., Yamaguchi, K., and Goto, Y., Dimethylsulfoxide-quenched hydrogen/deuterium exchange method to study amyloid fibril structure, *Biochim. Biophys. Acta* 1768, 1886–1899, 2007.

68. Mo, J., Todd, G.C., and Håkansson, K., Characterization of nucleic acid higher order structure by gas-phase H/D exchange in a quadrupole-FT-ICR mass spectrometer, *Biopolymers* 91, 256–264, 2009.

69. Woodward, C.K. and Rosenberg, A., Oxidized RNase as a protein model having no contribution to the hydrogen exchange rate from conformational restrictions, *Proc. Natl. Acad. Sci. USA* 66, 1067–1074, 1970.

70. Woodward, C.K. and Rosenberg, A., Studies of hydrogen exchange in proteins. V. The correlation of ribonuclease exchange kinetics with the temperature-induced transition, *J. Biol. Chem.* 246, 4105–4113, 1971.

71. Slobodian, E. and Fleisher, M., Effect of γ irradiation on the structure of ribonuclease, *Biochemistry* 5, 2192–2200, 1966.

72. Maier, C.S., Shcimerlik, M.I., and Deinzer, M.L., Thermal denaturation of *Escherichia coli* thioredoxin studied by hydrogen/deuterium exchange and electrospray ionization mass spectrometry: Monitoring a two-state protein folding transition, *Biochemistry* 38, 1136–1143, 1999.

73. Dzwolak, W., Kato, M., Shimizu, A., and Taniguchi, Y., Fourier-transform infrared spectroscopy study of the pressure-induced changes in the structure of the bovine α-lactalbumin: The stabilizing role of the calcium ion, *Biochim. Biophys. Acta* 1433, 45–55, 1999.

74. Babu, K.R. and Douglas, D.J., Methanol-induced conformations of myoglobin at pH 4.0, *Biochemistry* 39, 14702–14710, 2000.

75. Ikeuchi, Y., Nakagawa, K., Endo, T. et al., Pressure-induced denaturation of monomer β-lactoglobulin is partially irreversible: Comparison of monomer form (highly acidic pH) with dimer form (neutral pH), *J. Agric. Food Chem.* 49, 4052–4059, 2001.

76. Neira, J.L., González, C., Torion, C. et al., Three-dimensional solution structure and stability of thioredoxin m from spinach, *Biochemistry* 40, 15246–15256, 2001.

77. Mazon, H., Marcillat, O., Forest, E., and Vial, C., Denaturant sensitive regions in creatine kinase identified by hydrogen/deuterium exchange, *Rapid Commun. Mass Spectrom.* 19, 1461–1468, 2005.

78. Umezaki, T., Iimura, S., Noda, Y. et al., The confirmation of the denatured structure of pyrrolidone carboxyl peptidase under nondenaturing conditions: Difference in helix propensity of two synthetic peptides with single amino acid substitution, *Proteins* 71, 737–742, 2008.

79. Torta, F., Elviri, L., Careri, M. et al., Mass spectrometry and hydrogen/deuterium exchange measurements of alcohol-induced structural changes in cell retinol-binding protein type I, *Rapid Commun. Mass Spectrom.* 22, 330–336, 2008.

80. Ghaemmahami, S., Fitzgerald, M.C., and Oas, T.G., A quantitative, high-throughput screen for protein stability, *Proc. Natl. Acad. Sci. USA* 97, 8296–8301, 2000.

81. Powell, K.D. and Fitzgerald, M.C., Measurements of protein stability by H/D exchange and matrix-assisted laser desorption/ionization mass spectrometry using picomoles of material, *Anal. Chem.* 73, 3300–3304, 2001.

82. Dai, S.Y., Gardner, M.W., and Fitzgerald, M.C., Protocol for the thermodynamic analysis of some proteins using an H/D exchange- and mass spectrometry-based technique, *Anal. Chem.* 77, 693–697, 2005.

83. Hopper, E.D., Pittman, A.M.C., Fitzgerald, M.C., and Tucker, C.L., In vivo and in vitro examination of stability of primary hyperoxaluria-associated human alanine:glyoxylate aminotransferase, *J. Biol. Chem.* 283, 30493–30502, 2008.

84. Buija, J., Costa Vera, C., Ayala, E. et al., Conformational stability of adsorbed insulin studied with mass spectrometry and hydrogen exchange, *Anal. Chem.* 71, 3219–3225, 1999.

85. Baerga-Ortiz, A., Hughes, C.A., Mandell, J.G., and Komives, E.A., Epitope mapping of a monoclonal antibody against human thrombin by H/D exchange mass spectrometry reveals selection of a diverse sequence in a highly conserved protein, *Protein Sci.* 11, 1300–1308, 2002.

86. Tobler, S.A. and Fernandez, E.J., Structural features of interferon-γ aggregation revealed by hydrogen exchange, *Protein Sci.* 11, 1340–1352, 2002.

87. Turner, B.T. Jr. and Maurer, M.C., Evaluating the roles of thrombin and calcium in the activation of coagulation factor XIII using H/D exchange and MALDI-TOF MS, *Biochemistry* 41, 7947–7954, 2002.

88. Yan, X., Zhang, H., Watson, J. et al., Hydrogen/deuterium exchange and mass spectrometric analysis of a protein containing multiple disulfide bonds: Solution structure of recombinant macrophage colony stimulating factor-β (rhM-CSFβ), *Protein Sci.* 11, 2113–2124, 2002.

89. King, D., Lumpkin, M., Bergmann, C., and Orlando, R., Studying protein–carbohydrate interactions by amide hydrogen/deuterium exchange mass spectrometry, *Rapid Commun. Mass Spectrom.* 16, 1569–1574, 2002.

90. Hagen-Braun, C. and Tomer, K.B., Determination of protein-derived epitopes by mass spectrometry, *Expert Rev. Proteomics* 2, 745–756, 2005.

91. Andersen, T.E., Kirpekar, F., and Haselmann, K.F., RNA fragmentation in MALDI mass spectrometry studied by H/D-exchange: Mechanisms of general applicability to nucleic acids, *J. Am. Soc. Mass Spectrom.* 17, 1353–1368, 2006.

92. Kheterpal, I., Cook, K.D., and Wetzel, R., Hydrogen/deuterium exchange mass spectrometry analysis of protein aggregates, *Methods Enzymol.* 413, 140–166, 2006.

93. Olsen, O.H., Rand, K.D., Østergaard, H., and Persson, E., A combined structural dynamics approach identifies a putative switch in factor VIIa employed by tissue factor to initiate blood coagulation, *Protein Sci.* 16, 671–682, 2007.

94. Li, Y., Williams, T.D., Schowen, R.L., and Topp, E.M., Characterizing protein structure in amorphous solids using hydrogen/deuterium exchange with mass spectrometry, *Anal. Biochem.* 366, 18–28, 2007.

95. Li, Y., Williams, T.D., and Topp, E.M., Effects of excipients on protein conformation in lyophilized solids by hydrogen/deuterium exchange mass spectrometry, *Pharm. Res.* 25, 259–267, 2008.

96. Schuster, M.C., Ricklin, D., Papp, K. et al., Dynamic structural changes during complement C3 activation analyzed by hydrogen/deuterium exchange mass spectrometry, *Mol. Immunol.* 45, 3142–3151, 2008.

97. Bobat, C.E., Abzalimov, R.R., Houde, D. et al., Detection and characterization of altered conformations of protein pharmaceuticals using complementary mass spectrometry-based approaches, *Anal. Chem.* 80, 7473–7481, 2008.

98. Tsutsui, Y., Kuri, B., Sengupta, T., and Wintrode, P.L., The structural basis of serpin polymerization studied by hydrogen/deuterium exchange and mass spectrometry, *J. Biol. Chem.* 283, 30804–30811, 2008.

99. Coales, S.J., Tuske, S.J., Tomasso, J.C., and Hamuro, Y., Epitope mapping by amide hydrogen/deuterium exchange coupled with immobilization of antibody, on-line proteolysis, liquid chromatography and mass spectrometry, *Rapid Commun. Mass Spectrom.* 23, 639–647, 2009.

100. Houde, D., Arndt, J., Domeier, W. et al., Characterization of IgG1 conformation and conformational dynamics by hydrogen/deuterium exchange mass spectrometry, *Anal. Chem.* 81, 2644–2651, 2009.

101. Dürr, E. and Bosshard, H.R., A monoclonal antibody induces opening of a coiled coil. Global protection of amide protons from H/D exchange decreases by up to 1000-fold in antibody-bound triple-stranded coiled coil, *Eur. J. Biochem.* 249, 325–329, 1997.

102. Keire, D.A., Solomon, T.E., and Reeve, J.R. Jr., NMR evidence for different conformations of the bioactive region of rat CCK-8 and CCk-58, *Biochem. Biophys. Res. Commun.* 293, 1014–1020, 2002.

103. Fogle, J.L. and Fernandez, E.J., Amide hydrogen-deuterium exchange: A fast tool for screening protein stabilities in chromatography, *LC-GD North America*, June, 96–101, 2006.

104. Fogle, J.L., O'Connell, J.P., and Fernandez, E.J., Loading, stationary phase, and salt effects during hydrophobic interaction chromatography: α-Lactalbumin is stabilized at high loadings, *J. Chromatogr. A* 1121, 209–218, 2006.

105. Ueberbacher, R., Haimer, E., Hahn, R., and Jungbauer, A., Hydrophobic interaction chromatography of proteins—V. Quantitative assessment of conformational change, *J. Chromatogr. A* 1198, 154–163, 2008.

18 Use of Chemical Modification for the Conformational Analysis of Biopharmaceuticals

It is not the purpose of this chapter to present a comprehensive view of the use of chemical modification to study the conformation of proteins and other biopolymers. The reader is referred to several sources for more information of the use of chemical modification to study proteins, nucleic acids, and polysaccharides.[1–5]

The majority of biopharmaceuticals are proteins or protein conjugates; there are oligonucleotides such as aptamers and microRNAs, which are finding increasing application. Polysaccharides such as dextran and cellulose have been in use for some time. Oligonucleotides and polysaccharides do not present the same challenge for conformational analysis as presented by proteins and analytical problems associated with characterization and stability studies of these biopolymers[6,7] are usually easily solved by the various spectroscopic techniques discussed elsewhere in this book. Oligonucleotides and polysaccharides are, in general, more robust than proteins and do not present a similar stability problem.

Proteins are polymers composed of different monomer units such that the protein is considered a heteropolymer.[8–11] The solution chemistry of proteins includes the response of proteins to changes in solvent,[12–16] which can be measured by physical techniques and chemical modification as well as the intrinsic reactivity of functional groups on proteins, which may or may not be influenced by protein conformation and primary structure.

The chemical modification of a biopolymer results from the reaction of a chemical reagent with a functional group or groups. The reaction can be highly specific as with the modification of an active site residue such as the active site serine in trypsin or nonspecific such as reaction of a protein with acetic anhydride at alkaline pH. In general, reaction with nonspecific reagents such as acetic anhydride or hydroxyl radicals is used for conformational analysis. However, it is recognized that loss in conformational integrity (e.g., denaturation) eliminates the unique reactivity of an active serine residue referenced above. A variety of reagents are used to effect the modification of amino acid residues in proteins. A list of some of the more commonly used reagents is provided in Table 18.1.

Most reagents used for the chemical modification of biopolymers react with nucleophiles and the nucleophilicity is dependent on the state of protonation. The environment surrounding a given nucleophilic group in a biopolymer may differ from

TABLE 18.1

Reagents for the Conformational Analysis by Chemical Modification Molecules

Reagent	Specificity/Conditions	Molecular Weight	Refs.
Acetic anhydride	Lysine, α-amino groups, tyrosine hydroxyl; preferred reaction is at lysine; pH 8 or greater; reaction can be "driven" α-amino groups at pH less than 6.5. Avoid nucleophilic buffers such as Tris; hydrolysis of the reagent is an issue above pH 9.5. Acetic anhydride has been used for trace labeling in the study of protein conformation and more recently the deuterated derivative has been used in proteomics for differential isotope tagging	102.1	[17–21]
N-Acetylimidazole (1-acetylimidazole)	Tyrosine hydroxyl groups, lysine ε-amino groups, transient reaction at histidine; neutral pH	110.1	[22–26]
N-Bromosuccinimide	Modification of tryptophan with some oxidative side reactions; pH 4–6	178	[27–31]
2,3-Butanedione (diacetyl)	Modification of arginine residues; reversible reaction with the product stabilized by the presence of borate; reaction at alkaline pH	86.1	[32–36]
1,2-Cyclohexanedione	Modification of arginine in the presence of borate. At pH 7–9, the reaction is reversible; above pH 9, the reaction is irreversible with the formation of several products	112.1	[37–41]
Diethylpyrocarbonate	Modification of histidine residues in proteins with transient modification of tyrosine; possible reaction at amino groups. Disubstitution of histidine results in ring-opening	162.1	[42–46]
EDC [1-ethyl-3(3-dimethylaminopropyl)-carbodimide]; N-(3-dimethylaminopropyl)-N'-ethylcarbodiimide	Modification of carboxyl groups in proteins frequently with N-hydroxylsuccinimide. Used for zero-length cross-linking in proteins and for the coupling of proteins to matrices and for the preparation of protein conjugates	155.2	[47–51]
Ellman's reagent (5,5'-dithio-bis-nitrobenzoic acid [DTNB])	Modification and measurement of cysteine (sulfhydryl) groups in proteins	396.4	[52–56]

TABLE 18.1 (continued)
Reagents for the Conformational Analysis by Chemical Modification Molecules

Reagent	Specificity/Conditions	Molecular Weight	Refs.
Formaldehyde[e,f] (reductive methylation)	Formation of a Schiff base with a primary amine (e.g., ε-amino group of lysine), which is reduced with sodium borohydride or more often with sodium cyanoborohydride, which results in the formation, then in the case of lysine, of ε-methyllysine and ε-dimethyllysine. The modification is performed at alkaline pH; In a related reaction, the formylation of tryptophan occurs with formic acid under acidic conditions (HCOOH/HCl) which is reduced by base[g]	30.0	[57–61]
2-Hydroxy-5-nitrobenzyl bromide (Koshland's reagent)	Modification of tryptophan by alkylation of the indole ring. Reaction at acid pH (pH 2–6). Disubstitution can occur. This was one of the first "reporter groups"[h]	232.0	[62–66]
Methyl acetimidate	Modification of amino groups. Imido esters are the functional groups for a number of cross-linking agents such as dimethylsuberimidate.[k] One of the more interesting imido esters is methyl picolinimidate.[l] Reaction at pH 8–10	109.6 as HCl	[67–71]
Phenylglyoxal	Modification of arginine residues in proteins; reaction accelerated in the presence of bicarbonate buffers; reaction at alkaline pH. p-Hydroxyphenylglyoxal and p-nitrophenylglyoxal are useful derivates[q]	134.1 as hydrate	[34,72–75]
Tetranitromethane (TNM)	Nitration of tyrosine residues in proteins with nitration and possible cross-linking; also reacts with sulfhydryl groups; possible reaction with indole ring of tryptophan. Reaction at alkaline pH; does introduce a "reporter group" in proteins. The nitrotyrosine function can be reduced to aminotyrosine with sodium dithionite (sodium hydrosulfite). The modification with peroxynitrite is a similar reaction[s]	196	[76–80]

that surrounding another similar nucleophile. This lack of homogeneity provides for a variety of surface polarities around the various functional groups. For example, consider the effect of the addition of an organic solvent, ethyl alcohol, on the pK_a of acetic acid. In 100% H_2O, acetic acid has a pK_a of 4.70. The addition of 80% ethyl alcohol results in an increase of the pK_a to 6.9. In 100% ethyl alcohol, the pK_a of acetic acid is 10.3. The reader is directed to a study by García-Moreno and workers[80] for a listing of residues with marked changes in pK_a values resulting from microenvironmental influences. The reader is also directed to study by Hnízda and coworkers[81] on microenvironmental influences on the reactivity of lysine and histidine residues in proteins (lysozyme and human serum albumin). They concluded that while a chemical modification with a reagent such as hydroxyl radical, anhydride, or tetranitromethane can be an indication of surface accessibility, other factors also contribute to reactivity. Considering the importance of this information, it is surprising that there are not more studies in this area. Some 70 years ago, Richardson[82] concluded that lowering the dielectric constant decreases the acidity (increases the pK_a of carboxylic acids with little effect on the dissociation of protonated amino groups). These observations were confirmed by Duggan and Schmidt.[83] The increase in the pK_a of carboxyl groups in organic solvents has a favorable effect on transpeptidation reactions,[84,85] where the carboxyl groups is required to be protonated. The point here is that reactivity of a functional group on a protein or other biopolymer is a measure of conformation and conformation is critical to function.

Other factors that can influence the pK_a of a functional group in a protein include hydrogen binding with an adjacent functional group, the direct electrostatic effect of the presence of a charged group in the immediate vicinity of a potential nucleophile, and direct steric effects on the availability of a given functional group.

There is another consideration that can in a sense be considered either a cause or consequence of microenvironmental polarity. This has to do with the environment immediately around the residue modified. These are the "factors" that can cause a "selective" increase (or decrease) in reagent concentration in the vicinity of a potentially reactive species. The most clearly understood example of this is the process of affinity labeling.[86] The modification of most functional groups in a protein by a chemical reagent is a second-order with an observed linear relationship reagent concentration and rate (actually the rate is proportional to the square of the concentration of one reactant or to the product of the concentration of the two reactants). From a practical perspective, the relationship between reagent concentration and reaction rate is linear (Figure 18.1) with a second-order reaction. An affinity label shows saturation kinetics where a point is reached where an increase in reagent concentration does not increase the reaction rate (Figure 18.1). Affinity labels are generally designed to contain the structural features of substrates or a competitive inhibitor (or other ligand) and a reactive group (generally of low chemical reactivity to preclude nonspecific reaction with other protein functional groups).[86] There are examples where "non-specific reagents show saturation kinetics when reacting with a protein."[87–89]

While conformational change must be considered in the interpretation of the results of the site-specific modification of a protein,[90–92] it has been more frequent that site-specific chemical modification is used to assess conformational change in

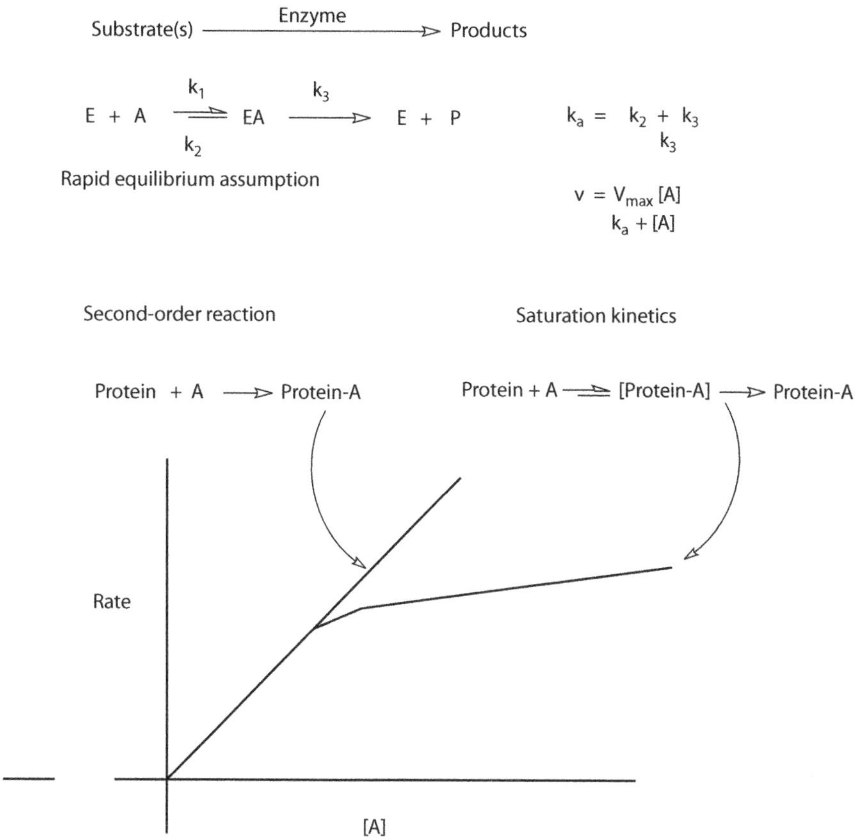

FIGURE 18.1   Saturation kinetics in the chemical modification of a biopolymer. The modification of a protein is usually a second-order reaction with the rate dependent on the concentration of both reactants. For practical purposes, the protein concentration is usually kept constant and the concentration of reagent varied and a straight line is obtained as shown in the figure as indicated (pseudo-first-order reaction). There are reactions where the reagent binds to the protein prior to the modification reaction resulting in saturation kinetics for the reaction as indicated in the figure (see Shen, W.C. and Colman, R.F., Cyanate modification of essential lysyl residues of the diphosphopyridine nucleotide-specific isocitrate dehydrogenase of pig heart, *J. Biol. Chem.* 250, 2973–2978, 1975; Hummel, C.F., Gerber, B.R., and Carty, R.P., Chemical modification of ribonuclease A with 4-arsono-2-nitrofluorobenzene, *Int. J. Protein Res.* 24, 1–13, 1984; Huynh, Q.K., Mechanism of inactivation of *Escherichia coli* 5-enolpyruvoylshikimate-3-phosphate synthase by *o*-phthalaldehyde, *J. Biol. Chem.* 265, 6700–6704, 1990). Denaturation usually eliminate the unique reactivity of a functional group as described above.

proteins.[93–102] The majority of studies use either lysine or cysteine as a target residue for modification.[103] The insertion of cysteine into recombinant proteins[104] via oligonucleotide-directed mutagenesis[105,106] has provided opportunity for the attachment of fluorescent probes to specific protein domains.[107] Fluorescent energy transfer (FRET) with covalently attached fluorescent probes is proving increasingly useful in the study of protein conformation.[108–113] The reaction of 1-dimethylaminonaphthalene-5-sulfonyl chloride (dansyl chloride) has been useful both in the structural analysis and amino group modification with proteins. Park and coworkers[114] used dansylation to improve sensitivity in peptide mass fingerprinting. Amoresano and coworkers[115] used dansyl chloride for the analysis of nitrotyrosine residues in proteins after reduction of the nitrotyrosyl residues to aminotyrosine. Cirulli and coworkers[116] used dansyl chloride for the labeling of bacterial surface proteins prior to mass spectrometric analysis.

Modification of protein amino groups with isothiocyanate derivatives of various dyes has proved to be an effective means of introducing structural probes into proteins at specific sites.[117] Fluorescein isothiocyanate has been used to modify cytochrome P-450 (reaction performed in 30 mM Tris, pH 8.0 containing 0.1% Tween 80; 2 h at 0°C in the dark),[118] actin (2 mM borate, pH 8.5; 3 h at ambient temperature then at 4°C for 16 h),[119] and ricin (pH 8.1, 6°C for 4 h).[120] The reader is directed to an elegant study on the effect of microenvironment on the fluorescence of arylaminophthalenesulfonates.[121]

Modification at tyrosine and carboxyl functional groups can also be useful for study of conformation. Modification with hydroxyl radicals or methylene carbine is not dependent on the presence of a specific functional group. Tetranitromethane (TNM) nitrates tyrosyl residues to yield the 3-nitro derivative, which markedly lowers the pK_a of the phenolic hydroxyl group (Figure 18.2). The modification of tyrosine with TNM is influenced by the ionization state of tyrosine, which is a function of the microenvironment around the residue. It is extremely useful for investigators to review early literature on the factors influencing tyrosine ionization in proteins[122–131] when considering the use of TNM as a conformational probe. The modification proceeds optimally at mildly alkaline pH. The rate of modification of N-acetyltyrosine is twice as rapid at pH 8.0 as at pH 7.0; it is approximately 10 times as rapid at pH 9.5 as at pH 7.0. Although the reaction of tetranitromethane with proteins is reasonably specific for tyrosine, oxidation of sulfydryl groups has been reported as has reaction with histidine, methionine, and tryptophan.[132–135] Reaction with TNM can also result in the covalent cross-linkage of tyrosyl residues resulting in inter- and intramolecular association of peptide chains.[136] Acidification of reaction mixtures tends to favor the cross-linkage reaction.[137] The extent of cross-linkage varies with the protein being studied. For example, reaction of pancreatic deoxyribonuclease with tetranitromethane results in extensive formation of dimer.[138] Early studies on the modification of tyrosyl residues by TNM was assessed by either spectrophotometric means or by amino acid analysis[139,140] but mass spectrometry is proving to be of increasing value in the analysis of nitrotyrosine in proteins.[141–143] Mass spectrometry permitted the identification of a dinitrotyrosine secondary to reaction with TNM. Antibodies to 3-nitrotyrosine in proteins have been developed and are useful not only for the analysis of purified proteins[144–146] and for identification on 2-D gel electrophoretograms.[147] Reduction with sodium dithionite (see below) improves the specificity for the detection of nitrotyrosine-containing proteins on Western blots; nitrotyrosine-positive bands are eliminated on reduction leaving the false-positive spots.[147,148]

Riordan and coworkers[133] suggested that the reactivity of tyrosyl residues in proteins with TNM was a measure of exposure to solvent. This concept of "free" and

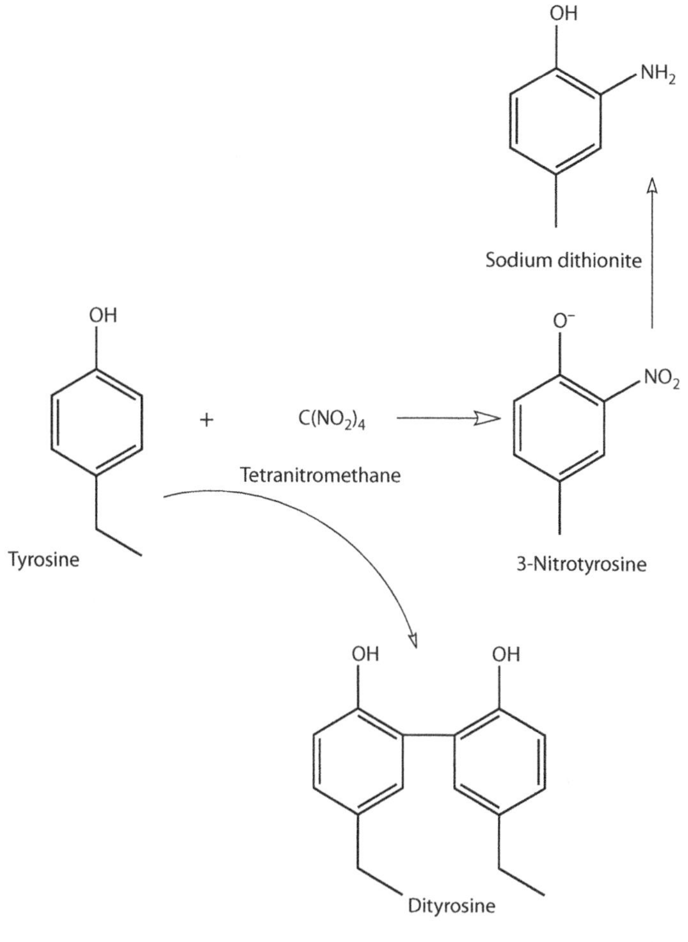

FIGURE 18.2   The reaction of tetranitromethane with tyrosine in proteins (see Ghesquière, B. and Gevaert, K., Improved tandem mass spectrometric characterization of 3-nitrotyrosine sites in peptides, *Rapid Commun. Mass Spectrom.* 20, 2885–2893, 2006; Lennon, C.W., Cox, H.D., Hennelly, S.P. et al., Probing structural differences in prion protein isoforms by tyrosine nitration, *Biochemistry* 46, 4850–4860, 2007; Lee, S.J., Lee, J.R., Kim, Y.H. et al., Investigation of tyrosine nitration and nitrosylation of angiotensin II and bovine serum albumin with electrospray ionization mass spectrometry, *Rapid Commun. Mass Spectrom.* 21, 2797–2804, 2007; Liu, H., Gaza-Bulseco, G., Chumsae, C. et al., Mass spectrometry analysis of *in vitro* nitration of a recombinant human IgG1 monoclonal antibody, *Rapid Commun. Mass Spectrom.* 22, 1–10, 2008).

"buried" residues was introduced earlier by the same group in their studies on the reaction of tyrosine with *N*-acetylimidazole in proteins (Figure 18.3).[149] In this model, "free" tyrosyl residues are considered to be in direct contact with solvent and have pK_a values between 9.5 and 10.5 while "buried" tyrosyl residues are relatively inaccessible to solvent and have pK_a values above 10.5. The general applicability of this

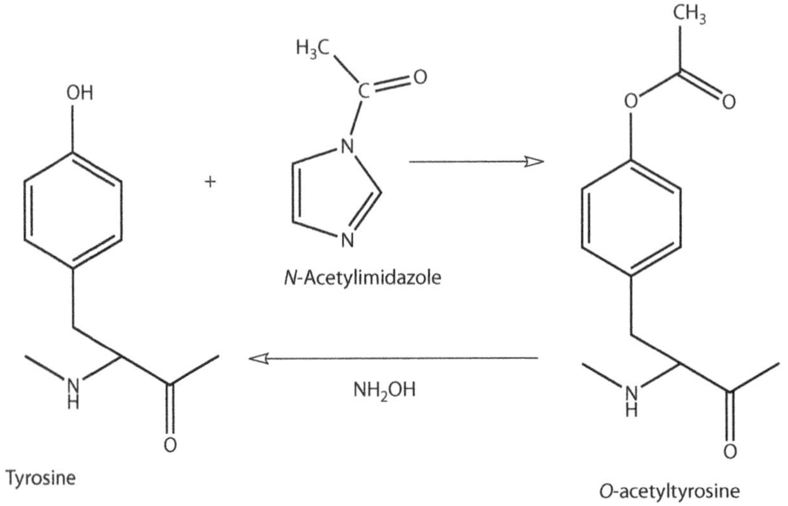

FIGURE 18.3 The reaction of *N*-acetylimidazole with tyrosine in proteins (see Martin, B.L., Wu, D., Jakes, S., and Graves, D.J., Chemical influences on the specificity of tyrosine phosphorylation, *J. Biol. Chem.* 265, 7108–7111, 1990; Silberring, J. and Nyberg, F., Analysis of tyrosine- and methionine-containing neuropeptides by fast atom bombardment mass spectrometry, *J. Chromatogr.* 562, 459–467, 1991; el Kabbaj, M.S. and Latruffe, N., Chemical reagents of polypeptides side chain: Relationships between properties and ability to cross the inner mitochondrial membranes, *Cell. Mol. Biol.* 40, 781–786, 1994; Zhang, F., Gao, F., Weng, J. et al., Structural and functional differentiation of three groups of tyrosine residues by acetylation of *N*-acetylimidazole in manganese stabilizing protein, *Biochemistry* 44, 719–725, 2005).

correlation between reactivity and solvent accessibility was challenged by Myers and Glazer[150] in studies on subtilisin. These investigators argued that tyrosyl residues in apolar locations are preferentially modified by TNM or *N*-acetylimidazole. It is not at all clear that reactivity correlates with solvent exposure or lack of solvent exposure.

Cysteine is (usually) the most powerful nucleophile in a protein and, as a result, is frequently the easiest to selectively modify with a variety of reagents; it also most frequently modified by reagents intended for other residues. Cysteine is the sulfur analogue of serine where the hydroxyl group is replaced with a sulfhydryl group. The reader is directed to an excellent review of thiols[151] for a more thorough discussion of both aliphatic and aromatic thiols. The bond dissociation energy for sulfhydryl groups is substantially less than that of the corresponding alcohol function, providing a basis for the increased acidity of sulfhydryl groups; for example, the pK_a for ethanethiol is 10.6 while it is 18 for ethanol. As a consequence, while the reaction of cysteine with chloroacetate is slow (5.3×10^{-3} M^{-1} min^{-1}); reaction with serine is nonexistent under the same conditions; reaction of a cysteine residue at an enzyme active site (papain) is some 30,000 times faster ($150 M^{-1}$ min^{-1}) than that of free cysteine at pH 6.0.[152] There is a single cysteine residue in blood coagulation factor VIII and in albumin but the presence of a free cysteine in biopharmaceutical proteins is rare.

Modification of cysteine residues proceeds via either a nucleophilic addition or displacement reaction with the thiolate anion as the nucleophile (Figure 18.4). The

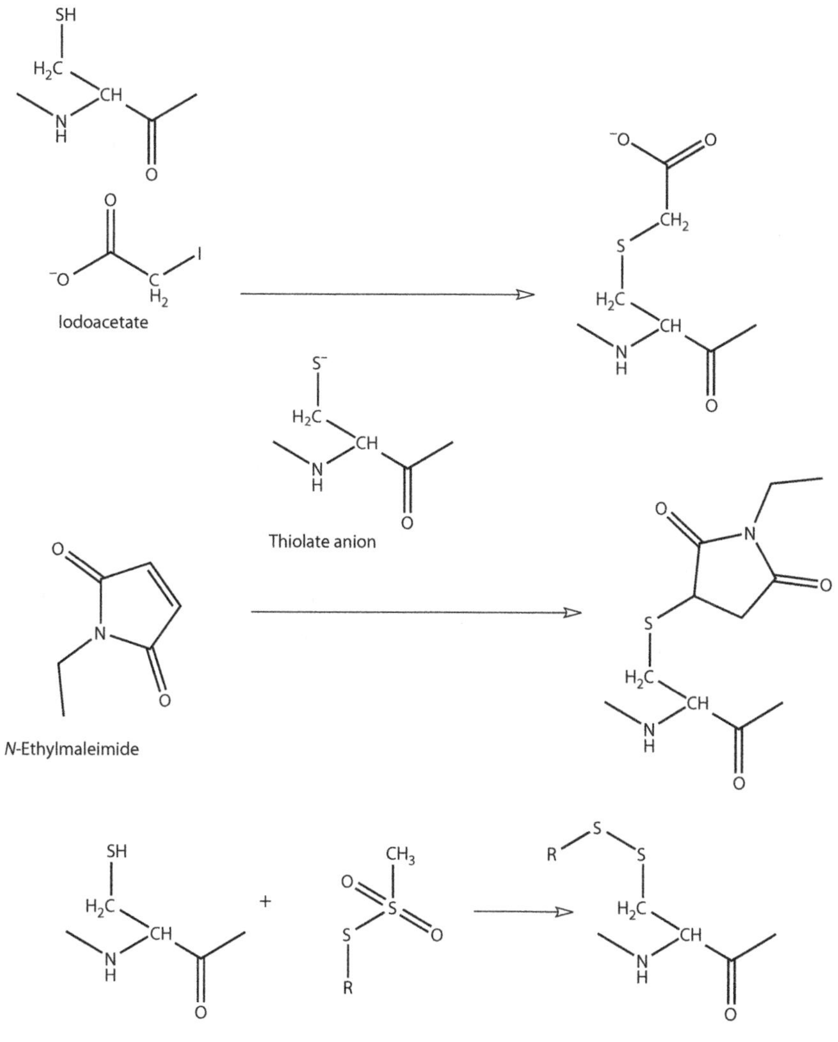

FIGURE 18.4 The modification of cysteine residues in proteins. Shows in modification of cysteine anion with iodoacetate (an alkylation reaction), the reaction with *N*-ethylmaleimide (a Michael addition reaction), and modification with an alkyl methanethiosulfonate (see Stable, E. and Finn, M.G., Chemical modification of viruses and virus-like particles, *Curr. Top. Microbiol. Immunol.* 327, 1–21, 2009; Wynn, R. and Richards, F.M., Chemical modification of protein thiols: Formation of mixed disulfides, *Methods Enzymol.* 251, 351–356, 1995; Carne, A.F., Chemical modification of proteins, *Methods Mol. Biol.* 32, 311–320, 1994).

reaction with the α-keto-haloalkyl compounds such as iodoacetate is an example of a nucleophilic displacement reaction while the reaction of maleimide is a nucleophilic addition to an olefin. This reaction is an example of a Michael reaction or Michael addition. In addition to the review cited above, there other reviews on sulfhydryl chemistry.[153–157] Local environment has a profound effect on the reactivity of cysteine residues in proteins. It has been shown[158] that local electrostatic potential modulates reactivity of individual cysteine residue in rat brain tubulin. Rat brain tubulin dimer contains 20 cysteine residues; 12 residues in the α-subunit and 8 in the β-subunit. The rates of reaction of the cysteine residues in rat brain tubulin were determined with a variety of reagents in 0.3 M MES, pH 6.9 containing 1.0 mM EGTA, and 1 mM MgCl$_2$ in the dark. The reagents evaluated included *syn*-monobromobimane, *N*-ethylmaleimide, iodoacetamide, and [5-((((2-iodoacetyl)amino) ethyl) amino) naphthalene-1-sulfonic acid] (AEDANS). Approximately 50% of the 20 sulfhydryl groups react equivalently with all reagents. Reaction is slower with iodoacetamide than with *N*-ethylmaleimide and a greater number of cysteine residues are modified with *N*-ethylmaleimide than with iodoacetamide. It is suggested that the difference in the rates of reaction is ascribed to the differences in the chemistry of the reaction of the two compounds with the thiolate ion with the reaction with iodoacetamide being a nucleophilic displacement while the reaction with *N*-ethylmaleimide is an addition reaction.

Ellman developed 5,5′-dithiobis(2-nitrobenzoic acid) (DTNB; Ellman's reagent),[159] which today is one of the more popular reagents for the modification and determination of the sulfhydryl group (Figure 18.5). Reaction with sulfhydryl groups in proteins results in the release of 2-nitro-5-mercaptobenzoic acid, which has a molar extinction coefficient of 13,600 M^{-1} cm^{-1} at 410 nm. Riddles and coworkers[160,161] studied the chemistry of DTNB in alkaline solutions as well as the reaction of DTNB with thiols. They concluded that the rate of reaction of DTNB was dependent on the pH and the pK_a of the thiol residue. The steric and electrostatic considerations, which have been discussed above, are also applicable to this reaction. At pH 7.0 and 25°C, model thiols and some protein thiols react rapidly but there are examples of protein thiols, which react more slowly ($t_{1/2} \geq 10$ min). There are several examples of the use of DTNB to study protein conformation.[162–165]

The modification of methionine in a native protein is generally accomplished with considerable difficulty. Since the dissociation of a proton from sulfur is unnecessary to generate the nucleophile, relatively specific derivatization by alkylating agents can be accomplished at low pH. Protonation of other nucleophiles such as cysteine or lysine restricts modification to methionine. The reversible alkylation of methionine by iodoacetate in dehydroquinase has been reported by Kleanthous and coworkers.[166] In this reaction, iodoacetate behaves kinetically as an affinity label with a K_i of 30 μM and a k_{inact} of 0.014 min^{-1}, pH 7.0 (50 mM potassium phosphate). There is no reaction with iodoacetamide. In a companion study, Kleanthous and Coggins[167] demonstrated that 2-mercaptoethanol treatment under alkaline conditions (0.5% ammonium bicarbonate, 37°C) could reverse modification at one of the two residues. If the modified protein is denatured, there is no reversal of modification at either residue. The results are interpreted in terms of the proximity of a positive charge (i.e., lysine) in close proximity to one of the two methionyl residues, which

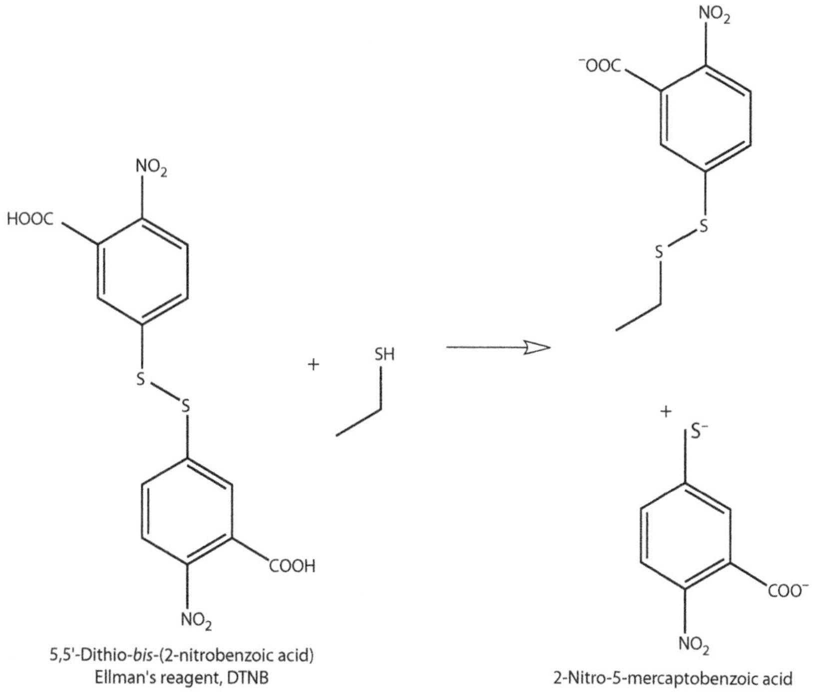

NO$_2$

$^-$OOC

$^-$OOC

S—S

+

S$^-$

COO$^-$

NO$_2$

HOOC

NO$_2$

S—S

S

+   SH   ⟶

COOH

NO$_2$

5,5'-Dithio-*bis*-(2-nitrobenzoic acid)
Ellman's reagent, DTNB

2-Nitro-5-mercaptobenzoic acid

FIGURE 18.5   The reaction of cysteine with Ellman's reagent (see Owusu-Apenten, R., Colorimetric analysis of protein sulfhydryl groups in milk: Applications and processing effects, *Crit. Rev. Food Sci. Nutr.* 45, 1–23, 2005; Landino, L.M., Mall, C.B., Nicklay, J.J. et al., Oxidation of 5-thio-2-nitrobenzoic acid, by the biological relevant oxidants peroxynitrite anion, hydrogen peroxide and hychlorous acid, *Nitric Oxide* 18, 11–18, 2008; Gergel, D. and Cederbaum, A.I., Interaction of nitric oxide with 2-thio-5-nitrobenzoic acid: Implications for the determination of free sulfhydryl groups by Ellman's reagent, *Arch. Biochem. Biophys.* 347, 282–288, 1997; Riener, C.K., Kada, G., and Gruber, H.J., Quick measurement of protein sulfhydryls with Ellman's reagent and with 4,4'-dithiopyridine, *Anal. Biochem. Chem.* 373, 266–276, 2002; Hansen, R.E., Østergaard, H., Nørgaard, P., and Winther, J.R., Quantification of protein thiols and dithiols in the picomolar range using sodium borohydride and 4,4'-dithiopyridine, *Anal. Biochem.* 363, 77–82, 2007; Giustarini, D., Dalle-Donne, I., Colombo, R. et al., Is ascorbate able to reduce disulfide bridges? A cautionary note, *Nitric Oxide* 19, 252–258, 2008).

(1) provides the basis for the affinity labeling and (2) provides the basis for the 2-mercaptoethanol-mediated reversal of modification. The observations that "non-specific" reagents such as iodoacetate can demonstrate saturation kinetics similar to those observed for affinity reagents[86] as described above. The ability to reverse the alkylation of methionine under relatively mild conditions as described above has resulted in the development of a clever affinity approach to the purification of methionine peptides. Several groups[168–170] have reported the isolation of methionine peptide from a mixtures by coupling to a bead containing a bromoacetyl function

under acidic conditions (e.g., 25% acetic acid); the peptide is subsequently released by elution with a reducing agent under alkaline conditions.

Oxidation of methionine (Figure 18.6) has been observed in the manufacture of recombinant proteins.[171–176] Conversion of methionine sulfoxide to methionine sulfone is essentially irreversible under common solvent conditions and requires more vigorous reagents such as performic acid.[177] The oxidation of methionine is used for protein surface mapping.[178] Reagents include hydroxyl radical, chloramine T,[179,180] and hydrogen peroxide.[181,182] The reaction of methionine with chloramine T can be followed spectrophotometrically.[181] The development t-butyl hydroperoxide by Keck[183] as a selective oxidizing agent for methionine in proteins represented a significant advance. Two methionine residues were oxidized in recombinant interferon with t-butyl hydroperoxide, while all five residues were oxidized to a varying extent by H_2O_2 under the same reaction conditions. Three methionine residues were oxidized in native tissue-type plasminogen activator; all five residues were oxidized to varying degrees in the presence of 8.0 M urea. t-Butyl hydroperoxide has been successfully used for recombinant human leptin (100 mM sodium borate, pH 9.0, room temperature)[184] and recombinant human granulocyte colony-stimulating factor (25 mM sodium acetate, pH 4.5, 25°C).[185]

Present approaches to the site-specific modification of arginyl residues in proteins used three reagents: phenylglyoxal (and derivatives such as p-hydroxyphenylglyoxal),[186] 2,3-butanedione,[187] and 1,2-cyclohexanedione.[188] A review of the literature suggests

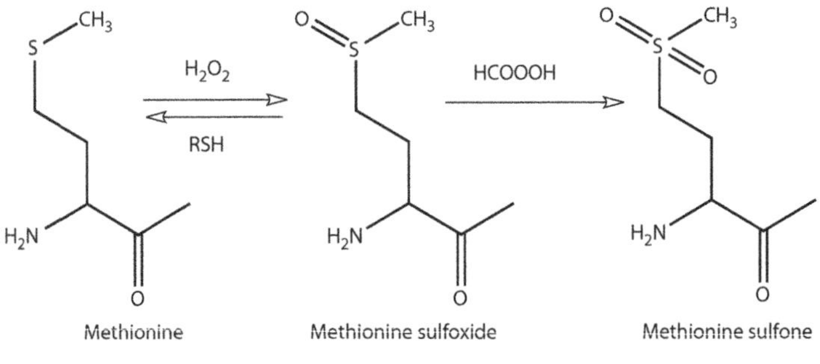

FIGURE 18.6 The oxidation of methionine residues in proteins. The reversible oxidation of methionine to methinone sulfoxide is shown as well as the subsequent further irreversible oxidation of methionine sulfoxide to methionine sulfone (see Jenkins, N., Modification of therapeutic proteins: Challenges and prospects, *Cryotechnology* 53, 121–125, 2007; Liu, H., Gaza-Bulseco, G., and Zhou, L., Mass spectrometry analysis of photo-induced methionine oxidation of a recombinant human monoclonal antibody, *J. Am. Soc. Mass Spectrom.* 20, 525–528, 2009; Takenawa, T., Yokota, A., Oda, M. et al., Protein oxidation during long storage: Identification of the oxidation sites in dihydrofoloate reductase from *Escherichia coli* through LC-MS and fragment studies, *J. Biochem.*, 145, 517, 2009; Pan, H., Chen, K., Chu, L. et al., Methionine oxidation in human IgG2 Fc decreases binding affinities to protein A and FcRn, *Protein Sci.* 18, 424–433, 2009; Jiang, H., Wu, S.L., Karger, B.L., and Hancock, W.S., Mass spectrometric analysis of innovator, counterfeit, and follow-on recombinant human growth hormone, *Biotechnol. Prog.* 25, 207–218, 2009).

that phenylglyoxal is the most extensively used reagent for the site-specific chemical modification of arginine in proteins. As with other site-specific chemical modifications of proteins, there has been increasing use of mass spectrometry to characterize the chemical modification of arginine in proteins.[189–191]

The use of phenylglyoxal (Figure 18.7) was developed by Takahashi[187] in 1968. The stoichiometry of the reaction involves the reaction of 2 mol of phenylglyoxal with 1 mol of arginine. The reaction of arginine with phenylglyoxal is greatly accelerated in bicarbonate–carbonate buffer systems.[192] The reaction of methylglyoxal with arginine is also enhanced by bicarbonate, while a similar effect is not seen with either glyoxal or 2,3-butanedione.

The use of carbodiimide-mediated modification[193,194] (Figure 18.8) is the most extensively used method for the modification of carboxyl groups in proteins. Carbodiimides react with protonated carboxyl groups yield an activated intermediate, most likely an acylisourea, which then reacts with a nucleophile such as an amine.[6,91] Carbodiimides are also used for zero-length cross-linking (Figure 18.9) of proteins between proximate lysine residues and carboxyl groups.[195–197] Zero-length cross-linking can be of value in the study of protein conformation.[198–203]

Other approaches, which have been advanced for surface labeling (conformational analysis) of proteins/biopolymers, include the photochemical labeling of tryptophan (Figure 18.9),[204] modification with diethylpyrocarbonate (Figure 18.10),[205] and formaldehyde cross-linking (Figure 18.11).[206] The reaction with formaldehyde can be quite complex[207,208] including the process of reductive methylation (Figure 18.12), which also could be used for surface labeling.[209,210]

Footprinting is a procedure for defining the surface area of a protein or nucleic acid, which is involved in interaction with ligand that could interact with another biopolymer or peptide/oligonucleotide. The process is based on the protection of a surface area from modification by rapidly reacting chemical reagent. The chemistry, which is used for footprinting, can also be used to define exposed area or conformation of biopolymers.

The chemistry used for footprinting is also used for the conformational analysis of biopolymers. The chemistry (e.g., nitration, carbodiimide, etc.) described in the earlier sections can be used for the footprinting/conformational analysis of biopolymers. The use of hydroxyl radical technology and trace labeling of lysine residues with anhydrides and amidates combined with mass spectrometric analysis appears to the most useful approach for accurate footprinting/conformational analysis. It is advised that these techniques yield exceptional data but require significant expertise; some of the other techniques discussed above (e.g., nitration, diethylpyrocarbonate) yield somewhat less sophisticated data but are somewhat more forgiving of experimental technique. Photooxidation and iodination are discussed below as there are similarities to free radical and carbine-based approaches.

Brian Hartley was the first to use the term fingerprinting in connection with the study of protein structure in a paper with Jim Brown in 1966.[211] This particular paper described the identification of disulfide bonds by diagonal paper electrophoresis. The concept of fingerprint arose from the pattern obtained from the 2-D (orthogonal) paper separation of peptides.[212] This technique developed by Virginia Richmond and Hartley in Seattle was a modification of an earlier procedure developed by Hans

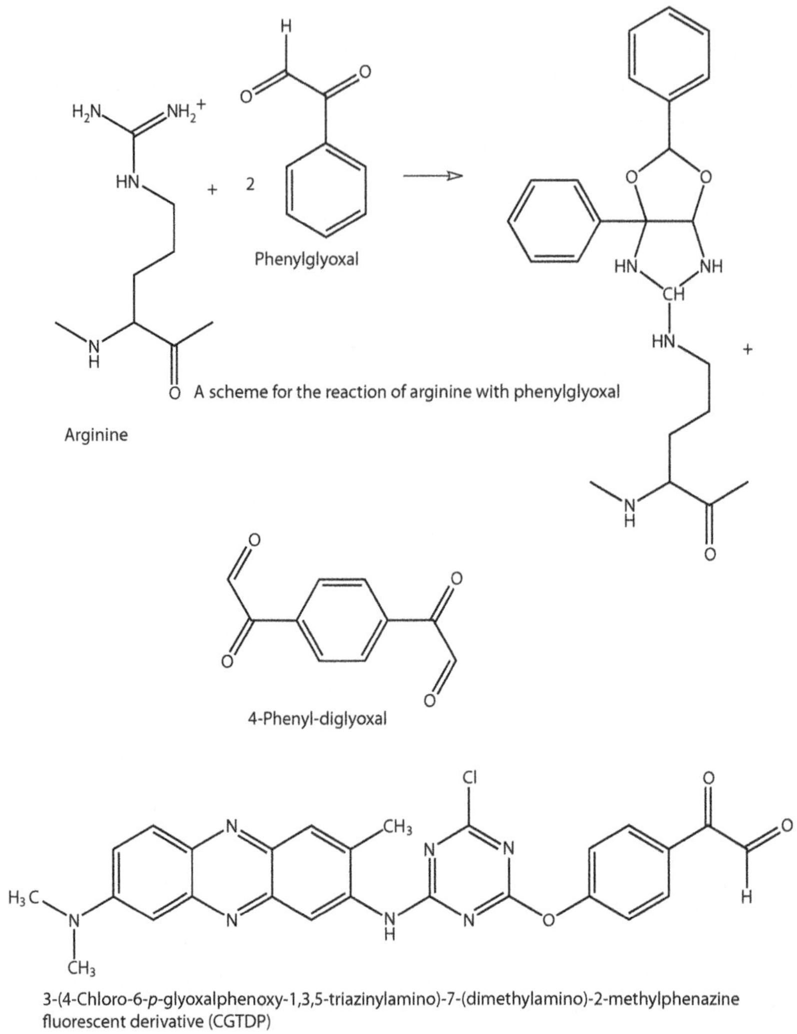

A scheme for the reaction of arginine with phenylglyoxal

4-Phenyl-diglyoxal

3-(4-Chloro-6-*p*-glyoxalphenoxy-1,3,5-triazinylamino)-7-(dimethylamino)-2-methylphenazine
fluorescent derivative (CGTDP)

FIGURE 18.7 The reaction of phenylglyoxal with arginine residues in proteins. Also is a
recently developed reagent for the cross-linking of arginine residues via phenylglyoxal chemistry (Zhang, Q., Crosland, E., and Fabris, D., Nested arg-specific bifunctional crosslinkers for
MS-based structural analysis of proteins and protein assemblies, *Anal. Chim. Acta* 627, 117–128,
2008) and a fluorescent derivative (Wang, S., Wang, X, Shi, W. et al., Detection of local polarity
and conformational changes at the active site of rabbit muscle creatine kinase with a new arginine-specific fluorescent probe, *Biochim. Biophys. Acta* 1784, 415–422, 2008). See also Saraiva,
M.A., Borges, C.M., and Florencio, M.H., Non-enzymatic model glycation reactions—A comprehensive study of the reactivity of a modified arginine with aldehydic and diketonic dicarbonyl
compounds by electrospray mass spectrometry, *J. Mass Spectrom.* 41, 755–770, 2006; Johans,
M., Milanesi, E., Franck, M. et al., Modification of permeability transition pore arginine(s) by
phenylglyoxal derivatives in isolated mitochondria and mammalian cells. Structure-function
relationships of arginine ligands, *J. Biol. Chem.* 280, 12130–12136, 2005.

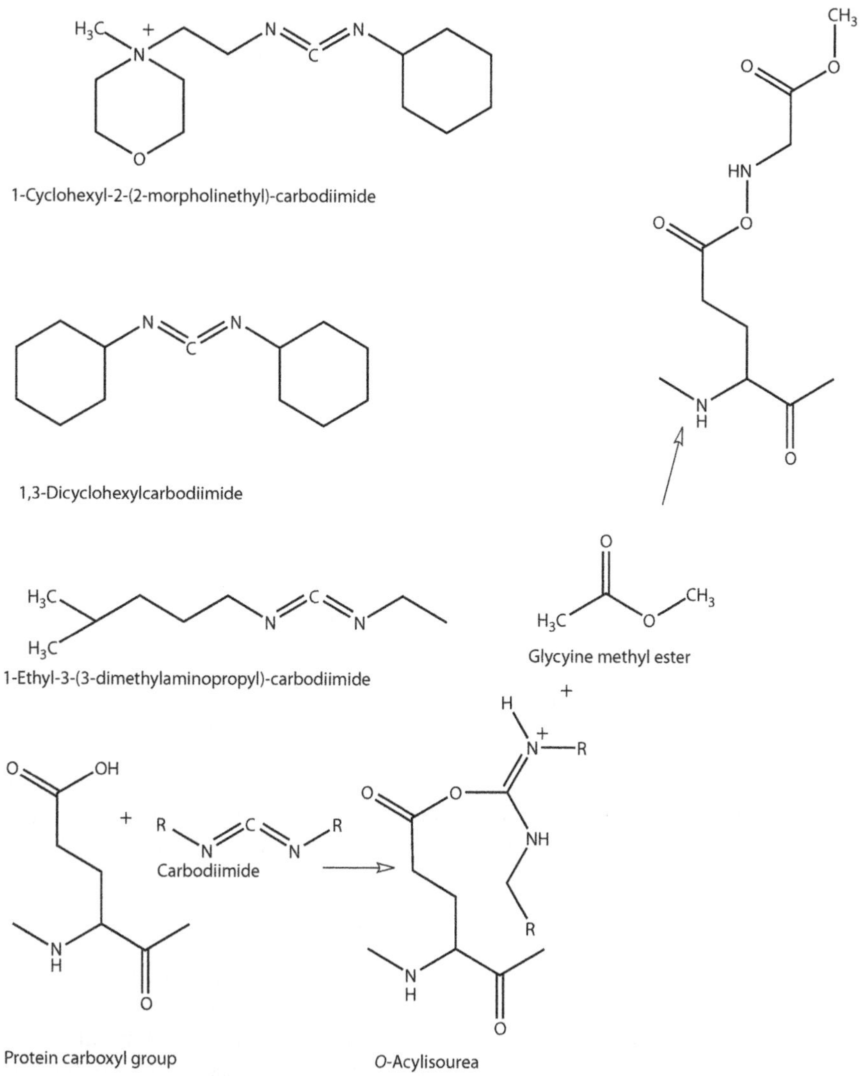

1-Cyclohexyl-2-(2-morpholinethyl)-carbodiimide

1,3-Dicyclohexylcarbodiimide

1-Ethyl-3-(3-dimethylaminopropyl)-carbodiimide

Glycyine methyl ester

Protein carboxyl group

Carbodiimide

O-Acylisourea

FIGURE 18.8 The use of carbodiimides for the modification of carboxyl groups in bio-polymers (see Bricker, A.L. and Belasco, J.G., Importance of a 5′ stem-loop for longevity of papAmRNA in *Escherichia coli, J. Bacteriol.* 181, 3587–3590, 1999; Bui, C.T., Rees, K., Lambrinakos, A. et al., Site-selective reactions of imperfectly matched DNA with small chemical molecules: Applications I mutation detection, *Bioorg. Chem.* 30, 216–232, 2002; Evran, S. and Telefoncu, A., Modification of porcine pancreatic lipoase with Z-proline, *Prep. Biochem. Biotechnol.* 35, 191–201, 2005; Boudet, C., Iliopoulos, I., Poncelet, O. et al., Control of the chemical cross-linking of gelatin by a thermosensitive polymer: Example of switchable reactivity, *Biomacromolecules* 6, 3073–3078, 2005).

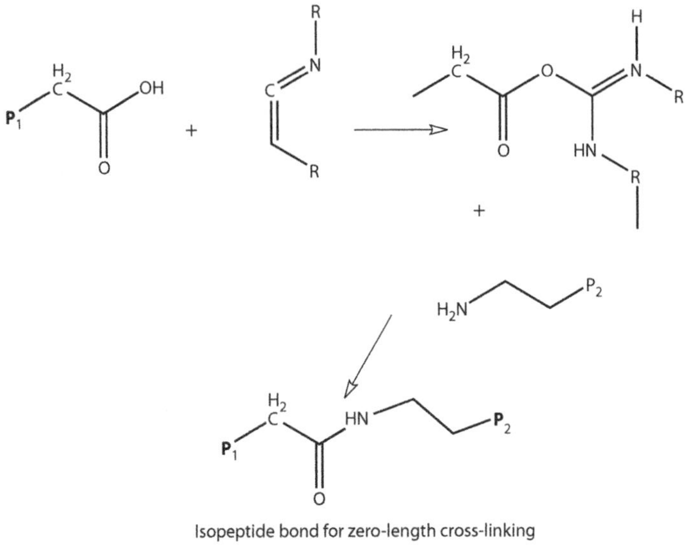

Isopeptide bond for zero-length cross-linking

FIGURE 18.9 Xero-length cross-linking in proteins. For additional information, see Novak, P. and Kruppa, G.H, Intra-molecular cross-linking of acidic residues for protein structure studies, *Eur. J. Mass Spectrom.* 14, 355–365, 2008; Maroufi, B., Ranjbar, B., Khajeh, K. et al., *Biochim. Biophys. Acta* 1784, 1043–1049, 2008; El-Shafey, A., Tolic, N., Young, M.M. et al., "Zero-length" cross-linking in solid state as an approach for analysis of protein-protein interactions, *Protein Sci.* 15, 429–440, 2006; Schmidt, A., Kalkof, S., Ihling, C. et al., Mapping protein interfaces by chemical cross-linking and Fourier transform ion cyclotron resonance mass spectrometry: Application to a calmodulin/adenyl cyclase 8 peptide complex, *Eur. J. Mass Spectrom.* 11, 525–534, 2005; Simons, B.L., King, M.C., Cyr, T. et al., Covalent cross-linking of proteins without chemical reagents, *Protein Sci.* 11, 1558–1564, 2002.

Neurath and colleagues.[213] These basic concepts have been extended into current protein science but using much more sophisticated analytical techniques such as UP-HPLC (ultraperformance-high performance liquid chromatography) and mass spectrometry.[214] Peptide mass fingerprinting is a technique used to identify proteins on the mass spectrometric analysis of digests of the protein analyte with a protease, usually trypsin. The masses of the various peptides obtained from mass spectrometric analysis are matched with a database providing accurate identification of the protein.[215–217] This technique has seen extensive use in proteomics for the identification of proteins.[218–224]

Footprinting is related to fingerprinting by terminology (anatomic) but refers to an entirely different concept. The term footprinting was used to describe an experimental approach where protection from enzymatic hydrolysis was used to define regions of DNA, which were bound protein or other ligands.[225–228] Subsequently, chemical modification with hydroxyl radicals has been used for DNA footprinting.[229–234] The reaction of hydroxyl radicals with nucleic acids[235] can lead to complex products (Figure 18.13).[236]

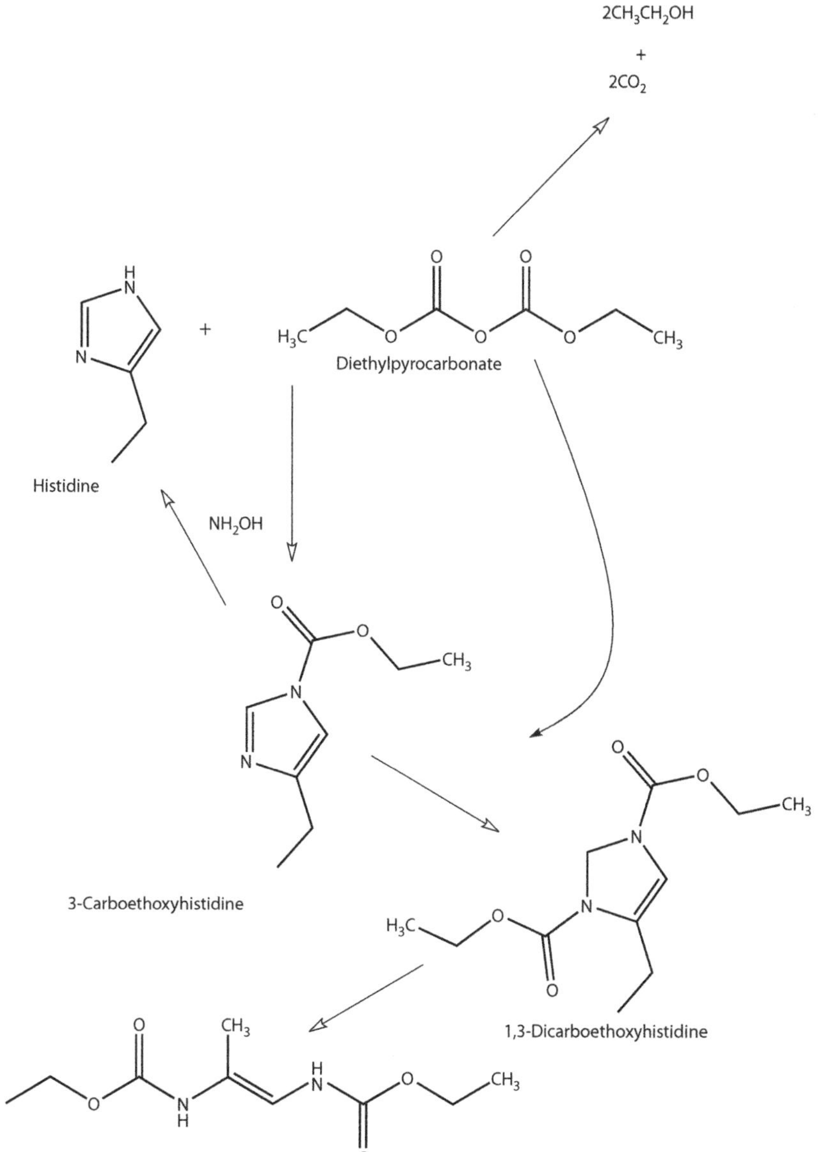

FIGURE 18.10 The use of diethylpyrocarbonate for surface mapping of proteins (see Mendoza, V.L. and Vachet, R.W., Protein surface mapping using diethylpyrocarbonate with mass spectrometric detection, *Anal. Chem.* 80, 2895–2904, 2008).

The definition of surface area by the footprinting of proteins with hydroxyl radicals can be considered an extension of much earlier work on the use of site-specific chemical modification to define interaction protein surface areas in macromolecular protein complexes.[237–241] As with nucleic acids, most work on protein footprinting

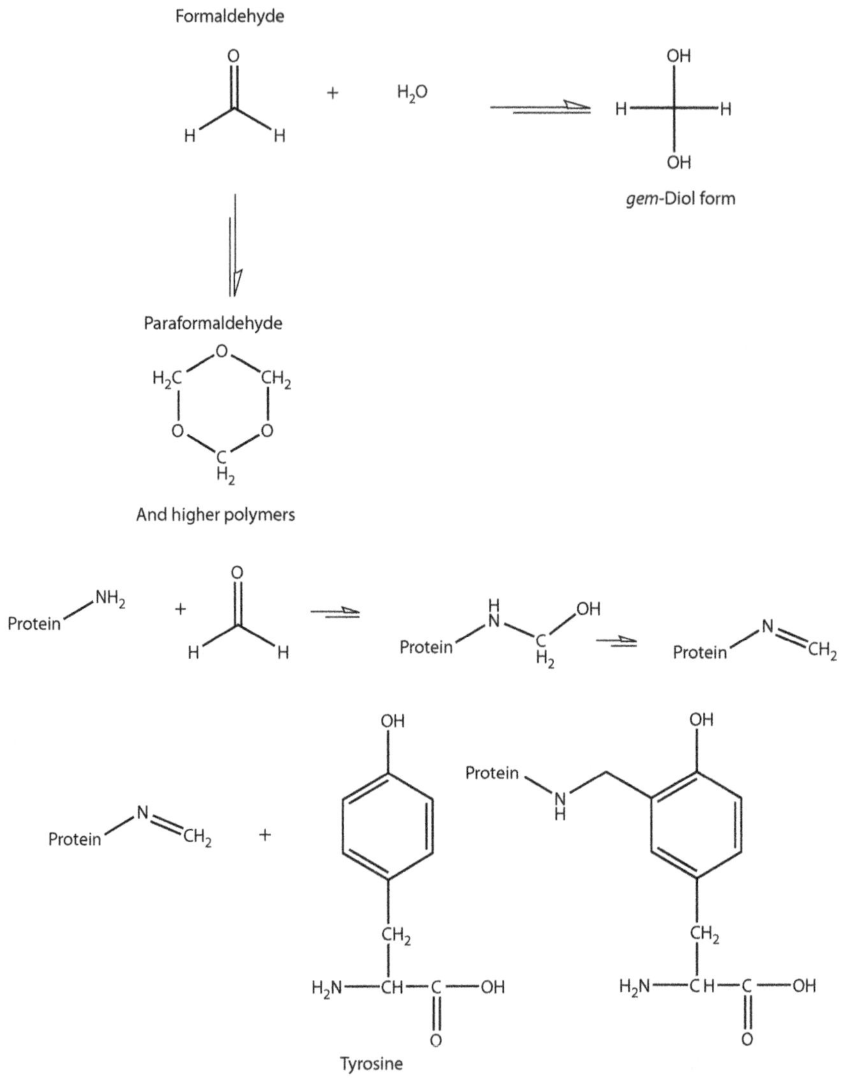

FIGURE 18.11 The use of formaldehyde for the cross-linking of protein (see Sutherland, B.W., Toews, J., and Kast, J., Utility of formaldehyde cross-linking and mass spectrometry in the study of protein-protein interactions, *J. Mass Spectrom.* 43, 699–715, 2008).

uses hydroxyl radicals, which are not dependent on the modification of a specific residue for modification of a polypeptide or polynucleotide.[242,243] Hydroxyl radicals (Figure 18.14) can be generated by radiolytic cleavage (synchrotron cleavage)[244–251] of water or photolysis of H_2O_2 or by the Fenton reaction (also from H_2O_2).[252–255] The Fenton reaction involved the reaction of ferrous iron with hydrogen peroxide to form

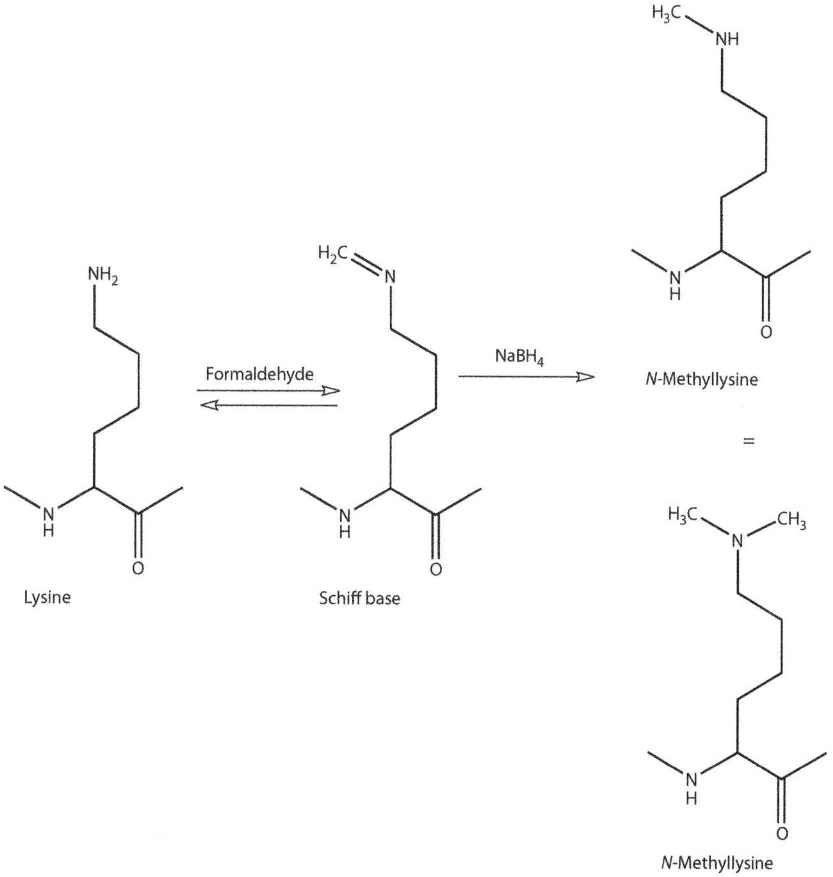

FIGURE 18.12 The use of reductive methylation to modify lysyl residues in proteins. The basic chemistry was developed by Means and coworkers (Means, G.E., Reductive alkylation of amino groups, *Methods Enzymol.* 47, 469, 1977). For applications directed at the study of biopolymer conformation, see Robinson, J.P., Picklesimer, J.B., and Puett, D., Tetanus toxin. The effect of chemical modifications on toxicity, immunogenicity, and conformation, *J. Biol. Chem.* 250, 7435–7442, 1975; Rice, R.H., Means, G.E., and Brown, W.D., Stabilization of bovine trypsin by reductive methylation, *Biochim. Biophys. Acta* 492, 316–321, 1977; Jentoft, J.E. and Rayford, R., Comparison of the Fc fragment from a human IgG1 and its CH2, pFc', and tFc' subfragments. A study using reductive methylation and ^{13}C NMR, *Biochemistry* 28, 3250–3257, 1989; Jentoft, J.E., Reductive methylatoin and carbon-13 nuclear magnetic resonance in structure–function studies of Fc fragment and its subfragments, *Methods Enzymol.* 203, 261–274, 1991: Abraham, S.J., Hobeisel, S. and Gaponenko, V., Detection of protein–ligand interactions by NMR using reductive methylation of lysine residues, *J. Biomol. NMR* 42, 143–148, 2008.

ferric ions, hydroxyl ion, and hydroxyl radical.[256] Iron is most frequently related to the generation of hydroxyl radical. Iron is also implicated in the generation of peroxides in cell culture media.[257] Copper ions can also generate hydroxyl radicals[258] and may be more effective than iron.[235] Hydroxyl radicals react with a variety of amino acids

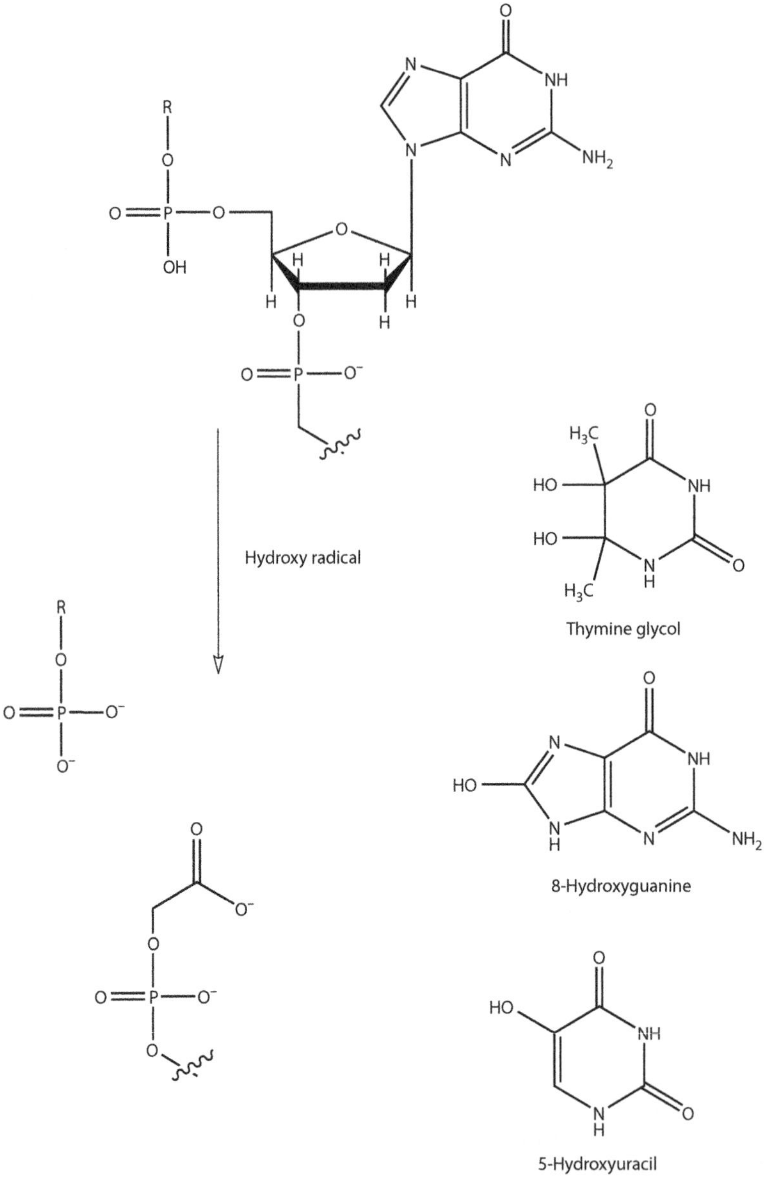

FIGURE 18.13 Some products for the hydroxyl radical modification of nucleic acids. Shown is the cleavage of DNA and some products identified from hydroxyl radicals (see Hertzberg, R.P. and Dervan, P.B., Cleavage of DNA with methidiumpropyl-EDTA-iron(II): Reaction conditions and product analyses, *Biochemistry* 23, 3934–3945, 1984; Jain, S.S. and Tullius, T.D., Footprinting protein–DNA complexes using the hydroxyl radical, *Nat. Protocols* 3, 1092–1100, 2008; Dizdaroglu, M., Aruoma, O.I., and Halliwell, B., Modification of bases in DNA by copper ion-1,10-phenanthroline complexes, *Biochemistry* 29, 8447–8451, 1990).

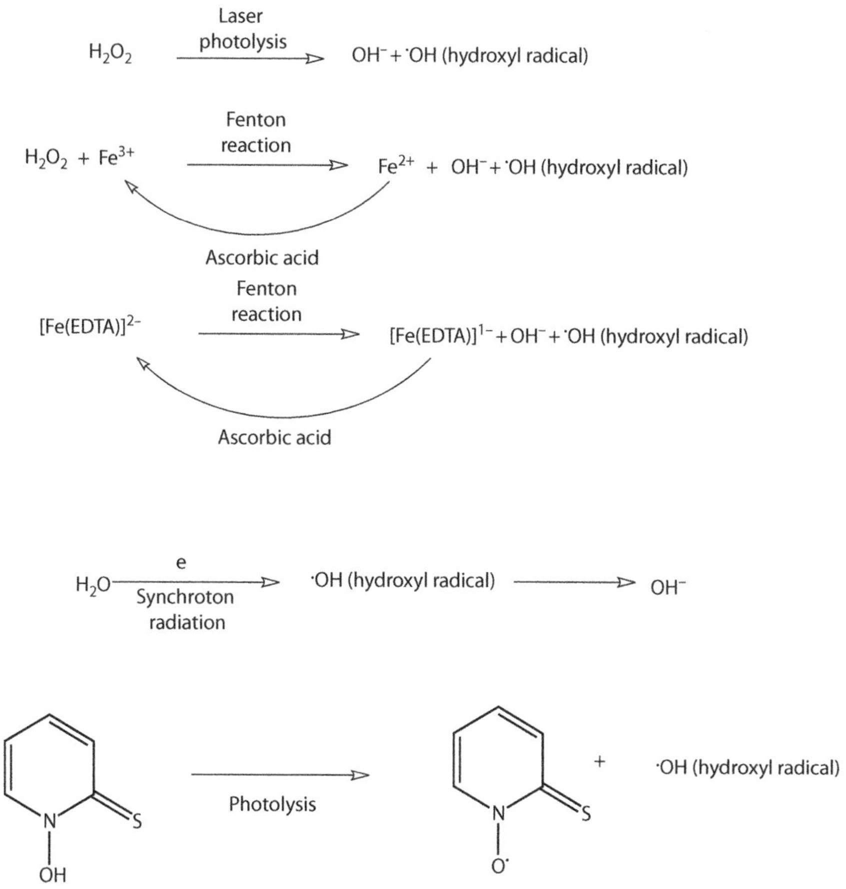

FIGURE 18.14 Mechanisms for the generation of hydroxyl radicals (see Green, M.J. and Hill, H.A.O., Chemistry of dioxygen, *Methods Enzymol.* 105, 3–22, 1984; Xu, G. and Chance, M.R., Hydroxyl radical-mediated modification of proteins as probes for structural proteomics, *Chem. Rev.* 107, 3514–3543, 2007; Jain, S.S. and Tullius, T.D., Footprinting protein–DNA complexes using the hydroxyl radical, *Nat. Protoc.* 3, 1092–1100, 2008; Watson, C., Janik, I., Zhuang, T. et al., Pulsed electron beam water radiolysis for submicrosecond hydroxyl radical protein footprinting, *Anal. Chem.*, 81, 2496, 2009). Also shown are the expected oxidation products from the reaction of hydroxyl radicals with proteins as well as generation from *N*-hydroxypyridine-2(1*H*)-thione (Chaulk, S.G., Pezacki, J.P., and MacMillan, A.M., Studies of RNA cleavage by photolysis of *N*-hydroxypyridine-2(1*H*)-thione. A new photochemical footprinting method, *Biochemistry* 39, 10448–10453, 2000).

in proteins,[259] resulting in a variety of products (Figure 18.15). Modification with hydroxyl radicals most often results in a mass gain of 14 (ketone), 16 (alcohol or sulfur oxygen), or 32 (multiple oxygen such as methionine sulfone).[248,260–263] Reactivity of individual amino acid residues is a function of accessibility.[243,251,259,260] The issue of structural change during modification must be considered and can be minimized by using very short (submicrosecond) periods of radiation.[251] As mentioned above,

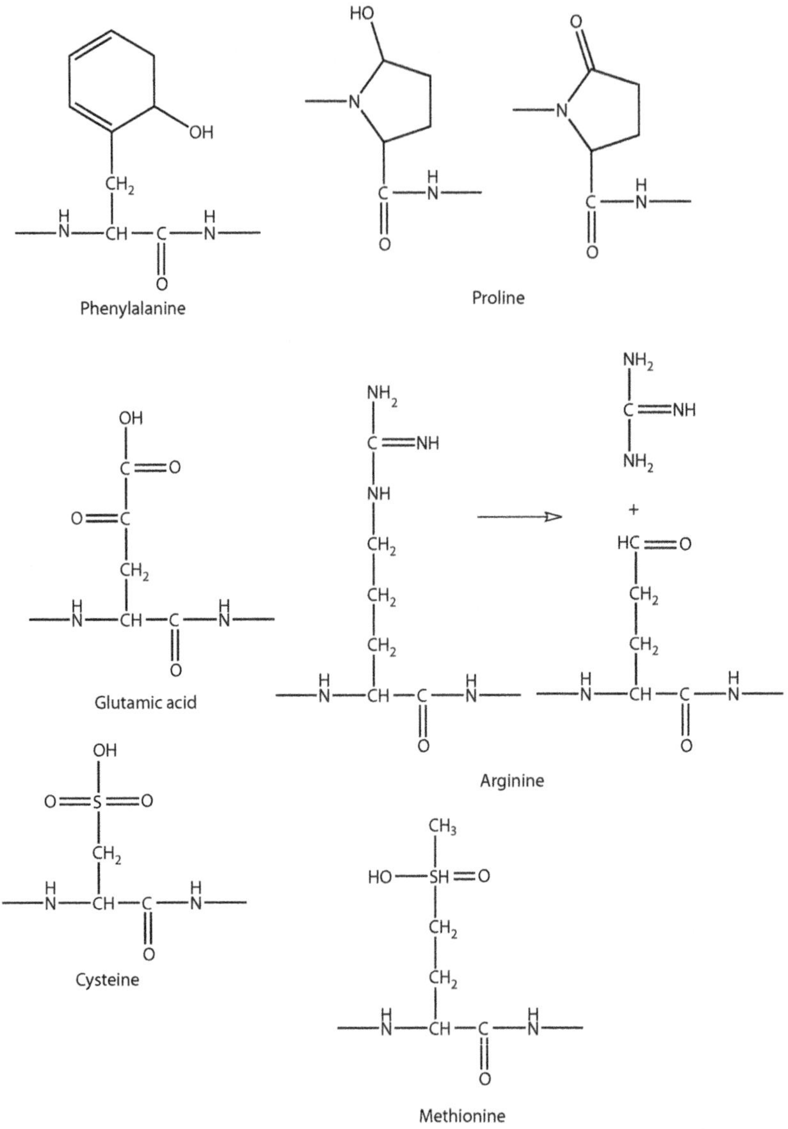

FIGURE 18.15 Products from the hydroxyl radical modification of proteins. Mass spectrometry analysis general results in mass gains of 16 or 18 reflecting either the addition of oxygen or hydroxyl group (see Maleknia, S.D., Brenowitz, M., and Chance, M.R., Millisecond radiolytic modification of peptides by synchrotron x-rays identified by mass spectrometry, *Anal. Chem.* 71, 3965–3973, 1999; Xu, G., Takamoto, K., and Chance, M.R., Radiolytic modification of basic amino acid residues in peptides: Probes for examining protein-protein interactions, *Anal. Chem.* 75, 6995–7007, 2003; Xu, G. and Chance, M.R, Radiolytic modification of acidic amino acid residues in peptides: Probes for examining protein–protein interactions, *Anal. Chem.* 76, 1213–1221, 2004; Xu, G. and Chance, M.R., Radiolytic modification of sulfur-containing amino acid residues in model peptides: Fundamental studies for protein footprinting, *Anal. Chem.* 77, 2437–2449, 2005).

hydroxyl radicals are promiscuous (Figure 18.15) but there is range of reactivity of specific amino acid residues ranging from cysteine (3.4. × $10^{10}\,M^{-1}\,s^{-1}$) to glycine ($1.7 \times 10^7\,M^{-1}\,s^{-1}$).[262] There are other reactivity species such as methylene that are used for rapid protein labeling.[264,265]

As with NMR, highly specialized equipment is required for the generation of hydroxyl radicals such that laboratory expertise is not trivial. The advantages of hydroxyl radical (or other short-lived highly reactive species such as methylene carbine) are definition of solvent-exposed area (conformation) and protein–ligand interactions.[243,244,246,247,266–270] Hydroxyl radical technology would appear to be most useful in the characterization of product and product interactions and not in the routine assay of critical process intermediates, active pharmaceutical ingredient, or final drug product.

Photooxidation (Figure 18.16) is a process by which radiation converts oxygen to oxygen radicals in a process mediated by a radiation receptor (dye) such as proflavin, Rose Bengal, or methylene blue. The oxygen radical then oxidizes amino acids and nucleic acid bases. While the reaction is inherently nonspecific[271–273] as with hydroxyl radical, specificity in the case of protein by binding of the dye to the protein. There are studies where photooxidation is used for the study of protein conformation.[274–277] Histidine would appear to the most sensitive amino acid to photooxidation with less reaction with methionine, cysteine, tyrosine, and tryptophan. Oxidation with chloramine T has been used as method for oxidizing methionine residues in peptides and proteins[180,278–285]; oxidation with chloramine T is suggested as an index of conformation.[180,285]

Iodination (Figure 18.17) is also a method for evaluating protein structure. There are a variety of chemical and enzymatic methods for iodinating proteins. Iodination can be assessed by incorporation of radioactivity or by mass spectrometry.[286,287] The use of mass spectrometry to identify sites of modification in derivatives obtained with nature ^{127}I isotope is preferable to the use of radioisotopes (^{125}I, ^{131}I). The radiolabeled derivatives are still used for some immunoassays and imaging procedures. Iodination is a measure of conformation[288–296] but the results can depend on the method.[297–301]

The use of acetic anhydride to modified amino groups in proteins (Figure 18.18) evolved into a much more sophisticated approach over the last decade as a result of advances in analytical technology. In particular, mass spectrometry has become extremely useful for the study of chemical modification in proteins.[302–305]

Suckau and colleagues[94] used modification with acetic anhydride to study the surface topology of hen egg white lysozyme. Modification was performed in 0.5 M NH_4CO_3, pH 7.0 [maintained by the addition of $NH_4OH(30\%\ NH_c)$ with a 10–10,000 molar excess (to amino groups)] or acetic anhydride for 30 min at 20°C. Analysis by mass spectrometry before and after tryptic hydrolysis permitted the identification of modified residues and the assignment of relative reactivity of the individual amino groups. This is an excellent paper that has been extensively cited since its publication in 1992 and continues to be of interest to investigators.[306]

Palczewski and colleagues[307] used reaction with acetic anhydride to study conformational changes in arrestin as well as the association of arrestin with P-Rho. The effect of light/dark cycles on reactivity with acetic anhydride were evaluated as well

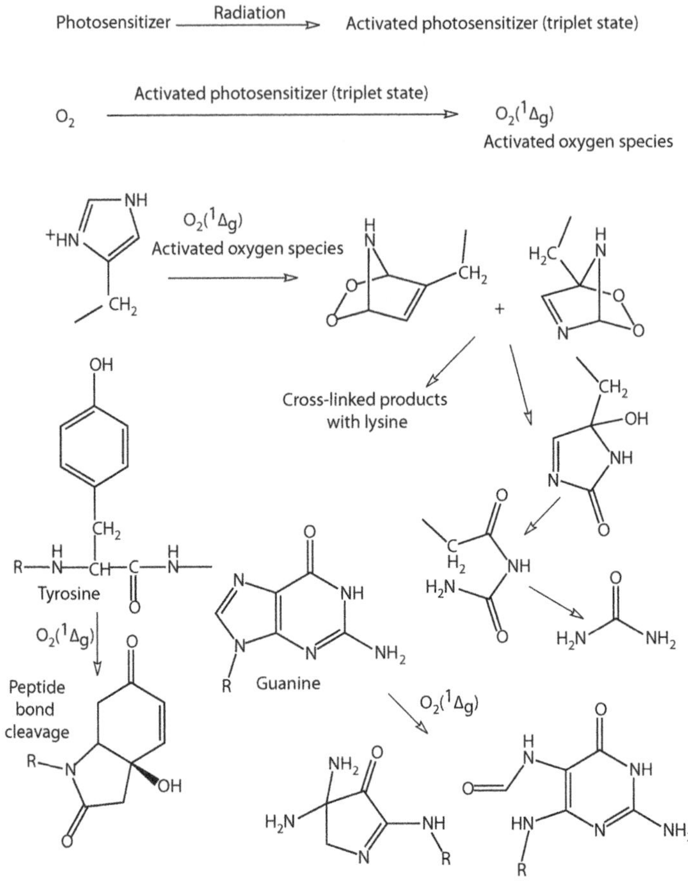

FIGURE 18.16 Photooxidation modification of biopolymers. Shown here is the sensitizer-mediated photooxidation of histidine and tyrosine, resulting in complex products frequently associated with cleavage of the peptide chain (Agon, V.V., Bubb, W.A., Wright, A. et al., Sensitizer-mediated photooxidation of histidine residues: Evidence for the formation of reactive side-chain peroxides, *Free Radic. Biol. Med.* 40, 468–710, 2006; Criado, S., Soltermann, A.T., Marioli, J.M., and Garcia, N.A., Sensitized photooxidation of di- and tripeptides of tyrosine, *Photochem. Photobiol.* 68, 453–458, 1998). The modification of tryptophan is also complex (Posadaz, A., Blasutti, A., Casale, C. et al., Rose Bengal-sensitized photooxidation of the dipeptides L-trytophanyl-L-phenylalanine, L-tryptophanyl-L-tyrosine and L-tryptophanyl-L-tryptophan: kinetics, mechanism, and photoproducts, *Photochem. Photobiol.* 80, 132–138, 2004). See also Jori, G., Galliazzo, G., Tamburro, A.M., and Scoffone, E., Dye-sensitized photooxidation as a tool for determining the degree of exposure of amino acid residues in proteins. The methionyl residues in ribonuclease A, *J. Biol. Chem.* 245, 3375–3383, 1970; Murachi, T., Tsudzuki, T., and Okumura, K., Photosensitized inactivation of stem bromolain. Oxidation of histidine, methionine, and tryptophan residues, *Biochemistry* 14, 249–255, 1975; Masson, J.F., Gauda, E., Mizaikoff, B., and Kranz, C., The interference of HEPES buffer during amperometric detection of ATP in clinical applications, *Anal. Bioanal. Chem.* 390, 2067–2071, 2008. Also shown is the photooxidation of guanine (Margolin, Y., Shafirovich, V., Geacintov, N.E. et al., DNA sequence context as a determinant of the quantity and chemistry of guanine oxidation produced by hydroxyl radicals and one-electron oxidants, *J. Biol. Chem.* 283, 35569–35578, 2008).

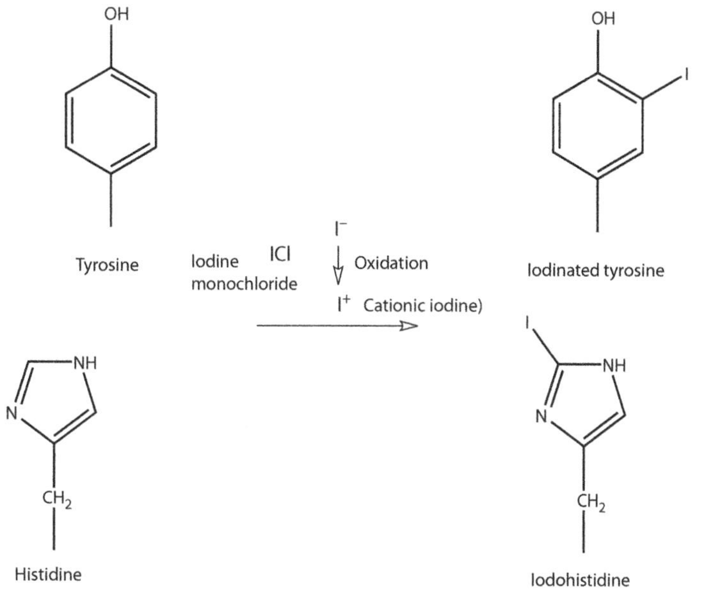

FIGURE 18.17 Iodination of biopolymers. The basic reaction is the generation of cationic iodine by oxidation of the iodide anion. The most common modification occurs at tyrosine with the formation of the monoiodinated derivative. Diiodination can occur with the formation of 3,5-diiodotyrosine. Iodination can also occur at histidinyl residues. Early work required the use of radioisotopic forms of iodine such as ^{131}I, which is still used in diagnostic applications. The development of mass spectrometry allows the use of the naturally occurring isotope, ^{127}I, for labeling protein (Ghosh, D., Erman, M., Sawicki, M. et al., Determination of a protein structure by iodination: The structure of iodinated acetylxyland esterase, *Acta Crystallogr. D Biol. Crystallogr.* 55, 779–784, 1999; Leinala, E.K., Davies, P.L., and Jia, Z., Elevated temperature and tyrosine iodination aid in the crystallization and structure determination of an antifreeze protein, *Acta Crystallogr. D Biol. Crystallogr.* 58, 1081–1083, 2002). Early work on iodination used chloramine T while more recent studies have used Iodogen or lactoperoxidase. Iodogen has the advantage of being insoluble and can be plated in the reaction vessel (microplate) prior to the iodination reaction (see Fraker, P.J. and Speck, J.C. Jr., Protein and cell membrane iodination with a sparingly soluble chloramine, 1,3,4,6-tetrachloro-3α,6α-diphenylglycoluril, *Biochem. Biophys. Res. Commun.* 80, 849–857, 1978; Salacinski, P.R.P., McLean, C., Sykes, J.E.C. et al., Iodination of proteins, glycoproteins, and peptides using a solid-phase oxidizing agent, 1,3,4,6-tetrachloro-3α, 6α-diphenyl glycouril (Iodogen), *Anal. Biochem.* 117, 136–146, 1981; Seevers, R.H. and Counsell, R.E., Radioiodination techniques for small organic molecules, *Chem. Rev.* 82, 575–590, 1982). Iodogen is also used for the iodination of nucleic acids (Piatysek, M.A., Jarmolowski, A., and Augustyniak, J., Iodogen mediated radiolabeling of nucleic acids, *Anal. Biochem.* 172, 356–359, 1988). Also shown is a reagent for the incorporation of iodine into proteins via the modification of lysine residues (Vaidyanathan, G. and Zalutsky, M.R., Synthesis of *N*-succinimidy 4-guanidinomehtyl-3-[*I] iodobenzoate: A radio-iodination agent for labeling internalizing proteins and peptides, *Nat. Protocols* 2, 282–286, 2007).

(*continued*)

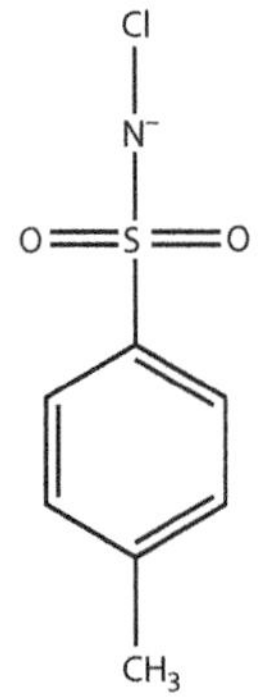

Chloramine-T

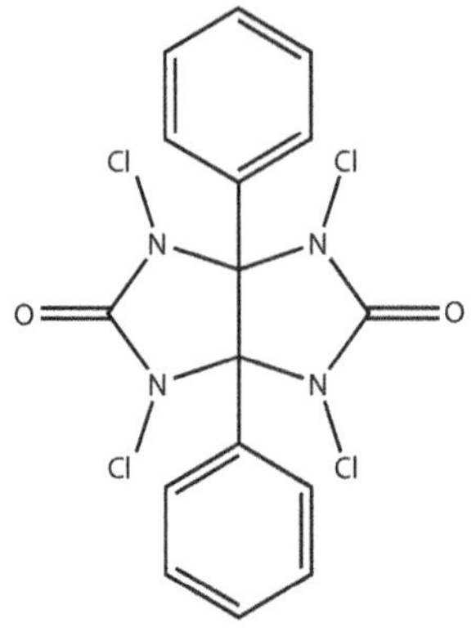

1,3,4,6-Tetrachloro-3a.6a-diphenylglycoluril
Iodogen

Oxidizing agents for iodination

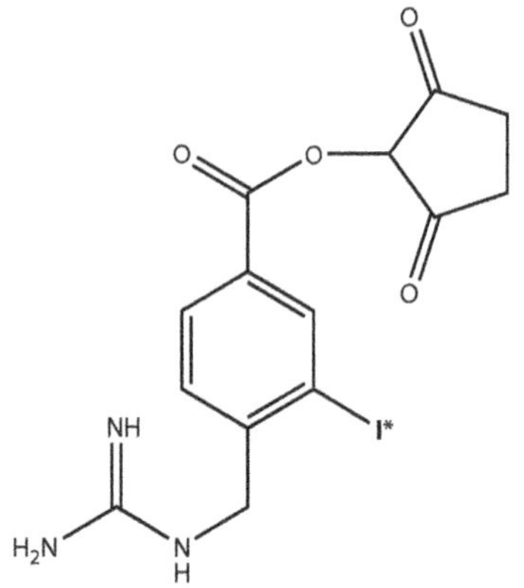

N-Succinimidyl-4-guanidinomethyl-
3-iodobenzoate

FIGURE 18.17 (continued)

as on the interaction of arrestin with P-Rho. The initial modification was performed with low levels (1–10 mM) of deuterated acetic anhydride at 0°C in 100 mM sodium borate, pH 8.5. This was followed by modification with higher concentrations of acetic anhydride (20 mM) in 100 mM sodium borate, pH 9.0 containing 6.0 M guanidine hydrochloride. The ratio of deuterated to protiated modification permitted the identification of residues "protected" from modification by interaction with P-Rho as well three lysine residues, which were more reactive as a result of the interaction. A similar approach was used by Scaloni and coworkers[308] to study the interaction of thyroid transcription factor 1 homeodomain with DNA. A 1- to 10-fold molar excess of acetic anhydride was added to free or DNA-complexed thyroid transcription factor 1 homeodomain in 20 mM Tris-HCl-75 mM KCl, pH 7.5 at 25°C for 10 min. The acetylated samples were subjected to cyanogen bromide cleavage in 70% formic acid at room temperature or 18 h in the dark. The fractions were separated by HPLC and subjected to analysis by mass spectrometry. D'Ambrosio and coworkers[101] have also used this differential modification approach to study the dimeric structure of porcine aminoacylase 1. Reaction with acetic anhydride was performed in 10 mM NH_4CO_3, 1 mM DTT, 1 MM $ZnCl_2$, pH 7.5 at 25°C for 10 min using a 100- to 5000-fold molar excess of reagent. Modification occurred readily at the amino terminus and at 8 of the 17 lysine residues. Calvete and coworkers[238] used a clever approach to identify the heparin-binding domain of bovine seminal plasma protein PDC-109. The PDC-109 protein was bound to heparin–agarose in 16.6 mM Tris- 50 mM NaCl–1.6 mM EDTA–0.025% NaN_3, pH 7.4. After washing the column to remove protein not bound to the matrix, the column was recycled at room temperature with the

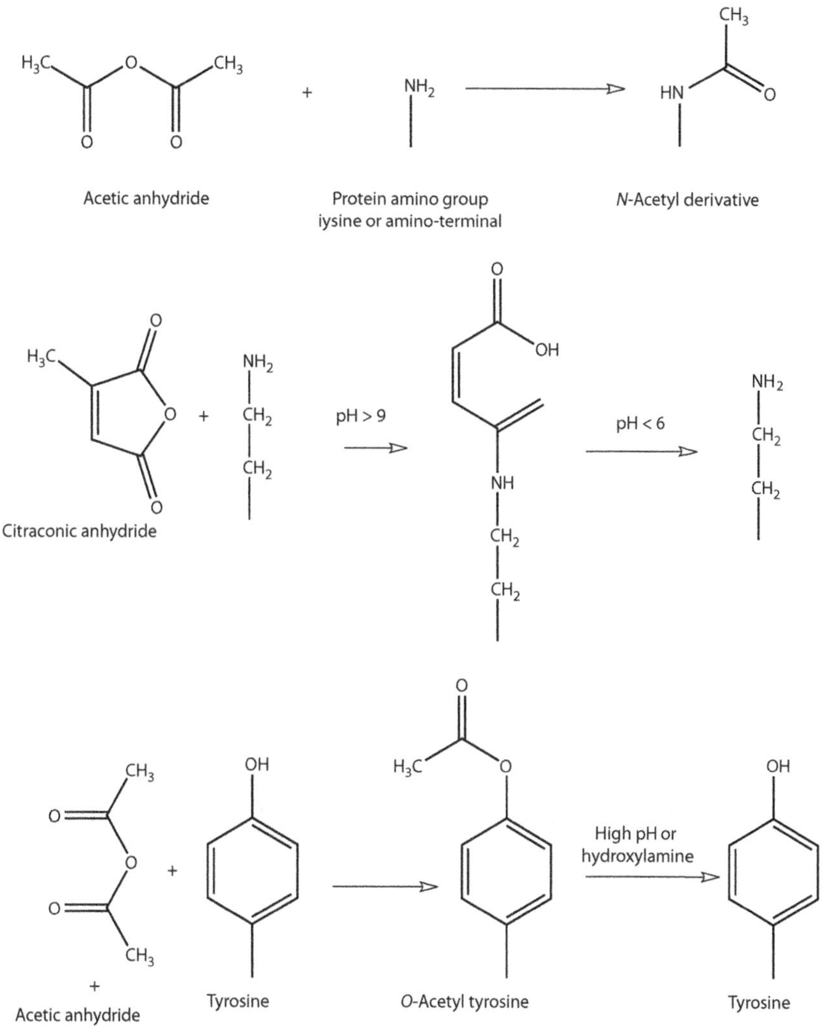

Acetic anhydride   Protein amino group   N-Acetyl derivative
                        iysine or amino-terminal

Citraconic anhydride

Acetic anhydride   Tyrosine   O-Acetyl tyrosine   Tyrosine

FIGURE 18.18 The reaction of acetic anhydride with proteins. This is well-known method for the modification of proteins which dates back to work by Fraekel-Conrat and coworkers (Fraekel-Conrat, H., Methods for investigating the essential groups for enzyme activity, *Methods Enzymol.* 4, 247–269, 1957; Fraenkel-Conrat, H. and Colloms, M., Reactivity of tobacco mosaic virus and its protein toward acetic anhydride, *Biochemistry* 6, 2740–2745, 1967). See also Riordan, J.F. and Vallee, B.L., Acetylation, *Methods Enzymol.* 11, 565–570, 1967; Higashimoto, Y., Sugishima, M., Sato, H. et al. Mass spectrometric identification of lysine residues of heme oxygenase-1 that are involved in its interaction with NADPH-cytochrome P450 reductase, *Biochem. Biophys. Res. Commun.* 367, 852–858, 2008. Also shown in the modification of protein by citraconic anhydride, which is reversible (see Hanai, R. and Wang, J.C., Protein footprinting by the combined use of reversible and irreversible lysine modifications, *Proc. Natl. Acad. Sci. USA* 91, 11904–11908, 1994).

application buffer containing acetic anhydride (25- to 1600-fold molar excess over protein lysine). A similar experiment was performed with 1,2-cyclohexanedione to study arginine modification. Six basic residues were protected from modification by binding to the heparin matrix. Zappacosta and coworkers[309] measured amino reactivity on Minobody, a small de novo-designed β-protein by reaction with a low concentration of acetic anhydride. The modification reaction was performed in 50mM NH_4CO_c, pH 7.5, at 25°C for 10min with a twofold molar excess (to amino groups) of acetic anhydride. Two lysine residues were highly reactive, one lysine and the α-amino group were less reactive, and one lysine residue was not modified. It was emphasized that the results are qualitative. However, it is still a clever, well-performed study, which provides considerable information.

Taralp and Kaplan[310] examined the reaction of acetic anhydride with lyophilized α-chymotrypsin in vacuo. α-Chymotrypsin was lyophilized from an unbuffered solution at pH 9.0 in one chamber in a reaction vessel. ^{3}H-Acetic anhydride was added to another compartment in the reaction vessel. The reaction vessel was evacuated and placed in an oven at 75°C. Several reaction vessels were used and removed at various time intervals for analysis. The proteins were then modified with ^{14}C-acetic anhydride and the ratio of ^{3}H to ^{14}C was used to determine the extent of modification. While complete modification of amino groups is achieved at pH 9.0 in aqueous solution, in the nonaqueous system, only 25% of the ε-amino groups and 90% of the α-groups were modified. It also appeared that mixed anhydrides formed with carboxyl groups on the protein surface. Kaplan and coworkers subsequently reported on the modification of amino groups in lyophilized proteins with iodomethane.[311] The pH memory effect describes the correlation between solution pH prior to freeze-drying (lyophilization) and functional group reactivity in the lyophilized state.[312] The ionization state of a given functional group in solution is maintained in the lyophilized state. Smith and coworkers[313] used reaction with acetic anhydride to determine the extent of posttranslational acetylation in histone H4. Histone H4 was modified with deuterated acetic anhydride (dried histone samples were suspected in deuterated glacial acid acetic and deuterated acetic anhydride added; the reaction mixture was allowed to stand for 6h at ambient temperature). The modified samples were subjected to mass spectrometric analysis. The extent of endogenous acetylation was determined from the ration of protiated to deuterated fragmentation ions. Hochleitner and coworkers[314] used reaction with acetic anhydride to define a discontinuous epitope in human immunodeficiency virus core protein p24. The protein was bound to the immobilized antibody and digested with endoprotease lys-C. Modification was then accomplished with acetic anhydride (10,000-fold molar excess of reagent, pH 7.8, 50mM NH_4CO_c, 20min) followed by a 100,000-fold excess of the hexadeuterated reagents under the same conditions; pH was maintained by the addition of 10% NH_4OH. Scholten and coworkers[315] used acetic anhydride and subsequent analysis with nanoC-MALDI tandem mass spectrometry to study the interaction of immunity protein Im9 and the DNase domain of Colicin E9.

There have been several recent studies which have used S-methylacetimidate (Figure 18.19) for the modification of lysine residues in protein conformation studies.[316–318] There is also a study that uses ethyl picolinimidate tetrafluoroborate[319]

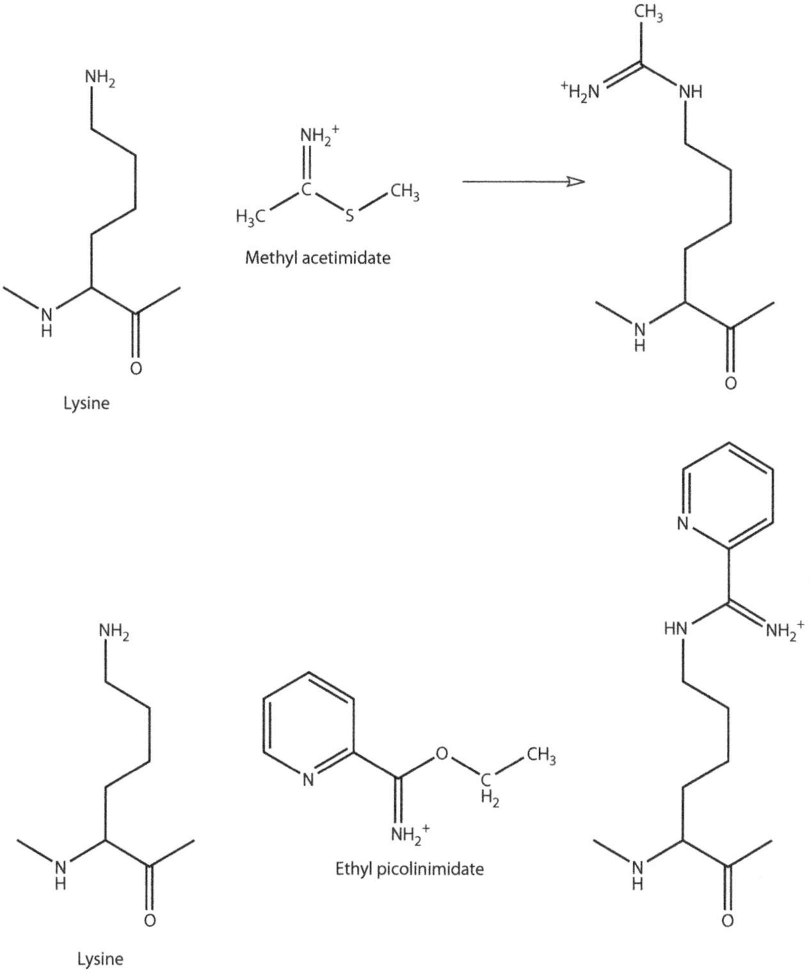

FIGURE 18.19 The amidation of proteins. Shown are two reagents for the amidation of proteins. This reaction is thought to be more specific than that with *O*-acyl derivatives (see Perham, R.N., The reactivity of functional groups as a probe for investigating the topography of tobacco mosaic virus. The use of mutants with additional lysine residues in the coat protein, *Biochem. J.* 131, 119–126, 1973; Janecki, D.J., Beardsley, R.L., and Reilly, J.P., Probing protein tertiary structure with amidation, *Anal. Chem.* 77, 7274–7281, 2005; Beardsley, R.I., Running, W.E., and Reilly, J.P., Probing the structure of the *Caulobacter crescentus* ribosome with chemical labeling and mass spectrometry, *J. Proteome Res.* 5, 2935–2946, 2006; Liu, X., Broshears, W.C., and Reilly, J.P., Probing the structure and activity of trypsin with amidation, *Anal. Biochem.* 367, 13–19, 2007; Kim, J.-S., Kim, J.-H., and Kim, H.-J., Matrix-assisted laser desorption/ionization signal enhancement of peptides by picolinamidination of amino groups, *Rapid Commun. Mass Spectrom.* 22, 495–502, 2008).

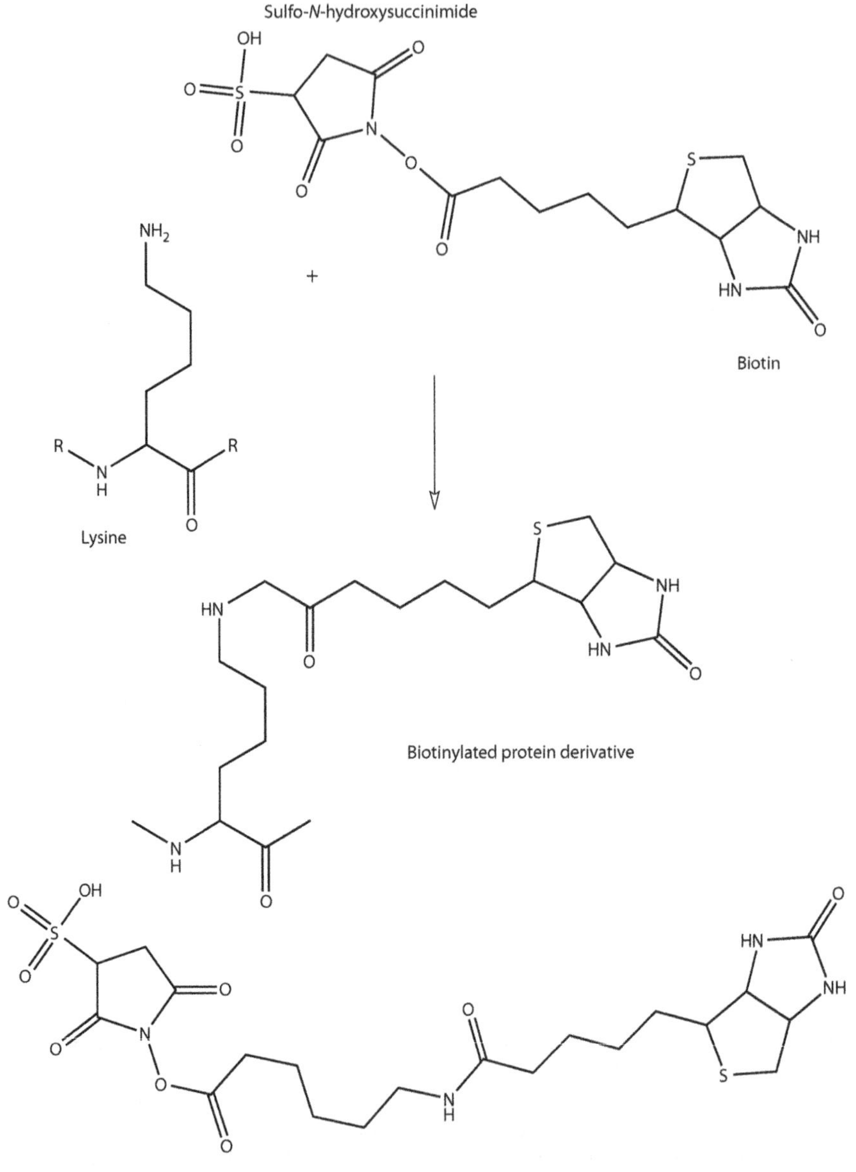

FIGURE 18.20 The reaction of proteins with *N*-hydroxysuccinimide derivatives (see Gabani, G., Augier, J., and Armengaud, J., Assessment of solvent residues accessibility using three sulfo-NHS-biotin reagents in parallel: Application to footprint changes of a methyl-transferase upon binding its substrate, *J. Mass Spectrom.* 43, 360–370, 2008).

(Figure 18.19), suggesting that the use of this reagent provides considerable signal enhancement over acetamidation or guanidination. One other recent study used sulfo-*N*-hydroxysuccinimide-biotin reagents (Figure 18.20) to measure solvent accessible residudes.[320]

REFERENCES

1. Lundblad, R.L., *Chemical Reagents for Protein Modification*, 3rd edn., CRC Press, Boca Raton, FL, November 2004.
2. Creighton, T.E., *Protein Function: A Practical Approach*, IRL Press at Oxford University Press, Oxford, U.K., 1997.
3. Mizuno, Y., *The Organic Chemistry of Nucleic Acids*, Elsevier, Amsterdam, the Netherlands, 1986.
4. Walker, J.M. ed., *The Protein Protocols*, Humana Press, Totowa, NJ, 1996.
5. Fraser-Reid, B.O., Tatsuta, K., and Thiem, J. eds., *Glycoscience: Chemistry and Chemical Biology I–III*, edn., Spinger-Verlag, Berlin, Germany, 2001.
6. Cabrera, J.C., Boland, A., Messiaen, J. et al., Egg box conformation of oligogalacturonides: The time-dependent stabilization of the elicitor-active conformation increases its biological activity, *Glycobiology* 18, 473–482, 2008.
7. Bolgiano, B., Mawas, F., Yost, S.E. et al., Effect of physico-chemical modification on the immunogenicity of *Haemophilus influenzae* type h oligosaccharide CMR_{197} conjugate vaccines, *Vaccine* 19, 3189–3200, 2001.
8. Cheung, J.K., Raverkar, P.S., and Truskett, T.M., Analytical model for studying how environmental factors influence protein conformational stability in solution, *J. Chem. Phys.* 125, 224903, 2006.
9. Kostareva, I. Hung, F., and Campbell, C., Purification of antibody heteropolymers using hydrophobic interaction chromatography, *J. Chromatogr.* 1177, 254–264, 2008.
10. Kallias, A., Bachmann, M., and Janke, W., Thermodynamics and kinetics of a Gō proteinlike heteropolymer model with two-state folding characteristics, *J. Chem. Phys.* 128, 055102, 2008.
11. Patel, B.A., Debenedetti, P.G., Stillinger, F.H., and Rossky, P.J., The effect of sequence on the conformational stability of a model heteropolymer in explicit water, *J. Chem. Phys.* 128, 175102, 2008.
12. Boström, M., Tavares, F.W., Finet, S. et al., Why forces between proteins follow different Hofmeister series for pK above and below pI, *Biophys. Chem.* 117, 217–224, 2005.
13. Xu, L.C., Vadillo-Rodriguez, V., and Logan, B.E., Residence time, loading force, pH, and ionic strength affect adhesion forces between colloids and biopolymer-coated surfaces, *Langmuir* 21, 7491–7500, 2005.
14. Sherrat, M.J., Baldock, C., Morgan, A., and Kielty, C.M., The morphology of adsorbed extracellular matrix assemblies is critically dependent on solution calcium concentration, *Matrix Biol.* 26, 156–166, 2007.
15. Hirano, A., Hamada, H., Okubo, T. et al., Correlation between thermal aggregation and stability of lysozyme with salts described by molar surface tension increment: An exceptional propensity of ammonium salts as aggregation suppressor, *Protein J.* 26, 423–433, 2007.
16. Ananthapadmanabhan, K.P., Lips, A., Vincent, C. et al., pH-induced alterations in stratum corneum properties, *Int. J. Cosmet. Sci.* 25, 103–112, 2003.
17. Giedroc, D.P., Puett, D., Sinha, S.K., and Brew, K., Calcium effects on calmodulin lysine reactivation, *Arch. Biochem. Biophys.* 252, 136–144, 1987.
18. Illy, C., Thielens, N.M., and Arlaud, G.J., Chemical characterization and location of ionic interactions involved in the assembly of the C1 complex of human complement, *J. Protein Chem.* 12, 771–781, 1993.
19. Che, F.Y. and Fricker, L.D., Quantitation of neuropeptides in Cpe(fat)Cpe(fat mice using differential isotopic tags and mass spectrometry, *Anal. Chem.* 74, 3190–3198, 2002.
20. Turner, B.T. Jr., Sabo, T.M., Wilding, D., and Maurer, M.C., Mapping of factor XIII solvent accessibility as a function of activation state using chemical modification methods, *Biochemistry* 43, 9755–9765, 2004.
21. Nam, H.W., Lee, G.Y., and Kim, Y.S., Mass spectrometric identification of K210 essential for rat malonyl-CoA decarboxylase catalysis, *J. Proteome Res.* 5, 1398–1406, 2006.

22. Scherer, H.J., Karthein, R., Strieder, S., and Ruf, H.H., Chemical modification of prostaglandin endoperoxide synthase by *N*-acetylimidazole. Effect on enzyme activities and EPR spectroscopic properties, *Eur. J. Biochem.* 205, 751–757, 1992.

23. Cymes, G.D., Igelesias, M.M., and Wolfenstein-Todel, C., Chemical modification of ovine prolactin with *N*-acetylimidazole, *Int. J. Pept. Protein Res.* 42, 33–28, 1993.

24. Vazeux, G., Iturrioz, X., Corvol, P., and Llorens-Cortes, C., A tyrosine residue essential for catalytic activity in aminopeptidase A, *Biochem. J.* 327, 883–889, 1997.

25. Pal, J.K., Bera, S.K., and Ghosh, S.K., Acetylation of α-crystallin with *N*-acetylimidazole and its influence upon the native aggregate and subunit reassembly, *Curr. Eye Res.* 19, 368–367, 1999.

26. Zhang, F., Gao, J., Weng, J. et al., Structural and functional differentiation of three groups of tyrosine residues by acetylation of *N*-acetylimidazole in manganese stabilizing protein, *Biochemistry* 44, 719–725, 2005.

27. McAllister, K.A., Marrone, L., and Clarke, A.J., The role of tryptophan residues in substrate binding to catalytic domains A and B of xylanase C from *Fibrobacter succinogenes* S85, *Biochim. Biophys. Acta* 1400, 342–352, 2000.

28. Takita, T., Nakagoshi, M., Inouye, K., and Tonomura, B., Changes observed in the amino acid activation reaction, *J. Mol. Biol.* 325, 677–685, 2003.

29. Sargisova, Y., Pierfederici, F.M., Scire, A. et al., Computational, spectroscopic, and resonant mirror biosensor analysis of the interaction of adenodoxin with native and tryptophan-modified NADPH-adrenodoxin reductase, *Proteins* 57, 302–310, 2004.

30. Faridmoayer, A. and Scaman, C.H., Binding residues and catalytic domain of soluble *Saccharomyces cerevisiae* processing α-glucosidase I, *Glycobiology* 15, 1341–1348, 2005.

31. Kumar, A., Tyagi, N.K., and Kinne, R.K., Ligand-mediated and conformational changes and positioning of tryptophans in reconstituted human sodium/*d*-glucose cotransporter (hSGLT1) probed by tryptophan fluorescence, *Biophys. Chem.* 127, 69077, 2007.

32. Leitner, A. and Lindner, W., Functional probing of arginine residues in proteins using mass spectrometry and an arginine-specific covalent tagging concept, *Anal. Chem.* 77, 4481–4488, 2005.

33. Foettinger, A., Leitner, A., and Lindner, W., Solid-phase capture and release of arginine peptides by selective tagging and boronate affinity chromatography, *J. Chromatogr. A* 1079, 187–196, 2005.

34. Saraiva, M.A., Borges, C.M., and Florencio, M.H., Reactions of a modified lysine with aldehydic and diketonic dicarbonyl compounds: An electrospray mass spectrometry structure/activity study, *J. Mass Spectrom.* 41, 216–228, 2006.

35. Holm, A., Rise, F., Sessler, N. et al., Specific modification of peptide-bound citrulline residues, *Anal. Biochem.* 352, 68–76, 2006.

36. Leitner, A., Amon, S., Rizzi, A., and Lindner, W., Use of the arginine-specific butanedione/phenylboronic acid tag for analysis of peptides and protein digests using matrix-assisted laser desorption/ionization mass spectrometry, *Rapid Commun. Mass Spectrom.* 21, 1321–1330, 2007.

37. Chang, L.S., Wu. P.F., Liou, J.C. et al., Chemical modification of arginine residues of *Notechis scutatus scutatus* notexin, *Toxicon* 44, 491–497, 2004.

38. Masuda, T., Ide, N., and Kitabatake, N., Structure–sweetness relationship in egg white lysozme: Role of lysine and arginine residues on the elicitation of lysozyme sweetness, *Chem. Senses* 30, 667–681, 2005.

39. Herrman, A., Svangard, E., Claeson, P. et al., Key role of glutamic acid for the cytotoxic activity of the cyclotide cycloviolacin O₂, *Chem. Mol. Life Sci.* 63, 235–245, 2006.

40. Schwartz, M.P., Barlow, D.E., Russell, J.N., Jr. et al., Semiconductor surface-induced 1,3-hydrogen shift: The role of covalent vs zwitterionic character, *J. Am. Chem. Soc.* 128, 11054–11061, 2006.

41. Daniel, J., Oh, T.J., Lee, C.M., and Kolattukudy, P.E., AccD6, a member of the Fas II locus, is a functional carboxyltranferase subunit of the acyl-coenzyme A carboxylase in *Mycobacterium tuberculosis, J. Bacteriol.* 189, 911–917, 2007.

42. Follmer, C. and Carlini, C.R., Effect of chemical modification of histidine on the copper-induced oligomerization of jack bean urease, *Arch. Biochem. Biophys.* 435, 15–20, 2005.

43. Colleluori, D.M., Reczkowski, R.S., Emig, F.A. et al., Probing the role of hyper-reactive histidine residue of arginase, *Arch. Biochem. Biophys.* 444, 15–26, 2005.

44. Runquist, J.A. and Miziorko, H., Functional contribution of a conserved mobile loop histidine of phosphoribulokinase, *Protein Sci.* 15, 837–842, 2006.

45. Wang, X.Y., Sun, M.L., Zhao, D.M., and Wang, M., Kinetics of inactivation of phytase (phy A) during modification of histidine residue by IAA and DEP, *Protein Pept. Lett.* 13, 565–570, 2006.

46. Nakanishi, N., Takeuchi, F., Okamoto, H. et al., Characterization of heme-coordinating histidyl residues of cytochrome b5 based on the reactivity with diethylpyrocarbonate: A mechanism for the opening of axial imidazole rings, *J. Biochem.* 140, 561–571, 2006.

47. Ghosh, M.K., Kildsig, D.O., and Mitra, A.K., Preparation and characterization of methotrexate-immunoglobulin conjugates, *Drug Des. Deliv.* 4, 13–25, 1989.

48. Shen, X., Lagergard, T., Yang, Y. et al., Preparation and preclinical evaluation of experimental group B streptococcus type III polysaccharide-cholera toxin B subunit conjugate vaccine for intranasal immunization, *Vaccine* 19, 850–861, 2000.

49. Hafemann, B., Ghofrani, K., Gattner, H.G. et al., Cross-linking by 1-ethyl-3-(3-dimethylaminopropyl)-carbodiimide (EDC) of a collagen/elastin membrane meant to be used as a dermal substitute: Effects on physical, biochemical and biological features in vitro, *J. Mater. Sci. Mater. Med.* 12, 437–446, 2001.

50. Zhang, R., Tang, M., Bowyer, A. et al., A novel pH- and ionic-strength-sensitive carboxy methyl dextran hydrogel, *Biomaterials* 26, 4677–483, 2005.

51. Li, D., He, Q., Cui, Y. et al., Immobilization of glucose oxidase onto gold nanoparticles with enhanced thermostability, *Biochem. Biophys. Res. Commun.* 355, 488–493, 2007.

52. Owusu-Apenten, R., Colorimetric analysis of protein sulfhydryl groups in milk: Applications and processing effects, *Crit. Rev. Food Sci. Nutr.* 45, 1–23, 2005.

53. Laragione, T., Gianazza, E., Tonelli, R. et al., Regulation of redox-sensitive exofacial protein thiols in CHO cells, *Biol. Chem.* 387, 1371–1376, 2006.

54. Landino, L.M., Koumas, M.T., Mason, C.E., and Alston, J.A., Ascorbic acid reduction of microtubule protein disulfides and its relevance to protein *S*-nitrosylation assays, *Biochem. Biophys. Res. Commun.* 340, 347–352, 2006.

55. de Araujo, A.D., Palomo, J.M., Cramer, J. et al., Diels-Alder ligation of peptides and proteins, *Chemistry* 12, 6095–6109, 2006.

56. Cliff, M.J., Alizadeh, T., Jelinska, C. et al., A thiol labelling competition experiment as a probe for sidechain packing in the kinetic folding intermediate of N-PGK, *J. Mol. Biol.* 364, 810–823, 2006.

57. Brubaker, G., Peng, D.Q., Somerlot, B. et al., Apolipoprotein A-1 lysine modification: Effects on helical content, lipid binding and cholesterol acceptor activity, *Biochim. Biophys. Acta* 1761, 64–72, 2006.

58. Fu, Q. and Li, L., Fragmentation of peptides with *N*-terminal dimethylation and imine/methylol adduction at the tryptophan side-chain, *J. Am. Soc. Mass Spectrom.* 17, 859–866, 2006.

59. Xu, J. and Bowden, E.F., Determination of the orientation of adsorbed cytochrome C on carboxyalkanethiol self-assembled monolayers by in situ differential modification, *J. Am. Chem. Soc.* 128, 6813–6822, 2006.

60. Hsu, J.L., Huang, S.Y., and Chen, S.H., Dimethyl multiplexed labeling combined with micro column separation and MS analysis for time course study in proteomics, *Electrophoresis* 27, 3652–3660, 2006.

61. Walter, T.S., Meier, C., Assenberg, R. et al., Lysine methylation as a routine rescue strategy for protein crystallization, *Structure* 14, 1617–1622, 2006.

62. Strohalm, M., Kodicek, M., and Pechar, M., Tryptophan modification by 2-hydroxy-5-nitrobenzyl bromide studied by MALDI-TOF mass spectrometry, *Biochem. Biophys. Res. Commun.* 312, 811–816, 2003.

63. Strohalm, M., Santrucek, J., Hynek, R., and Kodicek, M., Analysis of tryptophan surface accessibility in proteins by MALDI-TOF mass spectrometry, *Biochem. Biophys. Res. Commun.* 323, 1134–1138, 2004.

64. Jung, J.W., Kuk, J.H., Kim, K.Y. et al., Purification and characterization of exo-β-D-glucosaminidase from *Aspergillus fumigatus* S-26, *Protein Expr. Purif.* 45, 125–131, 2006.

65. Tashima, I., Yoshida, T., Asada, Y., and Ohmachi, T., Purification and characterization of a novel L-2-amino-Δ_2-thiazoline-4-carboxylic acid hydrolase from *Pseudomonas* sp. Strain ON-4a expressed in *E. coli*, *Appl. Microbiol. Biotechnol.* 72, 499–507, 2006.

66. Ma, S.F., Nishikawa, M., Yabe, Y. et al., Role of tyrosine and tryptophan in chemically modified serum albumin on its tissue distribution, *Biol. Pharm. Bull.* 29, 1926–1930, 2006.

67. Liu, X., Alexander, C., Serrano, J. et al., Variable reactivity of an engineered cysteine at position 338 in cystic fibrosis transmembrane conductance regulator reflects different chemical states of the thiol, *J. Biol. Chem.* 281, 8275–8285, 2006.

68. Audia, J.P., Roberts, R.A., and Winkler, H.H., Cysteine-scanning mutagenesis and thiol modification of the *Rickettsia prowazerkii* ATP/ADP translocase: Characterization of the TMs IV–VII and IX–XII and their accessibility to the aqueous translocation pathway, *Biochemistry* 45, 2648–2656, 2006.

69. Tombolato, F., Ferrarini, A., and Freed, J.H., Modeling the effects of structure and dynamics of the nitroxide side chain on the ESR spectra of spin-labeled proteins, *J. Phys. Chem. B Condens. Matter Mater. Surf. Interfaces Biophys.* 110, 26260–26271, 2006.

70. Karala, A.R. and Ruddock, L.W., Does *S*-methyl methanethiosulfonate trap the thiol-disulfide state of proteins?, *Antioxid. Redox Signal.* 9, 527–531, 2007.

71. Thonon, D., Jacques, V., and Desreux, J.F., A gadolinium triacetic monoamide DOTA derivative with a methanesulfonate anchor group. Relaxivity properties and conjugation with albumin and thiolated particles, *Contrast Media Mol. Imaging* 2, 24–34, 2007.

72. Johans, M., Milanesi, E., Franck, M. et al., Modification of permeability transition pore arginine(s) by phenylglyoxal derivatives in isolated mitochondria and mammalian cells. Structure-function relationship of arginine ligands, *J. Biol. Chem.* 280, 12130–12136, 2005.

73. Greig, N., Wyllie, S., Vickers, T.J., and Fairlamb, A.N, Trypanothione-dependent glyoxylase I in *Trypanosoma cruzi*, *Biochem. J.* 400, 217–223, 2006.

74. Ye, M. and English, A.M., Binding of polyaminocarboxylate chelators to the active-site copper inhibits the GSNO-reductase activity but not the superoxide dismutase activity of Cu, Zn-superoxide dismutase, *Biochemistry* 45, 12723–12732, 2006.

75. Santrucek, J., Strohalm, M., Kadlcik, V. et al., Tyrosine residue modification studied by MALDI-TOF mass spectrometry, *Biochem. Biophys. Res. Commun.* 323, 1151–1156, 2004.

76. Negrerie, M., Martin, J.L., and Njgiem, H.O., Functionality of nitrated acetylcholine receptor: The two-step formation of nitrotyrosines reveals their differential role in effector binding, *FEBS Lett.* 579, 2643–2647, 2005.

77. Carven, G.J. and Stern, L.J., Probing the ligand-induced conformational change in HLA-DR1 by selective chemical modification and mass spectrometric mapping, *Biochemistry* 44, 13625–13637, 2005.

78. Gruijthuijsen, Y.K., Grieshuber, I., Stocklinger, A. et al., Nitration enhances the allergic potential of proteins, *Int. Arch. Allergy Immunol.* 141, 265–275, 2006.
79. Ghesquiere, B., Goethals, M., van Damme, J. et al., Improved tandem mass spectrometric characterization of 3-nitrotyrosine sites in peptides, *Rapid Commun. Mass Spectrom.* 20, 2885–2893, 2006.
80. García-Moreno, B., Dwyer, J.J., GIttis, A.G. et al., Experimental measurement of the effective dielectric in the hydrophobic core of a protein, *Biophys. Chem.* 64, 211–224, 1997.
81. Hnízda, A., Šantrůček, J., Šanda, M. et al., Reactivity of histidine and lysine side-chains with diethylpyrocarbonate—A method to identify surface exposed residues in proteins, *J. Biochem. Biophys. Methods* 70, 1091–1097, 2008.
82. Richardson, G.M., The principle of formaldehyde, alcohol, and acetone titrations. With a discussion of the proof and implication of the zwitterionic conception, *Proc. R. Soc. Lond. B Biol. Sci.* 115, 121–141, 1934.
83. Duggan, E.L. and Schmidt, C.L.A., The dissociation of certain amino acids in dioxane–water mixtures, *Arch. Biochem.* 1, 453–471, 1943.
84. Canova-Davis, E. and Carpenter, F.H., Semisynthesis of insulin: Specific activation of arginine carboxyl group of the B chain of desoctapeptide-(B23-36)-insulin (Bovine), *Biochemistry* 20, 7053–7058, 1981.
85. Canova-Davis, E., Kessler, T.J., and Ling, V.T., Transpeptidation during the analytical proteolysis of proteins, *Anal. Biochem.* 196, 39–45, 1991.
86. Plapp, B.V., Application of affinity labeling for studying structure and function in enzymes, *Methods Enzymol.* 87, 469–499, 1982.
87. Sigrist, H., Kempf, C., and Zahler, P., Interaction of phenylisothiocyanate with human erythrocyte band 3 protein. I. Covalent modification and inhibition of phosphate transport, *Biochim. Biophys. Acta* 597, 137–144, 1980.
88. Church, F.C., Villanueva, G.B., and Griffith, M.J., Structure–function relationships in heparin cofactor II: Chemical modification of arginine and tryptophan and demonstration of a two-domain structure, *Arch. Biochem. Biophys.* 246, 175–184, 1986.
89. Basso, L.A. and Engel, P.C., Initial formation of a non-covalent enzyme–reagent complex during the inactivation of clostridial glutamate dehydrogenase by Ellman's reagent: Determination of the enzyme's dissociation constant for the binary complex with NAD+ from protection studies, *Biochim. Biophys. Acta* 1209, 222–226, 1994.
90. Rao, A.G. and Neet, K.E., Tryptophan residues of the gamma subunit of 7S nerve growth factor: Intrinsic fluorescence, solute quenching, and *N*-bromosuccinimide oxidation, *Biochemistry* 21, 6843–6850, 1982.
91. Davies, K.J. and Delsignore, M.E., Protein damage and degradation by oxygen radicals III. Modification of secondary and tertiary structure, *J. Biol. Chem.* 262, 9908–9913, 1987.
92. Okajima, T., Kawata, Y., and Hamaguchi, K., Chemical modification of tryptophan residues and stability changes in proteins, *Biochemistry* 29, 9168–9175, 1990.
93. Suckau, D., Mak, M., and Przybylski, M., Protein surface topology-probing by selective chemical modification and mass spectrometric peptide mapping, *Proc. Natl. Acad. Sci. USA* 89, 5630–5634, 1992.
94. Gettins, P.G.W., Fan, B., Crews, B.C., Turko, I.V., Olson, S.T., and Streusand, V.J., Transmission of conformational change from the heparin binding site to the reactive center of antithrombin, *Biochemistry* 32, 8385–8389, 1993.
95. Buechler, J.A., Vedvick, T.A., and Taylor, S.S., Differential labeling of the catalytic subunit of cAMP-dependent protein kinase with acetic anhydride: Substrate-induced conformational changes, *Biochemistry* 28, 3018–3024, 1989.
96. Mykkanen, H.M. and Wasserman, R.H., Reactivity of sulfhydryl groups in the brush-border membranes of chick duodena is increased by 1,2,5-dihydroxycholecalciferol, *Biochim. Biophys. Acta* 1033, 282–286, 1990.

97. Landfear, S.M., Evans, D.R., and Lipscomb, W.N., Elimination of cooperativity in aspartate transcarbamylase by nitration of a single tyrosine residue, *Proc. Natl. Acad. Sci. USA* 75, 2654–2658, 1978.

98. Kumar, G.K., Beegen, N., and Wood, H.G., Involvement of tryptophans at the catalytic site and subunit-binding domains of transcarboxylase, *Biochemistry* 27, 5972–5978, 1988.

99. Winkler, M.A., Fried, V.A., Merat, D.L., and Cheung, W.Y., Differential reactivities of lysines in calmodulin complexed to phosphatase, *J. Biol. Chem.* 262, 15466–15471, 1987.

100. Salhany, J.M., Sloan, R.L., and Cordes, K.S., The carboxyl side chain of glutamate 681 interacts with a chloride binding modifier site that allosterically modulates the dimeric conformational state of Band 3 (AE1). Implications for the mechanism of anion/proton cotransport, *Biochemistry* 42, 1589–1602, 2003.

101. D'Ambrosio, C., Talamo, C., Vitale, R.M. et al., Probing the dimeric structure of porcine aminoacylase 1 by mass spectrometric and modeling procedures, *Biochemistry* 42, 4430–4443, 2003.

102. Li, J. and Bigelow, D.J., Phosphorylation by cAMP-dependent protein kinase modulates the structural coupling between the transmembrane and cytosolic domains of phospholamban, *Biochemistry* 42, 10674–10682, 2003.

103. Waggoner, A., Covalent labeling of proteins and nucleic acids with fluorophores, *Methods Enzymol.* 246, 362–373, 1995.

104. Bech, L.M., Branner, S., Hastrup, S., and Breddam, K., Introduction of a free cysteine residue at position 68 in the subtilisin Savinase, based on homology with proteinase K, *FEBS Lett.* 297, 164–166, 1992.

105. Kunkel, T.A., Rapid and efficient site-specific mutagenesis without phenotypic selection, *Proc. Natl. Acad. Sci. USA* 82, 488–492, 1985.

106. Hogrefe, H.H., Cline, J., Youngblood, G.L., and Allen, R.M., Creating randomized amino acid libraries with the QuikChange® multi site-directed mutagenesis kit, *BioTechniques* 33, 1158–1165, 2002.

107. Karlin, A. and Akabas, M.H., Substituted-cysteine accessibility method, *Methods Enzymol.* 293, 123–145, 1998.

108. Ratner, V., Kahana, E., Eichler, M., and Haas, E., A general strategy for site-specific double labeling of globular proteins for kinetic FRET studies, *Bioconjug. Chem.* 13, 1163–1170, 2002.

109. Heyduk, T., Measuring protein conformational changes by FRET/LRET, *Curr. Opin. Biotechnol.* 13, 292–296, 2002.

110. Watrob, H.M., Pan, C.P., and Barkley, M.D., Two-step FRET as a structural tool, *J. Am. Chem. Soc.* 125, 7336–7343, 2003.

111. Rhoades, E., Gussakovsky, E., and Haran, G., Watching proteins fold one molecule at a time, *Proc. Natl. Acad. Sci. USA* 100, 3197–3202, 2003.

112. Franklin, J.G. and Leslie, J., Some enzymatic properties of trypsin after reaction with 1-dimethylaminonaphthalene-5-sulfonyl chloride, *Can. J. Biochem.* 49, 516–521, 1971.

113. Wagner, R., Podestá, F.E., González, D.H., and Andreo, C.S., Proximity between fluorescent probes attached to four essential lysyl residues in phosphoenolpyruvate carboxylase—A resonance energy transfer study, *Eur. J. Biochem.* 173, 561–568, 1988.

114. Park, S.J., Song, J.S., and Kim, H.J., Dansylation of tryptic peptides for increased sequence coverage in protein identification by matrix-assisted laser desorption/ionization time-of-flight mass spectrometric peptide mass fingerprinting, *Rapid Commun. Mass Spectrom.* 19, 3089–3096, 2005.

115. Amoresano, A., Chipappetta, G., Pucci, P. et al., Bidimensional tandem mass spectrometry for selective identification of nitration sites in proteins, *Anal. Chem.* 79, 2109–2017, 2007.

116. Cirulli, C., Marino, G., and Amoresano, A., Membrane proteins in *Escherichia coli* probed by MS3 mass spectrometry: A preliminary report, *Rapid Commun. Mass Spectrom.* 21, 2389–2397, 2007.

117. Haugland, R.P., Molecular probes, in *Handbook of Fluorescent Probes and Research Chemicals*, Molecular Probes, Inc., Eugene, OR, p. 37, 1989.

118. Tuls, J., Geren, L., and Millett, F., Fluorescein isothiocyanate specifically modifies lysine 338 of cytochrome P-450scc and inhibits adrenodoxin binding, *J. Biol. Chem.* 264, 16421–16425, 1989.

119. Miki, M., Interaction of Lys-61 labeled actin with myosin subfragment-1 and the regulatory proteins, *J. Biochem. (Tokyo)* 106, 651–655, 1989.

120. Bellelli, A., Ippoliti, R., Brunori, M. et al., Binding and internalization of ricin labelled with fluorescein isothiocyanate, *Biochem. Biophys. Res. Commun.* 169, 602–609, 1990.

121. Turner, D.C. and Brand, L., Quantitative estimation of protein binding site polarity. Fluorescence of *N*-arylaminonaphthalenesulfonates, *Biochemistry* 7, 3381–3390, 1968.

122. Donavan, J.W., The spectrophotometric titration of the sulfhydryl and phenolic groups of aldolase, *Biochemistry* 3, 67–74, 1964.

123. Weber, G. and Teale, F.W.J., Interaction of proteins with radiation, in *The Proteins*, ed. H. Neurath, Academic Press, New York, 1965.

124. Donovan, J.W., Changes in ultraviolet absorption produced by alteration of protein conformation, *J. Biol. Chem.* 244, 1961–1967, 1969.

125. Markland, F.S., Phenolic hydroxyl ionization in two subtilisins, *J. Biol. Chem.* 244, 694–700, 1969.

126. Mayberry, W.E. and Hockert, T.J., Kinetics of iodination. VI. Effect of solvent on hydroxyl ionization and iodination of L-tyrosine and 3-iodo-L-tyrosine, *J. Biol. Chem.* 245, 697–700, 1970.

127. Laws, W.R. and Shore, J.D., Spectral evidence for tyrosine ionization linked to a conformational change in liver alcohol dehydrogenase ternary complex, *J. Biol. Chem.* 254, 2582–2584, 1979.

128. Kuramitso, S., Hamaguchi, K., Miwa, S., and Nakashima, K., Ionization of the catalytic groups and tyrosyl residues in human lysozyme, *J. Biochem.* 87, 771–778, 1980.

129. Demchenko, A., *Ultraviolet Spectroscopy of Proteins*, Springer-Verlag, Berlin, Germany, 1981.

130. Kobayashi, J., Hagashijima, T., and Miyazawa, T., Nuclear magnetic resonance analyses of side chain conformations of histidine and aromatic acid derivatives, *Int. J. Pept. Protein Res.* 24, 40–47, 1984.

131. Poklar, N., Vesnaver, G., and Laponje, S., Studies by UV spectroscopy of thermal denaturation of beta-lactoglobulin in urea and alkylurea solutions, *Biophys. Chem.* 47, 143–151, 1993.

132. Sokolovsky, M., Harell, D., and Riordan, J.F., Reaction of tetranitromethane with sulfhydryl groups in proteins, *Biochemistry* 8, 4740–4745, 1969.

133. Riordan, J.F. and Vallee, B.L., Nitration with tetranitromethane, *Methods Enzymol.* 25, 515–521, 1972.

134. Cuatrecasas, P., Fuchs, S., and Anfinsen, C.B., The tyrosyl residues at the active site of staphylococcal nuclease. Modifications by tetranitromethane, *J. Biol. Chem.* 243, 4787–4798, 1968.

135. Sokolovsky, M., Fuchs, M., and Riordan, J.F., Reaction of tetranitromethane with tryptophan and related compounds, *FEBS Lett.* 7, 167–170, 1970.

136. Boesel, R.W. and Carpenter, F.H., Crosslinking during the nitration of bovine insulin with tetranitromethane, *Biochem. Biophys. Res. Commun.* 38, 678–682, 1970.

137. Nadeau, O.W., Traxler, K.W., and Carlson, G.M., Zero-length crosslinking of the beta subunit of phosphorylase kinase to the N-terminal half of its regulatory alpha subunit, *Biochem. Biophys. Res. Commun.* 251, 637–641, 1998.

138. Hugli, T.E. and Stein, W.H., Involvement of a tyrosine residue in the activity of bovine pancreatic deoxyribonuclease A, *J. Biol. Chem.* 246, 7191–7200, 1971.

139. Crow, J.P. and Ishiropoulos, H., Detection and quantitation of nitrotyrosine residues in proteins: In vivo marker of peroxynitrite, *Methods Enzymol.* 269, 185–194, 1996.

140. Greenacre, S.A.B. and Ischeriopoulos, H., Tyrosine nitration: Localisation, quantification, consequences for protein function and signal transduction, *Free Radic. Res.* 34, 541–581, 2001.

141. Schmidt, P., Youhnovski, N., Daiber, A. et al., Specific nitration at tyrosine 430 revealed by high resolution mass spectrometry as basis for redox regulation of bovine prostacyclin synthase, *J. Biol. Chem.* 278, 12813–12819, 2003.

142. Petersson, A.-S., Steen, H., Kalume, D.E. et al., Investigation of tyrosine nitration in proteins by mass spectrometry, *J. Mass Spectrom.* 36, 616–625, 2001.

143. Willard, B.B., Ruse, C.I., Keightley, J.A. et al., Site-specific quantitation of protein nitration using liquid chromatography/tandem mass spectrometry. *Anal. Chem.* 75, 2370–2376, 2003.

144. Daiber, A., Bachschmid, M., Kavakli, C. et al., A new pitfall in detecting biological end products of nitric oxide—Nitration, nitros(yl)ation and nitrite/nitrate artifacts during freezing, *Nitric Oxide* 9, 44–52, 2003.

145. Irie, Y., Saeki, M., Kamisaki, Y. et al., Histone H1.2 is a substrate for dinitrase, an activity that reduces nitrotyrosine immunoreactivity in proteins, *Proc. Natl. Acad. Sci. USA* 100, 5634–5639, 2003.

146. Nikov, G., Bhat, V., Wishnok, J.S. et al., Analysis of nitrated proteins by nitrotyrosine-specific affinity probes and mass spectrometry, *Anal. Biochem.* 320, 214–222, 2003.

147. Miyagi, M., Sakaguchi, H., Darrow, R.M. et al., Evidence that light modulates protein nitration in rat retina, *Mol. Cell. Proteomics* 1, 293–303, 2003.

148. Zu, Y., Strong, M., Huang, Z., and Beckman, J.S., Antibodies that recognize nitrotyrosine, *Methods Enzymol.* 269, 201–209, 1996.

149. Riordan, J.F., Wacker, W.E.C., and Vallee, B.L., *N*-Acetylimidazole: A reagent for determination of "free" tyrosyl residues of proteins, *Biochemistry* 4, 1758–1765, 1965.

150. Myers II B. and Glazer, A.N., Spectroscopic studies of the exposure of tyrosine residues in proteins with special reference to the subtilisins, *J. Biol. Chem.* 246, 412–419, 1971.

151. Ohno, A. and Oae, S., Thiols, in *Organic Chemistry of Sulfur*, ed. S. Oae, Plenum Press, New York, Chapter 4, 1977.

152. Sluyterman, L.A.A., The rate-limiting reaction in papain action as derived from the reaction of the enzyme with chloroacetic acid, *Biochim. Biophys. Acta* 151, 178–187, 1968.

153. Cecil, R. and McPhee, J.R., The sulfur chemistry of proteins, *Adv. Protein Chem.* 14, 255–389, 1959.

154. Liu, T.-Y., The role of sulfur in proteins, in *The Proteins*, 3rd edn., Vol. 3, eds. H. Neurath and R.L. Hill, Academic Press, New York, 1977.

155. Torchinsky, Y.M., *Sulfur in Proteins*, Pergamon Press, Oxford, U.K., 1981.

156. Creighton, T.E., Chemical nature of polypeptides, in *Proteins Structure and Molecular Principles*, W.H. Freeman and Company, New York, Chapter 1, 1983.

157. Modena, G., Paradisi, C., and Scorrano, G., Solvation effects on basicity and nucleophilicity, in *Organic Sulfur Chemistry. Theoretic and Experimental Advances*, eds. F. Bernardi, I.G. Csizmadia, and A. Mongini, Elsevier, Amsterdam, the Netherlands, 1985.

158. Britto, P.J., Knipling, L., and Wolff, J., The local electrostatic environment determines cysteine reactivity of tubulin, *J. Biol. Chem.* 277, 29018–29027, 2002.

159. Ellman, G.L., A colorimetric method for determining low concentrations of mercaptans, *Arch. Biochem. Biophys.* 74, 443–450, 1958.

160. Riddles, P.W., Blakeley, R.L., and Zerner, B., Ellman's reagent: 5,5′-Dithiobis (2-nitrobenzoic acid)—A reexamination, *Anal. Biochem.* 94, 75–81, 1979.

161. Riddles, P.W., Blakeley, R.L., and Zerner, B., Reassessment of Ellman's reagent, *Methods Enzymol.* 91, 49–60, 1983.
162. Fernandez-Diaz, M.D., Barsotti, L., Dumay, E., and Cheftel, J.C, Effects of electric fields on ovalbumin solutions and dialyzed egg white, *J. Agric. Food Chem.* 48, 2332–2339, 2000.
163. Helten, A. and Koch, K.W., Calcium-dependent conformational changes in guanylate cyclase-activting protein 2 monitored by cysteine accessibility, *Biochem. Biophys. Res. Commun.* 356, 687–692, 2007.
164. Okonjo, K.O., Bello, O.S., and Babalola, J.O., Transition of hemoglobin between two tertiary conformations: The transition constant differs significantly for the major and minor hemoglobins of the Japanese quail (*Corunix cortunix japonica*), *Biochim. Biophys. Acta* 1784, 464–471, 2008.
165. Jönsson, T.J., Ellils, H.R., and Poole, L.B., Cysteine reactivity and thiol-disulfide interchange pathways in AhpF and AhpC of the bacterial alkyl hydroperoxide reductase system, *Biochemistry* 46, 5709–5721, 2007.
166. Kleanthous, C., Campbell, D.G., and Coggins, J.R., Active site labeling of the shikimate pathway enzyme, dehydroquinase. Evidence for a common substrate binding site within dehydroquinase and dehydroquinate synthase, *J. Biol. Chem.*, 265, 10929–10934, 1990.
167. Kleanthous, C. and Coggins, J.R., Reversible alkylation of an active site methionine residue in dehydroquinase, *J. Biol. Chem.*, 265, 10935–10939, 1990.
168. Weinberger, S.R., Viner, R.J., and Ho, P., Tagless extraction-retentate chromatography: A new global protein digestion strategy for monitoring differential protein expression, *Electrophoresis* 23, 3182–3192, 2002.
169. Grunert, T., Pock, K., Buchacher, A., and Allmaier, G., Selective solid-phase isolation of methionine-containing peptides and subsequent matrix-assisted laser desorption mass spectrometric detection of methionine- and methionine-sulfoxide-containing tryptic peptides, *Rapid Commun. Mass Spectrom.* 17, 1815–1824, 2003.
170. Shen, M., Guo, L., Wallace, A. et al., Isolation and isotope labeling of cysteine- and methionine-containing tryptic peptides, *Mol. Cell. Proteomics* 2, 315–324, 2003.
171. Jensen, J.L., Kolvenbach, C., Roy, S., and Schöneich, C., Metal-catalyzed oxidation of bran-derived neurotrophic factor (BDNF): Analytical challenge for the identification of modified sites, *Pharm. Res.* 17, 190–196, 2000.
172. Duenas, E.T., Keck, R., De Vox, A. et al., Comparison between light induced and chemically induced oxidation of rhVEGF, *Pharm. Res.* 18, 1455–1460, 2001.
173. Shapiro, R.I., Wen, D., Levesque, M. et al., Expression of sonic hedgehog-Fc fusion protein in *Pichia pastoris*. Identification and control of post-translational, chemical, and proteolytic modifications, *Protein Expr. Purif.* 29, 272–283, 2003.
174. Wood, M.J., Prieto, J.H., and Komives, E.A., Structural and functional consequences of methionine oxidation in thrombomodulin, *Biochim. Biophys. Acta.* 1703, 141–147, 2005.
175. Bakhtiar, R. and Guan, Z., Electron dissociation mass spectrometry in characterization of peptides and proteins, *Biotechnol. Lett.* 28, 1047–1059, 2006.
176. Jenkins, N., Murphy, L., and Tyther, R., Post-translational modifications of recombinant proteins: Significance for biopharmaceuticals, *Mol. Biotechnol.* 39, 113–118, 2008.
177. Hirs, C.H.W., Performic acid oxidation, *Methods Enzymol.* 11, 197–199, 1967.
178. Sharp, J.S., Becker, J.M., and Hettich, R.L., Protein surface mapping by chemical oxidation: Structural analysis by mass spectrometry, *Anal. Biochem.* 313, 216–225, 2003.
179. Trout, G.E., The estimation of microgram amounts of methionine by reaction with chloroamine-T, *Anal. Biochem.* 93, 419, 1979.
180. Li, C., Takazaki, S., Jin, X. et al., Identification of oxidized methionine sites in erythrocyte membrane protein by liquid chromatography/electrospray mass spectrometry peptide mapping, *Biochemistry* 45, 12117–12124, 2006.

181. Corless, S. and Cramer, R., On-target oxidation of methionine residues using hydrogen peroxide for composition-restricted matrix-assisted laser desorption/ionization peptide mass-mapping, *Rapid Commun. Mass Spectrom.* 17, 1212–1215, 2003.

182. Caldwell, P., Luk, D.C., Weissbach, H., and Brot, N., Oxidation of the methionine residues of *Escherichia coli* ribosomal protein L12 decreases the protein's biological activity, *Proc. Natl. Acad. Sci. USA* 75, 5349, 1978.

183. Keck, R.G., The use of t-butyl hydroperoxide as a probe for methionine oxidation in proteins, *Anal. Biochem.* 236, 56–62, 1996.

184. Liu, J.L., Lu, K.V., Eris, T. et al., In vitro methionine oxidation of recombinant human leptin, *Pharm. Res.* 15, 632–640, 1998.

185. Lu, H.S., Fausset, P.R., Narhi, L.O. et al., Chemical modification and site-directed mutagenesis of methionine residues in recombinant human granulocyte colony-stimulating factor: Effect on stability and biological activity, *Arch. Biochem. Biophys.* 362, 1–11, 1999.

186. Takahashi, K., The reaction of phenylglyoxal with arginine residues in proteins, *J. Biol. Chem.* 243, 6171–6179, 1968.

187. Yankeelov, J.A. Jr., Mitchell, C.D., and Crawford, T.H., A simple trimerization of 2,3-butanedione yielding a selective reagent for the modification of arginine in proteins, *J. Am. Chem. Soc.* 90, 1664–1666, 1968.

188. Patthy, L. and Smith, E. L., Reversible modification of arginine residues. Application to sequence studies by restriction of tryptic hydrolysis to lysine residues, *J. Biol. Chem.* 250, 557–564, 1975.

189. Xu, G., Takamoto, K., and Chance, M.R., Radiolytic modification of basic amino acid residues in peptides: Probes for examining protein-protein interactions, *Anal. Chem.* 75, 6995–7007, 2003.

190. Leitner, A. and Linder, W., Probing of arginine residues in peptides and proteins using selective tagging and electrospray ionization mass spectrometry, *J. Mass Spectrom.* 38, 891–899, 2003.

191. Cotham, W.E., Metz, T.O., Ferguson, P.L. et al., Proteomic analysis of arginine adducts on glyoxal-modified ribonuclease, *Mol. Cell. Proteomics* 3, 1145–1153, 2004.

192. Cheung, S.-T. and Fonda, M.L., Reaction of phenylglyoxal with arginine. The effect of buffers and pH, *Biochem. Biophys. Res. Commun.* 90, 940–947, 1979.

193. Hoare, D.G. and Koshland, D.E. Jr., A method for the quantitative modification and estimation of carboxyl groups in proteins, *J. Biol. Chem.* 242, 2447–2453, 1967.

194. George, A.L. Jr. and Border, C.L. Jr., Essential carboxyl groups in yeast enolase, *Biochem. Biophys. Res. Commun.* 87, 59–65, 1979.

195. Khorana, H.G., The chemistry of carbodiimides, *Chem. Rev.* 53, 145–166, 1953.

196. Kunkel, G.R., Mehrabian, M., and Martinson, H.G., Contact-site cross-linking agents, *Mol. Cell. Biochem.* 34, 2–13, 1981.

197. Iwamoto, H., Oiwa, K., Suzuki, T., and Fujisawa, T., States of thin filament regulatory proteins as revealed by combining cross-linking/x-ray diffraction techniques, *J. Mol. Biol.* 317, 707–720, 2002.

198. Nadeau, O.W. and Carlson, G.M., Zero length conformation-dependent crosslinking of phosphorylase kinase subunits by transglutaminase, *J. Biol. Chem.* 269, 29670–29676, 1994.

199. Malencik, D.A. and Anderson, S.R., Dityrosine formation in calmodulin: Cross-linking and polymerization catalyzed by Arthromyces peroxidase, *Biochemistry* 35, 4375–4386, 1996.

200. Medvedeva, M.V., Kolobova, E.A., Huber, P.A.J. et al., Mapping of contact sites in the caldesmon-calmodulin complex, *Biochem. J.* 324, 255–262, 1997.

201. Usachenko, S.I. and Bradbury, E.M., Histone-DNA contacts in structure/function relationships of nucleosomes as revealed by crosslinking, *Genetics* 106, 103–115, 1999.

202. Juzmiene, D., Shapkina, T., Kirillov, S., and Wollenzien, P., Short-range RNA-RNA crosslinking methods to determine rRNA structure and interactions, *Methods* 25, 333–343, 2001.

203. Nogami, K., Wakabayashi, H., Ansong, C., and Fay, P.J., Localization of a pH-dependent, A2 subunit-interactive surface within the factor VIIIa A1 subunit, *Biochim. Biophys. Acta* 1701, 25–35, 2004.

204. Ladner, C.L., Turner, R.J., and Edwards, R.A., Development of indole chemistry to label tryptophan residues in protein for determination of tryptophan surface accessibility, *Protein Sci.* 16, 1204–1213, 2007.

205. Mendoza, V.L. and Vachet, R.W., Protein surface mapping using diethylpyrocarbonate with mass spectrometric detection, *Anal. Chem.* 80, 2895–2904, 2008.

206. Sutherland, B.W., Toews, J., and Kast, J., Utility of formaldehyde cross-linking and mass spectrometry in the study of protein-protein interactions, *J. Mass Spectrom.* 43, 699–715, 2008.

207. Krusemark, C.J., Ferguson, J.T., Wenger, C.D. et al., Global amine and acid functional group modification of proteins, *Anal. Chem.* 80, 713–720, 2008.

208. Toews, J., Rogalski, J.C., Clark, T.J., and Kast, J., Mass spectrometric identification of formaldehyde-induced peptide modifications under in vivo protein cross-linking conditions, *Anal. Chim. Acta* 618, 168–183, 2008.

209. Abraham, S.J., Kobayashi, T., Solaro, R.J., and Gaponenko, V., Differences in lysine pKa values may be used to improve NMR signal dispersion in reductively methylated proteins, *J. Biomol. NMR* 43, 239–246, 2009.

210. Rypniewski, W.R., Holden, H.M., and Rayment, I., Structural consequences of reductive methylation of lysine residues in hen egg white lysozyme: An X-ray analysis at 1.8-Å resolution, *Biochemistry* 32, 9851–9858, 1992.

211. Brown, J.R. and Hartley, B.S., Location of the disulphide bridges by diagonal paper electrophoresis. The disulphide bridges of bovine chymotrypsinogen A, *Biochem. J.* 101, 214–228, 1966.

212. Richmond, V. and Hartley, B.S., A two-dimensional system for the separation of amino-acids and peptides on paper, *Nature* 184, 1869–1870, 1959.

213. Dixon, G.H., Kaufmann, D.L., and Neurath, H., Amino acid sequence in the region of diisopropylphosphoryl binding in diisopropylphorphoryl-trypsin, *J. Biol. Chem.* 233, 1373–1381, 1958.

214. Mathesius, U., Imin, N., Chen, H. et al., Evaluation of protein reference maps for cross-species identification of protein by peptide mass fingerprinting, *Proteomics* 2, 1288–1303, 2002.

215. Pappin, D.J., Hojrup, P., and Bleasby, A.J., Rapid identification of proteins by peptide-mass fingerprinting, *Curr. Biol.* 3, 327–332, 1993.

216. Cottrell, J.S., Protein identification by peptide mass fingerprinting, *Pept. Res.* 7, 115–124, 1994.

217. Sutton, C.W., Pemberton, K.S., Cottrell, J.S. et al., Identification of myocardial proteins from two-dimensional gels by peptide mass fingerprinting, *Electrophoresis* 16, 308–316, 1995.

218. James, P., Quadroni, M., Carafoli, E., and Gonnet, G., Protein identification in DNA databases by peptide mass fingerprinting, *Protein Sci.* 3, 1347–1350, 1994.

219. Padliya, N.D. and Wood, T.D., Improved peptide mass fingerprinting matches via optimized sample preparation in MALDI mass spectrometry, *Anal. Chim. Acta* 627, 162–168, 2008.

220. Lee, S.W., Choi, J.P., Kim, H.J. et al., ASPMF: A new approach for identifying alternative splicing isoforms using peptide mass fingerprinting, *Biochem. Biophys. Res. Commun.* 377, 253–256, 2008.

221. Nicoli, R., Rudaz, S., Stella, C., and Veuthey, J.L., Trypsin immobilization on an ethylenediamine-based monolithic minidisk for rapid on-line peptide mass fingerprinting studies, *J. Chromatogr. A* 1216, 2695–2699, 2009.

222. Minagawa, H., Yamashita, T., Honda, M. et al., Comparative analysis of proteome and transcriptome in human hepatocellular carcinoma using 2D-DIGE and SAGE, *Protein J.* 27, 409–419, 2008.

223. Amini, A., Rapid identification of somatropin by peptide-mass fingerprinting, using MALDI-TOF mass spectrometry, *Phareur Sci. Notes* 2009, 11–16, 2009.

224. Wilson, N., Simpson, R., and Cooper-Liddell, C., Introductory glycosylation analysis using SDS-PAGE and peptide mass fingerprinting, *Methods Mol. Biol.* 534, 1–8, 2009.

225. Galas, D.J. and Schmitz, A., DNase footprinting: A simple method for the detection of protein-DNA binding specificity, *Nucleic Acids Res.* 5, 3157–3170, 1978.

226. Schreier, P.H., Davies, R.W., and Kotewicz, M.L., Protection against exonuclease III digestion—New way to investigate protein–DNA interactions, *FEBS Lett.* 109, 159–163, 1980.

227. Kirkegaard, K. and Wang, J.C., Mapping the topography of DNA wrapped around gyrase by nucleolytic and chemical probing of complexes of unique DNA sequences, *Cell* 23, 721–729, 1981.

228. Lutter, L.C., DNAS-II digestion of the nucleosome core—Precise locations and relative exposures of sites, *Nucleic Acids Res.* 9, 4251–4265, 1981.

229. Herzburg, R.P. and Dervan, P.B., Cleavage of DNA with methidiumpropyl-EDTA-Iron(II). Reaction conditions and product analyses, *Biochemistry* 23, 3934–3945, 1984.

230. Tullius, T.D., Dombroski, B.A., Churchill, M.E.A., and Kam, L., Hydroxyl modified footprinting: A high-resolution method for mapping protein–DNA contacts, *Methods Enzymol.* 155, 537–558, 1987.

231. Dixon, W.J., Hayes, J.J., Levin, J.R. et al., Hydroxyl radical footprinting, *Methods Enzymol.* 208, 380–413, 1991.

232. Brenowitz, M., Chance, M.R., Dhavan, G., and Takamoto, K., Probing the structural dynamics of nucleic acids by quantitative time-resolved and equilibrium hydroxyl radical "footprinting," *Curr. Opin. Struct. Biol.* 12, 648–653, 2002.

233. Hampshire, A.J., Rusling, D.A., Broughton-Head, V.J., and Fox, K.R., Footprinting: A method for determining the sequence selectivity, affinity and kinetics of DNA-binding ligands, *Methods* 42, 128–140, 2007.

234. Jain, S.S. and Tullius, T.D., Footprinting protein–DNA complexes using the hydroxyl radical, *Nat. Protoc.* 3, 1092–1100, 2008.

235. Aruoma, O.I., Halliwell, B., Gejewski, E., and Dizdaroglu, M., Copper-ion-dependent damage to the bases in DNA in the presence of hydrogen peroxide, *Biochem. J.* 273, 601–604, 1991.

236. Dizdaroglu, M., Aruoma, O.I., and Halliwell, B., Modification of bases in DNA by copper ion-1,10-phenanthroline complexes, *Biochemistry* 29, 8447–8451, 1990.

237. Nakata, H., Regan, J.W., and Lefkowitz, R.J., Chemical modification of α_2-adrenoreceptors. Possible role for tyrosine in the ligand binding site, *Biochem. Pharmacol.* 35, 4089–4094, 1986.

238. Calvete, J.J., Campanero-Rhodes, M.A., Raida, M., and Sanz, L., Characterisation of the conformational and quaternary structure-dependent heparin-binding region of bovine seminal plasma protein PDC-109, *FEBS Lett.* 444, 260–264, 1999.

239. Samata, A.K., Dutta, S., and Ali, E., Modification of sulfhydryl groups of interleukein-8 (IL-8) receptor impairs binding of IL-8 and IL-8-mediated chemotactic response of human polymorphonuclear neutrophils. *J. Biol. Chem.* 268, 6147–6153, 1993.

240. Ploug, M., Rahbek-Nielsen, H., Ellis, V. et al., Chemical modification of the urokinase plasminogen activator and its receptor using tetranitromethane. Evidence for the involvement of specific tyrosine residues in both molecules during receptor-ligand interaction, *Biochemistry* 34, 12524–12534, 1995.

241. Hobba, G.D., Forbes, B.E., Parkinson, E.J. et al., The insulin-like growth factor (IGF) binding site of bovine insulin-like growth factor binding protein-2 (bIGFBP-2) probed by iodination, *J. Biol. Chem.* 271, 30529–30536, 1996.

242. Edwards, A.M., Ruiz, M., Silva, E., and Lissi, E., Lysozyme modification by the Fenton reaction and gamma irradiation, *Free Radic. Res.* 36, 277–284, 2002.

243. Konermann, L., Tong, X., and Pan, Y., Protein structure and dynamics studied by mass spectrometry: H/D exchange, hydroxyl radical labeling, and related approaches, *J. Mass Spectrom.* 43, 1021–1036, 2008.

244. Sclavi, B., Woodson, S., Sullivan, M. et al., Following the folding of RNA with time-resolved synchrotron X-ray footprinting, *Methods Enzymol.* 295, 379–402, 1998.

245. Goshe, M.B., Chen, Y.H., and Anderson, V.E., Identification of the sites of hydroxyl radical reaction with peptides by hydrogen/deuterium exchange: Prevalence of reactions with the side chains, *Biochemistry* 39, 1761–1770, 2000.

246. Guan, J.Q., Almo, S.C., and Chance, M.R., Synchtrotron radiolysis and mass spectrometry: A new approach to research on the actin cytoskeleton, *Acc. Chem. Res.* 37, 221–229, 2004.

247. Aye, T.T., Low, T.Y., and Sze, S.K., Nanosecond laser-induced photochemical oxidation method for protein surface mapping with mass spectrometry, *Anal. Chem.* 77, 5814–5822, 2005.

248. Xu, G. and Chance, M.R., Radiolytic modification of sulfur-containing amino acid residues in model peptides: Fundamental studies for protein footprinting, *Anal. Chem.* 77, 2437–2449, 2005.

249. Sharp, J.S. and Tomer, K.B., Analysis of the oxidative damage-induced conformational changes of apo- and holocalmodulin by dose-dependent protein oxidative surface mapping, *Biophys. J.* 92, 1682–1692, 2007.

250. Liu, W. and Guo, R., Effects of Triton X-100 nanoaggregates on dimerization and antioxidant activity of morin, *Mol. Pharm.* 5, 588–597, 2008.

251. Watson, C., Janik, I., Zhuang, T. et al., Pulsed electron beam water radiolysis for submicrosecond hydroxyl radical protein footprinting, *Anal. Chem.* 81, 2496–2505, 2009.

252. Hampel, K.J. and Burke, J.M., Time-resolved hydroxyl-radical footprinting of RNA using Fe(II)-EDTA, *Methods* 23, 233–239, 2001.

253. Koppenol, W.H., The reaction of ferrous EDTA with hydrogen peroxide: Evidence against hydroxyl radical formation, *J. Free Radic. Biol. Med.* 1, 281–285, 1985.

254. Sclavi, B., Time-resolved footprinting for the study of the structural dynamics of DNA-protein interactions, *Biochem. Soc. Trans.* 36, 745–748, 2008.

255. Shcherbakova, I., Mitra, S., Beer, R.H., and Brenowitz, M., Fast Fenton footprinting: A laboratory-based method for the time-resolved analysis of DNA, RNA and proteins, *Nucleic Acids Res.* 34, e48, 2006.

256. Lloyd, R.V., Hanna, P.M., and Mason, R.P., The origin of the hydroxyl radical oxygen in the Fenton reaction, *Free Radic. Biol. Med.* 22, 885–888, 1997.

257. Wee, L.M., Long, L.H., Whiteman, M., and Halliwell, B., Factors affecting the ascorbate- and phenolic-dependent generation of hydrogen peroxide in Dulbecco's modified Eagles medium, *Free Radic. Res.* 37, 1123–1130, 2003.

258. Sutton, H.C. and Winterbourn, C.C., On the participation of higher oxidation states of iron and copper in Fenton reactions, *Free Radic. Biol. Med.* 6, 53–60, 1989.

259. Maleknia, S.D., Brenowitz, M., and Chance, M.R., Millisecond radiolytic modification of peptides by synchrotron x-rays identified by mass spectrometry, *Anal. Chem.* 71, 3965–3973, 1999.

260. Xu, G. and Chance, M.R., Hydroxyl radical-mediated modification of proteins as probes for structural proteomics, *Chem. Rev.* 107, 3514–3543, 2007.

261. Tong, X., Wren, J.C., and Konermann, L., Effects of protein concentration on the extent of x-ray-mediated oxidative labeling studied by electrospray mass spectrometry, *Anal. Chem.* 79, 6376–6382, 2007.

262. Xu, G. and Chance, M.R., Radiolytic modification of acidic amino acid residues in peptides: Probes for examining protein–protein interactions, *Anal. Chem.* 76, 1213–1221, 2004.

263. Xu, G., Takamoto, K., and Chance, M.R., Radiolytic modification of basic amino acids in peptides: Probes for examining protein-protein interactions, *Anal. Chem.* 75, 6995–7007, 1999.

264. Richards, F.M., Lamed, R., Wynn, R. et al., Methylene as a possible universal footprinting reagent that will include hydrophobic surface areas: Overview and feasibility: Properties of diazurine as a precursor, *Protein Sci.* 9, 2506–2517, 2000.

265. Nuss, J.E. and Alter, G.M., Denaturation of replication Protein A reveals an alterative conformation with intact domain structure and oligonucleotide binding activity, *Protein Sci.* 13, 1365–1378, 2004.

266. Loizos, N. and Darst, S.A., Mapping protein–ligand interactions by footprinting, a radical idea, *Structure* 6, 691–695, 1998.

267. Heyduk, T., Baichoo, N., and Heyduk, E., Hydroxyl radical footprinting of proteins using metal ion complexes, *Met. Ions Biol. Syst.* 38, 255–287, 2001.

268. Loizos, N., Mapping protein-ligand interactions by hydroxyl-radical protein footprinting, *Methods Mol. Biol.* 261, 199–210, 2004.

269. Kiselar, J.G., Janmey, P.A., Almo, S.C., and Chance, M.R., Structural analysis of gelsolin using synchrotron protein footprinting, *Mol. Cell. Proteomics* 2, 1120–1132, 2003.

270. West, G.M., Tang, L., and Fitzgerald, M.C., Thermodynamic analysis of protein stability and ligand binding using a chemical modification and mass spectrometry-based strategy, *Anal. Chem.* 89, 4175–4185, 2008.

271. Weil, L., Gordon, W.G., and Buchert, A.R., Photooxidation of amino acids in the presence of methylene blue, *Arch. Biochem. Biophys.* 33, 90–109, 1951.

272. Weil, L., On the mechanism of the photo-oxidation of amino acids sensitized by methylene blue, *Arch. Biochem. Biophys.* 110, 57–68, 1965.

273. Chatterjee, G.C. and Noltmann, E.A., Dye-sensitized photooxidation as a tool for the elucidation of critical amino acid residues in phosphoglucose isomerase, *Eur. J. Biochem.* 2, 9–18, 1967.

274. Weil, L., Seibles, T.S., and Herskovits, T.T., Photooxidation of bovine insulin sensitized by methylene blue, *Arch. Biochem. Biophys.* 111, 308–320, 1965.

275. Genov, N. and Idakieva, K., Photoreactivity of histidyl residues in subtilisins Novo and DY. Photooxidation of subtilisins, *Int. J. Pept. Protein Res.* 29, 368–373, 1987.

276. McDermott, M., Chiesa, R., Roberts, J.E., and Dillon, J., Photooxidation of specific residues in α-crystallin polypeptides, *Biochemistry* 30, 8653–8660, 1991.

277. Silverster, J.A., Timmins, G.S., and Davies, M.J., Photodynamically generated bovine serum albumin radicals: Evidence for damage transfer and oxidation at cysteine and tryptophan residues, *Free Radic. Biol. Med.* 24, 754–766, 1998.

278. Cascone, O., Biscoglio de Jimenez Bonino, M.J., and Santomé, J.A., Oxidation of methionine residues in bovine growth hormone by chloramine-T, *Int. J. Pept. Protein Res.* 16, 299–305, 1980.

279. Vogt, W., Zimmermann, B., Hesse, D., and Nolte, R., Activation of the fifth component of human complement, C5, without cleavage, by methionine oxidizing agents, *Mol. Immunol.* 29, 251–256. 1992.

280. Mihajlovic, V., Cascone, O., and Biscoglio de Jiménez Bonino, M.J., Oxidation of methionine residues in equine growth hormone by chloratmine-T, *Int. J. Biochem.* 25, 1189–1193, 1993.

281. Miles, A.M. and Smith, R.L., Functional methionines in the collagen/gelatin binding domain of plasma fibronectin: Effects of chemical modification by chloramine T, *Biochemistry* 32, 8168–8178, 1993.

282. Tomova, S., Cutruzzolà, F., Barra, D. et al., Selective oxidation of methionyl residues in the human recombinant secretroy leukocyte proteinase inhibitor. Effect of the inhibitor binding properties, *J. Mol. Recognit.* 7, 31–37, 1994.

283. Anraku, M., Yamasaki, K., Maruyama, T. et al., Effect of oxidative stress on the structure and function of human serum albumin, *Pharm. Res.* 18, 632–639, 2001.

284. Anraku, M., Kragh-Hansen, U., Kawai, K. et al., Validation of the chloramine-T induced oxidation of human serum albumin as a model for oxidative damage in vivo, *Pharm. Res.* 20, 684–692, 2003.

285. Santarelli, L.C., Wassef, R., Heinemann, S.H., and Hoshi, T., Three methionine residues located within the regulator of conductance for K$^+$(RCK) domains confer oxidative sensitivity to large-conductance Ca^{2+}-activated K$^+$ channels, *J. Physiol.* 571, 329–348, 2006.

286. Silberring, J. and Nyberg, F., Analysis of tyrosine- and methionine-containing neuropeptides by fast atom bombardment mass spectrometry, *J. Chromatogr.* 562, 469–467, 1991.

287. Rosenfeld, R., Philo, J.S., Haniu, M. et al., Sites of iodination in recombinant human brain-derived neurotrophic factor and its effect on neurotrophic activity, *Protein Sci.* 2, 1664–1674, 1993.

288. Villiers, C.L., Chesne, S., Lacroix, M.B. et al., Structural features of the first component of human complement, C1, as revealed by surface iodination, *Biochem. J.* 203, 185–191, 1982.

289. Callaway, J.E., Ho, Y.S., and DeLange, R.J., Accessibility of tyrosyl residues altered by the formation of the histone 2A/2B complex, *Biochemistry* 24, 2692–2697, 1985.

290. Maly, P. and Lüthi, C., The binding sties of insulin-like growth factor I (IGF I) to type I IGF receptor and to a monoclonal antibody. Mapping by chemical modification of tyrosine residues, *J. Biol. Chem.* 263, 7068–7072, 1988.

291. Illy, C., Thielens, N.M., Gagnon, J., and Arlaud, G.J., Effect of lactoperoxidase-catalyzed iodination on the Ca^{2+}-dependent interactions of human C1s. Location of the iodination sites, *Biochemistry* 30, 7135–7141, 1991.

292. Frantzen, F., Heggli, D.E., and Sudrehagen, E., Radiolabelling of human haemoglobin using the ^{125}I-Boton-Hunter reagent is superior to oxidative iodination of for conservation of the native structure of labelled protein, *Biotechnol. Appl. Biochem.* 22, 161–167, 1995.

293. Mund, M., Weise, C., Franke, P. et al., Mapping of exposed surfaces of the nicotinic acetylcholine receptor by identification of iodinated tyrosine residues, *J. Protein Chem.* 16, 161–170, 1997.

294. Kung, C.K. and Goldwasser, E., A probable conformational difference between recombinant and urinary erythropoietins, *Proteins* 28, 94–98, 1997.

295. Saboori, A.M., Rose, N.R., Bresler, H.S. et al., Iodination of human thyroglobulin (Tg) alters its immunoreactivity. I. Iodination alters multiple epitopes of human Tg, *Clin. Exp. Immunol.* 113, 297–302, 1998.

296. Ghosh, D., Erman, M., Sawicki, M. et al., Determination of a protein structure by iodination: The structure of iodinated acetylxylan esterase, *Acta Crystallogr. D Biol. Crystallogr.* 55, 779–784, 1999.

297. Thean, E.T., Comparison of specific radioactivities of human α-lactalbumin iodinated by three different methods, *Anal. Biochem.* 188, 330–334, 1990.

298. Kienhuis, C.B., Heuvel, J.J., Ross, H.A. et al., Six methods for direct radioiodination of mouse epidermal growth factor compared: Effect of nonequivalence in binding behavior between labeled and unlabeled ligand, *Clin. Chem.* 37, 1749–1755, 1991.

299. Kuo, B.S., Nordblom, G.D., and Wright, D.S., Perturbation of epidermal growth factor clearance after radioiodination and its implications, *J. Pharm. Sci.* 86, 290–296, 1997.

300. Sobal, G., Resch, U., and Sinzinger, H., Modification of low-density lipoprotein by different radioiodination methods, *Nucl. Med. Biol.* 31, 381–388, 2004.

301. Vergote, V., Bodé, S., Peremans, K. et al., Analysis of iodinated peptides by LC-DAD/ESI ion trap mass spectrometry, *J. Chromatogr. B Anal. Technol. Biomed. Life Sci.* 850, 213–220, 2007.

302. Glish, G.L. and Vacher, R.W., The basics of mass spectrometry in the twenty-first century, *Nat. Rev. Drug Discov.* 2, 140–150, 2003.

303. Fligge, T.A., Kast, J., Bruns, K., and Przybylski, M., Direct monitoring of protein-chemical reactions utilizing nanoelectrospray mass spectrometry, *J. Am. Soc. Mass Spectrom.* 10, 112–118, 1999.

304. Bennett, K.L., Smith, S.V., Lambrecht, R.M. et al., Rapid characterization of chemically-modified proteins by electrospray mass spectrometry, *Bioconjug. Chem.* 7, 16–22, 1996.

305. Jahn, O., Hofmann, B., Brauns, O. et al., The use of multiple ion chromatograms in on-line HPLC-MS for the characterization of post-translational and chemical modifications of proteins, *Int. J. Mass Spectrom.* 214, 37–51, 2002.

306. Yeboah, F.K., Alli, I., Yaylayan, V.A. et al., Effect of limited solid-state glycation on the conformation of lysozyme by ESI-MSMS peptide mapping and molecular modeling, *Bioconjug. Chem.* 15, 27–34, 2004.

307. Ohguro, H., Palczewski, K., Walsh, K.A., and Johnson, R.S., Topographic study of arrestin using differential chemical modification and hydrogen/deuterium exchange, *Protein Sci.* 3, 2428–2434, 1994.

308. Scaloni, A., Monti, M., Acquaviva, R. et al., Topology of the thyroid transcription factor 1 homeodomain-DNA complex, *Biochemistry* 38, 64–72, 1999.

309. Zappacosta, F., Ingallinella, P., Scaloni, A. et al., Surface topology of Minibody by selective chemical modification and mass spectrometry, *Protein Sci.* 6, 1901–1909, 1997.

310. Taralp, A. and Kaplan, H., Chemical modification of lyophilized proteins in nonaqueous environments, *J. Protein Chem.* 16, 183–193, 1997.

311. Vakos, H.T., Kaplan, H., Black, B. et al., Use of the pH memory effect in lyophilized proteins to achieve preferential methylation of α-amino groups, *J. Protein Chem.* 19, 231–237, 2000.

312. Govindarajan, R., Chatterjee, K., and Gatlin, L., Impact of freeze-drying on ionization of sulfonephthalein probe molecules in trehalose-citrate system, *J. Pharm. Sci.* 95, 498–510, 2006.

313. Smith, C.M., Gafken, P.R., Zhang, Z. et al., Mass spectrometric quantification of acetylation at specific lysines within the amino-terminal tail of histone H4, *Anal. Biochem.* 316, 23–33, 2003.

314. Hochleitner, E.O., Borchers, C., Parker, C. et al., Characterization of a discontinuous epitope of the human immunodeficiency virus (HIV) core protein p24 by epitope excision and differential chemical modification followed by mass spectrometric peptide mapping analysis, *Protein Sci.* 9, 487–496, 2000.

315. Scholten, A., Visser, N.F.C., van den Heuvel, R.H.H., and Heck, A.J.R., Analysis of protein-protein interaction surfaces using a combination of efficient lysine acetylation and nanoLC-MALDI-MS/MS applied to the E9:Im9 bacteriotoxin-immunity protein complex, *J. Am. Soc. Mass Spectrom.* 17, 983–994, 2006.

316. Janecki, D.J., Beardsley, R.L., and Reilly, J.P., Probing protein tertiary structure with amidination, *Anal. Chem.* 77, 7274–7251, 2005.

317. Beardsley, R.L., Running, W.E., and Reilly, J.P., Probing the structure of the *Caulobacter crescentus* ribosome with chemical labeling and mass spectrometry, *J. Proteome Res.* 5, 2935–2946, 2006.
318. Liu, X., Broshears, W.C., and Reilly, J.P., Probing the structure and activity of trypsin with amidination, *Anal. Biochem.* 307, 13–19, 2007.
319. Kim, J.-S., Kim, J.-H., and Kim, H.-J., Matrix-assisted laser desorption/ionization signal enhancement of peptides by picolinamidination of amino groups, *Rapid Commun. Mass Spectrom.* 22, 495–502, 2008.
320. Gabant, G., Augier, J., and Armengaud, J., Assessment of solvent residue accessibility using three sulfo-NHS-biotin reagent in parallel: Application to footprint changes of a methyltransferase upon binding its substrates, *J. Mass Spectrom.* 43, 360–370, 2008.

19 Use of Immunology to Characterize Biopharmaceutical Conformation

Proteins/polypeptides, oligonucleotides/polynucleotides, and oligosaccharides/polysaccharides all can elicit an immunological response. Immunological responses are more common against proteins and protein conjugates but antibody responses to polynucleotides and polysaccharides have been demonstrated. Autoantibodies to double-stranded DNA are present in system lupus erythrematosis[1] and there are several polysaccharide-based vaccines.[2–4]

The majority of biopharmaceutical products are proteins or protein conjugates and, as with the rest of this work, the emphasis in this section will be directed toward the study of protein products. There are a number of reports on adverse reactions to protein therapeutics, which are immunological in nature.[5–11] This has resulted in considerable interest in the development of assays to assess the immunogenicity of biopharmaceutical products.[12–16] It is therefore not unreasonable to use immunological reactivity as a measure of conformational integrity of biopharmaceutical products both in bioequivalence evaluation[17–23] and for changes in manufacturing.[24–26] It is noted that some biotherapeutic proteins are antibodies, which can be immunogenic.[7,10,27] Failure to maintain immunologic identity (avoidance of immunogenicity) is essential for both biosimilar products and for assuring the safety of manufacturing/ formulation changes.

An epitope or antigenic determinant is defined as the object that reacts with an antibody to form an antigen–antibody complex. A hapten is a small molecule such as organic molecule such as trinitrophenyl or small saccharide, which is an antigenic determinant and usually requires a carrier to be immunogenic. A hapten can react with its antibody in the absence of carrier[28]; the interaction of a hapten and its antibody can be studied by equilibrium dialysis.[29–32] Antibodies against oligosaccharides and polysaccharides are usually directed against a linear determinant consisting of one to four monosaccharide units and not sensitive to conformation. A protein can have one or more antigenic determinants, which may be linear (continuous) or discontinuous (conformational). A linear determinant is a sequence of amino acids in a protein. A discontinuous epitope consists of amino acids, which are from disparate regions of a protein molecule[33–36]; a discontinuous epitope is sensitive to the conformation of a protein and is lost on denaturation.[37–40] There are also examples of epitopes being formed as a result of conformational change such as those as the metal ion-dependent epitopes in various proteins.[41–45]

Granted that the immunochemistry of a biopharmaceutical is of importance in the characterization for comparability studies both for the evaluation of changes in manufacturing process as well as the study of biosimilar products, the question is then one of selection of technology to answer the question. An example is provided from the development of recombinant forms of human blood coagulation factor VIII (FVIII).[46] First, factor VIII is used as chronic drug, not an acute treatment drug. Chronic use means continuing exposure to the biopharmaceutical, maximizing the opportunity for adverse reactions including the development of antibodies. Factor VIII is used for the treatment of hemophilia A; approximately 10% of the hemophilia A patients develop inhibitors (antibodies) to factor VIII.[47,48] The possibility of increased inhibitor response to recombinant factor VIII was a major consideration in the development of this product. Esmon and colleagues[49] prepared polyclonal antibody to recombinant factor VIII in rabbits. This antibody preparation was evaluated for the presence of antibodies, which would bind to recombinant factor VIII and not plasma factor VIII using competitive immunoassay and immunoadsorption. A factor VIII protein with the B-domain deleted did develop specific antibodies. These antibodies appeared to be directed against the deletion site and not the extensive carbohydrate found in the B-domain of the native protein. Other studies did show that there are at least four distinct epitopes in human FVIII.[50] While preclinical studies suggest the immunologic identity, there is continued surveillance of the hemophilia treatment population,[51,52] indicating that while it is necessary to establish immunological identity in preclinical studies and have successful clinical studies prior to licensure, phase IV surveillance will likely be required for biosimilar products[17,23,53] and useful for changes in the manufacturing process.[52]

The challenge then is to select immunochemistry that will be useful in defining the immunogenicity of a biopharmaceutical product. Immunogenicity is defined as the capability of eliciting an immune response. Immunogenicity is a desirable quality for a vaccine but not useful for a biopharmaceutical product. An interesting example is provided by allergoids where the primary amino groups on an allergen such as pollen protein are modified with reagent such as formaldehyde[54] or organic acid anhydrides.[55] Modification with formaldehyde converts an allergen such as those derived from pollen to an allergoid. The allergoid no longer reacts with IgE for the development of an allergic response but maintains immunogenicity provides a mechanism for tolerization.[56] There is a continuing debate regarding the value of chemically modified allergoids versus pollen-based vaccines.[57,58]

Immunological assays are based on the interaction of an antibody and antigen. These assays usually involve binding of antibody and/or antigen to solid matrix providing the basis for enzyme-linked immunosorbant assay (ELISA) technology or surface plasmon resonance (SPR).[21] SPR assays are also referred to as label-free assays.[59–67] SPR assays rely on the binding of an analytic to a target protein or other macromolecule.[68] The target or capture macromolecule is covalently coupled (Figure 19.1) to carboxymethylated dextran bound to a gold surface. The binding of the analyte to the capture molecule is measured by the change in the angle of reflected light. SPR allows the study the kinetics of binding as well as the specific

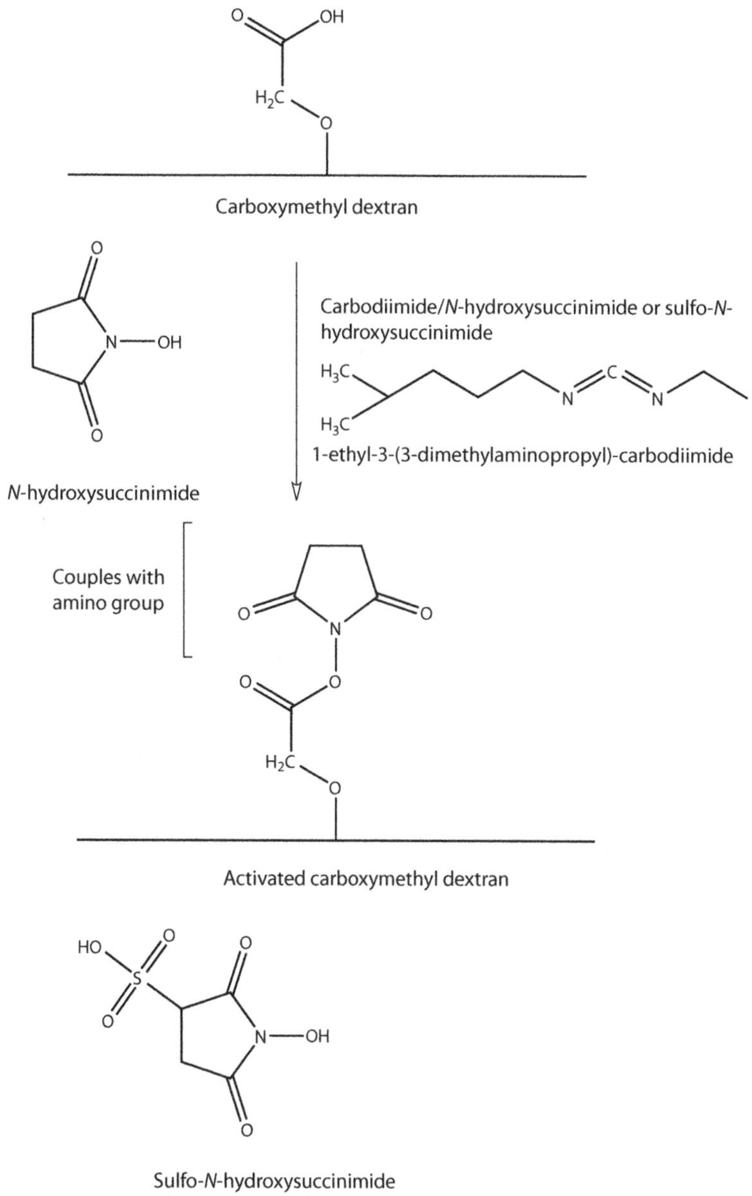

Carboxymethyl dextran

N-hydroxysuccinimide

Carbodiimide/N-hydroxysuccinimide or sulfo-N-hydroxysuccinimide

1-ethyl-3-(3-dimethylaminopropyl)-carbodiimide

Couples with amino group

Activated carboxymethyl dextran

Sulfo-N-hydroxysuccinimide

FIGURE 19.1 Activated carboxymethyl dextran for SPR. Carboxymethyl dextran is activated with carbodiimide/NHS to form the succinimide derivative of the carboxymethyl group which can be coupled with amine groups (see Johnsson, B., Löfås, S., and Lindquist, G., Immobilization of proteins to a carboxymethyldextran-modified gold surface for biospecific interaction analysis in SPR sensors, *Anal. Biochem*. 198, 268–277, 1991; Johnsson, B., Löfås, S., Lindquist, G. et al., Comparison of methods for immobilization to carboxymethyl dextran sensor surfaces by analysis of the specific activity of monoclonal antibodies, *J. Mol. Recognit.* 8, 125-131, 1995).

identification of an analytic in a complex biological fluid.[69] While the greatest use has been for the measurement of antibody/antigen, heparin was immobilized to study specific binding sites for this substance on fibrinogen.[70] There are few studies comparing ELISA assays and SPR technology but one recent study suggests comparable sensitivity.[71] Another study used SPR and ELISA to evaluate the immunogenicity of pegylated *Escherichia coli* and pegylated *Erwinia* asparaginases.[72] Patient and control sera were evaluated for the presence of antibody using SPR with the *E. coli* or *Erwinia* proteins covalently coupled to a carboxymethylated dextran matrix. SPR was shown to be more sensitive than an ELISA method in detected antibodies to the pegylated proteins in patient samples. Sandwich technology where a second antibody is added after the initial SPR coupling has been shown to improve sensitivity.[73–76] One of these studies[73] observed that while SPR was useful for the measurement of recombinant human blood coagulation factor VIII, it was not useful for plasma-derived human protein.

There have been several studies where SPR technology is used to monitor in-process biopharmaceutical production.[77–81] SPR is useful in the development of biopharmaceuticals in which the development of specific binding is a major consideration.[82] SPR is more rapid than ELISA procedures and is a promising technology, but does not yet have the thorough track record of ELISA technologies. SPR does have the advantage of providing real-time data, which would be of great importance for the manufacturing process. ELISA and SPR technologies have similar sensitivity with limit of detection (LOD) of approximately 0.05 ng; the range of quantitation is usually as high ng/low μg level. LOD assays are of most importance in the measurement of impurities and contaminants.[83–86] Selected applications of SPR to study of biological products are shown in Table 19.1.

The sensitivity of ELISA technologies can be markedly enhanced by a change in detection technology such as seen with fluorogenic substrates[98] or chemiluminescent detection.[99] A recent study[100] on the use of chemiluminescence detected human growth hormone at less that 0.05 ng/L. A mixture of two monoclonal antibodies was used for the capture; a separate monoclonal labeled with acridinium (*N*-hydroxysuccinimide, NHS).

Each detection technology should be evaluated to determine performance in a particular system.[101] ELISA-based systems also have the advantage of using technologies easily adapted to microarray/multiplexing platforms[102–104]; favorable comparability studies for microarray and ELISA microplate systems have been performed.[105,106] ELISA assays are sensitive (LOD) to the nanogram quantities and quantitation in the low microgram levels.[107–111]

ELISA assays are designed to measure quantity bound rather than association/dissociation rates. ELISA assays can measure the appearance of epitopes (neoantigenicity)[112–120] and the disappearance of epitopes.[121–130] ELISA assays can either directly analyze a biopolymer bound to a matrix or with a competitive assay; ELISA assays can also measure a protein or other biopolymer bound a capture material (capture antibody) where there is another specific biopolymer with a signal, which will recognize the bound biopolymer (sandwich format). This can present a problem where conformation is of importance; binding to a matrix can alter the conformation

TABLE 19.1
Use of SPR for the Characterization of Biopolymers[a]

Analyte	Capture Material[b]	System	Reference
IL-5	Soluble domain of human IL-5 receptor expressed in *Drosophila*	The soluble domain of human IL-5 (single transmembrane domain and a cytoplasmic tail) coupled to carboxymethyl dextran using carbodiimide/NHS chemistry (Figure 19.1). This provides for the preparation of an activated form of the carboxymethyl dextran, which will couple to biopolymers with free amino groups. This system provides a system for the laboratory evaluation of receptor agonists and antagonists.	[87]
Lipoprotein lipase	Heparan sulfate, heparin, heparin fragments	Heparan sulfate, heparin/heparin disaccharides previously modified with biotin were coupled to streptavidin-modified carboxymethyl dextran (streptavidin was coupled to carboxymethyl dextran using sulfo-NHS chemistry). These matrices were used to study the structure of lipoprotein lipase.	[88]
Human extracellular superoxide dismutase	Heparin	Biotinylated heparin was bound to streptavidin or avidin covalently bound to a carboxymethyl dextran matrix. This matrix was used to study the structural features of superoxide dismutase important for binding to heparin.	[89]
Irradiated ovalbumin	Monoclonal antibodies to ovalbumin	Monoclonal antibodies were coupled to the carboxymethyl dextran with carbodiimide/NHS chemistry and used to study conformation change in ovalbumin subjected to gamma-irradiation	[89a]
Thrombin substrates[c]	Anhydrothrombin	Anhydrothrombin was coupled to carboxymethyl dextran (presumably with carbodiimide/NHS chemistry). Analysis was performed using a IAsys single-channel resonant mirror biosensor.	[90]
Blood coagulation factor VIII preparations[d]	von Willebrand factor[e]	Purified von Willebrand factor was coupled to carboxymethyl dextran surface via amino groups[f]	[91]
von Willebrand factor[g]	Various collagen preparations with bovine serum albumin as a control	The various collagen preparations and the bovine serum albumin were coupled to the carboxymethyl dextran matrix via carbodiimide/NHS chemistry	[92]

(continued)

TABLE 19.1

Use of SPR for the Characterization of Biopolymers[a]

Analyte	Capture Material[b]	System	Reference
Erythropoietin glycosylation variants[h]	Erythropoietin receptor–Fc chimera	The erythropoietin receptor–Fc chimera is bound to protein A coupled to carboxymethyl dextran with carbodiimide/NHS chemistry. This system provides a useful approach to evaluation of changes in biopharmaceutical composition on receptor binding.	[93]
IgG2a and Fab and F(ab)$_2$ fragments in the presence and absence of hapten[i]	Protein A and protein G	Biotinylated protein A bound to streptavidin previous coupled to carboxymethyl dextran.	[94]
Human complement C1q	Mouse prion protein	Mouse prion protein was bound to carboxymethyl dextran using carbodiimide/NHS chemistry. Human complement C1q binds to this immobilized form of mouse prion protein but not to the soluble form.	[94a]
Lectins[j]	Neoglycopeptides (oxime linked glycoprobes)	Open and ring forms of an oxime-linked glycopeptides. Through use of a secondary amine, the aglycon-bound monosaccharide is forced into a ring form. The peptides were coupled via the amino-terminal to the carboxymethyldextran matrix using carbodiimide/NHS chemistry[k].	[95]
Native or denatured insulin or citrate synthase	α-Crystallin	α-Crystallin was bound to the carboxymethyl dextran surface with carbodiimide/NHS chemistry. The native forms of insulin or citrate synthase did not bind while the denatured forms bound.	[96]
Aptamers[l]	Mouse prion protein or "capture DNA"	Mouse prion protein was bound to the carboxymethyl dextran via carbodiimide/NHS coupling. A capture DNA oligonucleotide was labeled on the 3′-end and bound to a streptavidin sensor surface.	[97]

[a] Most of the SPR studies cited herein used the Biocore system developed by Pharmacia (see *Optical Biosensors Present and Future*, eds. F.S. Ligler and C.A. Rowe, Elsevier, Amsterdam, the Netherlands, 2002 for general discussion). The reactive matrix (carboxymethyl dextran) is bound to a gold layer. The capture substance is bound to activated dextran matrix. The sample is allowed to flow over the bound capture substance. The amount of material (mass) bound to the capture substance chances the angle of reflected light incident on the opposing surface of the gold layer. SPR measures mass, not numbers of bound materials.

[b] The term capture substance refers to the material coupled the matrix. This material could be an antibody, a ligand, or a receptor/receptor homolog. It is analogous to the capture antibody used in solid-phase immunoassays (Englebienne, P., *Immune and Receptor Assays in Theory and Practice*, CRC Press, Boca Raton, FL, 2000; König, T. and Skerra, A., Use of an albumin-binding domain for the selective immobilization of recombinant capture antibody fragments on ELISA plates, *J. Immunol. Methods* 218, 73–83, 1998).

[c] The thrombin substrates used in this study included fibrinogen, factor VIII, protein C, and factor XIII.

[d] Various factor VIII preparations included a recombinant B-domain–deleted form and two plasma-derived preparations; one missing the entire B-domain(designated as low-molecular weight form) and another containing varying amounts of the B-domain (designated as high-molecular-weight form). The various preparations showed almost identical association kinetics. These studies were performed as part of the preclinical characterization of a new therapeutic product for the treatment of hemophilia A.

[e] Human von Willebrand factor was prepared from a commercial factor VIII concentrate and was not characterized with respect to multimer content (Lind, P., Larsson, K., Spira, J. et al., Novel forms of B-domain-deleted recombinant factor VIII molecules. Construction and biochemical characterization, *Eur. J. Biochem.* 232, 19–27, 1995). The quality of vWF does vary in various factor VIII concentrates (Fricke, W.A. and Yu, M.Y., Characterization of von Willebrand factor in factor VIII concentrates, *Am. J. Hematol.* 31, 41–45, 1989).

[f] Presumably via carbodiimide/NHS chemistry.

[g] von Willebrand factor purified from human plasma cryoprecipitate and said to contain the normal multimer distribution.

[h] The variants differed in sialic acid content as well as a nonglycosylated form.

[i] These experiments were evaluated the conformation change cause in antibody on binding hapten. Affinity was reduced in the presence of hapten.

[j] *Erythrina cristagalli* agglutinin (ECA); *Ricinus communis* agglutinin (RCA); *Tricium vulgaris* agglutinin (WGA).

[k] Binding (K_a) enhanced minimally fourfold with the ring form.

[l] These aptamers undergo a conformation change of binding to their target and then bind to a capture DNA sequence which may be in solution or bound to the sensor surface.

of a biopolymer,[94a,131,132] which can be detected with ELISA-like technology.[133–136] Zamarron and colleagues identified monoclonal antibodies specific for fibrinogen bound to a surface.[137] These antibodies also inhibited fibrin polymerization, suggesting that the epitopes exposed on surface adsorption were also exposed on the conversion of fibrinogen to fibrin. It has been possible to avoid this problem by using a specific peptide tag to bind to the matrix[136] or by coating the surface with protein A, which will then bind the capture antibody.[147] However, the use of ELISA-based systems to assess conformation has been limited as it is not practical to measure rates of association. There are several examples where it has been possible to use ELISA technology and these are presented in Table 19.2.

TABLE 19.2
The Application of ELISA to Analysis of Conformational Change in Biopolymers

Analyte	Antibody	System	References
Factor IX	Purified rabbit antibody	Use of ELISA to demonstrate presence of antibodies specific for calcium-stabilized epitopes on blood coagulation factor IX. This provided initial support for the existence of a calcium ion-stabilized conformer of factor IX; this is a critical product quality attribute for current therapeutic products.	[45]
β-Lactoglobulin	Multiple monoclonal antibodies	Noncompetitive and competitive ELISA assay were used to study the reaction of β-lactoglobulin with monoclonal antibodies. Two antibodies bound more tightly to the native conformation while two other antibodies bound more tightly to the reduced, carboxymethylated protein. The antibodies were used to study the thermal denaturation process of β-lactoglobulin. These antibodies were used in a subsequent study of the refolding of β-lactoglobulin.	[138,139]
Potato virus X protein	Multiple monoclonal antibodies	A blocking ELISA was used to study antibody-induced conformational change in proteins.	[140]
Outer membrane protein P1.16 from *Neisseria meningitides*	Two monoclonal antibodies	ELISA used to monitor stability of monoclonal antibodies.	[141]
β-Lactoglobulin conjugated to carboxymethyl dextran	Monoclonal antibodies	ELISA used to measure local conformation changes in the β-lactoglobulin–carboxymethyl dextran conjugate. The β-lactoglobulin conjugate demonstrated reduced immunogenicity in several mouse models.	[142]

TABLE 19.2 (continued)
The Application of ELISA to Analysis of Conformational Change in Biopolymers

Analyte	Antibody	System	References
Monoclonal antibodies	His-tagged Nha antiporter	His-tagged NhaA is bound to NTA (Ni^{3+}-nitriloacetic acid). A competitive assay was used to measure a pH-dependent conformation change in Nha.	[143]
Calreticulin	C1q	No interaction between calreticulin and C1q in solution; when C1q is bound to a polystyrene surface, calreticulin binds to the C1q. A mouse anticalreticulin polyclonal followed by a alkaline phosphatase-goat antimouse IgG is used for detection It is suggested that calreticulin binds an altered form of C1q such as when bound to immunoglobulins thus serving a catabolic function.	[144]
Monoclonal antibodies	Troponin C	Divalent-induced conformational changes demonstrated by using monoclonal antibodies. Troponin C was bound to a microplate in the presence or absence of the divalent cation; binding to the microtiter plate. Conformational change was detected with the addition of monoclonal antibodies specific for the metal ion-dependent epitopes.	[145]
EG95 Hybrid vaccine	Polyclonal antibodies	Polyclonal antibodies against the conformationally dependent epitope (nonlinear) on the EB95 hydatid vaccine were screened with phage library to identify peptide sequences. This identified four conformational epitopes, one of which (E100) were then used to purify specific antibody from the polyclonal material. These antibodies can be used in ELISA assays or Western Blots for quality control.	[146]

There are other methods to assess the interaction of antibody with antigen. Most are not as sensitive as ELISA or SPR.[148,149] Such techniques as light scattering[150,151] and immunoaffinity electrophoresis[152] might be of value since there is no matrix dependence; as long as a matrix is involved, it is possible that artifact could be introduced into the analysis.

REFERENCES

1. Ghirardello, A., Villalta, D., Morozzi, G. et al., Evaluation of current methods for the measurement of serum anti-double stranded DNA antibodies, *Ann. N.Y. Acad. Sci.* 1109, 401–406, 2007.

2. Destefano, F., Pfeifer, D., and Nohynek, H., Safety profile of pneumococcal conjugate vaccines: Systematic review of pre- and post-licensure data, *Bull. World Health Organ.* 86, 373–380. 2008.

3. Smith, M.J., Meningococcal tetravalent conjugate vaccine, *Expert Opin. Biol. Ther.* 8, 1941–1946, 2008.

4. Schaffer, A.C. and Lee, J.C, Staphylococcal vaccines and immunotherapies, *Infect. Dis. Clin. North Am.* 23, 153–171, 2009.

5. Gupta, S., Indelicato, S.R., Jethwa, V. et al., Recommendations for the design, optimization, and qualification of cell-based assays used for the detection of neutralizing antibody responses elicited to biological therapeutics, *J. Immunol. Methods* 321, 1–18, 2007.

6. Wadhwa, M., Bird, C., Dilger, P. et al., Strategies for detection, measurement and characterization of unwanted antibodies induced by therapeutic biologicals, *J. Immunol. Methods* 278, 1–17, 2003.

7. Geng, D., Shankar, G., Schantz, A. et al., Validation of immunoassays used to assess immunogenicity to therapeutic monoclonal antibodies, *J. Pharmaceut. Biomed. Anal.* 39, 364–375, 2005.

8. Frost, H., Antibody-mediated side effects of recombinant proteins, *Toxicology* 209, 155–160, 2005.

9. Barbosa, M.D.F.S. and Celis, E., Immunogenicity of protein therapeutics and interplay between tolerance and antibody responses, *Drug. Discov. Today* 12, 674–681, 2007.

10. Pendley, C., Schantz, A., and Wagner, C., Immunogenicity of therapeutic antibodies, *Curr. Opin. Mol. Ther.* 5, 172–179, 2003.

11. Wang, J., Lozier, J., Johnson, G. et al., Neutralizing antibodies to therapeutic enzymes: Considerations for testing, prevention and treatment, *Nat. Biotechnol.* 26, 901–908, 2008.

12. Chirino, A.J., Ary, M.L., and Marshall, S.A., Minimizing the immunogenicity of protein therapeutics, *Drug Discov. Today* 9, 82–90, 2004.

13. Hermeling, S., Crommelin, D.J.A., Schellekens, H., and Jiskoot, W., Structure–immunogenicity relationships of therapeutics proteins, *Pharm. Res.* 21, 897–903, 2004.

14. Thorpe, R. and Swanson, S.J., Current methods for detecting antibodies against erythropoietin and other recombinant proteins, *Clin. Diag. Lab. Immunol.* 12, 28–39, 2005.

15. Kessler, M., Goldsmith, D., and Schellekens, H., Immunogenicity of biopharmaceuticals, *Nephrol. Dial. Transplant.* 21(Suppl 5), v9–v12, 2006.

16. Qin, C.J., Rodilla, R., Morgan, S.J. et al., Investigation of the immunogenicity of a protein drug using equilibrium dialysis and liquid chromatography tandem mass spectrometry detection, *Anal. Chem.* 77, 5529–5533, 2005.

17. Schellekens, H., Bioequivalence and the immunogenicity of biopharmaceuticals, *Nat. Rev. Drug Discov.* 1, 457–462, 2002.

18. Bertolotto, A., Deisenhammer, F., Gallo, P., and Sorensen, P.S., Immunogenicity of interferon β: Differences among products, *J. Neurol.* 251(Suppl 2), 15–24, 2004.

19. DeFelippis, M.R. and Larimore, F.S., The role of formulation in insulin comparability assessments, *Biologicals* 34, 49–54, 2006.

20. Shankar, G., Shores, E., Wagner, C., and Mire-Sluis, A., Scientific and regulatory considerations on the immunogenicity of biologics, *Trends Biotechnol.* 24, 274–280, 2006.

21. Baumann, A., Preclinical development of therapeutic biologics, *Expert Opin. Drug Discov.* 3, 289–297, 2008.

22. Cohen, B.A., Oger, J., Gagnon, A., and Giovannoni, G., The implications of immunogenicity for protein-based multiple sclerosis therapeutics, *J. Neurol. Sci.* 275, 7–17, 2008.

23. Schellekens, H., Recombinant human erythropoietin, biosimilars and immunogenicity, *J. Nephrol.* 21, 497–502, 2008.

24. Hemeling, S., Crommelin, D.J.A., Schellekens, H., and Jiskoot, W., Structure–immunogenicity relationships of therapeutic proteins, *Pharm. Res.* 21, 897–903, 2004.

25. Chirino, A.J. and Mire-Sluis, A., Characterizing biological products and assessing comparability following manufacturing changes, *Nat. Biotechnol.* 22, 1385–1391, 2004.

26. Doyle, J.W., Johnson, G.L., Eshhar, N., and Hammond, D., The use of rabbit polyclonal antibodies to assess neoantigenicity following viral reduction of an alpha-1-proteinase inhibitor preparation, *Biologicals* 34, 199–207, 2006.

27. Schneider, C.K. and Kalinke, U., Toward biosimilar monoclonal antibodies, *Nat. Biotechnol.* 26, 905–990, 2008.

28. Berzofsky, J.A. and Berkower, I.J., Immunogenicity and antigenic structure, in *Fundamental Immunology*, 5th edn, ed. W.E. Paul, Lippincott Williams & Wilkins, Philadelphia, PA, Chapter 21, 2003.

29. Geffard, M., Sequela, P., and Buijs, R.M., Immunorecognition of anti-serotonin antibodies by using a radiolabeled ligand, *Neurosci. Lett.* 50, 217–222, 1984.

30. Li, C.K., ELISA-based determination of immunological binding constants, *Mol. Immunol.* 22, 321–327, 1985.

31. Delcros, J.G., Clement, S., Thomas, V. et al., Differential recognition of free and covalently bound polyamines by the monoclonal anti-spermine antibody SPM8-2, *J. Immunol. Methods* 185, 191–198, 1995.

32. Feng, X., Pak, R.H., Kroger, L.A. et al., New anti-Cu-TETA and anti-Y-DOTA monoclonal antibodies for potential use in the pre-targeted delivery of radiopharmaceuticals to tumor, *Hybridoma* 17, 125–132, 1998.

33. Binder, M., Otto, F., Mertelsmann, R. et al., The epitope recognized by rituximab, *Blood* 108, 1975–1975, 2006.

34. Haste Anderson, P., Nielson, M., and Lund, O., Prediction of residues in discontinuous B-cell epitopes using protein 3D structures, *Protein Sci.* 15, 2558–2567, 2006.

35. Rapberger, R., Lukaas, A., and Mayer, B., Identification of discontinuous antigenic determinants on proteins based on shape complementarities, *J. Mol. Recognit.* 20, 113–121, 2007.

36. Lisova, O., Hardy, F., Petit, V. et al., High-affinity IgE recognition of a conformational epitope of the major respiratory allergen Phl p 2 as revealed by X-ray crystallography, *J. Immunol.* 182, 2141–2151, 2009.

37. Wright, K.E., Salvato, M.S., and Buchmeier, M.J., Neutralizing epitopes of lymphocytic choriomeningitis virus are conformational and require both glycosylation and disulfide bonds for expression, *Virology* 171, 417–426, 1989.

38. Moore, J.P. and Ho, D.D., Antibodies to discontinuous or conformationally sensitive epitopes on the gp120 glycoprotein of human immunodeficiency virus type 1 are highly prevalent in sera of infected humans, *J. Virol.* 67, 863–875, 1993.

39. Lee, N., Ahn, B., Jung, S.B. et al., Conformation-dependent antibody response to *Pseudomonas aeruginosa* outer membrane proteins induced by immunization in humans, *FEMS Immunol. Med. Microbiol.* 27, 79–85, 2000.

40. Munkonda, M.N., Pelletier, J., Ivanenkov, V.V. et al., Characterization of a monoclonal antibody as the first specific inhibitor of human NTP diphophohydrolase-3: Partial characterization of the inhibitory epitope and potential applications, *FEBS J.* 276, 479–496, 2009.

41. Lee, C.C., Ko, T.P., Chou, C.C. et al., Crystal structure of infectious bursal disease virus VP2 subviral particle at 2.6 Å resolution: Implications in virion assembly and antigenicity, *J. Struct. Biol.* 155, 74–86, 2006.

42. Takebe, M., Soe, G., Kohno, I. et al., Calcium ion-dependent monoclonal antibody against human fibrinogen: Preparation, characterization, and application to fibrinogen purification, *Thromb. Haemost.* 73, 662–667, 1995.

43. Persson, E. and Petersen, L.C., Structurally and functionally distinct Ca^{2+} binding sites in the γ-carboxyglutamic acid-containing domain of factor VIIa, *Eur. J. Biochem.* 234, 293–300, 1995.
44. Church, W.R., Boulanger, L.L., Messier, T.L., and Mann, K.G., Evidence for a common metal ion-dependent transition in the 4-carboxyglutamic acid domains of several vitamin K-dependent proteins, *J. Biol. Chem.* 264, 17882–17887, 1989.
45. Liebman, H.A., Limentani, S.A., Furie, B.C., and Furie, B., Immunoaffinity purification of factor IX (Christmas factor) by using conformation-specific antibodies directed against the factor IX-metal complex, *Proc. Natl. Acad. Sci. USA* 82, 3879–3883, 1985.
46. Kingdon, H.S. and Lundblad, R.L., An adventure in biotechnology: The development of haemophilia A therapeutics—From whole-blood transfusion to recombinant DNA to gene therapy, *Biotechnol. Appl. Biochem.* 35, 141–148, 2002.
47. ter Avest, P.C., Fischer, K., Mancuso, M.E. et al., Risk stratification for inhibitor development at first treatment for severe hemophilia A: A tool for clinical practice, *J. Thromb. Haemost.* 6, 2048–2054, 2008.
48. Mauser-Bunschoten, E.P., Fransen Van de Putte, D.E., and Schutgens, R.E., Co-morbidity in the aging haemophilia patient: The down side of increased life expectancy, *Haemophilia* 15, 853–863, 2009.
49. Esmon, P.C., Kuo, H.S., and Fournal, M.A., Characterization of recombinant factor VIII and a recombinant factor VIII deletion mutant using a rabbit immunogenicity model system, *Blood* 76, 1593–1600, 1990.
50. Leyte, A., Mertens, K., Distel, B. et al., Inhibition of human coagulation factor VIII by monoclonal antibodies. Mapping of functional epitopes. Mapping of epitopes with the use of recombinant factor VIII fragments, *Biochem. J.* 263, 187–194, 1989.
51. Tarantino, M.D., Collins, P.W., Hay, C.R. et al., Clinical evaluation of an advanced category antihaemophilic factor prepared using a plasma/albumin-free method: Pharmacokinetic, efficacy, and safety, *Haemophilia* 10, 428–437, 2004.
52. Blanchette, V.S., Shapiro, A.D., Liesner, R.J. et al., Plasma and albumin-free recombinant factor VIII: Pharmacokinetics, efficacy and safety in previously treated pediatric patients, *J. Thromb. Haemost.* 6, 1319–1326, 2008.
53. Schellekens, H. and Casadevall, N., Immunogenicity of recombinant human proteins: Causes and consequences, *J. Neurol.* 251(Suppl 2), II4–II9, 2004.
54. Norman, P.S., Lichtenstein, L.M., and March, D.G., Studies on allergoids from naturally occurring allergens. IV. Efficacy and safety of long-term allergoid treatment of ragweed pollen hay fever, *J. Allergy Clin. Immunol.* 68, 460–470, 1981.
55. Ćirković, T.D., Bukilica, M.N., Gavrović, M.D. et al., Physicochemical and immunologic characterization of low molecular-weight allergoids of *Dactylis glomerata* pollen proteins, *Allergy* 54, 128–134, 1999.
56. Vrtala, S., Focke-Tejkl, M., Swoboda, I. et al., Strategies for converting allergens into hypoallergenic vaccine candidates, *Methods* 32, 313–320, 2004.
57. Henmar, H., Lund, G., Lund, L. et al., Allergenicity, immunogenicity and dose-relationship of three intact allergen vaccines and four allergoid vaccines for subcutaneous grass pollen immunotherapy, *Clin. Exp. Immunol.* 153, 316–323, 2008.
58. Carnés, J., Himly, M., Gallego, M. et al., Detection of allergen composition and in vivo immunogenicity of depigmented allergoids of *Betula alba*, *Clin. Exp. Allergy* 39, 426–434, 2009.
59. Luppa, P.B., Sokoll, L.J., and Chan, D.W., Immunosensors—Principles and applications to clinical chemistry, *Clin. Chim. Acta* 314, 1–26, 2001.
60. Vetter, D., Chemical microarrays, fragment diversity, label-free imaging by plasmon resonance—A chemical genomics approach, *J. Cell Biochem. Suppl.* 39, 79–84, 2002.

61. Cooper, M.A., Label-free screening of bio-molecular interactions, *Anal. Bioanal. Chem.* 377, 834–842, 2003.
62. Englebienne, P., Van Hoonacker, A., and Verhas, M., Surface plasmon resonance: Principles, methods and applications in biomedical sciences, *Spectroscopy* 17, 255–273, 2003.
63. Yu, X., Xu, D., and Cheng, Q., Label-free detection methods for protein microarrays, *Proteomics* 6, 5493–5503, 2006.
64. Campbell, C.T. and Kim, G., SPR microscopy and its applications to high-throughput analyses of biomolecular binding events and their kinetics, *Biomaterials* 28, 2380–2392, 2007.
65. Visser, N.F. and Heck, A.J., Surface plasmon resonance mass spectrometry in proteomics, *Expert Rev. Proteomics* 5, 425–433, 2008.
66. Piliarik, M., Vaisocherová, H., and Homola, J., Surface plasmon resonance biodensity, *Methods Mol. Biol.* 503, 65–88, 2009.
67. Zourob, M., Elwary, S., Fan, X. et al., Label-free detection with the resonant mirror biosensor, *Methods Mol. Biol.* 503, 89–138, 2009.
68. Malmqvist, M., Surface plasmon resonance for detection and measurement of antibody–antigen affinity and kinetics, *Curr. Opin. Immunol.* 5, 282–287, 1993.
69. Stubenrauch, K., Wessels, U., and Lenz, H., Evaluation of an immunoassay for human-specific quantitation of therapeutic antibodies in serum samples from non-human primates, *J. Pharm. Biomed. Anal.* 49, 1003–1008, 2009.
70. Raut, S. and Gaffney, P.J., Interaction of heparin with fibrinogen using surface plasmon resonance technology: Investigation of heparin binding site on fibrinogen, *Thromb. Res.* 81, 503–509, 1996.
71. Vaisocherová, H., Faca, V.M., Taylor, A.D. et al., Comparative study of SPR and ELISA methods based on analysis of CD166/ALCAM levels in cancer and control human sera, *Biosens. Bioelectron.* 24, 2143–2148, 2009.
72. Avramis, V.I., Avramis, E.V., Hunter, W., and Long, M.C., Immunogenicity of native or pegylated *E. coli* and *Erwinia* asparaginases assessed by ELISA and surface plasmon resonance (SPR-biacore) assays of IgG antibodies (Ab) in sera from patients with acute lymphoblastic leukemia (ALL), *Anticancer Res.* 29, 299–302, 2009.
73. McCormick, A.N., Leach, M.E., Savidge, G., and Alhaq, A., Validation of a quantitative SPR assay for recombinant factor FVIII, *Clin. Lab. Haematol.* 26, 57–64, 2004.
74. Wang, L., Cole, K.D., Peterson, A. et al., Monoclonal antibody selection for interleukin-4 quantitation using suspension arrays and forward-phase protein microarrays, *J. Proteome Res.* 6, 4720–4727, 2007.
75. Arima, Y., Teramura, Y., Takiguchi, H. et al., Surface plasmon resonance and surface plasmon field-enhanced fluorescence spectroscopy for sensitive detection of tumor markers, *Methods Mol. Biol.* 503, 3–20, 2009.
76. Ladd, J., Lu, H., Taylor, A.D. et al., Direct detection of carcinoembryonic antigen autoantibodies in clinical human serum samples using a surface plasmon resonance sensor, *Colloids Surf. B. Biointerfaces* 70, 1–6, 2009.
77. Van Regenmortel, M.H., Use of biosensors to characterize recombinant proteins, *Dev. Biol. Stand.* 83, 143–151, 1994.
78. Van Regenmortel, M.H., Binding measurements as surrogate biological assays: Surface plasmon resonance biosensors for characterizing vaccine components, *Dev. Biol. Stand.* 103, 69–74, 2000.
79. Thillaivinayagalingam, P., Newcombe, A.R., O'Donovan, K. et al., Detection and quantification of affinity ligand leaching with specific antibody fragment concentration within chromatographic fractions using surface plasmon resonance, *Biotechol. Appl. Biochem.* 48, 179–188, 2007.
80. Jacquemart, R., Chavane, N., Durocher, Y. et al., At-line monitoring of bioreactor protein production by surface plasmon resonance, *Biotechnol. Bioeng.* 100, 184–188, 2008.

81. Mandenius, C.F., Wang, R., Aldén, A. et al., Monitoring of influenza virus hemaggluti-nin in process samples using weak affinity ligands and surface plasmon resonance, *Anal. Chim. Acta* 623, 66–75, 2008.

82. Huber, A., Demartis, S., and Neri, D., The use of biosensor technology for the engineer-ing of antibodies and enzymes, *J. Mol. Recognit.* 12, 198–216, 1999.

83. Edevåg, G., Eriksson, M., and Granström, M., The development and standardization of an ELISA for ovalbumin determination in influenza vaccines, *J. Biol. Stand.* 14, 223–230, 1986.

84. Lombardo, S., Inampudi, P., Scotton, A. et al., Development of surface swabbing proce-dures for a cleaning validation program in a biopharmaceutical manufacturing facility, *Biotechnol. Bioeng.* 48, 513–519, 1995.

85. Spanggord, R.J., Wu, B., and Sun, M., Development and application of an analytical method for the determination of squalene in formulations of anthrax vaccine adsorbed, *J. Pharm. Biomed. Anal.* 29, 183–193, 2002.

86. Li, Y., Song, C., Zhang, K. et al., Establishment of a highly sensitive sandwich enzyme-linked immunosorbent assay specific for ovomucoid from hen's egg white, *J. Agric. Food Chem.* 56, 337–342, 2008.

87. Morton, T.A., Bennett, D.B., Appelbaum, E.R. et al., Analysis of the interaction between human interleukin-5 and the soluble domain of its receptor using a surface plasmon resonance biosensor, *J. Mol. Recognit.* 7, 47–55, 1994.

88. Lookene, A., Chevreuil, O., Østergaard, P., and Olivecrona, G., Interaction of lipoprotein lipase with heparin fragments and heparan sulfate: Stoichiometry, stabilization, and kinetics, *Biochemistry* 35, 12155–12163, 1996.

89. Lookene, A., Stenlund, P., and Tibell, L.A.E., Characterization of heparin binding of human extracellular superoxide dismutase, *Biochemistry* 39, 230–236, 2000; (a) Masuda, T., Yasumoto, K., and Kitabatake, N., Monitoring the irradiation-induced conformational changes of ovalbumin by using monoclonal antibodies and surface plasmon resonance, *Biosci. Biotechnol. Biochem.* 64, 710–716, 2000.

90. Hosokawa, K., Ohnishi, T., Shima, M. et al., Preparation of anhydrothrombin and characterization of its interaction with natural thrombin substrates, *Biochem. J.* 354, 309–313, 2001.

91. Sandberg, H., Almstedt, A., Brandt, J. et al., Structural and functional characteristics of the B-domain-deleted recombinant factor VIII protein, r-VIII SQ, *Thromb. Haemost.* 85, 93–100, 2001.

92. Li, F., Maoke, J.L., and McIntire, L.V., Characterization of von Willebrand factor inter-action with collagens in real time using surface plasmon resonance, *J. Biomed. Eng.* 30, 1107–1116, 2002.

93. Darling, R.J., Kuchibohotla, U., Glaesner, W. et al., Glycosylation of erythropoietin affects receptor binding kinetics: Role of electrostatic interactions, *Biochemistry* 41, 14524–14531, 2002.

94. Sagawa, T., Oda, M., Morii, H. et al., Conformational changes in the antibody constant domains upon hapten-binding, *Mol. Immunol.* 42, 9–18, 2005; (a) Blanquet-Grossard, F., Thielens, N.M., Vendrely, C. et al., Complement protein C1q recognizes a conforma-tionally modified form of the prion protein, *Biochemistry* 44, 4349–4356, 2005.

95. Jiménez-Castells, C., de la Torre, B.G., Andrea, D., and Gutiérez-Gallego, R., Neo-glycopeptides: The importance of sugar core conformation in oxime-linked glyco-probes for interaction studies, *Glyconj. J.* 25, 879–887, 2008.

96. George, D.F., Bilek, M.M.M., and McKenzie, D.R., Detecting and exploring partially unfolded states of proteins using a sensor with a chaperone bound to its surface, *Biosens. Bioelectron.* 24, 963–969, 2008.

97. Ogasawara, D., Hachiya, N.S., Kaneko, K. et al., Detection system based on the con-formational change in an aptamers and its application to simple bound/free separation, *Biosens. Bioelectron.* 24, 1372–1376, 2009.

98. Meng, Y., High, K., Antonello, J. et al., Enhanced sensitivity and precision in an enzyme-linked immunosorbent assay with fluorogenic substrates compared with commonly used chromogenic substrates, *Anal. Biochem.* 345, 227–236, 2005.

99. Bi, S., Zhou, H., and Zhang, S., Multilayers enzyme-coated carbon nanotubes as biolabel for ultrasensitive chemiluminescence immunoassay for cancer biomarker, *Biosens. Bioelectron.* 24, 2961–2966, 2009.

100. Bidlngmaier, M., Suhr, J., Ernst, A. et al., High-sensitivity chemiluminescence immunoassays for detection of growth hormone doping in sports, *Clin. Chem.* 55, 445–453, 2009.

101. Kadkhodayan, S., Elliot, L.O., Mausisa, G. et al., Evaluation of assay technologies for the identification of protein–peptide interaction antagonists, *Assay Drug. Dev. Technol.* 5, 501–513, 2007.

102. Schweitzer, B. and Kingsmore, S.F., Measuring proteins on microarrays, *Curr. Opin. Biotechnol.* 13, 14–19, 2002.

103. Gonzalez, R.M., Seurynck-Servoss, S.L., Crowley, S.A. et al., Development and validation of sandwich ELISA microarrays with minimal assay interference, *J. Proteome Res.* 7, 2406–2414, 2008.

104. Pickering, J.W., Hoopes, J.D., Groll, M.C. et al., A 22-plex chemiluminescent microarray for pneumococcal antibodies, *Am. J. Clin. Pathol.* 128, 23–31, 2007.

105. Lebrun, S.J. and VanRenterghem, B., Performance characteristics of colorimetric protein microarrays compared to ELISA, *Assay Drug. Dev. Technol.* 4, 197–202, 2006.

106. Pang, S., Smith, J., Onley, D. et al., A comparability study of the emerging protein array platforms with established ELISA procedures, *J. Immunol. Methods* 302, 1–12, 2005.

107. Borg, L., Kristiansen, J., Christensen, J.M. et al., Evaluation of accuracy and uncertainty of ELISA assays for the determination of interleukin-4, interleukin-5, interferon-γ, and tumor necrosis factor-α, *Clin. Chem. Lab. Med.* 40, 509–519, 2002.

108. Choi, D.H., Katahura, Y., Matsuda, R. et al., Validation of a method for predicting the precision, limit of detection and range of quantitation in competitive ELISA, *Anal. Sci.* 23, 215–218, 2007.

109. Liang, M., Klakamp, S.L., Funelas, C. et al., Detection of high- and low-affinity antibodies against a monoclonal antibody using various technology platforms, *Assay Drug Dev. Technol.* 5, 655–662, 2007.

110. Morishita, N., Kamiya, K., Matsumoto, T. et al., Reliable enzyme-linked immunosorbent assay for the determination of soybean protein in processed foods, *J. Agric. Food Chem.* 56, 6818–6824, 2008.

111. Bhogal, H.S., Snodgrass, M., McLaws, L.J. et al., A suspension array immunoassay for the toxin stimulant ovalabumin, *J. Immunoassay Immunochem.* 30, 119–134, 2009.

112. Mollnes, T.E., Lea, T., Frøland, S.S. et al., Quantification of the terminal complement complex in human plasma by an enzyme-linked immunosorbent assay based on monoclonal antibodies against a neoantigen of the complex, *Scand. J. Immunol.* 22, 197–202, 1985.

113. Bekisz, T.E. and Brown, E.J., A mouse monoclonal antibody that reacts specifically with immune-complexed human IgG, *Mol. Immunol.* 22, 1225–1230, 1985.

114. Nugent, D.J., Kunicki, T.J., Berglund, C. et al., A human monoclonal autoantibody recognizes a neoantigen on glycoprotein IIa expressed on stored and activated platelets, *Blood* 70, 16–22, 1987.

115. Kusunoki, Y., Takekoshi, Y., and Nagasawa, S., Using polymerized C9 to produce a monoclonal antibody against a neoantigen of the human terminal complement complex, *J. Pharmacobiodyn.* 13, 454–460, 1990.

116. Accardo-Palumbo, A., Triolo, G., Casiglia, D. et al., Two-site ELISA for quantification of the terminal C5b-9 complement complex in plasma. Use of monoclonal and polyclonal antibodies against a neoantigen of the complex, *J. Immunol. Methods* 163, 169–172, 1993.

117. MacGregor, I.R., McLaughlin, L.F., MacGregor, M.C. et al., No detectable alternations in immunogenicity following terminal severe dry–heat treatment of high-purity factor VIII (Liberate) and factor IX (HP9)) concentrates, *Vox Sang.* 69, 319–327, 1995.
118. Philippou, H., Adami, A., Amersey, R.A. et al., A novel specific immunoassay for plasma two-chain factor VIIa: Investigation of FVIIa levels in normal individuals and in patients with acute coronary syndromes, *Blood* 89, 767–775, 1997.
119. Docena, G.H., Benítez, P., Fernández, R., and Fossati, C.A., Identification of allergenic proteins in condoms by immunoenzymatic methods, *Ann. Allergy Asthma Immunol.* 85, 77–83, 2000.
120. Hammel, M., Sfyroera, G., Pyrpassopoulos, S. et al., Charactizatoin of Ehp, a secreted complement inhibitory protein from *Staphylococcus aureus, J. Biol. Chem.* 282, 30051–30061, 2007.
121. Brown, M.A., Stenberg, L.M., Persson, U. et al., Identification and purification of vitamin K-dependent proteins and peptides with monoclonal antibodies specific for γ-carboxyglutarmyl (Gla) residues, *J. Biol. Chem.* 275, 19795–19802, 2000.
122. Matsuura, E., Inagaki, J., Kasahara, H. et al., Proteolytic cleavage of β_2-glycoprotein I: Reduction of antigenicity and the structural relationship, *Int. Immunol.* 12, 1183–1192, 2000.
123. Brett, G.M., Mills, E.N., Bacon, J. et al., Temperature-dependent binding of monoclonal antibodies to *C. hordein, Biochim. Biophys. Acta* 1594, 17–26, 2002.
124. Kim, Y.B., Han, D.P., Cao, C., and Cho, M.W., Immunogenicity and ability of variable loop-deleted human immunodeficiency virus type 1 envelope glycoproteins to elicit neutralizing antibodies, *Virology* 305, 124–137, 2003.
125. Mine, Y., Sasaki, E., and Zhang, J.W., Reduction of antigenicity and allergenicity of genetically modified egg white allergen, ovormucoid third domain, *Biochem. Biophys. Res. Commun.* 302, 133–137, 2003.
126. Buongiorno, A.M., Sagratella, E., Morellin, S. et al., Two polyclonal antisera detect different AGE epitopes in human plasma samples, *Immunol. Lett.* 85, 243–249, 2003.
127. Kamal, N., Chowhury, S., Madan, T. et al., Tryptophan residue is essential for immunoreactivity of a diagnostically relevant peptide epitope of *A. fumigatus, Mol. Cell. Biochem.* 275, 223–231, 2005.
128. Prabhakaran, V., Rajshekhar, V., Murrell, K.D., and Oommen, A., Conformation-sensitive immunoassays improve the serodiagnosis of solitary cysticercus granuloma in Indian patients, *Trans. R. Soc. Trop. Med. Hyg.* 101, 570–577, 2007.
129. Siddiqui, N.I., Yigzaw, Y., Préux, G. et al., Involvement of glycans in the immunological cross-reaction between α-macroglobulin and hemocyanin of the gastropod *Helix pomatia, Biochimie* 91, 508–516, 2009.
130. Hino, S., Matsubara, T., Urisu, A. et al., Periodate-resistant carbohydrate epitopes recognized by IgG and IgE antibodies from some of the immunized mice and patients with allergy, *Biochem. Biophys. Res. Commun.* 380, 632–637, 2009.
131. Prokopowicz, M., Banecki, B., Lukasiak, J., and Pryzyjazny, A., The measurement of conformational stability of proteins adsorbed on siloxanes, *J. Biomater. Sci. Polym. Ed.* 14, 103–118, 2003.
132. Lee, V.A., Craig, R.G., Filisko, F.E., and Zand, R., Microcalorimetry of the adsorption of lysozyme onto polymeric substrates, *J. Colloid Interface Sci.* 288, 6–13, 2005.
133. Hollander, Z. and Katchalski-Katzir, E., Use of monoclonal antibodies to detect conformational alternations in lactate dehydrogenase isoenzyme 5 on heat denaturation and on adsorption to polystyrene plates, *Mol. Immunol.* 23, 927–933, 1986.
134. Horbett, T.A. and Lew, K.R., Residence time effects on monoclonal antibody binding to adsorbed fibrinogen, *J. Biomater. Sci. Polym. Ed.* 6, 15–33, 1994.

135. Butler, J.E., Navarro, P., and Lü, E.P., Comparative studies on the interaction of proteins with a polydimethylsiloxane elastomer. II. The comparative antigenicity of primary and secondarily adsorbed IgG1 and IgG2a and their non-adsorbed counterparts, *J. Mol. Recognit.* 10, 52–62, 1997.

136. Kumada, Y., Zhao, C., Ishimura, R. et al., Protein-protein interaction analysis using an affinity peptide tag and hydrophilic polystyrene plate, *J. Biotechnol.* 128, 354–361, 2007.

137. Zamarron, C., Ginsberg, M.H., and Plow, E.F., Monoclonal antibodies specific for a conformationally altered state of fibrinogen, *Thromb. Haemost.* 64, 41–46, 1990.

138. Cho, E.W., Lee, M.K., Kim, K.L., and Hahm, K.S., Binding kinetics of monoclonal antibody using antigen-β-galactosidase hybrid protein: Application to measurement of peptide antigenicity, *J. Immunoassay* 16, 349–363, 1995.

139. Kaminogawa, S., Shimizu, M., Ametani, A. et al., Monoclonal antibodies as probes for monitoring the denaturation process of bovine β-lactoglobulin, *Biochim. Biophys. Acta* 998, 50–56, 1989.

140. Hattori, M., Ametani, A., Katakura, Y. et al., Unfolding/refolding studies on bovine β-lactoglobulin with monoclonal antibodies as probes. Does a renatured protein completely refold?, *J. Biol. Chem.* 268, 22414–22419, 1993.

141. Cepica, A., Yason, C., and Ralling, G., The use of ELISA for detection of antibody-induced conformational change in a viral protein and its intermolecular spread, *J. Virol. Methods* 28, 1–13, 1990.

142. Jiskoot, W., Beurvery, E.C., de Koning, A.A. et al., Analytical approaches to the study of monoclonal antibody stability, *Pharm. Res.* 7, 1234–1241, 1990.

143. Hattori, M., Nagasawa, K., Ohgata, M. et al., Reduced immunogenicity of β-lactoglobulin by conjugation with carboxymethyl dextran, *Bioconjug. Chem.* 11, 84–93, 2000.

144. Venturi, M., Rimon, A., Gerchman, Y. et al., The monoclonal antibody 1F6 identifies a pH-dependent conformational change in the hydrophilic NH_2 terminus of NhaA Na^+/H^+ antiporter of *Escherichia coli*, *J. Biol. Chem.* 275, 4734–4742, 2000.

145. Steine, A., Jørgensen, C.S., Laursen, I., and Houen, G., Interaction of C1q with the receptor calreticulin requires a conformational change in C1q, *Scand. J. Immunol.* 59, 485–495, 2004.

146. Jin, J.P., Chong, S.M., Hossain, M.M., Microtiter plate monoclonal antibody epitope analysis of Ca^{2+} and Mg^{2+}-induced conformational changes in troponin C, *Arch. Biochem. Biophys.* 466, 1–7, 2007.

147. Read, A.J., Casey, J.L., Coley, A.M. et al., Isolation of antibodies specific to a single conformation-dependent antigenic determinant on the EG95 hydatid vaccine, *Vaccine* 27, 1024–1031, 2009.

148. Butler, J.E., Antibody–antigen and antibody–hapten reactions, in *Enzyme Immunoassay*, ed. E.T. Maggio, CRC Press, Boca Raton, FL, 1980.

149. Gosling, J.P., A decade of development in immunoassay methodology, *Clin. Chem.* 36, 1408–1427, 1990.

150. Whicher, J.T., Price, C.P., and Spencer, K., Immunonephelometric and immunoturbidimetric assays for proteins, *Crit. Rev. Clin. Lab. Sci.* 18, 213–260, 1983.

151. Price, C.P., Spender, K., and Whicher, J., Light-scattering immunoassay of specific proteins: a review, *Ann. Clin. Biochem.* 20, 1–14, 1983.

152. Guzman, N.A., Immunoaffinity capillary electrophoresis applications of clinical electrophoresis of clinical and pharmaceutical relevance, *Anal. Bioanal. Chem.* 378, 37–39, 2004.

20 Use of Limited Proteolysis to Study the Conformation of Proteins of Biotechnological Interest

Limited proteolysis has been used for the study protein conformation for at least 60 years.[1] Linderstrom-Lange[1] observed in 1939 that while native proteins were slightly susceptible to proteolysis, reversible denaturation increased the susceptibility to proteolysis. Lineweaver and Hoover[2] described the action of papain on native and urea-denatured proteins. The rate of digestion increased as much as a 100-fold on denaturation of the substrate; the rate increase depended on the protease and the protein substrate. Bernheim and colleagues[3] showed that while the rate of proteolysis showed the greatest increase with fully denatured proteins, a substantial but lesser increase was still observed with a renatured protein as compared to the native protein. Thus it was demonstrated that a reversibly denatured protein can be differentiated from native by limited proteolysis. A year later in 1943[4] Erickson and Neurath reported that while a reversibly denatured proteins retained native epitopes, it was less immunogenic. Much of this early work has been reviewed by Putnam[5] in 1953. Somewhat later, Mihalyi[6] presented a comprehensive review of the proteolysis of proteins with an extensive discussion of the role of conformation. The susceptibility/rate of hydrolysis of a peptide bonds is dependent on (1) the amino acids in the scissile peptide bond and the sequence of amino acids surrounding the scissile peptide bonds (primary structure effects) and (2) the environment around the peptide bond (long-range effects), which is a function of the secondary and tertiary structure providing the environment around the scissile peptide bond. It is this latter consideration that is most important in the use of limited proteolysis for the study of protein conformation. The use of mass spectrometry for the analysis of the products of limited proteolysis has provided a significant advance in the use of this tool.[7-11]

It should be noted that a regulatory protease is more sensitive to primary structure effects than a digestive enzyme[12,13]; a digestive enzyme such as trypsin is more useful as a conformational probe[14-22] since the purpose is to identify peptide bonds which become exposed as a result of conformational change. Since many biological products are participate in highly regulated processes such as hemostasis or immunomodulation, regulatory proteolysis is a critical consideration.[23] Thrombin is an example of a regulatory protease that is a biopharmaceutical product[24–26] as are some products for fibinolysis.[27–29] Activated protein C is another example of a therapeutic regulatory protease.[30] Blood coagulation factor IX is activated by the process of regulatory proteolysis; the correct cleavage products obtained with factor XIa is a quality product attribute. Factor VIII is also controlled by the process of regulatory proteolysis and digestion with thrombin is used in the process of quality control.[31]

Bovine pancreatic ribonuclease A (RNAase A) is resistant to tryptic hydrolysis at 23°C. Rupley and Scheraga[32] demonstrated that RNAase A was susceptible to chymotryptic hydrolysis at 50°C. The rate of hydrolysis was inhibited by phosphate and citrate; such polyanions had previous been demonstrated to stabilize RNAase A toward urea denaturation. The temperature (50°C) is at the bottom end of a conformational change as measured by optical rotation[33]; the transition temperature for RNAase A was 61.9°C in water and 66.1°C in deuterium oxide. Winchester and coworkers[34] determined a T_m of 60.5°C by differential scanning calorimetry (DSC). Alcohols lowered the transition temperature as determined by difference spectroscopy[35] and increased susceptibility to proteolysis.[36] Other studies using limited proteolysis extended these observations for the effect of temperature on RNAase conformation as assessed with limited proteolysis.[37–40] These studies provided early support for the use of limited proteolysis to study conformational change in proteins. Pecher and Arnold[41] used proteinase K to study the stability of RNAse A derivatives with additional disulfide bonds, which had been inserted via protein engineering. This group also used limited proteolysis (proteinase K or subtilisin) to study the stability of RNAse A in trifluorethanol.[42] Earlier studies[43] from this group used limited proteolysis to study the thermal stability differences between RNAse A and the glycosylated form, RNAse B. Klink and Raines[44] studied variant form of RNAse A with limited proteolysis (proteinase K) and DSC. Conformational stability as assessed by DSC and limited proteolysis is a determinant of cytotoxity. Tsai and coworkers[45] compared protein fragmentation determined by computational cutting with the actual fragment produced by limited proteolysis. They used RNAse A, apomyoglobin, cytochrome c, and α-lactalbumin as model proteins. There was more computational cost than solution chemistry cost but there is enough consistency to use the approach to study protein folding.

Ribonuclease S (RNase-S),[46] discovered in the laboratory of the late Fred Richards at Yale University, is one of the most studied protein derivatives obtained via limited proteolysis. Cleavage occurs between Alanine-20 and Serine-21 (numbering from the amino terminal). The cleaved protein is active with the S-protein and S-peptide bound by noncovalent interactions. The protein and peptide can be separated and recombined. The structure of RNase-S is well understood.[47] RNAse A and the RNase S derivative continue to be useful as well understood models in solution protein chemistry.[48]

The limited proteolysis of immunoglobulins has been of great interest since the observations of Porter on the hydrolysis of an antibody directed against ovalabumin.[49] He observed that a fragment could be prepared from an antibody preparation directed against ovalbumin by proteolysis with papain. This fragment prevented the precipitation of ovalbumin by the native antibody preparation. He concluded that an antibody consisted of two general regions also as suggested by Pauling. Concern about the purity of the reagents in the earlier studies prompted a subsequent study in 1958,[50] confirming the early studies providing support for the structure of IgG composed of the Fc domain and F(ab) fragments.[51] Nisonoff and colleagues[52] used pepsin to digest IgG yielding the F(ab)$_2$, which had the properties of a dimer of F(ab). As discussed by Kolar and Capra, these results provided an understanding of the function of the IgG protein where the F(ab) region is the variable portion of the molecule and Fc region the constant portion. Karlsson and colleagues[53] provided an early synthesis of these findings in combination with results obtained from the study of Bence-Jones proteins to provide a functional framework for IgG function. Subsequent work showed that cleavage by papain occurs proximal from the "hinge" disulfide bonds yielding monomer F(ab) derivative; cleavage with pepsin is distal from the "hinge" disulfide bonds yielding the F(ab')$_2$ dimer.[54] These various observations have provided basis for the use of Fab/Fab' and F(ab')$_2$ fragments for therapeutic and diagnostic use[55-59] and subsequent development of engineered antibodies.[60-63] There are also early studies on the effect of temperature on immunoglobulin structure and limited proteolysis.[64-67] Fab fragments have slightly reduced affinity when compared to the parent IgG but retain epitopic specificity.[68-73]

Immunoaffinity chromatography is used for the purification of proteins and other biopolymers in the production of biopharmaceuticals.[74-76] One of the challenges in immunoaffinity chromatography is balancing affinity and specificity in binding the desired product. In general, a monoclonal antibody is the preferred element for attachment to a matrix for reasons of specificity, affinity, and reasonable production. Polyclonal antibodies may have increased degeneracy, higher affinity, and sustained production and are therefore less preferable. In a little recognized work, Recktenwald and coworkers[77] reduced the affinity of a polyclonal matrix by proteolysis, permitting the use of the polyclonal antibody for the affinity matrix. This is consistent with the earlier observations of Erickson and Neurath[4] on the reduction of antibody affinity but not specificity by proteolysis.

The reader is directed to several recent reports on the development of limited proteolysis for the study of protein conformation.[45,78-84] Hubbard and Beynon[80] presented examples of the use of limited proteolysis for surface mapping similar to the footprinting described in Chapter 18, identification of domain structure and elucidation of conformational change including denaturation and folding pathways. Another excellent review of the application of limited proteolysis to protein conformation is provided by Fontana and colleagues.[81] Park and Marqusee[82] have introduced pulse proteolysis as a technique for evaluating protein stability.

Some selected examples of the use of limited proteolysis for the study of proteins, which have specific interest for biopharmaceuticals, are shown in Table 20.1.

TABLE 20.1

Some Examples of the Application of Limited Proteolysis for the Study of Biopharmaceutical Protein Conformation

Protein	Enzyme	Study	Ref.
Human IL-4	Glu-C (*Staphylococcus aureus*)	Hydrolysis of human interleukin-4 with Glu-C protease from *S. aureus* resulted in two derivative forms, both of which were inactive. Derivative I was full-length with cleavage at Glu-26 and Glu-103. Derivative II extended from Glu-20 to Glu-103 with peptide bond cleavage at Glu-26.	[85]
Human vWF	Trypsin	von Willebrand factor (vWF) interaction with blood platelets depends on binding to exposed vascular endothelium. Ristocetin also promotes the interaction of vWF with blood platelets. The presence of ristocetin also causes a difference in the cleavage pattern with trypsin, suggesting that the interaction of vWF with blood platelets requires a conformational change.	[86]
Holo- and apo-α-lactalbumin	Trypsin	Limited proteolysis by trypsin is used to study the effect of ethanol on the conformation of α-lactalbumin. Holo-α-lactalbumin was resistant to proteolysis in the absence of ethanol; the presence of 20% ethanol increased susceptibility to proteolysis. Apo-α-lactalbumin was degraded by trypsin in the absence of ethanol; the presence of ethanol decreased proteolysis. CD and intrinsic fluorescence were also used to evaluate the effect of ethanol on α-lactalbumin.	[19]
Influenza A virus M1 protein	Bromolain	Surface mapping of the M1 protein globule.	[87]
Glucose oxidase	Subtilisin	Limited proteolysis was used to evaluate the effect of monovalent cation on protein conformation. The presence of 2.0 M NaCl protected the enzyme from proteolysis. Other techniques demonstrated an effect on enzymatic activity and tertiary structure.	[88]
Human growth hormone; interferon-α2b	Trypsin	Limited proteolysis with trypsin was using for surface mapping of the binding of Cibacron blue F3G-A to human growth hormone and interferon-α2b. These investigators used time-resolved proteolysis[a] to identify binding sites for Cibacron blue F3G-A on these two polypeptides.	[89]
Bovine serum albumin	Trypsin, chymotrypsin	Surface mapping of changes in protein conformation occurring on binding to silica particles. Study also includes the effect of solvent on the binding to silica particles as assessed by limited proteolysis.	[90]

TABLE 20.1 (continued)

Some Examples of the Application of Limited Proteolysis for the Study of Biopharmaceutical Protein Conformation

Protein	Enzyme	Study	Ref.
Cytochrome *c*	Proteinase K	Limited proteolysis by proteinase K was used to evaluate conformational change occurring on the binding of the equine apoprotein to heme. The formation of the holoprotein occurring on the binding of the apoprotein to heme is associated with an increase in resistance to proteolysis. This is consistent with other studies showing a transition from an open or unfolded conformation to a folded conformation occurring on the transition from the apoprotein to holoprotein.	[91]

[a] Time-resolved proteolysis is a technique where samples are removed from a reaction mixture as a function of time and subjected to analysis by, for example, mass spectrometry (see Tao, L., Kiefer, S.E., Xie, D. et al., Time-resolved limited proteolysis of mitogen-activated protein kinase-activated protein kinase-2 determined by LC/MS only, *J. Am. Soc. Mass Spectrom.* 19, 841–854, 2008).

REFERENCES

1. Linderstrom-Lange, K., Globular proteins and proteolytic enzymes, *Proc. R. Soc. B* 127, 17–18, 1939.
2. Lineweaver, H. and Hoover, S.P., A comparison of the action of crystalline papain on native and urea-denatured proteins, *J. Biol. Chem.* 137, 325–335, 1941.
3. Bernheim, E., Neurath, H., and Erickson, J.O., The denaturation of proteins and its apparent reversal IV. Enzymatic hydrolysis of native, denatured, and apparently irreversibly denatured proteins, *J. Biol. Chem.* 144, 259–264, 1942.
4. Erickson, J.O. and Neurath, H., Antigenic properties of native and regenerated horse serum albumin, *J. Exp. Med.* 78, 1–8, 1943.
5. Putnam, F.W., Protein denaturation, in *The Proteins*, Vol. 1, Part B, eds. H. Neurath and K. Bailey, Academic Press, New York, pp. 807–892, Chapter 9, 1953.
6. Mihalyi, E., *Application of Proteolytic Enzymes to Protein Structure Studies*, 2nd edn., CRC Press, West Palm Beach, FL, 1978.
7. Villanueva, J., Villegas, V., Querol, E. et al., Protein secondary structure and stability determined by combining exoproteolysis and matrix-assisted laser desorption/ionization time-of-flight mass spectrometry, *J. Mass Spectrom.* 37, 974–984, 2002.
8. Stroh, J.G., Loulakis, P., Lanzetti, A.J., and Xie, J., LC-mass spectrometry analysis of N- and C-terminal boundary sequences of polypeptide fragments by limited proteolysis, *J. Am. Soc. Mass Spectrom.* 16, 38–45, 2005.
9. Gao, X., Bain, K., Bonanno, J.E. et al., High-throughput limited proteolysis/mass spectrometry for protein domain elucidation, *J. Struct. Funct. Genomics* 6, 129–134, 2005.
10. Person, M.D., Shen, J., Traner, A. et al., Protein fragment domains identified using 2D gel electrophoresis/MALDI-TOF, *J. Biomol. Tech.* 17, 145–156, 2006.
11. Breuker, K., Jin, M., Han, X. et al., Top-down identification and characterization of biomolecules by mass spectrometry, *J. Am. Soc. Mass Spectrom.* 19, 1045–1053, 2008.

12. Neurath, H., Proteolytic enzymes, past and present, *Fed. Proc.* 44, 2907–2913, 1985.
13. Friedrich, P. and Bozóky, Z., Digestive versus regulatory proteases: On calpain action in vivo, *Biol. Chem.* 386, 609–612, 2005.
14. Egelund, R., Petersen, T.E., and Andreasen, P.A., A serpin-induced extensive proteolytic susceptibility of urokinase-type plasminogen activator implicates distortion of the proteinase pocket and oxyanion hole in the serpin inhibitory mechanism, *Eur. J. Biochem.* 268, 673–685, 2001.
15. Reid, J., Kelly, S.M., Watt, K. et al., Conformational analysis of the androgen receptor amino-terminal domain involved in transactivation. Influence of structure-stabilizing solutes and protein-protein interactions, *J. Biol. Chem.* 277, 22079–20086, 2002.
16. Varne, A., Muthukumaraswamy, K., Jatiani, S.S., and Mittal, R., Conformational analysis of the GTP-binding protein MxA using limited proteolysis, *FEBS Lett.* 516, 129–132, 2002.
17. Bito, R., Shikano, T., and Kawabata, H., Isolation and characterization of denatured serum albumin from rats with endotoxicosis, *Biochim. Biophys. Acta* 1646, 100–111, 2003.
18. Stiuso, P., Marabotti, A., Facchiano, A. et al., Assessment of the conformational features of vasoactive intestinal peptide in solution by limited proteolysis experiments, *Biopolymers* 81, 110–119, 2006.
19. Wehbi, Z., Pérez, M.D., Dalgalarronda, M. et al., Study of ethanol-induced conformational changes of holo and apo alpha-lactalbumin by spectroscopy and limited proteolysis, *Mol. Nutr. Food Res.* 50, 34–43, 2006.
20. Manea, M., Mezo, G., Hudecz, F., and Przybylski, M., Mass spectrometric identification of the trypsin cleavage pathway in lysyl-proline containing oligotuftsin peptides, *J. Pept. Sci.* 13, 227–236, 2007.
21. Liu, H., Gaza-Bulseco, G., Xiang, T., and Chumsae, C., Structural effect of deglycosylation and methionine oxidation of a recombinant monoclonal antibody, *Mol. Immunol.* 45, 701–708, 2008.
22. Kathir, K.M., Ibrahim, K., Rajalingam, D. et al., S100A13–lipid interactions—Role in the non-classical release of the acidic fibroblast growth factor, *Biochim. Biophys. Acta* 1768, 3080–3089, 2007.
23. Reich, E., Rifkin, D.B., and Shaw, E. eds., *Proteases and Biological Control*, Cold Spring Harbor Press, Cold Spring Harbor, NY, 1975.
24. Kumar, V. and Chapman, J.R., Whole blood thrombin: Development of a process for intraoperative production of human thrombin, *J. Extra Corpor. Technol.* 39, 18–23, 2007.
25. Weaver, F.A., Lew, W., Granke, K. et al., A comparison of recombinant thrombin to bovine thrombin as a hemostatic ancillary in patients undergoing peripheral arterial bypass and arteriovenous graft procedures, *J. Vasc. Surg.* 47, 1266–1273, 2008.
26. Clark, J., Crean, S., and Reynolds, M.W., Topical bovine thrombin and adverse events: A review of the literature, *Curr. Med. Res. Opin.* 24, 2071–2087, 2008.
27. Rubin, R.N., Fibrinolysis and its current usage, *Clin. Ther.* 5, 211–222, 1983.
28. Milligan, K.S., Tissue-type plasminogen activator: A new fibrinolytic agent, *Heart Lung* 16, 69–74, 1987.
29. Davydov, L. and Cheng, J.W., Tenecteplase: A review, *Clin. Ther.* 23, 982–997, 2001.
30. Schoots, I.G., Levi, M., van Vliet, A.K. et al., Inhibition of coagulation and inflammation by activated protein C or antithrombin reduces intestinal ischemia/reperfusion injury in rats, *Crit. Care Med.* 32, 1375–1383, 2004.
31. Sandberg, H., Almstedt, A., Brandt, J. et al., Structural and functional characteristics of the B-domain-deleted recombinant factor VIII protein, r-VIII SQ, *Thromb. Haemost.* 85, 93–100, 2001.
32. Rupley, J.A. and Scheraga, H.A., Digestion of ribonuclease A with chymotrypsin and trypsin at high temperatures, *Biochim. Biophys. Acta* 44, 191–193, 1960.

33. Hermans, J. Jr. and Scheraga, H.A., The thermally induced configurational change of ribonuclease I water and deuterium, *Biochim. Biophys. Acta* 36, 534–535, 1959.
34. Winchester, B.G., Mathias, A.P., and Robin, B.R., Study of the thermal denaturation of ribonuclease A by differential thermal analysis an susceptibility to proteolysis, *Biochem. J.* 117, 299–307, 1970.
35. Schrier, E.E. and Scheraga, H.A., The effect of aqueous alcohol solutions on the thermal transition of ribonuclease, *Biochim. Biophys. Acta* 64, 406–408, 1962.
36. Ooi, T. and Scheraga, H.A., Structural studies of ribonuclease XIV. Tryptic hydrolysis of ribonuclease in propyl alcohol solution, *Biochemistry* 3, 1209–1213, 1964.
37. Ooi, T. and Scheraga, H.A., Structural studies of ribonuclease XII. Enzymic hydrolysis of active tryptic modifications of ribonuclease, *Biochemistry* 3, 641–647, 1964.
38. Ooi, T. and Scheraga, H.A., Structural studies of ribonuclease 13. Physicochemical properties of tryptic modification of ribonuclease, *Biochemistry* 3, 648–652, 1964.
39. Ooi, T., Rupley, J.A., and Scheraga, H.A., Structural studies of ribonuclease VIII. Tryptic hydrolysis of ribonuclease A at elevated temperatures, *Biochemistry* 3, 432–437, 1963.
40. Rupley, J.A. and Scheraga, H.A., Structural studies of ribonuclease VII. Chymotryptic hydrolysis of ribonuclease at elevated temperatures, *Biochemistry* 2, 421–431, 1963.
41. Pecher, P. and Arnold, U., The effect of additional disulfide bonds on the stability and folding of ribonuclease A, *Biophys. Chem.* 141, 21–28, 2009.
42. Köditz, J., Arnold, U., and Ulbrich-Hofmann, R., Dissecting the effect of trifluorethanol on ribonuclease A. Subtle structural changes detected by nonspecific proteases, *Eur. J. Biochem.* 269, 3931–3837, 2002.
43. Arnold, U., Schierhorn, A., and Ulrich-Hofmann, R., Modification of the unfolding region in bovine pancreatic ribonuclease and its influence on the thermal stability and proteolytic fragmentation, *Eur. J. Biochem.* 259, 470–475, 1999.
44. Klink, T.A. and Raines, R.T., Conformational stability is a determinant of ribonuclease A cytotoxicity, *J. Biol. Chem.* 275, 17463–17467, 2000.
45. Tsai, C.J., Polverino de Laureto, P., Fontana, A., and Nussinov, R., Comparison of protein fragments identified by limited proteolysis and by computational cutting of proteins, *Protein Sci.* 11, 1753–1770, 2002.
46. Richard, F.M. and Vithayathil, P.J., The preparation of subtilisin-modified ribonuclease and the separation of the peptide and protein components, *J. Biol. Chem.* 234, 1459–1465, 1959.
47. Kim, E.E., Varadarajan, R., Wycoff, H.W., and Richards, F.M., Refinement of the crystal structure of ribonuclease S. Comparison with and between the various ribonuclease A structures, *Biochemistry* 31, 12304–12314, 1992.
48. Marshall, G.R., Feng, J.A., and Kuster, D.J., Back to the future: Ribonuclease A, *Biopolymers* 90, 259–277, 2008.
49. Porter, R.R., The formation of a specific inhibitor by hydrolysis of rabbit antiovalbumin, *Biochem. J.* 46, 479–484, 1950.
50. Porter, R.R., Separation and isolation of fractions of rabbit gamma-globulin containing the antibody and antigen combining sites, *Nature* 182, 670–671, 1958.
51. Kolar, G.R. and Capra, J.D., Immunoglobulins: Structure and function, in *Fundamental Immunology*, 5th edn., ed. W.E. Paul, Lippincott Williams & Wilkins, Philadelphia, PA, pp. 47–68, Chapter 3, 2003.
52. Nisonoff, A., Wissler, F.D., and Lipman, L.N., Properties of the major components of a peptic digest of rabbit antibody, *Science* 132, 1770–1771, 1960.
53. Karlsson, F.A., Peterson, P.A., and Berggård, I., Properties of halves of immunoglobulin light chains, *Proc. Natl. Acad. Sci. USA* 64, 1257–1263, 1969.
54. Butler, J.E. and Kennedy, N., The differential enzyme susceptibility of bovine immunoglobulin G_1 and immunoglobulin G_2 to pepsin and papain, *Biochim. Biophys. Acta* 535, 125–137, 1978.

55. Cresswell, C., Newcombe, A.R., Davies, S. et al., Optimal conditions for the papain digestion of polyclonal ovine IgG for the production of biotherapeutic Fab fragments, *Biotechnol. Appl. Biochem.* 42, 163–168, 2005.
56. Colburn, W.A., Specific antibodies and Fab fragments to alter the pharmacokinetics and reverse the pharmacologic/toxicologic effects of drugs, *Drug Metab. Rev.* 11, 223–262, 1980.
57. Flanagan, R.J. and Jones, A.L., Fab antibody fragments: Some applications in clinical toxicology, *Drug. Saf.* 27, 1115–1133, 2004.
58. Bourne, T., Fossati, G., and Nesbitt, A., A PEGylated Fab' fragment against tumor necrosis factor for the treatment of Crohn disease: Exploring a new mechanism of action, *BioDrugs* 22, 331–337, 2008.
59. Jazayeri, J.A. and Carroll, G.J., Fc-based cytokines: Prospects for engineering superior therapeutics, *BioDrugs* 22, 11–26, 2008.
60. Molnár, E., Prechl, J., Isaák, A., and Erdei, A., Targeting with scFv: Immune modulation by complement receptor specific constructs, *J. Mol. Recognit.* 16, 318–323, 2003.
61. Holliger, P. and Hudson, P.J., Engineered antibody fragments and the rise of single domains, *Nat. Biotechnol.* 23, 1126–1136, 2005.
62. Filpula, D., Antibody engineering and modification technologies, *Biomol. Eng.* 24, 201–215, 2007.
63. Benhar, I., Design of synthetic antibody libraries, *Exp. Opin. Biol. Ther.* 7, 763–779, 2007.
64. Seon, B.-K., Roholt, O.A., and Pressman, D., Differences in the enzymatic digestability of the variable and constant halves of Bence-Jones protein with the temperature, *J. Biol. Chem.* 247, 2151–2155, 1972.
65. Seon, B.-K. and Pressman, D., Fragment from the constant portion of IgG obtained by peptic digestion at high-temperatures, *J. Immunol.* 113, 1190–1198, 1974.
66. Edmundson, A.B., Ely, K.R., Abola, E.E. et al., Rotational allosterism and divergent evolution of domains in immunoglobulin light-chains, *Biochemistry* 14, 3953–3961, 1974.
67. Pascual, D.W. and Clem, L.W., Low temperature pepsin proteolysis. An effective procedure for mouse IgM F(ab')$_2$ fragment production, *J. Immunol. Methods* 146, 249–255, 1992.
68. Pitner, J.B., Beyer, W.F., Venetta, T.M. et al., Bivalency and epitope specificity of a high-affinity IgG3 monoclonal antibody to the *Streptococcus gropu* A carbohydrate antigen. Molecular modeling of a Fv fragment, *Carbohydr. Res.* 324, 17–29, 2000.
69. Orchekowski, R.P., Plescia, J., Altieri, D.C. et al., $\alpha_M\beta_2$ (Cd11b/CD18, Mac-1) integrin activation by a unique monoclonal antibody to α_MI domain that is divalent cation-sensitive, *J. Leukoc. Biol.* 68, 641–649, 2000.
70. Nason, E.L., Wetzel, J.D., Mukherjee, S.K. et al., A monoclonal antibody specific for reovirus outer-capsid protein σ3 inhibits signal-mediated hemagglutination by steric hindrance, *J. Virol.* 75, 6625–6234, 2001.
71. Khawli, L.A., Biela, B.H., Hu, P. et al., Stable, genetically engineered F(ab')$_2$ fragments of chimeric TNT-3 expressed in mammalian cells, *Hybrid. Hybridom.* 21, 11–18, 2002.
72. Blake II R.C., Delahanty, J., Khosraviani, M. et al., Allosteric binding properties of a monoclonal antibody and its Fab fragment, *Biochemistry* 42, 497–508, 2003.
73. Todorova-Balvay, D., Pitiot, O., Bourhim, M. et al., Immobilized metal-ion affinity chromatography of human antibodies and their proteolytic fragments, *J. Chromatogr.* 808, 57–62, 2004.
74. Griffith, M., Ultrapure plasma factor VIII produced by anti-F VIII$_c$ immunoaffinity chromatography and solvent/detergent viral inactivation. Characterization of the Method M process and Hemofil M antihemophilic factor (human), *Ann. Hematol.* 63, 131–137, 1991.

75. Jurlander, B., Thim, L., Klausen, N.K. et al., Recombinant activated factor VII (rFVIIa): Characterization, manufacturing, and clinical development, *Semin. Thromb. Hemost.* 27, 373–384, 2001.
76. Blank, K., Lindner, P., Diefenbach, B. et al., Self-immobilizing recombinant antibody fragments for immunoaffinity chromatography: Generic, parallel, and scalable protein purification, *Protein Expr. Purif.* 24, 313–322, 2002.
77. Recktenwald, A., Vernet, T., Storer, A.C., and Ziomek, E., Reduction of strong lipase-polyclonal antibodies binding by limited proteolysis, *Anal. Biotech.* 226, 31–34, 1995.
78. Król, M., Roterman, I., Piekarska, B. et al., Local and long-range structural effects caused by the removal of the N-terminal polypeptide fragment from immunoglobulin L chain lamba, *Biopolymers* 69, 189–200, 2003.
79. Hubbard, S.J., Eisensenger, F., and Thornton, J.M., Modeling studies of the change in conformation required for cleavage of limited proteolytic sites, *Protein Sci.* 3, 757–768, 1994.
80. Hubbard, S. and Beynon, R.J., Proteolysis of native proteins as a structural probe, in *Proteolytic Enzyme, A Practical Approach*, 2nd edn., eds. R.J. Benyon and J. Bond, Oxford University Press, Oxford, U.K., pp. 233–264, Chapter 10, 2001.
81. Fontana, A., de Laureto, P.P., Spolaore, B. et al., Probing protein structure by limited proteolysis, *Acta Biochim. Pol.* 51, 299–321, 2004.
82. Park, C. and Marqusee, S., Pulse proteolysis: A simple method for quantitative determination of protein stability and ligand binding, *Nat. Methods* 2, 207–212, 2005.
83. Breitender, H. and Mills, E.N., Molecular properties of food allergens, *J. Allergy Clin. Immunol.* 155, 14–23, 2005.
84. Reyda, M.R., Dippold, R., Dotson, M.E., and Jarrett, J.T., Loss of iron–sulfur clusters from biotin synthase as a result of catalysis promotes unfolding and degradation, *Arch. Biochem. Biophys.* 471, 32–41, 2008.
85. Le, H.V., Seelig, G.F., Syto, R. et al., Selective proteolytic cleavage of recombinant human interleukin-4. Evidence for a critical role of the C-terminus, *Biochemistry* 30, 9576–9582, 1991.
86. Kang, M., Wilson, L., and Kermode, J.C., Evidence from limited proteolysis of a ristocetin-induced conformational change in human von Willebrand factor that promotes its binding to platelet glycoprotein Ib-IX-V, *Blood Cells Mol. Dis.* 40, 433–443, 2008.
87. Kordyvkova, L.V., Serebryakava, M.V., Polyakov, V.Y. et al., Influenza A virus M1 protein structure probed by in situ limited proteolysis with bromolain, *Protein Pept. Lett.* 15, 922–930, 2008.
88. Ahmed, A., Akhtar, M.S., and Bhakuni, V., Monovalent cation-induced conformational change in glucose oxidase leading to stabilization of the enzyme, *Biochemistry* 40, 1945–1955, 2001.
89. Sutkeviciute, I., Serekaite, J., and Bumelis, V.A., Analysis of Cibacron blue F3G-A interaction with therapeutic proteins by MALDI-TOF mass spectrometry, *Biomed. Chromatogr.* 22, 1001–1007, 2008.
90. Larsericsdotter, H., Oscarsson, S., and Buijs, J., Structure, stability, and orientation of BSA adsorbed to silica, *J. Colloid Interface Sci.* 289, 26–35, 2005.
91. Spolaore, B., Bermejo, R., Zambonin, M., and Fontana, A., Protein interactions leading to conformational changes monitored by limited proteolysis: Apo form and fragments of horse cytochrome c, *Biochemistry* 40, 9460–9468, 2001.

21 Other Technologies for the Characterization of Conformational Change in Biopharmaceuticals

In the context of the current discussion, a biological therapeutic product refers to a peptide/protein, oligonucleotide/polynucleotide, oligosaccharide/polysaccharide, or conjugate product thereof (e.g., pegylated protein, peptide–nucleic acid, polysaccharide–protein complex). While the term biopharmaceutical is used on occasion, it is noted that this term does not necessarily have a meaning based on composition but perhaps more of a marketing definition.[1] The term biopharmaceutical is used in the current work more as a "shortcut" rather than a definitive term.

The thrust of this book has been on the description of methods useful for studying conformation during stability studies and/or the comparison of a "follow-on" or biosimilar product with an existing product. This has been a lively area of interest with erythropoietin (EPO).[2,3] The major techniques discussed in this are listed in Table 21.1. The use of these various techniques for product characterization is contained within the overall discussion of strategy below.

This chapter is divided into two sections. The first section is concerned with a brief discussion of techniques not covered in one of the preceding chapters. The second section contains a discussion of strategy for the characterization of biopharmaceutical products as defined above.

There are several techniques which should be briefly mentioned.

CRYSTALLOGRAPHIC ANALYSIS

The author received his graduate education in the Department of Biochemistry at the University of Washington in Seattle. This was in the days before the proliferation of specialty departments and subsequent talent dilution not unlike that seen in professional sports teams[19,20] and in business.[21] The basic concept there is that there is only so much great athletic talent; the point is that there are only a few really bright people (the author is not among this number but has been privileged to have met a number along the way). The Department of Biochemistry faculty included Hans Neurath, the two Eddies (Fischer and Krebs—later Nobel laureates secondary to an intense discussion following a student presentation some thirty years previous). Milt Gordon, Don Hanahan, Earl Davie, Frank Huennekens, Bill Rutter, and Joe Kraut. Joe Kraut

TABLE 21.1
Some Examples of Therapeutic Enzymes[a]

Enzyme	Therapeutic Target	Assay and Units of Activity	Refs.
Thrombin	Hemostatic agent; fibrinogen clotting; platelet aggregation	NIH Unit; International Unit—both based the "clotting" of fibrinogen	[4–6]
DNase (Pulmozyme®)	High-molecular DNA, which promotes pathogen colonization in pulmonary abscesses; cystic fibrosis	The original assay described by Kunitz[b] was based on hyperchromicity.[c] More recent assays have used the hydrolysis of the methyl green complex with DNA[d,e]	[7–9]
Glucocerebrosidase; imiglucerase (Cerezyme®) β-D-glucosyl-N-acylsphingosine glucohydrolase	Glucosylceramide (glucocerebroside)	A unit of enzyme activity is amount of enzyme that catalyzes the hydrolysis of one micromole of p-nitrophenyl-β-D-glucopyranoside per minute at 37°C	[10–12]
Blood coagulation factor VIIa	Factor VIII inhibitor-bypass activity (FEIBA); hemostatic agent	VIIa is measured in FEIBA units (units of factor VIII inhibitor bypassing activity). More recently, factor VIIa is measured with respect to an International Standard[f]	[13–15]
Tissue plasminogen activator (tPA); Alteplase (Activase®)	Activation of plasminogen to plasmin in therapeutic fibrinolysis		[16–18]

[a] See also Vellard, M., The enzyme as drug: Application of enzymes as pharmaceuticals, *Curr. Opin. Biotechnol.* 14, 444–450, 2003.

[b] Kunitz, M., Crystalline desoxyribonuclease I. Isolation and general properties. Spectrophotometric method for the measurement of desoxyribonuclease activity, *J. Gen. Physiol.* 33, 349–360, 1950.

[c] Plapp, B.V., Moore, S., and Stein, W.H., Activity of bovine pancreatic desoxyribonuclease A with modified amino groups, *J. Biol. Chem.* 246, 939–945, 1971.

[d] Sinicropi, D., Baker, D.L., Prince, W.S. et al., Colorimetric determination of DNase I activity with a DNA-methyl green substrate, *Anal. Biochem.* 222, 351–358, 1994.

[e] Lichtinghagen, R., Determination of Pulmozyme® (dornase alpha) stability using a kinetic colorimetric DNase I activity assay, *Eur. J. Pharm. Biopharm.* 63, 365–368, 2006.

[f] CBER, Summary Basis of Approval for Novoseven®; http://www.fda.gov/cber/sba/viianov032599S.pdf

was the resident crystallographer who soon moved south to warmer weather. None of the graduate students chose to work with Joe, not because he was not a nice guy but you had to be extremely clever in those days to do crystallography. In today's work, we have moved to high-throughput crystallographic analysis,[22–26] which is important for discovery research and product development but does not appear to add value for the measure of critical product attributes, subsequent product release criteria, or stability-indicating assays. It is not intended to suggest that today's crystallographers are not as bright as Joe Kraut and Steve Freer, but the analytical process does seem a bit easier with today's tools. X-ray crystallography, together with NMR spectroscopy, are the primary tools used in structural proteomics.[27–29] As noted above, these technologies will assist in the discovery characterization phase of biopharmaceutical research. There is particular interest in the use of x-ray crystallographic analysis for vaccine development.[30–37]

SMALL-ANGLE NEUTRON SCATTERING AND SMALL-ANGLE X-RAY SCATTERING

Small-angle neutron scattering (SANS) combined with deuterium labeling can provide information about the position of components within a complex (e.g., monomers in a quaternary complex).[38,39] Neutron scattering does require sophisticated instrumentation,[40,41] presenting a challenge for routine use. Neutrons can penetrate into condensed matter and have been used for harder materials such as metals[42] and plastics.[43] Its use has been increasing in biological systems.[44,45] SANS can distinguish between nucleic acid and proteins in complexes.[46–48]

Gabel and coworkers[49] have reviewed the early work on the use of SANS to study protein conformation. SANS was used to study the filament structures of polymers formed with RecA proteins from *Escherichia coli* and *Pseudomonas aeruginosa*.[50] Magani and coworkers used SANS to study the effect of temperature on lysozyme-trehalose, lysozyme-sucrose, trehalose, and sucrose solutions.[51] This study provides insight into the role of trehalose and sucrose as excipients in biopharmaceutical formulation. Wood and colleagues[52] used SANS to measure the dynamics (temperature) of RNAse A in dry and hydrated power forms, concentrated and dilute solutions. The results were intended to provide information on a well-understood protein, which can be taken to other systems. Marconi and coworkers[53] performed a similar study using SANS to study lysozyme in a hydrated power and solution. SANS was used to study the effect of pH on the structure of bovine serum albumin in solution and bound to negatively charged mica.[54]

Small-angle x-ray scattering (SAXS) is similar to SANS but provides more precise information about shape.[39] SAXS is complementary to x-ray crystallography in providing solution structure data on conformation, domain relationships, and folding at a resolution of 10–50 Å.[55] Pilz and coworkers used SAXS to measure conformational change on the interaction of tetra-alanine and anti-poly(D-alanyl) antibodies.[56] Subsequent work by this group with the Fab' and (Fab')$_2$ fragments using SAXS did not show a conformational change.[57] SAXS was used to study the degradation of polycarbonate urethanes by cholesterol esterase.[58] A vaccine delivery system consisting of polyelectrolyte complexes based on chitosan or dextran sulfate was characterized by SAXS.[59] SAXS was also used to characterize standard allergens and vaccines[60] as well as hydrogels and aerogels.[61–63]

EQUILIBRIUM DIALYSIS

Equilibrium dialysis is a technique for determining the association constant for a small molecule such as a dye or drug and a large molecule such as a protein. The technique is based on the ability of the small molecule to pass freely through a semipermeable membrane (dialysis membrane) while the large molecule is excluded from passage.

Klotz and coworkers published the first description of an equilibrium dialysis experiment in 1946.[64] These experiments used cellophane bags prepared from commercial sausage casings and measured the binding of methyl orange to bovine serum albumin. Teresi and Luck[65] subsequently used equilibrium dialysis to measure the association of bovine serum albumin with some aromatic carboxylic acids and nitrophenolates. The protein was placed inside the dialysis bag, which was in turn placed in a tube of a solution containing a defined amount of ligand. A control dialysis bag containing only solvent was placed in another tube with identical ligand concentration. A series of tubes was used containing a varying concentration of ligand. Measurement of the differences in concentration allows calculation of a binding constant. Other techniques, such as gel filtration (Hummel–Dreyer), provide data in a shorter period of time but equilibrium dialysis remains competitive[66] because of expense, requirement of technique expertise, and ability to perform experiments in plasma.

Ji and workers[67] used equilibrium dialysis to measure the interaction between an antibody and a low-molecular-weight (10 kDa) protein drug. This technique allows the measurement of free protein drug in plasma. Equilibrium dialysis has proved useful for the study of the binding of metal ions to prothrombin[68–70] and blood coagulation factor X,[71,72] the binding of metal ions and small molecules to albumin,[73–77] the binding of calcium ions to C-reactive protein,[78] drug binding to cardiac troponin C,[79] divalent cation binding to blood coagulation factor VIII,[80] and the binding of several amphiphilic penicillins to myoglobin.[81]

ENZYME KINETICS

There are several biopharmaceuticals that are enzymes, as shown in Table 21.1. This list is quite limited and some examples relevant to the overall discussion in this chapter are presented. In addition, there is considerable interest in the therapeutic potential of ribozymes[82–84] and catalytic antibodies.[85–90] There is a commonality among these diverse products in that the conventional method of expressing the potency of these materials is units/mg of biopolymer. A unit of enzyme activity was defined by the IUBMB (International Union of Biochemistry and Molecular Biology) as $1\,\mu mol$ of substrate min^{-1} under specified assay conditions.[91] More recent, the term katal (kat) has been adopted by IUPAC (International Union of Pure and Applied Chemistry) as unit of enzyme activity.[92] A katal is defined as $1\,mol$ substrate s^{-1}. A unit of enzyme activity as defined above is 16.667×10^{-9} kcat.[93] Irrespective of nomenclature, the activity of an enzyme is a function of affinity for substrate, ability convert substrate to product, and ability to release product and is expressed by the following equation[94,95]:

$$v = V[S]/K_m + [S] \qquad (21.1)$$

where

v is the observed velocity
V is V_{max}, $[S]$ is substrate concentration
K_m is the Michaelis–Menten constant

This is a simple reaction such as that represented by hydrolysis where the second reactant, water, is in overwhelming molar excess. While it is clear that the measurement of biological activity and subsequent expression in units is a critical quality attribute,[96] it is not clear that changes in kinetic constants such as V (V_{max}), k_{cat}, kat, or K_m (K_S) are useful as quality attributes.

The use of enzyme kinetic constants for the assessment of biopharmaceutical enzyme conformation first requires the accurate measurement of enzyme activity. This is not an overwhelming issue but does require going back to basics.[97–99] First, solvent conditions must be well-described and must be constant for intra- and interlaboratory comparison. Selection of buffer can be a major issue.[100] An example is the dependence of K_m and V_{max} values of alkaline phosphatase on buffer.[101] Tris has been demonstrated to have an effect on a variety of enzymes.[102–107]

Yoshioka and colleagues have studied the denaturation of β-galactosidase as part of a study on the prediction of stability for protein drugs.[108] The model suggested the conversation of native enzyme to an intermediate denatured form which then undergoes aggregation. Size-exclusion chromatography was used to distinguish between the various forms. Loss of activity corresponded to the formation of aggregated protein and not the intermediate denatured form. Samoshina and Samoshin[109] have suggested the use of the ratio of Michaelis constants (MCR) for comparing enzymes using β-galactosidases as a model. In another study, Nguyen and coworkers[110] compared the use of o-nitrophenyl-β-galactopyranoside and lactose as substrates for two β-galactosidases with the nitrophenyl substrate the more sensitive. Bhattacharya and Bhattacharya[111] studied the effect of sodium dodecyl sulfate (SDS) on bromelain structure and enzymatic activity. The loss of activity (benzyloxycarbonyl-arg-arg-7-aminocoumarin) was consistent with the loss of secondary structure; the decrease in V_{max} occurred at lower SDS concentration than did any change in K_m (K_m actually decreased with increasing SDS concentration). Adenosine deaminase is a therapeutic enzyme.[112] Shu and Frieden[113] showed that enzyme activity of adenosine deaminase is lost more rapidly than measured conformational change on denaturation with guanidine hydrochloride. Use of site-specific ^{19}F NMR showed that a change in the environment of one of the four tryptophan residues correlated with the loss of enzyme activity.

The second issue concerns the validity of "nonbiological" substrates as surrogate substrates for the biological/therapeutic activity of biopharmaceutical enzymes. Thrombin is an excellent example of this issue. Thrombin is one of the oldest biological therapeutic products[4] and was used originally (and currently) as a hemostatic agent and more recently as a component in fibrin sealant,[114] which is a combination product. The biological or therapeutic effect of thrombin is based on the conversion of fibrinogen to fibrin and somewhat less on the aggregation of platelets, which involves the cleavage of a platelet membrane receptor. The conversion of fibrinogen to fibrin, the formation of a fibrin clot, involves the cleavage of the A(α) chain of fibrinogen with the release of fibrinopeptide A and later the cleavage of the B(β) chain with release of fibrinopeptide B. The kinetic parameters for these reactions are well-understood.[115,116] The activity of thrombin as measured for biological potency as a licensed product is based on the clotting of fibrinogen and is expressed in NIH units or International units.[117] In either case, the measurement is the clotting of a standard fibrinogen solution determined by visual observation or mechanical measurement. Thrombin can also hydrolyze peptide nitroanilide and ester substrates, which provide data inconsistent with fibrinogen clotting.[118-124] Thus, in the case of thrombin, there is no effective substitute for fibrinogen in the measurement of therapeutic thrombin activity; chromogenic substrates may be effectively used in diagnostic tests.

Hyaluronidase,[125] earlier known as spreading factor,[126-129] is being used in cosmetic surgery and as a adjunct for drug delivery.[130-132] Hyaluronidase degrades hyaluronic acid, a high-molecular-weight glycosaminoglycan found in connective tissue and joint fluids[133] by hydrolyzing β-N-acetylhexosaminic bonds.[134] The assay methods are complex but can provide good data[134]; other assay methods are being developed.[135-139] A microplate-based assay has been developed.[140] Enzyme activity is used as a measure of molecular integrity of hyaluronidase.[141,142]

Another example is provided by carboxyl proteases.[143] Fujiwara and colleagues studied the effect of pressure on the catalytic activity and conformation of two pepstatin-sensitive carboxyl proteases, porcine pepsin and proteinase A from Baker's yeast, and two pepstatin-insensitive carboxyl proteases, pseudomonapepsin and xanthomonapepsin. Activity was determined with an octapeptide fluorescent substrate and acid-denatured myoglobin. Increasing pressure causes a decrease in k_{cat}/K_m for pseudomonapepsin and xanthomonapeptide while greater lost of activity was observed with pepsin and proteinase A from Baker's yeast; there was little effect on the hydrolysis of acid-denatured myoglobin by any of the four proteinases. Change in conformation was observed with fourth derivative spectroscopy. The point here is that there is a difference in response between the "biological" substrate and synthetic substrate.

Larner and colleagues showed that the chain length of the oligosaccharide acceptor influenced the activity of glycogen synthase.[144] These investigators used the metrics of $S_{0.5}$ and V_{max}. $S_{0.5}$ is a value analogous to K_M in that its substrate concentration is at $\frac{1}{2}V_{max}$. $S_{0.5}$ is usually used for cooperative enzyme systems[145-152] but would also be useful for the study of enzymes where there is substrate heterogeneity. In this case, it would be important to standardize the substrate. The assay for the presence of prekallikrein activator (activated Hageman factor, factor XIIa) is used for the evaluation of plasma protein products.[153] While current technology uses a peptide nitroanilide substrate[154] with an international standard.[155] To the best of my knowledge, this synthetic substrate has not been validated with the biological assay based on kinin release[156]: this assay (biological assay) required the preparation of a standardized crude substrate from plasma as a source of prekallikrein.[157]

The bulk of the data would suggest that smaller, synthetic substrates while providing for a more facile assay, do not necessarily measure the biological (therapeutic) activity of a biopharmaceutical enzyme. The use of such a synthetic substrate should be validated against the biological substrate. The use of concepts such as $S_{0.5}$ and V_{max} instead of K_M (K_S) and k_{cat} may be more valuable in the characterization of biopharmaceutical enzyme.

REFERENCES

1. Radar, R.A., (Re)defining biopharmaceutical, *Nat. Biotechnol.* 26, 743–751, 2008.
2. Combe, C., Tredree, R.L., and Schellekens, H., Biosimilar epoetins: An analysis based on recently implemented European medicines evaluation agency guidelines on comparability of biopharmaceutical proteins, *Pharmacotherapy* 25, 954–962, 2005.
3. Sörgel, F., Thyroff-Friesinger, U., Vetter, A. et al., Bioequivalence of HX575 (recombinant human epoetin alfa) and a comparable epoetin alfa after multiple subcutaneous administration, *Pharmacology* 93, 122–130, 2009.
4. Lundblad, R.L., Bradshaw, R.A., Gabriel, D. et al., A review of the therapeutic uses of thrombin, *Thromb. Haemost.* 91, 851–860, 2004.
5. Cheng, C.M., Meyer-Massetti, C., and Kayser, S.R., A review of three stand-alone topical thrombins for surgical hemostasis, *Clin. Ther.* 31, 32–41, 2009.
6. Anderson, C.D., Bowman, L.J., and Chapman, W.C., Topical use of recombinant human thrombin for operative hemostasis, *Expert Opin. Biol. Ther.* 9, 133–137, 2009.
7. Ayvazian, J.H., Johnson, A.J., and Tillett, W.S., The use of parenterally administered pancreatic desoxyribonuclease as an adjunct in the treatment of pulmonary abscesses, *Am. Rev. Tuberc.* 76, 1–21, 1957.
8. Bryson, H.M. and Sorkin, E.M., Dornase alfa. A review of its pharmacological properties and therapeutic potential in cystic fibrosis, *Drugs* 48, 894–906, 1994.
9. Suri, R., The use of human deoxyribonuclease (rhDNase) in the management of cystic fibrosis, *BioDrugs* 19, 135–144, 2005.
10. Pentchev, P.G., Brady, R.O., Hibbert, S.R. et al., Isolation and characterization of glucocerebrosidase from human placental tissue, *J. Biol. Chem.* 248, 5256–5261, 1973.
11. Morales, L.E., Gaucher's disease: A review, *Ann. Pharmacother.* 30, 381–388, 1996.
12. Cerezyme® NDA 20-362/S-053, Center for Drug Evaluation and Research, FDA, http://www.fda.gov/cder/foi/label/2002/20367s53lbl.pdf
13. Hedner, U., Recombinant coagulation factor VIIa: From the concept to clinical application in hemophilia treatment in 2000, *Semin. Thromb. Hemost.* 26, 363–366, 2000.
14. Jurlander, B., Thim, L., Klausen, N.K. et al., Recombinant activated factor VII (rFVIIa): Characterization, manufacturing, and clinical development, *Semin. Thromb. Hemost.* 27, 373–384, 2001.
15. Monroe, D.M., Further understanding of recombinant activated factor VII mode of action, *Semin. Hematol.* 45 (2 Suppl 1), S7–S11, 2008.
16. Karlan, B.Y., Clark, A.S., and Littlefield, B.A., A highly sensitive chromogenic microtiter plate assay for plasminogen activators, *Biochim. Biophys. Res. Commun.* 142, 147–154, 1987.
17. Christodoulides, M. and Boucher, D.W., The potency of tissue-type plasminogen activator (TPA) determined with chromogen and clot-lysis assays, *Biologicals* 18, 103–111, 1990.
18. Koley, K., Owen, W.G., and Machovich, R., Dual effect of synthetic plasmin substrates on plasminogen activation, *Biochim. Biophys. Acta* 1247, 239–245, 1995.
19. Bradbury, J.C., What really ruined baseball, *New York Times*, April 2, 2007.
20. Zimbalist, A., Baseball in the twenty-first century, in *Stee-Rike Four*, ed. D.R. Marburger, Greenwood Publishing, Santa Barbara, CA, Chapter 13, 1997.
21. Frank, J., Perverse outcomes of intense competition in the popular arts and its implications on product quality, *J. Cult. Econ.* 32, 215–224, 2008.
22. Lamzin, V.S. and Perrakis, A., Current state of automated crystallographic data analysis, *Nat. Struct. Biol.* 7 (Suppl), 978–981, 2000.
23. Delucas, L.J., Hamrick, D., Cosenza, L. et al., Protein crystallization: Virtual screening and optimization, *Prog. Biophys. Mol. Biol.* 88, 285–309, 2005.

24. Cianci, M., Antonyuk, S., Bliss, N. et al., A high-throughput structural biology/proteomics beamline at the SRS on a new multipole wiggler, *J. Synchrotron Radiat.* 12, 455–466, 2005.

25. Hiraki, M., Kato, R., Nagai, M. et al., Development of an automated large-scale protein-crystallization and monitoring system for high-throughput protein-structure analysis, *Acta Crystallogr. D Biol. Crystallogr.* 62, 1058–1065, 2006.

26. Baumes, L.A., Moliner, M., and Corma, A., Design of a full-profile-matching solution for high-throughput analysis of multiphase samples through power x-ray diffraction, *Chemistry* 15, 4258–4269, 2009.

27. Sali, A., Glaerser, R., Earnest, T., and Baumister, W., From words to literature in structural proteomics, *Nature* 422, 216–225, 2003.

28. Jung, J.W. and Lee, W., Structure-based functional discovery of proteins: Structural proteomics, *J. Biochem. Mol. Biol.* 37, 28–34, 2004.

29. Shin, J., Lee, W., and Lee, W., Structural proteomics by NMR spectroscopy, *Expert Rev. Proteomics* 5, 589–601, 2008.

30. Hashiguchi, T., Kajikawa, M., Maita, N. et al., Crystal structure of measles virus hemagglutinin provides insight into effective vaccines, *Proc. Natl. Acad. Sci. USA* 104, 19535–19540, 2007.

31. Patarroyo, M.E., Cifuentes, G., and Rodríguez, R., Structural characterization of sporozite components for a multistage, multi-epitope, anti-malarial vaccine, *Int. J. Biochem. Cell. Biol.* 40, 543–557, 2008.

32. Bartlam, M., Xue, X., and Rao, Z., The search for a structural basis for therapeutic intervention against the SARS coronavirus, *Acta Crystallogr. A.* 64, 204–213, 2008.

33. Buchanan, M.S., Carroll, A.R., Wessling, D. et al., Clavatadine A, a natural product with selective recognition and irreversible inhibition of factor XIa, *J. Med. Chem.* 51, 3583–3587, 2008.

34. Ng, A.K., Zhang, H., Tan, K. et al., Structure of the influenza virus A H5N1 nucleoprotein: Implications for RNA binding, oligomerization, and vaccine design, *FASEB J.* 22, 3638–3647, 2008.

35. Vulliez-Le Normand, B., Saul, F.A., Phalipon, A. et al., Structures of synthetic *O*-antigen fragments from serotype 2a *Shigella flexneri* in complex with a protective monoclonal antibody, *Proc. Natl. Acad. Sci. USA* 105, 9976–9981, 2008.

36. Caines, M.E., Zhu, H., Vuckovic, M. et al., The structural basis for T-antigen hydrolysis by *Streptococcus pneurmoniae*: A target for structure-based vaccine design, *J. Biol. Chem.* 283, 31279–31283, 2008.

37. Ekiert, D.C., Bhabba, G., Elsliger, M.A. et al., Antibody recognition of a highly conserved influenza virus epitope, *Science* 324, 246–251, 2009.

38. Engelman, D.M. and Moore, P.B., Determination of quaternary structure by small angle neutron scattering, *Annu. Rev. Biophys. Bioeng.* 4, 219–241, 1975.

39. Neylon, C., Small angle neutral and x-ray scattering in structural biology: Recent examples from the literature, *Eur. Biophys. J.* 37, 531–541, 2008.

40. Bucknall, D.G., Neutron scattering in analysis of polymers and rubbers, in *Encyclopedia of Analytical Chemistry*, ed. R.A. Meyers, John Wiley & Sons, Chichester, U.K., 2000.

41. Roe, R.-J., *Methods of X-Ray and Neutron Scattering in Polymer Sciences*, Oxford University Press, Oxford, U.K., 2006.

42. Lebedev, V., Didenko, V., Lapin, A. et al., Small-angle neutron scattering investigation of plastically deformed stainless steel, *J. Appl. Crystallogr.* 36, 629–631, 2003.

43. Wiyatno, W., Fuller, G.G., Pople, J.A. et al., Component stress–strain behavior and small-angle neutron scattering investigation of stereoblock elastomeric polypropylene, *Macromolecules* 36, 1178–1187, 2003.

44. Fitter, J., Gutberlet, T., and Katseras, J., eds., *Neutron Scattering in Biology*, Springer-Verlag, Berlin, Germany, 2006.

45. Teixeira, S.C.M., Zaccai, G., Ankner, J. et al., New sources and instrumentation for neutrons in biology, *Chem. Phys.* 345, 133–151, 2008.

46. Capel, M.S., Engelman, D.M., Freeborn, B.R. et al., A complete mapping in the small ribosomal subunit of *Escherichia coli*, *Science* 238, 1403–1406, 1987.

47. Svergyun, D.I. and Nierhaus, K.H., A map of protein-rRNA distribution in the 70 S *Escherichia coli* ribosome, *J. Biol. Chem.* 275, 14432–14439, 2000.

48. Petoukhov, M.V. and Svergun, D.I., Joint use of small-angle X-ray and neutron scattering to study biological macromolecules in solution, *Eur. Biophys. J.* 35, 567–576, 2006.

49. Gabel, F., Bicout, D., Lehnert, U. et al., Protein dynamics studied by neutron scattering, *Q. Rev. Biophys.* 35, 327–367, 2002.

50. Lebedev, D.V., Baitin, D.M., Islamov, A.Kh. et al., Analytical model for determination of parameters of helical structures in solution by small angle scattering: Comparison of RecA structures by SANS, *FEBS Lett.* 537, 182–186, 2003.

51. Magani, S., Romeo, G., and Telling, M.T.F., Temperature dependence of protein dynamics as affected by sugars: A neutron scattering study, *Eur. Biophys. J.* 36, 685–691, 2007.

52. Wood, K., Caronna, C., Fouquet, P. et al., A benchmark for protein dynamics: Ribonuclease A measured by neutral scattering in a large wavevector-energy transfer range, *Chem. Phys.* 345, 305–314, 2008.

53. Marconi, M., Cornicchi, E., Onori, G., and Paciaroni, A., Comparative study of protein dynamics in hydrated powers and in solutions: A neutron scattering investigation, *Chem. Phys.* 345, 224–229, 2008.

54. Li, Y., Lee, J., Lai, J. et al., Effects of pH on the interactions and conformations of bovine serum albumin: Comparison between chemical force microscopy and small-angle neutron scattering, *J. Phys. Chem. B* 112, 3797–3806, 2008.

55. Putnam, C.D., Hammel, M., Hura, G.L., and Tainer, J.A., X-ray solution scattering (SAXS) combined with crystallography and computation: Defining accurate macromolecular structures, conformations and assemblies in solution, *Q. Rev. Biophys.* 40, 191–285, 2007.

56. Pilz, I., Kratky, O., Light, A., and Sela, M., Shape and volume of anti-poly-(D-alanyl) antibodies in the presence and absence of tetra-D-alanine followed by small-angle x-ray scattering, *Biochemistry* 12, 4998–5005, 1973.

57. Pilz, I., Kratky, O., Light, A., and Sela, M., Shape and volume of fragments Fab' and (Fab')$_2$ of anti-poly(D-alanyl) antibodies in the presence and absence of tetra-D-alanine as determined by small-angle x-ray scattering, *Biochemistry* 14, 1326–1333, 1975.

58. Tang, Y.W., Labow, R.S., and Santerre, J.P., Enzyme-induced biodegradation of polycarbonate–polyurethane: Dependence on hard-segment chemistry, *J. Biomed. Mater. Res.* 57, 597–611, 2001.

59. Drogoz, A., Munier, S., Verrier, B. et al., Towards biocompatible vaccine delivery systems: Interactions of colloidal PECs based on polysaccharides with HIV-1 p24 antigen, *Biomacromolecules* 9, 583–591, 2008.

60. Chapman, M.D., Ferreira, F., Vaillalba, M. et al., The European Union CREATE project: A model for international standardization of allergy diagnostics and vaccines, *J. Allergy Clin. Immunol.* 122, 882–889, 2008.

61. Tamon, H. and Ishizaka, H., SAXS study on gelation process in preparation of resorcinol-formaldehyde aerogel, *J. Colloid Interface Sci.* 206, 577–582, 1998.

62. Loizou, E., Weisser, J.T., Dundigalla, A. et al., Structural effects of crosslinking a biopolymer hydrogel derived from marine mussel adhesive protein, *Macromol. Biosci.* 6, 711–718, 2006.

63. Schack, N.B., Oliveira, C.L., Young, N.W. et al., Oxygen diffusion in cross-linked, ethanol-swollen poly(vinyl alcohol) gels: Counter-intuitive results reflect microscopic heterogeneities, *Langmuir* 25, 1148–1153, 2009.

64. Klotz, I.M., Walker, F.M., and Pivan, R.B., The binding of organic ions by proteins, *J. Am. Chem. Soc.* 68, 1486–1490, 1946.
65. Teresi, J.D. and Luck, J.M., The combination of organic anions with serum albumin. VI. Quantitative studies by equilibrium dialysis, *J. Biol. Chem.* 174, 653–661, 1948.
66. Berger, G. and Girault, G., Macromolecule-ligand binding studied by the Hummel and Dreyer method: Current status of the methodology, *J. Chromatogr. B* 797, 51–61, 2003.
67. Ji, Q.C., Rodila, R., Morgan, S.J. et al., Investigation of the immunogenicity of a protein drug using equilibrium dialysis and liquid chromatography tandem mass spectrometry detection, *Anal. Chem.* 77, 5529–5533, 2005.
68. Deerfield II D.W., Olson, D.L., Berkowitz, P. et al., Relative affinity of Ca(II) and Mg(II) ions for human and bovine prothrombin and fragment 1, *Biochem. Biophys. Res. Commun.* 144, 520–527, 1987.
69. Monroe, D.M., Deerfield II D.W., Olson, D.L. et al., Calcium ion binding to human and bovine factor X, *Blood Coagul. Fibrinolysis* 1, 633–640, 1990.
70. Huh, N.W., Berkowitz, P., Hiskey, R.G., and Pedersen, L.G., Determination of strontium binding to macromolecules, *Anal. Biochem.* 198, 391–393, 1991.
71. Morita, T., Issacs, B.S., Esmon, C.T., and Johnson, A.E., Derivatives of blood coagulation factor IX contain a high affinity Ca^{2+}-binding site that lack γ-carboxyglutamic acid, *J. Biol. Chem.* 259, 5698–5704, 1984.
72. Morita, T. and Kisiel, W., Calcium binding to a human factor IXa derivatived lacking γ-carboxyglutamic acid: Evidence for two high-affinity sites that do not involve β-hydroxyaspartic acid, *Biochem. Biophys. Res. Commun.* 130, 841–847, 1985.
73. Bos, O.J., Fischer, M.J., Wilting, J., and Janssen, L.H., Drug-binding and other physicochemical properties of a large tryptic and a large peptic fragment of human serum albumin, *Biochim. Biophys. Acta* 953, 37–47, 1988.
74. Fischer, M.J., Bos, O.J., van der Linden, R.F. et al., Steroid binding to human serum albumin and fragments thereof. Role of protein conformation and fatty acid content, *Biochem. Pharmacol.* 45, 2411–2416, 1993.
75. Brée, F., Urien, S., Nguyen, P. et al., Human serum albumin conformational changes as induced by tenoxicam and modified by simultaneous diazepam binding, *J. Pharm. Pharmacol.* 45, 1050–1053, 1993.
76. Yamasaki, K., Maruyama, T., Yoshimoto, K. et al., Interactive binding to the principal ligand binding sites of human serum albumin: Effect of the neutral-to-base transition, *Biochim. Biophys. Acta* 1432, 313–323, 1999.
77. Shen, X., Liang, H., Guo, J.H. et al., Studies on the interaction between Ag^+ and human serum albumin, *J. Inorg. Biochem.* 95, 124–130, 2003.
78. Mullenix, M.C. and Mortensen, R.F., Calcium ion binding regions in C-reactive protein: Location and regulation of conformational changes, *Mol. Immunol.* 31, 615–622, 1995.
79. Kleerekoper, Q. and Putkey, J.A., Drug binding to cardiac troponin C, *J. Biol. Chem.* 274, 23932–23939, 1999.
80. Wakabayashi, H., Zhen, Z., Schmidt, K.M., and Fay, P.J., Mn^{2+} binding to factor VIII subunits and its effect on cofactor activity, *Biochemistry* 42, 145–153, 2003.
81. Taboada, P., Fernández, Y., and Mosquera, V., Interactions of two amphiphilic penicillins with myoglobin in aqueous buffered solutions: A thermodynamic and spectroscopy study, *Biomacromolecules* 5, 2201–2211, 2004.
82. Sioud, M., Nucleic acid enzymes as a novel generation of anti-gene agents, *Curr. Mol. Med.* 1, 575–588, 2001.
83. Sullenger, B.A. and Gilboa, E., Emerging clinical applications of RNA, *Nature* 418, 252–258, 2002.
84. Alvarez-Salas, L.M., Nucleic acids as therapeutic agents, *Curr. Top. Med. Chem.* 8, 1379–1404, 2008.

85. Haney, M. and Kosten, T.R., Therapeutic vaccines for substance dependence, *Expert Rev. Vaccines* 3, 11–18, 2004.
86. Xu, Y., Hixon, M.S., Yamamoto, N. et al., Antibody-catalyzed anaerobic destruction of methamphetamine, *Proc. Natl. Acad. Sci. USA* 104, 3681–3686, 2007.
87. Planque, S., Nishiyama, Y., Taguchi, H. et al., Catalytic antibodies to HIV: Physiological role and potential clinical utility, *Autoimmun. Rev.* 7, 473–479, 2008.
88. Wójcik, T. and Kieć-Konowicz, K., Catatlyic activity of certain antibodies as a potential tool for drug synthesis and for directed prodrug therapies, *Curr. Med. Chem.* 15, 1606–1615, 2008.
89. Taguchi, H., Planque, S., Nishiyama, Y. et al., Catalytic antibodies to amyloid beta peptide in defense against Alzheimer disease, *Autoimmun. Rev.* 7, 391–397, 2008.
90. Nieri, P., Donadio, E., Rossi, S. et al., Antibodies for therapeutic uses and the evolution of biotechniques, *Curr. Med. Chem.* 16, 753–779, 2009.
91. Racker, E. for the National Academy of Sciences-National Research Council, International Unit of Enzyme Activity, *Science* 128, 19–20, 1958.
92. Dybkaer, R., Unit "katal" for catalytic activity (IUPAC technical support), *Pure Appl. Chem.* 73, 927–931, 2001.
93. Dybkaer, R., The tortuous road to the adsorption of katal for the expression of catalytic activity by the General Conference on Weights and Measures, *Clin. Chem.* 48, 586–590, 2002.
94. Dixon, M. and Webb, E.C., *Enzymes*, Academic Press, New York, 1964.
95. Purich, D. ed., *Contemporary Enzyme Kinetics and Mechanism*, Academic Press, New York, 1981.
96. ICHQ8 Pharmaceutical Development Revision 1, International Conference on Harmonization, http://www.fda.gov/cber/gdlns/ichq8pharmann.htm
97. Dixon, M. and Webb, E.C., *Enzymes*, Academic Press, New York, Chapter 2, 1964.
98. Purich, D. ed., *Contemporary Enzyme Kinetics and Mechanism,* Academic Press, New York, 1983.
99. Engel, P.C. ed., *Enzymology LabFax*, Bios/Academic Press, Oxford, U.K., 1996.
100. Lundblad, R.L., *Biochemistry and Molecular Biology Compendium*, Taylor & Francis/CRC, Boca Raton, FL, 2007.
101. Stinson, R.A., Kinetic parameters for the chemical substrate, and enzyme and substrate stability, vary with the phosphoacceptor in alkaline phosphatase catalysis, *Clin. Chem.* 39, 2993–2997, 1982.
102. Pavlic, M., The inhibitory effect of Tris on the activity of cholinesterase, *Biochim. Biophys. Acta* 139, 133–137, 1967.
103. Ogilvie, J.W. and Whitaker, S.C., Reaction of Tris with aldehydes. Effect of Tris on reactions catalyzed by homoserine dehydrogenase and glyceraldehyde-3-phosphate dehydrogenase, *Biochim. Biophys. Acta* 445, 525–536, 1976.
104. Kasche, V. and Zöllner, B., Tris(hydroxymethyl)methylamine is acylated when it reacts with acyl-chymotrypsin, *Hoppe Seylers Z. Physiol. Chem.* 363, 531–534, 1982.
105. Trivić, S., Leskovac, V., Zeremski, J. et al., Influence of Tris(hydroxymethyl)aminomethane on kinetic mechanism of yeast alcohol dehydrogenase, *J. Enzyme Inhib.* 13, 57–68, 1998.
106. Gordon-Weeks, R., Koren'kov, V.D., Steele, S.H., and Leigh, R.A., Tris is a competitive inhibitor of K^+ activation of the vacuolr H^+-pumping pyrophophatase, *Plant Physiol.* 114, 901–905, 1997.
107. Quan, L., Wei, D., Jiang, X. et al., Resurveying the Tris buffer solution: The specific interaction between tris(hydroxymethyl)aminomethane and lysozyme, *Anal. Biochem.* 378, 144–150, 2008.
108. Yoshioka, S., Aso, Y., Izutsu, K.-i., and Kojima, S., Is stability prediction possible for protein drugs? Denaturation kinetics of β-galactosidase in solution, *Pharm. Res.* 11, 1721–1725, 1994.

109. Samoshina, N.M. and Samoshin, V.V., The Michaelis constants ratio for two substrates with a series of fungal (mould and yeast) β-galactosidases, *Enzyme Microbiol. Technol.* 36, 239–251, 2005.
110. Nguyen, T.-H., Splechtna, B., Steinbock, M. et al., Purification and characterization of two novel β-galactosidases from *Lactobacillus reuteri, J. Agric. Food Chem.* 54, 4989–4998, 2006.
111. Bhattacharya, R. and Bhattacharya, D., Resistance of bromelain to SDS binding, *Biochim. Biophys. Acta* 1794, 698–708, 2009.
112. Haag, R. and Kratz, F., Polymer therapeutics: Concepts and applications, *Angew. Chem. Int. Ed. Engl.* 45, 1198–1215, 2006.
113. Shu, Q. and Frieden, C., Relation of enzyme activity to local/global stability of murine adenosine deaminase: ^{19}F NMR studies, *J. Mol. Biol.* 345, 599–610, 2005.
114. Spotnitz, W.D. and Prabhu, R., Fibrin sealant tissue adhesive—Review and update, *J. Long Term Eff. Med. Implants* 15, 245–270, 2005.
115. Lewis, S.D., Shields, P.P., and Shafer, J.A., Characterization of the kinetic pathway for liberation of fibrinopeptides during assembly of fibrin, *J. Biol. Chem.* 260, 10192–10199, 1985.
116. De Cristofaro, R. and Castagnola, M., Kinetics aspects of release of fibrinopeptides AP and AY by human α-thrombin, *Haemostasis* 21, 85–90, 1991.
117. Gaffney, P.J. and Edgell, T.A., The International and "NIH" units for thrombin—How do they compare? *Thromb. Haemost.* 74, 900–903, 1995.
118. Ronwin, E., The relationship between the peptidase, esterase and clotting activity of thrombin, *Acta Haematol.* 23, 129–139, 1960.
119. Lundblad, R.L. and Harrison, J.H., The differential effect of tetranitromethane on the proteinase and esterase activity of bovine thrombin, *Biochem. Biophys. Res. Commun.* 45, 1344–1349, 1971.
120. Lottenberg, R., Hall, J.A., Fenton II J.W., and Jackson, C.W., The action of thrombin on peptide *p*-nitroanilide substrates: Hydrolysis of Tos-gly-pro-arg-pNA and D-phe-pip-arg-pNA by human α and γ and bovine α and β-thrombins, *Thromb. Res.* 28, 313–332, 1982.
121. Jenny, N.S., Lundblad, R.L., and Mann, K.G., Thrombin, in *Hemostasis and Thrombosis. Basic Principles and Clinical Practice*, 5th edn., eds. R.W. Colman, V.J. Marder, A.W. Cloves, J.N. George, and S.Z. Goldhaber, Lippincott William & Wilkins, Philadelphia, PA, pp. 193–213, Chapter 10, 2007.
122. Rijkers, D.T., Hemker, H.C., and Tesser, G.I., Synthesis of peptide *p*-nitroanilides mimicking fibrinogen- and hirudin-binding to thrombin. Design of slow reacting thrombin substrates, *Int. J. Pept. Protein Res.* 48, 182–193, 1996.
123. Meddahi, S., Bara, L., Fessi, H., and Samama, M.M., Standard measurement of clot-bound thrombin by using a chromogenic substrate for thrombin, *Thromb. Res.* 114, 51–56, 2004.
124. Nowak, G., Lange, U., Wiesenburg, A., and Bucha, E., Measurement of maximum thrombin generation capacity in blood and plasma using the thrombin generation assay (THROGA), *Semin. Thromb. Hemost.* 33, 508–514, 2007.
125. Menzel, E.J. and Farr, C., Hyaluronidase and its substrate hyaluronan: Biochemistry, biological activities and therapeutic uses, *Cancer Lett.* 131, 3–11, 1998.
126. Bergquist, S. and Packalen, T., The bacteriological origin of a spreading factor present in the nasal secretion, *Acta Pathol. Microbiol. Scand.* 25, 255–258, 1948.
127. Woodin, A.M., Hyaluronidase as a spreading factor in the cornea, *Br. J. Ophthalmol.* 34, 375–379, 1950.
128. Bensley, S.H. Histological studies of the reactions of cells and intercellular substances of loose connective tissue to the spreading factor of testicular extracts, *Ann. N.Y. Acad. Sci.* 52, 983–988, 1950.

129. Farrar, G.E. Jr., The spreading factor, *Clin. Ther.* 11, 705–706, 1989.
130. Girish, K.S. and Kemparaju, K., The magic glue hyaluronan and its eraser hyaluronidase, a biological overview, *Life Sci.* 80, 1921–2007, 2007.
131. Frost, G.I., Recombinant human hyaluronidase (rHuPH20): An enabling platform for subcutaneous drug and fluid administration, *Expert Opin. Drug Deliv.* 4, 427–440, 2007.
132. Anwer, K., Formulations for DNA delivery via electroporation in vivo, *Methods Mol. Biol.* 423, 77–89, 2008.
133. Gandhi, N.S. and Mancera, R.L., The structure of glycosaminoglycans and their interactions with proteins, *Chem. Biol. Drug. Des.* 72, 455–482, 2008.
134. Vercruysse, K.P., Lauwers, A.R., and Demeester, J.M., Absolute and empirical determination of the enzymatic activity and kinetic investigation of the action of hyaluronidase on hyaluronan using viscosimetry, *Biochem. J.* 305, 153–160, 1995.
135. Kinoshita, M., Okino, A., Oda, Y. et al., Anomalous migration of hyaluronic acid oligomers in capillary electrophoresis: Correlation to susceptibility to hyaluronidase, *Electrophoresis* 22, 3458–3465, 2001.
136. Courel, M.N., Maingonnat, C., Transchepain, F. et al., Importance of hyaluronan length in a hyaladherin-based assay for hyaluronan, *Anal. Biochem.* 302, 285–290, 2002.
137. He, D., Zhou, A., Wei, W. et al., A new study of the degradation of hyaluronic acid by hyaluronidase using quartz crystal impedance, *Talanta* 53, 1021–1029, 2001.
138. Schulze, C., Bittorf, T., Walzel, H. et al., Experimental evaluation of hyaluronidase activity in combination with specific drugs applied in clinical techniques of interventional pain management and local anaesthesia, *Pain Physician* 11, 877–883, 2008.
139. Yavav, G., Prasad, R.L., Jha, B.K. et al., Evidence for inhibitory interaction of hyaluronan-binding protein 1 (HABP1/p32/gClqR) with *Streptococcus pneumonia* hyaluronidase, *J. Biol. Chem.* 284, 3897–3905, 2009.
140. Frost, G.I. and Stern, R., A micro-titer assay for hyaluronidase activity not requiring specialized reagents, *Anal. Biochem.* 251, 263–269, 1997.
141. Gupta, G.S. and Sharma, P.K., Molecular inactivation of testicular hyaluronidase in solid state after proton irradiation: A study based on target size, substrate binding and thermodynamic analysis of heat denaturation, *Indian J. Biochem. Biophys.* 32, 266–271, 1995.
142. Maksiemenko, A.V., Schechilina, Y.V., and Tischenko, E.G., Resistance of dextran-modified hyaluronidase to inhibition by heparin, *Biochemistry* 66, 456–463, 2001.
143. Fujiwara, S., Kunugi, S., Oyama, H., and Oda, K., Effects of pressure on the activity and spectroscopic properties of carboxyl proteinases. Apparent correlation of pepstatin-insensitivity and pressure response, *Eur. J. Biochem.* 268, 645–655, 2001.
144. Larner, J., Takeda, Y., and Hizukuri, S., The influence of chain size and molecular weight on the kinetic constants for the span glucose to polysaccharide for rabbit muscle glycogen synthase, *Mol. Cell. Biochem.* 12, 131–136, 1976.
145. Neet, K.E., Cooperativity in enzyme function: Equilibrium and kinetic aspects, in *Contemporary Enzyme Kinetics and Mechanism*, ed. D.L. Purich, Academic Press, New York, Chapter 11, 1983.
146. Lawlis, V.B. and Roche, T.E., Regulation of bovine kidney α-ketoglutarate dehydrogenase complex by calcium ions and adenine nucleotides. Effects on $S_{0.5}$ for α-ketoglutarate, *Biochemistry* 28, 2512–2518, 1981.
147. Tippett, P.S. and Neet, K.E., Interconversions between different sulfhydryl-related kinetic states in glucokinase, *Arch. Biochem. Biophys.* 222, 285–298, 1982.
148. Richards, E.W., Hamm, M.W., and Otto, D.A., The effect of palmitoyl-CoA binding to albumin on the apparent kinetics behavior of carnitine palmitoyltransferase I, *Biochim. Biophys. Acta* 1076, 23–28, 1991.
149. Suganuma, T., Maeda, Y., Kitahara, K., and Nagahama, T., Study of the action of human salivary α-amylase on 2-chloro-4-nitrophenyl α-maltotrioside in the presence of potassium thiocyanate, *Carbohydr. Res.* 303, 219–227, 1997.

150. Buchbinder, J.L., Luong, C.B., Browner, M.F. et al., Partial activation of muscle phosphorylase by replacement of serine 14 with acidic residues at the site of regulatory phosphorylation, *Biochemistry* 36, 8039–8044, 1997.

151. Akowski, J.P. and Bauerle, R., Steady-state kinetics and inhibitor binding of 3-deoxy-D-arabino-heptulosonate-7-phosphate synthase (tryptophan sensitive) from *Escherichia coli*, *Biochemistry* 36, 15817–15822, 1997.

152. Swieca, A., Rybakowska, I., Nagel-Starczynowska, G. et al., AMP-deaminase from human term placenta, *Mol. Cell. Biochem.* 252, 363–367, 2003.

153. Tanaka, K., Sawatani, K., Dias, G.A. et al., High quality human immunoglobulin G purified from Cohn fractions by liquid chromatography, *Braz. J. Med. Biol. Res.* 33, 27–30, 2000.

154. Snape, T.J., Griffith, D., Vallet, L., and Wesley, E.D., The assay of prekallikrein activator in human blood products, *Dev. Biol. Stand.* 44, 115–120, 1979.

155. Kerry, P.J., Curtis, A.D., Paton, C.J., and Thomas, D.P., Standardization of prekallikrein activator (PKA): The 1st British reference preparation of PKA, *Br. J. Haematol.* 52, 275–281, 1985.

156. Tankersley, D.L., Fournel, M.A., and Schroeder, D.D., Kinetics of activation of prekallikrein activator, *Biochemistry* 19, 3121–3127, 1980.

157. Lundblad, J.L., In vitro assay for prekallikrein activator (PKA), *Dev. Biol. Stand.* 44, 107–114, 1979.

22 Development of an Experimental Approach for the Study of the Conformation of a Biological Therapeutic Product

First, any strategy must fit within the regulatory guidelines for drug and biological products[1-3] and represent solid science.[4-9] Second, a strategy must be realistic with respect to goals. Finally, the goals need to be clearly defined.

There are two major issues to address in the planning stage. The first is the selection of the sample and reference preparation. In the case where a manufacturing change in being evaluated, then the logical reference preparation is the active pharmaceutical ingredient immediately before formulation and transition to final drug product (final drug product). Characterization of biosimilar products presents a more significant problem. While it is possible to obtain the final approved drug product, there is concern that this does not provide a satisfactory reference preparation.[10] There is, of course, an emerging literature on this subject.[11-14] Whatever the decision, it is critical that it be driven by scientific, not marketing considerations. The second question is a clear definition of the expected outcome. Since the primary consideration is the assurance of the efficacy and safety of the product, the experimental design should be focused toward developing data in support of the bioequivalence of the drug substance. The rest of this brief chapter will focus on this second question. I wish to note the following before continuing: First, there is no precise regulatory guidance as what compromises a satisfactory consideration; second, there is likely considerable subjectivity reflecting my biases. Based on my personal observations, I have the following general comments:

- There is no such thing as faster, better, cheaper.
- Pressure from senior management cannot reduce cost.
- The technique you used is graduate school/post-doc/last job/heard about from your friend/etc. is not necessarily the best technique.
- The newest technology is not necessarily the best technology for your problem.

341

- If you do not have direct experience with the technology, it is best to outsource.
- Time is your greatest enemy.
- Documentation from day 1 is critical.
- There is no single correct approach.

Given the above considerations, upfront time is well spent. As such, you want to select techniques that will be complementary not duplicative. For example, circular dichroism (CD), Fourier transform infrared (FTIR), and H/D exchange mostly provide information about secondary structure. It would be more useful to combine CD, intrinsic fluorescence, and differential scanning calorimetry (DSC). Thus, you need to spend time considering resources before embarking on the program. In addition, it is essential to assure total documentation of the process including the research plan. There is limited case history available for guidance in this work. It is useful to consider some earlier works[5,15,16] on the development of biotechnology products. There is recent article on the comparability of human erythropoietin obtained from different manufacturers.[16a]

It should be obvious that your first consideration is the biological activity of the biopharmaceutical. The choice of assay is critical as you want a assay that measure true biological activity, most likely then a bioassay.[17–20] This assay may well be one selected for stability studies. It might also be the surrogate assay for measuring concentration in the subject after administration. If you are lucky, the same assay may serve all three (and addition) functions. For example, the assay for blood coagulation factor VIII[21] can serve all three functions. More often, as is the case with cytokines, a cell-based assay might be used for comparability and stability but an immunoassay used for measurement of product after administration.[22] This is also an issue with monoclonal antibody therapeutics.[23–30] My experience would suggest that the assay selected for stability-indicating is the best choice. This, of course, is easy for the organization addressing comparability in product resultant from changes in manufacturing.

The use of biological assays for biosimilar products is somewhat more challenging. While there is (usually) no access to the final drug substance (active pharmaceutical ingredient), the assay systems used are generally known. The goal here is to develop an assay for biological activity, which can be used both a quality attribute for the biopharmaceutical and as a stability-indicating assay. It is most likely that an immunoassay (either enzyme-linked immunosorbent essay [ELISA] or surface plasmon resonance based) will be used to study pharmacokinetics.[31–35] It is important to remember that these assays must reflect clinical effectiveness and must be structured to address the pathophysiology. In the case of factor VIII assays mentioned above, there is generally a correlation between the ability to correct the defective clotting of hemophilia A plasma in a test tube and the ability of a material to treat a bleeding diathesis in a patient. However, there is concern that a surrogate end point is not necessarily correlated with clinical outcome.[36] A particularly noteworthy example was the disastrous clinical trial of anti-CD28 antibody, TBN1412, in 2006.[37,38] Prediction of safety issues is a challenge with protein biopharmaceuticals.[39] Some adverse reactions such as a severe insulin allergy after a percutaneous transluminal

TABLE 22.1

Comparison of Various Technical Approaches to the Study of Biological Polymer Characterization

Technique	Amount of Material Required[a]	Comment	Disadvantages
Native electrophoresis	Microgram	Good for demonstrating protein–protein interaction and protein–ligand (i.e., protein–heparin) interaction	Destructive
Differential scanning calorimetry	Microgram to milligram	Excellent for stability; sensitive to conformational changes; useful for formulation development	High skill level required
Diffusion and light scattering	Microgram to milligram	Good for aggregation; sensitive to gross conformational change	High skill level required
Intrinsic fluorescence	Microgram to milligram	Measurement of tertiary structure; dependent on content of tryptophan	
HPLC	Microgram	Reverse phase is the dominant technology—dependent on conformational transition in initial solvent	Destructive; skill required in interpretation
UV-VIS	Microgram to milligram	Tertiary structure; derivative spectroscopy can be useful	
Infrared	Low milligram	Secondary structure	
Raman	milligram	Secondary structure	High skill level required
CD and ORD	Microgram to milligram	Secondary structure	High skill level required
Nuclear magnetic resonance	Milligram	Secondary and tertiary structure	High skill level required
Dye-binding/ fluorescent labeling	Microgram	Tertiary structure	Destructive
Hydrogen–deuterium exchanges	Microgram	Secondary structure	Destructive; Mass spectrometry required for analysis
Chemical footprinting/trace labeling	Microgram	Tertiary structure	Destructive; Mass spectrometry required for analysis
Immunogenicity	Microgram	Critical; not optional	Destructive
Enzyme kinetics	Microgram	Global structure; useful for enzymes acting on large substrates.	Destructive; skill required for data analysis
Limited proteolysis	Microgram	Tertiary structure	Destructive; Mass spectrometry required for analysis

[a] Approximation only.

coronary angioplasty and administration of protamine[40] would be extremely difficult to predict.

Immunological reactivity is a consideration equal importance to biological activity. Unwanted immunogenicity is likely the primary cause for adverse reactions in biopharmaceutical protein products.[26,41–52] Chapter 20 discusses immunogenicity issues and biopharmaceutical conformation. One problem is that there is no animal model that is an entirely a reliable predictor of immunological performance in human subjects. It is of interest that immunogenicity is discussed as a possible mechanism for adverse reaction with the use of recombinant feline erythropoietin in cats with erythropoietin-dependent nonregenerative anemia.[53] The reader is directed to two studies that are of direct relevance to the immunogenicity of biopharmaceuticals.[49,54] Lysozyme has been used as a model for the immunochemistry of proteins[55] and there is a useful literature.[56–60]

Selection of activity and immunogenicity as two keys assays for comparability is relatively easy because of (1) the high importance of these attributes and (2) the requirement for modest amounts of material. The selection of the various physicochemical assays poses a bit more of a problem. First, the combinations of assays should yield more than just the sum of assays; in other words, the assays should not necessarily be redundant. For example, it is not clear that both FTIR and CD measurements would not be duplicative. Second, since the sample is valuable material, selection should balance material used with the value of data obtained. If I were to chose three assays, I would chose CD, DSC, and intrinsic fluorescence. CD would provide information about secondary structure and some information on tertiary structure, intrinsic fluorescence for tertiary structure, and DSC as a global stability estimate; DSC is also useful in formulation development. There are other combinations that are likely to be useful. Table 22.1 contains a list of technologies with advantages and disadvantages based on my biases.

REFERENCES

1. ICH Q6: Specifications: Test Procedures and Acceptance Criteria for Biotechnological/Biological Products; http://www.ich.org/LOB/media/MEDIA432.pdf
2. ICH Q8(R1) Pharmaceutical Development; http://www.ich.org/LOB/media/MEDIA 4986.pdf
3. ICH Q5E: Comparability of Biotechnological/Biological Products Subject to Changes in Their Manufacturing Process; http://www.ich.org/LOB/media/MEDIA1196.pdf
4. Miller, J.M. and Crowther, J.B. eds., *Analytical Chemistry in a GMP Environment*, John Wiley & Sons, Inc., New York, 2000.
5. Wang, Y.J. and Pearlman, R. eds., *Stability and Characterization of Protein and Peptide Drugs. Case History*, Plenum Press, New York, 1993.
6. Kozlowski, S. and Swann, P., Current and future issues in the manufacturing and development of monoclonal antibodies, *Adv. Drug. Deliv. Rev.* 58, 707–722, 2006.
7. Egrie, J.C. and Browne, J.K., Development and characterization of novel erythropoiesis stimulating protein (NESP), *Nephrol. Dial. Transplant* 16(Suppl 3), 3–13, 2001; (a) Chirino, A.J. and Mire-Sluis, A., Characterizing biological products and assessing comparability following manufacturing changes, *Nat. Biotechnol.* 22, 1383–1391, 2004.
8. Beck, A., Wagner-Rousset, E., Bussat, M.C. et al., Trends in glycosylation, glycoanalysis and glycoengineering of therapeutic antibodies and Fc-fusion proteins, *Curr. Pharm. Biotechnol.* 9, 482–501, 2008.

9. Ondo, M., Reducing the immunogenicity of protein therapeutics, *Curr. Drug Targets* 10, 131–139, 2009.
10. Heavner, G.A., Arakawa, T., Philo, J.S. et al., Protein isolated from biopharmaceutical formulations cannot be used for comparative studies: Follow-up to "a case study using Epoeitn Alpha from Epogen and EPREX", *J. Pharm. Sci.* 96, 3214–3225, 2007.
11. Green, J.D., Tsang, L., and Cavagnaro, J.A., "Generic" or "follow-on" biologics: Scientific considerations and safety issues, *Expert Opin. Biol. Ther.* 3, 1019–1022, 2003.
12. Duconge, J., The case of biotech-derived product equivalence: Much ado about nothing? *Curr. Clin. Pharmacol.* 1, 147–156, 2006.
13. Hoppe, W. and Berghout, A., Biosimilar somatropin: Myths and facts, *Horm. Res.* 69, 29–30, 2008.
14. Kuhlmann, M. and Covic, A., The protein science of biosimilars, *Nephrol. Dial. Transplant* 21(Suppl 5), v4–v8, 2006.
15. Walsh, G., *Proteins Biochemistry and Biotechnology*, John Wiley & Sons Ltd., Chichester, U.K., 2002.
16. Stoll, R.E., Ball, D.J., Burchiel, S.W. et al., Case history of a biotechnology product: Toxicological protocol, design and results, *Prog. Clin. Biol. Res.* 235, 173–182, 1987; (a) Deechongkit, S., Aoki, K.H., Park, S.S., and Kerwin, B.A., Biophysical comparability of the same protein form different manufacturers: A case study using Epoetin Alfa from Epogen® and Eprex®, *J. Pharm. Sci.* 95, 1931–1943, 2006.
17. Thorpe, R., Wadhwa, M., Page, C., and Mire-Sluis, A., Bioassays for the characterization and control therapeutic cytokines: Determinations of potency, *Dev. Biol. Stand.* 97, 61–71, 1999.
18. Lebron, J.A., Wolf, J.J., Kaplanski, C.V., and Ledwith, B.J., Ensuring the quality, potency and safety of vaccines during preclinical development, *Expert Rev. Vaccines* 4, 855–866, 2005.
19. Mire-Sluis, A.R., Progress in the use of biological assays during the development of biotechnology products, *Pharm. Res.* 18, 1239–1246, 2001.
20. Indelicato, S.R., Bradshaw, S.L., Chapman, J.S., and Weiner, S.H., Evaluation of standard and state of the art analytical technology-bioassays, *Dev. Biol.* (Basel) 122, 103–114, 2005.
21. Lundblad, R.L., Kingdon, H.S., Mann, K.G., and White, G.C., Issues with the assay for factor VIII activity in plasma and factor VIII concentrates, *Thromb. Haemost.* 84, 842–948, 2000.
22. Piscitelli, S.C., Reiss, W.G., Figg, W.D., and Petros, W.P., Pharmacokinetic studies with recombinant cytokines. Scientific issues and practical considerations, *Clin. Pharmacokinet.* 32, 368–381, 1997.
23. Murano, G., FDA perspective on specification for biotechnology products—From IND to PLA, *Dev. Biol. Stand.* 91, 3–13, 1997.
24. Gossett, K.A., Narayan, P.K., and Williams, D.M., Flow cytometry in the preclinical development of biopharmaceuticals, *Toxicol. Pathol.* 27, 32–37, 1999.
25. Goel, S., Mani, S., and Perez-Soler, R., Tyrosine kinase inhibitors: A clinical perspective, *Curr. Oncol. Rep.* 4, 9–19, 2002.
26. Pendley, C., Schantz, A., and Wagner, C., Immunogenicity of therapeutic monoclonal antibodies, *Curr. Opin. Mol. Ther.* 5, 172–179, 2003.
27. Harding, J. and Burtness, B., Cetuximab: An epidermal growth factor receptor chimeric human-murine monoclonal antibody, *Drugs Today* (Barc) 41, 107–127, 2005.
28. Pegram, M. and Ngo, D., Application and potential limitations of animal models utilized in the development of trastuzumab (Herceptin): A case study, *Adv. Drug Deliv. Rev.* 58, 723–734, 2006.
29. McEarchern, J.A., Smith, L.M., McDonagh, C.F. et al., Preclinical characterization of SGN-70, a humanized antibody directed against CD70, *Clin. Cancer Res.* 14, 7763–7772, 2008.

30. Jamieson, D., Cresti, N., Verril, M.W. et al., Development and validation of cell-based ELISA for the quantification of trastuzmab in human plasma, *J. Immunol. Methods* 345, 106–111, 2009.

31. Furie, R., Stohl, W., McCune, W.J. et al., Belimumab study group. Biologic activity and safety of belimumab, a neutralizing anti-B-lymphocyte stimulator (BLyS) monoclonal antibody: Phase I trial in patients with systemic lupus erythematosus, *Arthritis Res. Ther.* 10, R109, 2008.

32. Rahmati, K., Lernmark, A., Becker, C. et al., A comparison of serum and EDTA plasma in measurement of glutamic acid decarboxylase autoantibodies (GADA) and autoantibodies to islet antigen-2 (IA-2A) using the RSR radioimmunoassay (RIA) and enzyme linked immunosorbent assay (ELISA) kits, *Clin. Lab.* 54, 227–235, 2008.

33. Ray, C.A., Patel, V., Shih, J. et al., Application of multi-factorial design of experiments to successfully optimize immunoassays for robust measurements of therapeutic proteins, *J. Pharm. Biomed. Anal.* 49, 311–318, 2009.

34. Call, S.K., Kasow, K.A., Barfield, R. et al., Total and active rabbit antithymocyte globulin (rATG; Thymoglobulin) pharmacokinetics in pediatric patients undergoing unrelated donor bone marrow transplantation, *Biol. Blood Marrow Transplant.* 15, 274–278, 2009.

35. Braund, R., Hook, S.M., Greenhill, N. et al., Distribution of fibroblast growth factor-2 (FGF-2) within model excisional wounds following topical application, *J. Pharm. Pharmacol.* 61, 193–200, 2009.

36. Grimes, D.A. and Schulz, K.F., Surrogate end points in clinical research: Hazardous to your health, *Obstet. Gynecol.* 105, 1114–1118, 2005.

37. Suntharalingam, G., Perry, M.R., Ward, S. et al., Cytokine storm in a phase 1 trial of the anti-CD28 monoclonal antibody TGN1412, *N. Engl. J. Med.* 355, 1018–1028, 2006.

38. St. Clair, E.W., The calm after the cytokine storm: Lessons from the TGN1412 trial, *J. Clin. Invest.* 118, 1344–1347, 2008.

39. Haller, C.A., Cosenza, M.E., and Sullivan, J.T., Safety issues specific to clinical development of protein therapeutics, *Clin. Pharmacol. Ther.* 84, 624–627, 2008.

40. Wang, C., Ding, Z.-Y., Shu, S.-Q. et al., Severe insulin allergy after percutaneous transluminal coronary angioplasty, *Clin. Ther.* 31, 569–574, 2009.

41. Konrad, M., Immunogenicity of proteins administered to humans for therapeutic purposes, *Trends Biotechnol.* 7, 175–179, 1989.

42. Porter, S., Human immune response to recombinant human proteins, *J. Pharm. Sci.* 90, 1–11, 2001.

43. Schellekens, H., Bioequivalence and the immunogenicity of biopharmaceuticals, *Nat. Rev. Drug Discov.* 1, 457–462, 2002.

44. Wadhwa, M., Bird, C., Dilger, P. et al., Strategies for detection, measurement and characterization of unwanted antibodies induced by therapeutic biologicals, *J. Immunol. Methods* 278, 1–17, 2003.

45. Hermeling, S., Crommelin, D.J.A., Schellekens, H., and Jiskoot, W., Structure-immunogenicity relationships of therapeutic proteins, *Pharm. Res.* 21, 897–903, 2004.

46. Chirino, A.J., Ary, M.L., and Marshall, S.A., Minimizing the immunogenicity of protein therapeutics, *Drug Discov. Today* 9, 82–90, 2004.

47. Thorpe, R. and Swanson, S.J., Current methods for detecting antibodies against erythropoietin and other recombinant proteins, *Clin. Diagn. Lab. Immunol.* 12, 28–39, 2005.

48. Geng, D., Shankar, G., Schantz, A. et al., Validation of immunoassays used to assess immunogenicity to therapeutic monoclonal antibodies, *J. Pharm. Biomed. Anal.* 39, 364–375, 2005.

49. Doyle, J.W., Johnson, G.L., Eshhar, N., and Hammond, D., The use of rabbit polyclonal antibodies to assess neoantigenicity following viral reduction of an alpha-1-proteinase inhibitor preparation, *Biologicals* 34, 199–207, 2006.

50. Kessler, M., Goldsmith, D., and Schellekens, H., Immunogenicity of biopharmaceuticals, *Nephrol. Dial. Transplant* 21(Suppl 5), V9–V12, 2006.
51. Gupta, S., Indelicato, S.R., and Jethwa, V., Recommendations for the design, optimization, and qualification of cell-based assays used for the detection of neutralizing antibody responses elicited to biological therapeutics, *J. Immunol. Methods* 321, 1–18, 2007.
52. Wang, J., Lozier, J., Johnson, G. et al., Neutralizing antibodies to therapeutic enzymes: Considerations for testing, prevention and treatment, *Nature Biotechnol.* 26, 901–908, 2008.
53. Randolph, J.E., Scarlett, J.M., Stokol, T. et al., Expression, bioactivity and clinical assessment of recombinant feline erythropoietin, *Am. J. Vet. Res.* 65, 1355–1366, 2004.
54. Esmon, P.C., Kuo, H.S., and Fournel, M.A., Characterization of recombinant factor VIII and a recombinant factor VIII deletion mutant using a rabbit immunogenicity model system, *Blood* 76, 1593–1600, 1990.
55. Arnon, R., Immunochemistry of Lysozyme, in *Immunochemistry of Enzymes and Their Antibodies*, ed. M.R.J. Salton, John Wiley & Sons, New York, pp. 1–29, Chapter 1, 1977.
56. Ibrahimi, I.M., Eder, J., Prager, E.M. et al., Effect of a single amino acid substitution on the antigenic specificity of the loop region of lysozyme, *Mol. Immunol.* 17, 37–46, 1980.
57. Hirayama, A., Fujio, H., Dohi, Y. et al., Preparations and immunological characterization of 2 lysozyme derivatives dinitrophenylated at lysine-33 and lysine-96, respectively, *J. Biochem.* 89, 963–974, 1981.
58. Smithgill, S.J., Wilson, A.C., Potter, M. et al., Mapping the antigenic epitope for a monoclonal-antibody against lysozyme, *J. Immunol.* 128, 314–322, 1982.
59. Smithgill, S.J., Lavoie, T.G., and Mainhart, C.R., Antigenic regions defined by monoclonal-antibodies correspond to structural domains of avian lysozyme, *J. Immunol.* 133, 384–393, 1984.
60. Amit, A.G., Mariuzza, R.A., Phillips, S.E.V., and Poijak, R.J., 3-Dimensional structure of an antigen–antibody complex at 6-Å resolution, *Nature* 313, 156–158, 1985.

Index

Printed and bound by CPI Group (UK) Ltd, Croydon, CR0 4YY

18/10/2024

01776266-0007